AF775166

INVISCID INCOMPRESSIBLE FLOW

INVISCID INCOMPRESSIBLE FLOW

JEFFREY S. MARSHALL
Department of Mechanical Engineering and
Iowa Institute of Hydraulic Research
University of Iowa
Iowa City, Iowa

A Wiley-Interscience Publication

JOHN WILEY & SONS, INC.

New York / Chichester / Weinheim / Brisbane / Singapore / Toronto

This book is printed on acid-free paper. ♾

Published simultaneously in Canada.

Library of Congress Cataloging-in-Publication Data

Marshall, Jeffrey S.
Inviscid incompressible flow / Jeffrey S. Marshall.
p. cm.
Includes index.
ISBN 0-471-37566-7 (cloth : alk. paper)
1. Fluid dynamics. I. Title.
TA357 .M36 2001
620.1′064—dc21 00-049617

10 9 8 7 6 5 4 3 2 1

CONTENTS

PREFACE xi

1 INTRODUCTION 1

1.1 Role of Viscosity in High-Reynolds-Number Flows / 3
1.2 Inviscid Flows / 6
Bibliography / 7

2 VECTORS AND TENSORS 9

2.1 Cartesian Index Notation / 9
2.2 Kronecker Delta and Permutation Tensor / 11
2.3 Vector and Tensor Operations / 12
2.4 Vector and Tensor Transformations / 14
2.5 Symmetric and Skew-Symmetric Tensors / 15
2.6 Vector Identities / 16
2.7 Integral Identities / 18
Bibliography / 22
Problems / 23

3 KINEMATICS OF FLUID MOTION 24

3.1 Descriptions of Fluid Motion / 24
3.2 Deformation Measures / 28

3.3 Stretch Rate and Vorticity / 29
3.4 Dilatation / 33
3.5 Pathlines, Streaklines, Streamlines, and Vortex Lines / 36
3.6 Transport Theorem / 40
Bibliography / 42
Problems / 42

4 LAWS OF FLUID DYNAMICS 44

4.1 Mass Conservation / 44
4.2 Momentum Conservation / 46
4.3 Moment of Momentum Conservation / 49
4.4 Kinetic Energy Transport Theorem / 51
4.5 Constraint of Incompressibility / 53
4.6 Navier-Stokes Equation and Some Simple Solutions / 55
Bibliography / 59
Problems / 59

5 DYNAMICS OF DISCONTINUITY SURFACES 63

5.1 Extended Transport Theorem for a Region Containing a Discontinuity Surface / 63
5.2 Jump Conditions across a Surface of Discontinuity / 66
5.3 Surface Tension / 68
5.4 Boundary Conditions for Fluid Flow / 73
Bibliography / 74
Problems / 75

6 VELOCITY REPRESENTATIONS AND ASSOCIATED THEOREMS 76

6.1 Irrotational (Lamellar) and Related Flows / 76
6.2 Incompressible (Solenoidal) Flows / 79
6.3 Flows That Are Both Incompressible and Irrotational / 81
6.4 Helmholtz Representation Theorem / 83
6.5 Biot-Savart Law / 86
6.6 Far-Field Asymptotic Form for Velocity in a Flow Extending to Infinity / 88
Bibliography / 90
Problems / 90

7 VORTICITY TRANSPORT THEOREMS 92

7.1 Vorticity Transport Equation / 92
7.2 Motion of Vortex Lines / 95
7.3 Preservation of Circulation / 97
7.4 Vorticity Measures and Invariants / 98
7.5 Vorticity Invariants in Two-Dimensional Flows / 105
7.6 Vorticity Transport and Invariants in Viscous Flows / 107
Bibliography / 109
Problems / 110

8 PRESSURE THEOREMS 114

8.1 Pressure Poisson Equation / 114
8.2 Bernoulli Theorem / 115
8.3 Boundary Integral Equation for Pressure / 116
8.4 Special Results for Steady, Two-Dimensional Flows / 119
Bibliography / 120
Problems / 120

9 TWO-DIMENSIONAL POTENTIAL FLOWS 122

9.1 Analogy between Potential Flows and Analytic Functions of a Complex Variable / 122
9.2 Some Basic Potential Flows / 124
9.3 Superposition of Basic Flows / 128
9.4 Circle Theorem / 133
9.5 Method of Images / 136
9.6 Conformal Transformation of Fluid Flows / 139
9.7 Transformation $\zeta = z^n$ / 144
9.8 Joukowski Transformation / 148
9.9 Schwarz-Christoffel Equation / 151
9.10 Free-Streamline Theory / 157
Bibliography / 162
Problems / 163

10 FORCES ON BODIES IN TWO-DIMENSIONAL FLOWS 166

10.1 Integration in the Complex Plane / 166
10.2 Blasius Force and Moment Laws / 168
10.3 Lagally's Theorem / 172

10.4 D'Alembert's Paradox and the Kutta-Joukowski Lift Law / 173
10.5 Application to Two-Dimensional Airfoils / 175
Bibliography / 180
Problems / 180

11 TWO-DIMENSIONAL FLOWS WITH VORTICITY **183**

11.1 Systems of Point Vortices / 183
11.2 Conformal Transformation of Uniform-Vorticity Patches / 186
11.3 Contour Integration Method for Uniform-Vorticity Patches / 191
11.4 Discrete-Vortex Numerical Method: Basic Method / 193
11.5 Discrete-Vortex Numerical Method: Acceleration Techniques / 197
11.6 Vortex Sheets / 203
11.7 Vortex Sheet Representation of Flat-Plate Airfoils / 209
11.8 Flow with Uniform Background Vorticity / 212
Bibliography / 215
Problems / 216
Computational Projects / 218

12 THREE-DIMENSIONAL POTENTIAL FLOWS **219**

12.1 Governing Equations / 219
12.2 Basic Potential Flows / 221
12.3 Some Axisymmetric Flows with Immersed Bodies / 225
12.4 Sphere Theorems / 231
12.5 Slender-Body Theory / 235
12.6 Source and Doublet Sheet Boundary-Integral Methods / 238
12.7 Vortex Sheet Boundary-Integral Method / 243
12.8 Forces Induced by Singularities / 246
12.9 Added Mass and Buoyancy Forces / 249
Bibliography / 255
Problems / 256
Computational Projects / 258

13 AXISYMMETRIC VORTEX FLOWS **260**

13.1 General Theory / 260
13.2 Thin-Core Vortex Rings / 265
13.3 Hill's Spherical Vortex / 271
13.4 Axisymmetric Contour Dynamics / 273
13.5 Steady Axisymmetric Flows / 277
13.6 Waves of Variable Core Area / 279

13.7 Plug-Flow Model / 283
13.8 Axisymmetric Discrete-Vortex Method / 288
Bibliography / 292
Problems / 293
Computational Projects / 295

14 VORTEX TUBES 296

14.1 Velocity Field Induced by a Curved Vortex Filament / 297
14.2 Cut-Off Model for a Vortex Tube / 298
14.3 Local-Induction Approximation / 303
14.4 Bending Waves on a Vortex Tube / 306
Bibliography / 311
Problems / 312
Computational Projects / 313

15 INTERFACIAL WAVE MOTION 314

15.1 Internal Waves in Layered Media / 314
15.2 Linear Wave Theory / 316
15.3 Capillary and Gravity Waves / 319
15.4 Particle Displacement in a Wave Field / 320
15.5 Wave Energy and Group Velocity / 322
15.6 Boundary-Integral Method for Nonlinear Interfacial Waves / 325
Bibliography / 327
Problems / 327

16 STABILITY OF FLUID FLOWS 330

16.1 General Concepts / 331
16.2 Stability of an Elliptical Vortex Patch in a Straining Flow / 332
16.3 Stability of Two-Dimensional Point Vortex Arrays / 337
16.4 Interfacial Instabilities / 343
16.5 Capillary Instability of a Liquid Jet / 346
16.6 Centrifugal Instability / 348
16.7 Stability of Parallel Shear Flows / 352
16.8 Three-Dimensional Instability of a Vortex Pair / 358
Bibliography / 361
Problems / 363

APPENDIX A COMMON EXPRESSIONS IN ORTHOGONAL CURVILINEAR COORDINATE SYSTEMS 367

INDEX 373

PREFACE

This book gives an introduction to the dynamics of inviscid, incompressible fluids. By omitting effects inherent to viscosity and compressibility, the book examines fluid flow at its most fundamental level, while retaining the essential nonlinear processes that make fluid mechanics of enduring interest to physicists, mathematicians, and engineers. The book is intended as a text for a beginning graduate-level course on fluid mechanics, which in many institutions of higher learning deals principally, and in some cases entirely, with inviscid flows. Unlike more classical texts on inviscid fluids, the book also contains detailed coverage of vorticity transport phenomena in both two- and three-dimensional spaces and of a variety of computational methods for inviscid flows. The later chapters of the book might also be used as a text for more advanced fluid dynamics courses, for instance on vortex dynamics, or as a professional reference.

The first chapter discusses the importance of inviscid flows and examines modifications to ideal flow behavior in real fluids caused by viscous forces in high Reynolds number laminar and turbulent flows. Chapter 2 provides background on vector and tensor analysis, which is used extensively in the remainder of the text. Chapters 3–5 give a rigorous introduction to the continuum mechanics of fluid flows, where at this stage both viscosity and compressibility are included in the governing equations. Aspects of kinematics that are important to the study of fluid motion are developed in Chapter 3, including flow lines, stretching and rotation of fluid elements, and the transport theorem. Chapter 4 introduces mass and momentum conservation and derives local and control-volume forms of the conservation laws. Incompressibility is introduced as a constraint form of the compressible flow theory. Chapter 5 deals with discontinuity surfaces, which may occur either internal to a flow or along boundaries of the flow domain. A modified form of the transport theorem is developed and used to derive discontinuity jump conditions, which are applied to obtain boundary conditions for different flows. The effects of surface tension at a material interface are also discussed.

Inviscid flows are typically solved using solution methods, both analytical and computational, that are not based directly on the continuity and momentum conservation equations, but rather on theorems that are derived from these equations. These general theorems are developed in Chapters 6–8. Chapter 6 derives vector representation theorems that, when applied to the velocity field, yield a relationship between the vorticity and rate of dilatation fields and the velocity that these fields generate. Chapter 7 develops the vorticity transport equation and examines the various laws governing vorticity transport in two- and three-dimensional flows. A discussion of how these laws are modified for viscous fluids is given at the end of Chapter 7. Chapter 8 covers theorems associated with the pressure field, both with and without vorticity present.

There exist a number of solution methods for two-dimensional inviscid flows that cannot be extended to three-dimensional spaces. These two-dimensional solution methods are described in Chapters 9–11. Chapter 9 examines the analogy between analytic functions of a complex variable and potential flow solutions and shows how complex variable theory can be used to simplify flow solutions and to transform one solution into other flow solutions. Chapter 10 applies the complex variable analogy to develop theorems for forces and moments acting on a body immersed in a potential flow field. The dynamics of vortex patches and sheets in two-dimensional flows is examined in Chapter 11, including development of different computational methods for two-dimensional vorticity transport.

Solutions methods for three-dimensional flows are covered in Chapters 12–14; many of these methods can be applied to two-dimensional flows as a special case. Chapter 12 examines three-dimensional potential flows, including analytic solutions for axisymmetric flows, an approximate solution method for flow past slender bodies, and computational methods of the boundary-integral type for general potential flows. Chapter 13 provides an in-depth discussion of axisymmetric vorticity dynamics, including analytic solutions for vortex rings, computational methods for vorticity transport in axisymmetric flows, and approximate methods for solution of axisymmetric wave propagation on a vortex core. Chapter 14 describes different approximations used to model the motion of thin-core vortex tubes.

The final two chapters are devoted to perturbations of equilibrium solutions and ensuing stability issues. Chapter 15 covers the dynamics of wave motions on an internal fluid interface and at a free surface. Solutions of the linear wave equations are examined, and a computational method is described for nonlinear wave motions. Chapter 16 covers stability of inviscid flows, progressing from vortex systems governed by ordinary differential equations, to interfacial wave instabilities, and finally to instabilities of rotating flows and shear flows.

The text is self-contained and requires only standard background in college calculus and differential equations. All necessary topics from more advanced mathematics are reviewed in the text. The book should be of interest to students of any discipline dealing with fluid motion, including mechanical, civil, and chemical engineering, applied mathematics, physics, and earth sciences. The text provides enough detail in the derivations and sufficient number of examples to allow students to follow easily, along with end-of-chapter problems to aid in instruction. Computational problems are given at the end of chapters dealing with numerical methods that are suitable for more long-term student projects. Subroutines for many of the computational methods described in the text can be downloaded from the Internet site of the Iowa Institute of Hydraulic Research (IIHR), which can be accessed at www.iihr.uiowa.edu.

There are a number of individuals and institutions that have made invaluable contributions to the development of this book. I would first like to acknowledge my instructors and advisers at the University of California, Los Angeles, and the University of California, Berkeley. My Ph.D. adviser, P. M. Naghdi, was particularly instrumental in illustrating to me the beauty of rigorous mechanics, and any Berkeley graduate will recognize his imprint in Chapters 2–5 of the book. I have also learned a great deal from the colleagues with whom I have associated closely during my career, including M. R. Dhanak and T. C. Su at Florida Atlantic University and V. C. Patel at the University of Iowa. I would particularly like to acknowledge the contributions of J. R. Grant, with whom I have collaborated closely for over ten years on a journey of discovery in vortex dynamics and computational methods, and of T. L. Doligalski, without whose financial support of my research and personal encouragement this work would not have been possible. The first nine chapters of the book were written during a series of summers spent at the Naval Undersea Warfare Center in Newport, Rhode Island, under support from the ASEE/Navy Summer Faculty Research Program. Most of Chapters 10–16 were written during a one-semester sabbatical at the Institut de Mécanique des Fluides de Toulouse (IMFT) in Toulouse, France. I express thanks to H. Boisson, A. Giovannini, and P. Brancher for hosting my visit to IMFT and making me feel welcome there, and to the French Ministry of Education for providing financial support. Undoubtedly, the individuals having the most direct impact on this book are the students in the inviscid flow course that I have taught over many years at Iowa, and I offer them my most sincere thanks for correcting numerous errors and omissions in early drafts and for encouraging me to write the book in the first place.

CHAPTER 1

INTRODUCTION

Earth, air, fire, and water—the four basic elements of the ancient Greek world—are each vital to human existence. Air is the medium by which oxygen is transported to our lungs, surging through our twisting pulmonary passageways with each gasped breath. Air is the domain of flight, of soaring birds and buzzing insects and the screaming vehicles of man. Air transports our words, allowing us to communicate with each other. Air and water together regulate the temperature of the earth's atmosphere, blocking out harmful solar emissions and surrounding us in blanketing warmth. They determine our weather, be it a tranquil summer day, a spring shower, a winter blizzard, or a roaring fall hurricane, and over time they control the climate in which we must strive to survive. Water is also vital to human life, transporting nourishment and waste products as it courses through our veins and regulating our body temperature as it evaporates from our skin. Much of the world depends upon oceans and rivers as a medium of transport for goods and a source of bounteous food. The ability to utilize fire is one of the major human characteristics distinguishing us from other animals. Fire cooks our food, warms our bodies, propels our vehicles, hardens and forms our metals. Even the earth is not stagnant and unmoving. Water percolates through its pores, providing us with drink and sustaining the plants that feed us and provide us with shelter. Oil gushes from its fissures, providing fuel for automobiles and engine lubrication. With too much water, the earth forms giant slides, burying houses and roads. Over geologic time scales the earth's motion forms our surroundings, be they from violent volcanic explosions to the subtle drifting of the earth's crust in response to the ever-present convective stirring of the mantle.

Nearly all human endeavors must in some way deal with restrictions imposed by fluid transport. This observation is obvious in the aerospace and marine transport industries, but equally true in other industries in which the importance of fluid flow may not be as apparent. For instance, the major limitation to the smallness, and hence speed, of modern computer chips is the restriction imposed by convective heat transfer in cooling electrical components. As computer chips become increasingly compact, the heating rates become higher and the passageways available for cooling become more confined. Biomedical applications must deal with a host of fluid transport issues, including oxygen supply to the lungs, pumping of blood and the associated transport of nutrients throughout the body, digestion of foods, and human reproduction. On a cellular level, transport of matter, such as viruses or nutrients, over cell boundaries controls the body's ability to fight diseases and heal wounds. Major medical crises, such as strokes and lung cancer, are related to transport of particulate matter either in the blood stream or in the pulmonary system, whereas other emergencies, such as heart attacks or hemorrhaging, are caused by inability to maintain a continual fluid flow. Agriculture must continually deal with fluid transport issues, such as water supply to crops, pesticide distribution, heating of livestock, and soil moisture during harvesting operations. Material processing, which is so vital for modern technological advances, deals with a host of fluid processes, ranging from metal casting to liquid spray coating.

Aside from its importance in diverse applications, fluid dynamics has had a central influence in development of much of modern science and mathematics. A fluid view of elementary matter dates back to ancient Greece, where Anaxagoras (500–428 BC) proposed that all matter consists of a fluid continuum whose basic component is a vortex. This continuum view competed with the atomic (or particulate) view of Democritus (460–370 BC), in which matter is formed of small particles immersed in a fluid and ordered by the action of vortices. The fluid theory of matter, again based on vortices, was later taken up by Descartes (1596–1650) to explain the suspension of celestial bodies and by Kelvin (1824–1907), who proposed that all matter is constructed of a set of "vortex atoms" that exist in the ether. Despite the errors that are now apparent in these concepts, the fluid/particle analogies of matter proposed by these early philosophers spurred development of the science of mechanics and of objective scientific processes to test these models. For instance, Kelvin's quest to uncover the structure of vortex atoms resulted in discovery of many of the basic laws and phenomena associated with vorticity transport in inviscid fluids. The wave/particle models used in modern physics to describe the theory of light and quantum models of elementary particles, as well as the quantum vortices in liquid helium II (London, 1954) and the "magnetic vortices" in high-temperature superconducting materials (Tinkham, 1975), have their base in the fluid/particle models of matter developed in ancient times.

Fluid dynamics has also had a large impact on development of modern applied mathematics. For instance, linear partial differential equations are usually categorized as hyperbolic, parabolic, or elliptic, where the classic examples for these three categories are the wave equation, the heat equation, and the Laplace equation, all of which play a prominent role in description of fluid systems. Two important paradigms for development of the theory of characteristics of nonlinear hyperbolic partial differential equations (Courant and Hilbert, 1962) are wave propagation in compressible gas dynamics and free-surface oscillations in shallow water layers.

Kolmogorov's advancements in stochastic analysis found immediate application in his models of the energy cascade process of turbulent flows (Hunt et al., 1991). One of the earliest investigations of deterministic chaos was performed by Lorentz (1963) using a model of atmospheric transport, and further investigations into chaotic systems have been spurred by the need to formulate better models for turbulent flows. Fractal geometry is also commonly exhibited by fluid systems, as illustrated by many of the examples given in the book by Mandelbrot (1977).

Despite its importance to many fields, many centuries of study, and widespread use of advanced supercomputing systems, scientists and engineers specializing in fluid dynamics are still far from able to reliably predict most fluid flow problems. The principal difficulty lies in the inherent nonlinearity of the equations governing fluid flow. This nonlinearity arises from the fluid inertia and is responsible for instabilities and eventual transition of the flow to a turbulent state. Turbulent flows span an enormous range of length scales, with the ratio of the largest to the smallest scale exceeding a factor of 10^5 in many marine and aerospace engineering applications and a factor of 10^8 for flow in the earth's oceans and atmosphere. This wide range of length scales makes direct computational solution of the governing equations impossible for all but a few rather academic cases at low Reynolds numbers, thus requiring the use of models, often combined with empiricism, to truncate the mathematics to a manageable system. The situation is made yet more difficult by the fact that many natural and industrial processes involve particle or droplet transport, phase change, and chemical reactions that influence the fluid momentum transport, introducing still broader scales and breaking down self-similar behavior of the turbulence over these scales.

Because of its many practical applications, its use as a paradigm to stimulate new mathematical methods and physical models in diverse other fields, and its intrinsic beauty and difficulty, fluid dynamics remains an active field of both engineering and fundamental scientific inquiry. In this introductory chapter, we examine the role of viscous forces in fluid flows using simple scaling arguments and then explore the relevance of inviscid flow theory to real fluid flow problems.

1.1 ROLE OF VISCOSITY IN HIGH-REYNOLDS-NUMBER FLOWS

All real fluid flows are viscous, with the possible exception of quantum fluids such as liquid helium II below the transition temperature. One measure of the importance of viscous effects is given by the flow Reynolds number, $\text{Re} = \rho UL/\mu$, which is the ratio of the order of magnitude of the characteristic inertial stress ρu^2 to the viscous stress $\mu(\partial u/\partial y)$. Here ρ and μ are the fluid density and viscosity, and U and L are characteristic velocity and length scales of the problem. When the Reynolds number is $O(1)$ or less, the viscous forces are important everywhere in the flow. This type of flow is often called *creeping flow*, and it is commonly observed in situations with extremely low velocities and high viscosities, such as certain liquid metal melts or convective circulation in the earth's mantle, and in flows with very small length scales, such as locomotion of microorganisms, dispersion of particulate matter in two-phase flow, and flow within microfluidic devices. For most problems on a human scale, the Reynolds number is quite large. For instance, a man standing in a mild 10-m/s breeze has a Reynolds number of about 3×10^5. For a baseball thrown by a major league pitcher, the Reynolds number is about 2×10^5. Water flow in a 2.5-cm-diameter bathroom supply pipe has Reynolds number (based on pipe diameter) of about 2×10^4, whereas a modest river might have Reynolds number (based on width) of 5×10^7. An automobile driving at highway speeds has a Reynolds number of about 1×10^7, and the wing of a commercial jet airplane has Reynolds number (based on chord length) of about 2×10^8. The Reynolds numbers of major atmospheric or oceanic features are quite large; for instance, for a typical atmospheric low-pressure system $\mathrm{Re} \sim 10^{12}$ and for an oceanic Gulf Stream ring eddy $\mathrm{Re} \sim 5 \times 10^{10}$.

For flows with Reynolds number much larger than unity, viscous forces will be of importance only in regions with small length scales or over very long convective time scales. This statement can be made more precise by the following argument. Let us consider a steady-state flow with length scale L characteristic of the flow geometry (e.g., the body diameter or channel width), and assume that viscous forces are important only within a small layer $\mathfrak{R}$ with length scale δ in the cross-stream direction, as illustrated in Figure 1.1. The inertial force acting on the region $\mathfrak{R}$ has order of magnitude $\rho U^2 \delta L$, which is simply the inertial stress times the cross-sectional area δL of $\mathfrak{R}$. The viscous shear force acting on the lateral surfaces of $\mathfrak{R}$ has order of magnitude $\mu U L^2/\delta$. Equating these two forces and solving for the length scale ratio gives

$$\frac{\delta}{L} = O(\mathrm{Re}^{-1/2}). \qquad (1.1.1)$$

As the Reynolds number increases, the thickness of the region in which viscous forces can be important correspondingly decreases compared to the length scale L that characterizes the flow as a whole. The estimate (1.1.1) is characteristic of the ratio of thickness of a laminar viscous boundary layer along a body surface to the body diameter (Figure 1.2), where the formation of the boundary layer is necessary in order to satisfy the no-slip condition for the tangential velocity component, which cannot in general be satisfied in inviscid flow theory. Viscous forces can also be important in internal regions within a fluid flow, such as in the vortex reconnection

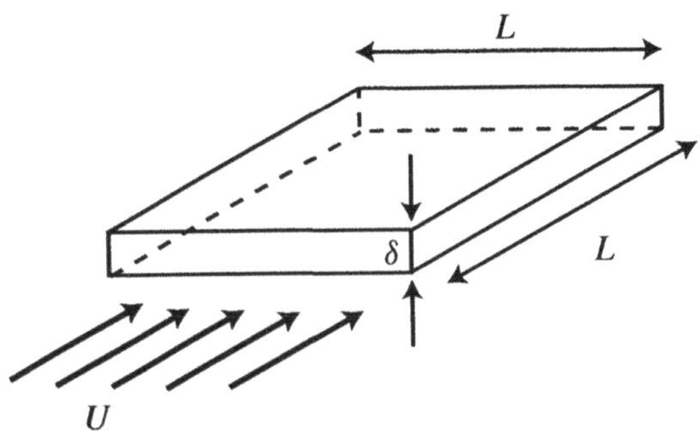

Figure 1.1 Schematic of a thin viscous layer $\mathfrak{R}$ with thickness δ and lateral dimension L.

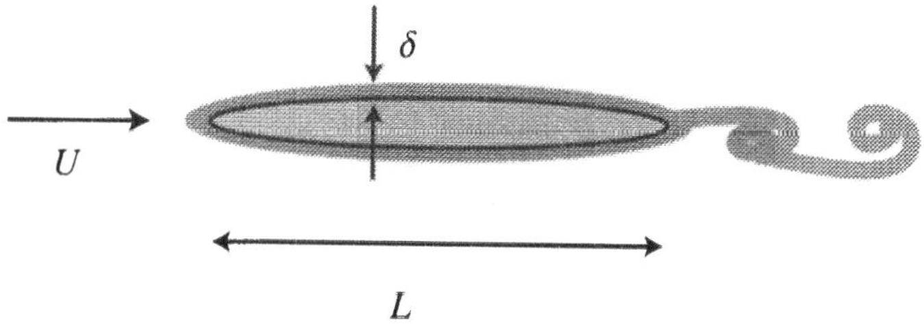

Figure 1.2 Boundary layer of thickness δ formed around a body of length L.

problem shown in Figure 1.3. In this problem, two vortex tubes with vorticity of opposite sign collide due to inviscid instability, deforming the vortex cores as they are driven together. Viscous forces become important in a thin layer in-between the impacting tubes, in which viscous cross diffusion causes cancellation of vorticity between the two structures, allowing the vortex lines within each tube to be cut and to reconnect to vortex lines in the opposing tube to form loops.

Viscous forces are important everywhere in a turbulent flow, but only at small scales of motion. We equate the flow length scale L to the turbulence integral length scale, which is characteristic of the eddy size containing the most energetic turbulent fluctuations, and let U be proportional to the square root of the turbulent kinetic energy. Viscous dissipation is important on the smallest scale of turbulent motion, denoted by δ. Following the reasoning of Kolmogorov (see Hunt et al., 1991), we assume that the net rate of energy dissipation in the turbulent flow is controlled by the rate at which energy cascades from larger scales to smaller scales of motion. The dissipation scale δ (called the *Kolmogorov scale*) is thus assumed to vary as a function of the average dissipation rate per unit mass ε and the kinematic viscosity $\nu \equiv \mu/\rho$, such that from elementary dimensional analysis $\delta = O(\nu^3/\varepsilon)^{1/4}$. The energy dissipation rate ε is assumed to be proportional to the turbulent kinetic energy U^2 divided by the large eddy turnover time L/U, such that $\varepsilon = O(U^3/L)$. Substituting into the expression for the Kolmogorov length scale δ gives the ratio of the smallest to the

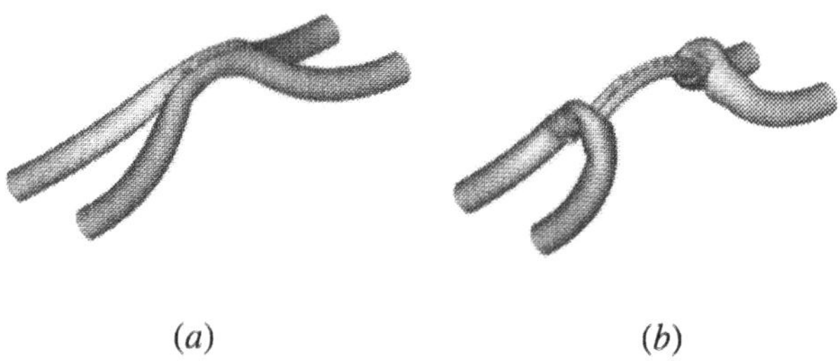

Figure 1.3 Two colliding vortex tubes (*a*) just prior to and (*b*) following vortex reconnection. The computation is performed using a pseudospectral method with vortex Reynolds number $\mathrm{Re} = \Gamma/\nu = 1500$. The vortex surfaces are identified using the λ_2 method of Jeong and Hussain (1995).

largest length scales of a turbulent flow as

$$\frac{\delta}{L} = O(\mathrm{Re}^{-3/4}). \tag{1.1.2}$$

Turbulent flows are characterized by high Reynolds numbers and a broad range of length scales, ranging from scales characteristic of the large-scale flow geometry to the dissipation scale. Viscous effects are negligible at all but the smallest scales of this range.

If both viscous and inertial forces are restricted to act over the same length scale L, then the viscous forces must act much more slowly than the convective forces in a high-Reynolds-number flow. The ratio of viscous and convective time scales, defined by $T_V = \rho L^2/\mu$ and $T_C = L/U$, respectively, is given by

$$\frac{T_V}{T_C} = O(\mathrm{Re}). \tag{1.1.3}$$

The estimate (1.1.3), for instance, is characteristic of the ratio of viscous decay time to turnover time of a vortex patch.

1.2 INVISCID FLOWS

In view of the length scale estimates presented in the previous section, it might be supposed that inviscid flow can be obtained as a limit of actual viscous flows as the Reynolds number increases to infinity. Actually, this is not quite the case. There are certain observed properties of viscous flows that, no matter how high the Reynolds number, occur in violation of inviscid flow theory. For instance, while the turbulent boundary layer about a blunt body grows progressively thinner as the Reynolds number increases (with the body size held fixed), the location of boundary layer separation in the rear of the body is approximately independent of Reynolds number. Even if the Reynolds number is increased indefinitely, the boundary layer will still separate, ejecting vorticity from the body, whereas inviscid flow theory requires vorticity generated on the body surface to remain on the surface. As another example, we note that many turbulent flows are observed to adopt a self-similar state, in which the large scales of motion are not influenced by a change in Reynolds number (Tennekes and Lumley, 1972). In such flows, energy dissipation is controlled by the cascade of energy from the large scales, such that with increasing Reynolds number the energy dissipation occurs at progressively smaller scales but at nearly the same overall dissipation rate. In an inviscid flow, of course, no energy dissipation is possible.

If inviscid flow theory can never provide a complete model of actual fluid flow, even in the high-Reynolds-number limit, the reader might reasonably ask, why study it? Why not proceed directly to viscous flows, or better yet to turbulent flows? The reason is that fluid flows are hard to solve, many of them much too hard even with our fastest computers without invoking some form of ad hoc model. Inviscid flow theory, if employed judiciously and with a clear understanding of when it is valid and when it must be replaced by more general flow theories, offers considerable simplification to fluid flow problems. In particular, the differential equation governing momentum transport in inviscid flows possesses only first-order spatial derivatives instead of the second-order derivatives in the full viscous flow momentum equation. This seemly minor difference has important implications: it allows integration of the momentum equation to obtain a solution for pressure in irrotational flows and it makes possible an exact integral of the vorticity transport equation for all inviscid flows, which can be used to derive a number of fundamental invariant quantities. These simplifications make possible a wide range of powerful analytical and computational solution methods for inviscid flows. For many inviscid flows, the initial conditions and the inlet and outlet conditions are such that one can argue that vorticity must vanish completely. In such *irrotational* (or *potential*) flows, the velocity field can be found by solution of a simple linear equation (the Laplace equation) in regions with no vorticity. This condition can be applied, for instance, to flow past airfoils (if one ignores the thin downstream wake), for which the inviscid theory can be used to obtain accurate solutions for the airfoil lift. In other flows, the vorticity can be confined to discrete points or a thin sheet. The evolution of the vorticity can be followed using simple computational methods, where the velocity field anywhere in the fluid can be obtained by solution of an integral over the vorticity field. Vortex sheet models, for instance, are commonly used for solution of interfacial or free-surface wave fields. Even in flows with distributed vorticity support, the invariant properties of the inviscid theory can be used to show that the vorticity support is time independent, such that on points where the vorticity initially vanishes, it must always vanish. This fact leads to efficient and accurate numerical methods in which the computational points follow the evolving vorticity field.

Although inviscid flow theory may not be applicable to all parts of a fluid flow, except in special circumstances, it does apply to most regions of high-Reynolds-number flows. Thus for laminar flow past a body, the potential flow field external to the boundary layer and the evolution of vortical structures in the body wake can be efficiently modeled using inviscid flow theory. In turbulent flow, both the large-scale instabilities leading to turbulence and the broad turbulent inertial range are controlled by inviscid effects. Even in situations where a more refined viscous flow solution is desirable, usually requiring a time-consuming physical experiment or computational solution, the comparatively simple results attainable from inviscid flow theory can be used to interpret and provide physical intuition for many aspects of the experimental or computational results.

BIBLIOGRAPHY

Courant, R., and D. Hilbert (1962). *Methods of Mathematical Physics*, Vol. II: *Partial Differential Equations*, John Wiley & Sons, New York (reprinted 1989).

Jeong, J., and F. Hussain (1995). "On the identification of a vortex," *Journal of Fluid Mechanics* **285**, 69–94.

Hunt, J.C.R., O.M. Phillips, and D. Williams (1991). *Turbulence and Stochastic Processes: Kolmogorov's Ideas 50 Years On*, Cambridge University Press, London (also *Proceedings of the Royal Society of London A* **434**, 1–240, 1991).

Kelvin, Lord (1867). "On vortex atoms," *Philosophical Magazine* **34**, 15–24 (reprinted *Mathematical and Physical Papers*, Vol. 4, Cambridge University Press, Cambridge, 1910).

London, F. (1954). *Superfluids*, Vol. II: *Macroscopic Theory of Superfluid Helium*, John Wiley & Sons, New York (reprinted Dover Publications, New York, 1964).

Lorentz, E.N. (1963). "Deterministic non-periodic flow," *Journal of Atmospheric Science* **20**, 130–141.

Lugt, H.J. (1983). *Vortex Flow in Nature and Technology*, John Wiley & Sons, New York.

Mandelbrot, B.B. (1977). *The Fractal Geometry of Nature*, W.H. Freeman and Company, New York.

Robinson, J.M. (1968). *An Introduction to Early Greek Philosophy*, Houghton Mifflin, Boston.

Tennekes, H., and J.L. Lumley (1972). *A First Course in Turbulence*, MIT Press, Cambridge, MA.

Tinkham, M. (1975). *Introduction to Superconductivity*, McGraw-Hill, New York.

Tokaty, G.A. (1971). *A History and Philosophy of Fluid Mechanics*, G.T. Foulis & Company, Oxfordshire, England (reprinted Dover Publications, New York, 1994).

Truesdell, C.A. (1954). *The Kinematics of Vorticity*, Indiana University Press, Bloomington, IN.

CHAPTER 2

VECTORS AND TENSORS

The variables describing the motion of a fluid consist of scalar quantities such as density and pressure, vector quantities such as velocity and acceleration, and quantities forming second-order tensors, such as velocity gradient and stress. As we shall see, vector-valued quantities can be considered as first-order tensors and scalars can be considered as zero-order tensors. Before one can fully appreciate the mathematical structure of the theory of fluid motion, it is first necessary to develop some background in the theory of tensors.

2.1 CARTESIAN INDEX NOTATION

In this text, we employ a common notation in which scalar variables are written in italics and vector- and tensor-valued variables are written in boldface type. We will generally write vectors with Greek or lowercase Roman letters and second-order tensors with capital Roman letters. Although a tensor can be defined without regard to a particular system of coordinates, it is often desirable to refer to the *components* of a tensor with respect to a specific coordinate system. For simplicity, we shall always refer the components of tensors to a Cartesian coordinate system, unless stated otherwise. A brief summary of various vector operations in non-Cartesian coordinate systems is given in an appendix.

The three base vectors of a Cartesian coordinate system are denoted by $\{\mathbf{e}_1, \mathbf{e}_2, \mathbf{e}_3\}$. Cartesian coordinate systems are orthogonal, such that the scalar product of any two different base vectors is zero, and the base vectors have unit magnitudes. The components of a vector $\mathbf{v}$ with respect to this coordinate system are scalars $\{v_1, v_2, v_3\}$ such that

$$\mathbf{v} = v_1\mathbf{e}_1 + v_2\mathbf{e}_2 + v_3\mathbf{e}_3 = \sum_{i=1}^{3} v_i\mathbf{e}_i. \tag{2.1.1}$$

Writing out the various terms of an equation can become very tedious when dealing with complex vector and tensor expressions, which are found frequently in the laws governing fluid motion, and often obscures the basic physical significance of these terms. While use of the summation symbol $\sum$, as in (2.1.1), allows us to avoid writing out each term, this notation also becomes tedious during long calculations. In this text, we instead follow the common implicit summation convection, by which (2.1.1) is written as

$$\mathbf{v} = v_i\mathbf{e}_i, \tag{2.1.2}$$

and the repetition of the index i in (2.1.2) implies summation over all values that the index may take. Index values typically cover the set $\{1, 2, 3\}$ in a three-dimensional space and $\{1, 2\}$ in a two-dimensional space. The *summation convention* can be stated as follows: *whenever an index is repeated in a given term, sum over all values of the index.* In this convention, quantities separated by a plus or minus sign or an equality are considered different terms, but quantities separated by a product are in the same term.

Example 2.1.1. Write out the products $a_i b_i$ and $a_i b_i c_j$ using the summation convention.

SOLUTION

$$a_i b_i = a_1 b_1 + a_2 b_2 + a_3 b_3, \tag{2.1.3a}$$

$$a_i b_i c_j = a_1 b_1 c_j + a_2 b_2 c_j + a_3 b_3 c_j. \tag{2.1.3b}$$

A repeated index, over which summation is performed, is called a *dummy index* and a nonrepeated index is called a *free index*. For instance, i is a dummy index in equations (2.1.3), but j is a free index in (2.1.3b). Because equations (2.1.3) are written in terms of the components of vectors, rather than the vector and tensor quantities themselves, we say that such equations are written using *index notation*. When using index notation, two precautions must be kept in mind. The first is that a dummy index may appear only twice in any given term. To repeat the dummy index more than once in a given term, e.g., $a_i b_i c_i$, has no meaning in the summation convention and can lead to ambiguities and erroneous results. The second precaution is that the same free indices must appear in *every* term in an equation. For instance, the equation $a_i = 3 + b_i$ has no meaning since there is no free index i associated with the first term on the right-hand side.

The basis of all tensors of order 2 or higher can be constructed from the set of Cartesian base vectors $\mathbf{e}_i$ using the tensor product, denoted by $\otimes$. The tensor product of two vectors is a second-order tensor, and more generally the tensor product of n vectors is an nth order tensor. For instance, for a second-order tensor $\mathbf{A}$ with components A_{ij}, we can write

$$\mathbf{a} = A_{ij}\mathbf{e}_i \otimes \mathbf{e}_j = A_{ij}\mathbf{e}_i\mathbf{e}_j. \tag{2.1.4}$$

The combination of two vectors written next to each other, such as $\mathbf{uv}$ or $\mathbf{e}_1\mathbf{e}_2$, is called a *dyad* and is an alternative notation for the tensor product. Expanding (2.1.4) using the summation convention gives

$$\begin{aligned}\mathbf{a} = {} & A_{11}\mathbf{e}_1\mathbf{e}_1 + A_{12}\mathbf{e}_1\mathbf{e}_2 + A_{13}\mathbf{e}_1\mathbf{e}_3 \\ & + A_{21}\mathbf{e}_2\mathbf{e}_1 + A_{22}\mathbf{e}_2\mathbf{e}_2 + A_{23}\mathbf{e}_2\mathbf{e}_3 \\ & + A_{31}\mathbf{e}_3\mathbf{e}_1 + A_{32}\mathbf{e}_3\mathbf{e}_2 + A_{33}\mathbf{e}_3\mathbf{e}_3.\end{aligned} \tag{2.1.5}$$

The summation convention also applies to the indices of tensors, such that, for instance, in the equation

$$A_{ij}b_j = A_{i1}b_1 + A_{i2}b_2 + A_{i3}b_3, \tag{2.1.6}$$

i is a free index and j is a dummy index.

2.2 KRONECKER DELTA AND PERMUTATION TENSOR

We now introduce two special tensors which are useful in a large number of vector operations. The first tensor is the identity tensor $\mathbf{I}$, whose components δ_{ij} are called the *Kronecker delta* and are defined such that δ_{ij} equals unity if $i = j$ and vanishes otherwise. Multiplication of the components a_i of any vector by the Kronecker delta δ_{ij} yields a change in the index as follows:

$$a_i\delta_{ij} = a_1\delta_{1j} + a_2\delta_{2j} + a_3\delta_{3j} = a_j. \tag{2.2.1}$$

This operation, called *contraction*, is simply the indicial form of the identity $\mathbf{I}\cdot\mathbf{a} = \mathbf{a}$. Using the summation convention again, we note that $\delta_{ii} = \delta_{11} + \delta_{22} + \delta_{33} = 3$. From the definition of δ_{ij} and the orthogonality property of the Cartesian base vectors, it follows that

$$\mathbf{e}_i \cdot \mathbf{e}_j = \delta_{ij}, \tag{2.2.2}$$

where the centered dot denotes the scalar product. In a Cartesian coordinate system, we further find that if $\mathbf{x}$ denotes the position vector with components x_i, then

$$\frac{\partial x_i}{\partial x_j} = \delta_{ij}. \tag{2.2.3}$$

Another useful tensor is the *permutation tensor*, the components of which are denoted by ε_{ijk} and are defined such that ε_{ijk} equals 1 if (i, j, k) are cyclic, ε_{ijk}

equals -1 if (i, j, k) are anticyclic, and ε_{ijk} vanishes if any of (i, j, k) are the same. The permutation symbol ε_{ijk} denotes the components of a third-order tensor and thus has three free indices. The value of ε_{ijk} is unchanged by any cyclic change of its indices (i.e., $\varepsilon_{ijk} = \varepsilon_{kij} = \varepsilon_{jki}$) and its sign is changed whenever two indices are interchanged. The permutation symbol is related to the vector product of Cartesian base vectors by

$$\mathbf{e}_i \times \mathbf{e}_j = \varepsilon_{ijk}\mathbf{e}_k. \tag{2.2.4}$$

A useful identity that relates the Kronecker delta and permutation symbol is

$$\varepsilon_{ijk}\varepsilon_{imn} = \delta_{jm}\delta_{kn} - \delta_{jn}\delta_{km}, \tag{2.2.5}$$

which can be proven by direct expansion.

Example 2.2.1. Prove that $\varepsilon_{ijk}\varepsilon_{ijk} = 6$.

SOLUTION. Use (2.2.5) to write

$$\varepsilon_{ijk}\varepsilon_{ijk} = \delta_{jj}\delta_{kk} - \delta_{jk}\delta_{kj} = \delta_{jj}\delta_{kk} - \delta_{jj} = (3 \cdot 3) - 3 = 6,$$

where we recall that $\delta_{ii} = 3$ and have used the contraction operation (2.2.1) to write $\delta_{jk}\delta_{kj} = \delta_{jj}$.

2.3 VECTOR AND TENSOR OPERATIONS

It is often easier to manipulate a complicated expression involving vectors, tensors, and their derivatives by writing the expression in terms of the components of the vectors and tensors. It is therefore desirable to be fluent at the process of transforming combinations of vectors and tensors into their components, and vice versa. Letting **u** and **v** denote two arbitrary vectors with components u_i and v_i, the scalar and vector products of **u** and **v** can be written as

$$\mathbf{u} \cdot \mathbf{v} = u_i v_i, \tag{2.3.1}$$

$$\mathbf{u} \times \mathbf{v} = \varepsilon_{ijk} u_j v_k \mathbf{e}_i. \tag{2.3.2}$$

If **A** and **B** are arbitrary second-order tensors, with components A_{ij} and B_{ij}, we can write

$$\mathbf{A} \cdot \mathbf{B} = A_{ij} B_{jk} \mathbf{e}_i \otimes \mathbf{e}_k, \tag{2.3.3}$$

$$\mathbf{A} : \mathbf{B} = A_{ij} B_{ji}. \tag{2.3.4}$$

The single centered dot is used to denote repetition of the innermost index of **A** and **B** such that the product is also a second-order tensor. A double dot denotes that the outermost index of **A** and **B** is also repeated such that the product is a scalar. Different notations for products of tensors are sometimes used in the literature.

The *transpose* of a tensor **A**, denoted with a superscript T, can be formed simply by switching the order of the indices of the components of **A** such that

$$\mathbf{A}^{\mathrm{T}} = A_{ji}\mathbf{e}_i \otimes \mathbf{e}_j. \tag{2.3.5}$$

Example 2.3.1. Use index notation to show that

$$(\mathbf{A} \cdot \mathbf{B})^{\mathrm{T}} = \mathbf{B}^{\mathrm{T}} \cdot \mathbf{A}^{\mathrm{T}}. \tag{2.3.6}$$

SOLUTION. Using (2.3.3) and (2.3.5) gives

$$\begin{aligned}
(\mathbf{A} \cdot \mathbf{B})^{\mathrm{T}} &= (A_{ij}B_{jk}\mathbf{e}_i \otimes \mathbf{e}_k)^{\mathrm{T}} \\
&= A_{kj}B_{ji}\mathbf{e}_i \otimes \mathbf{e}_k \\
&= (B_{ji}\mathbf{e}_i \otimes \mathbf{e}_j) \cdot (A_{k\ell}\mathbf{e}_\ell \otimes \mathbf{e}_k) \\
&= \mathbf{B}^{\mathrm{T}} \cdot \mathbf{A}^{\mathrm{T}}.
\end{aligned}$$

The *trace* of a tensor **A** is the sum of all diagonal components of **A**, or

$$\mathrm{tr}(\mathbf{A}) = A_{ii}. \tag{2.3.7}$$

The *determinant* of **A** can be related to its components by the identity

$$\det(\mathbf{A}) = \varepsilon_{ijk}A_{1i}A_{2j}A_{3k}, \tag{2.3.8}$$

or alternatively by

$$\varepsilon_{qmn}\det(\mathbf{A}) = \varepsilon_{ijk}A_{iq}A_{jm}A_{kn}. \tag{2.3.9}$$

Example 2.3.2. Show that if **A** and **B** are second-order tensors, then

$$\det(\mathbf{A} \cdot \mathbf{B}) = \det(\mathbf{A})\det(\mathbf{B}). \tag{2.3.10}$$

SOLUTION. Using (2.3.3), (2.3.8), and (2.3.9), we have

$$\begin{aligned}
\det(\mathbf{A} \cdot \mathbf{B}) &= \det(A_{ik}B_{kj}\mathbf{e}_i \otimes \mathbf{e}_j) \\
&= \varepsilon_{ijk}A_{1q}B_{qi}A_{2m}B_{mj}A_{3n}B_{nk} \\
&= (A_{1q}A_{2m}A_{3n})(\varepsilon_{ijk}B_{qi}B_{mj}B_{nk}) \\
&= (A_{1q}A_{2m}A_{3n})\varepsilon_{qmn}\det(\mathbf{B}) \\
&= \det(\mathbf{A})\det(\mathbf{B}).
\end{aligned}$$

The inner product of a vector **u** and a tensor **A** satisfies

$$\mathbf{u} \cdot \mathbf{A} = \mathbf{A}^{\mathrm{T}} \cdot \mathbf{u} = u_j A_{ji}\mathbf{e}_i. \tag{2.3.11}$$

The *scalar triple product* of three vectors **a**, **b**, and **c** is defined by

$$[\mathbf{abc}] \equiv (\mathbf{a} \times \mathbf{b}) \cdot \mathbf{c} = \varepsilon_{ijk} a_i b_j c_k. \tag{2.3.12}$$

The scalar triple product is invariant to a cyclic change in the vectors **a**, **b**, and **c**.

A variety of differential operations with vectors and tensors are commonly written in terms of the *del operator* ∇, which is a vector-valued operator defined by

$$\nabla \equiv \mathbf{e}_i \frac{\partial}{\partial x_i}. \tag{2.3.13}$$

For instance, the *divergence* of a vector **u** can be written as

$$\nabla \cdot \mathbf{u} = \frac{\partial u_i}{\partial x_i}. \tag{2.3.14}$$

The *curl* of **u** is defined by

$$\nabla \times \mathbf{u} = \varepsilon_{ijk} \frac{\partial u_k}{\partial x_j} \mathbf{e}_i. \tag{2.3.15}$$

The *gradients* of a scalar ϕ or of a vector **u** are defined by

$$\nabla \phi = \frac{\partial \phi}{\partial x_i} \mathbf{e}_i, \qquad \nabla \mathbf{u} = \frac{\partial u_j}{\partial x_i} \mathbf{e}_i \otimes \mathbf{e}_j. \tag{2.3.16}$$

The *directional derivative* of a scalar ϕ in the direction of a unit vector **a** is defined by

$$\mathbf{a} \cdot \nabla \phi = a_i \frac{\partial \phi}{\partial x_i}. \tag{2.3.17}$$

The *Laplacian* of a scalar ϕ is a scalar-valued operator ∇^2 defined by

$$\nabla^2 \phi \equiv \nabla \cdot \nabla \phi = \frac{\partial^2 \phi}{\partial x_i \, \partial x_i}, \tag{2.3.18}$$

and the Laplacian of a vector is defined similarly.

2.4 VECTOR AND TENSOR TRANSFORMATIONS

A second-order tensor **A** is said to be *orthogonal* if the transpose $\mathbf{A}^T$ of **A** is equal to its inverse $\mathbf{A}^{-1}$. It follows that if **A** is orthogonal, then $\mathbf{A} \cdot \mathbf{A}^T = \mathbf{A}^T \cdot \mathbf{A} = \mathbf{I}$. Applying the identity (2.3.10), and noting that the determinant of the transpose $\mathbf{A}^T$ is the same as the determinant of **A** for any tensor **A**, it follows that the determinant of any orthogonal tensor **A** must equal ± 1. The components A_{ij} of an orthogonal tensor satisfy the relation $A_{ij}A_{kj} = \delta_{ik}$.

Let v_i be the components of some vector **v** and D_{ij} be the components of some second-order tensor **D** with respect to a set of Cartesian base vectors $\mathbf{e}_i$. Suppose now that we wish to write **v** and **D** in terms of their components v'_i and D'_{ij} with respect to a different set of Cartesian base vectors $\mathbf{e}'_i$, where $\mathbf{e}_i$ and $\mathbf{e}'_i$ differ by a rotation. Let **A** denote the *direction cosine tensor*, with components A_{ij}, defined by

$$\mathbf{e}'_i = A_{ij}\mathbf{e}_j, \tag{2.4.1}$$

where we note that **A** is orthogonal. Writing **v** and **D** in terms of their components and using (2.4.1) give

$$\mathbf{v} = v_i\mathbf{e}_i = v'_i\mathbf{e}'_i = v'_i A_{ij}\mathbf{e}_j, \tag{2.4.2}$$

$$\mathbf{D} = D_{ij}\mathbf{e}_i \otimes \mathbf{e}_j = D'_{ij}\mathbf{e}'_i \otimes \mathbf{e}'_j = D'_{ij}A_{ik}A_{jm}\mathbf{e}_k \otimes \mathbf{e}_m. \tag{2.4.3}$$

From (2.4.2) and (2.4.3), the vector and tensor components in the two coordinate systems are found to be related by

$$v_j = v'_i A_{ij}, \qquad D_{km} = D'_{ij}A_{ik}A_{jm}. \tag{2.4.4}$$

Multiplying the vector and tensor transformation equations in (2.4.4) by A_{kj} and $A_{nk}A_{pm}$, respectively, and making use of the orthogonality properties of A_{ij} give the components in the primed system in terms of those in the unprimed system as

$$v'_k = A_{kj}v_j, \qquad D'_{np} = A_{nk}A_{pm}D_{km}. \tag{2.4.5}$$

The transformation relationships (2.4.4) and (2.4.5) are satisfied if and only if the numbers v_i and D_{ij} are components of tensors. We can in general define a *tensor* **T** of order N as some quantity whose components $T_{i_1 i_2 \cdots i_N}$ transform according to

$$T'_{i_1 i_2 \cdots i_N} = A_{i_1 j_1}A_{i_2 j_2} \cdots A_{i_N j_N} T_{j_1 j_2 \cdots j_N} \tag{2.4.6}$$

under a rotation of the coordinate system (Sokolnikoff, 1951).

2.5 SYMMETRIC AND SKEW-SYMMETRIC TENSORS

A second-order tensor **A** with components A_{ij} is said to be *symmetric* if $\mathbf{A} = \mathbf{A}^{\mathrm{T}}$, or $A_{ij} = A_{ji}$. The tensor is said to be *skew-symmetric* if $\mathbf{A} = -\mathbf{A}^{\mathrm{T}}$, or $A_{ij} = -A_{ji}$. A symmetric second-order tensor has six independent components and a skew-symmetric second-order tensor has three independent components. The diagonal components of a skew-symmetric tensor are all zero. By way of example, the Kronecker delta δ_{ij} is symmetric, so that $\delta_{ij} = \delta_{ji}$. The permutation symbol ε_{ijk} is skew-symmetric in any two of its indices, so that $\varepsilon_{ijk} = -\varepsilon_{jik} = -\varepsilon_{ikj}$. We now present two theorems related to symmetric and skew-symmetric tensors.

Theorem 2.5.1. If **A** is an arbitrary second-order symmetric tensor and **B** is an arbitrary second-order skew-symmetric tensor with components A_{ij} and B_{ij}, respectively, then the scalar product of **A** and **B**, written as **A** : **B** or $A_{ij}B_{ji}$, vanishes.

Theorem 2.5.2. Any second-order tensor **C** can be decomposed as the sum of a symmetric tensor **A** and a skew-symmetric tensor **B**, where

$$\mathbf{C} = \mathbf{A} + \mathbf{B}, \qquad \mathbf{A} = \tfrac{1}{2}(\mathbf{C} + \mathbf{C}^{\mathrm{T}}), \qquad \mathbf{B} = \tfrac{1}{2}(\mathbf{C} - \mathbf{C}^{\mathrm{T}}). \tag{2.5.1}$$

The proof of the first theorem can be obtained by applying the definitions of symmetric and skew-symmetric tensors as follows: $A_{ij}B_{ji} = A_{ji}B_{ji} = -A_{ji}B_{ij}$. Since the dummy indices can be given any label, the indices i and j in the above result can be switched to obtain $A_{ij}B_{ji} = -A_{ij}B_{ji}$, which can only be satisfied if the product $A_{ij}B_{ji}$ vanishes. The proof of the second theorem follows from substitution using (2.5.1).

Any second-order skew-symmetric tensor **B** can be associated with a unique *axial vector* **w** whose components are given by

$$B_{ij} = \varepsilon_{ijk} w_k, \tag{2.5.2}$$

where both **B** and **w** contain three independent components. Multiplying (2.5.2) by $\varepsilon_{ij\ell}$ and using (2.2.5) give

$$\varepsilon_{ij\ell} B_{ij} = \varepsilon_{ij\ell}\varepsilon_{ijk} w_k = (\delta_{jj}\delta_{\ell k} - \delta_{jk}\delta_{j\ell}) w_k = 3w_\ell - w_\ell = 2w_\ell,$$

or

$$w_k = \tfrac{1}{2}\varepsilon_{ijk} B_{ij}. \tag{2.5.3}$$

Taking the vector product of any arbitrary vector **a** with the vector **w**, defined by (2.5.3), and using (2.2.5), it readily follows that

$$\mathbf{a} \times \mathbf{w} = \mathbf{B} \cdot \mathbf{a}. \tag{2.5.4}$$

2.6 VECTOR IDENTITIES

Familiarity with vector identities is necessary in proving a variety of fundamental theorems in fluid dynamics. Some of the more frequently used vector identities, written in terms of an arbitrary scalar ϕ and arbitrary vectors **a**, **b**, and **c**, are given in Table 2.1. While the identities in Table 2.1 can be proved without regard to any particular coordinate system, it is a useful exercise to rewrite these identities in their component forms and prove them (for Cartesian coordinates) using the properties of the Kronecker delta and the permutation tensor and the theorems on symmetric and skew-symmetric tensors given in the previous sections. A few examples are given below and proofs of the rest of the identities in Table 2.1 are left as exercises.

TABLE 2.1 A Collection of Useful Vector Identities

$\nabla \times \nabla\phi = 0$	(2.6.1)
$\nabla \cdot (\nabla \times \mathbf{a}) = 0$	(2.6.2)
$\nabla^2\mathbf{a} = \nabla(\nabla \cdot \mathbf{a}) - \nabla \times (\nabla \times \mathbf{a})$	(2.6.3)
$\nabla \cdot (\phi\mathbf{a}) = \phi(\nabla \cdot \mathbf{a}) + \mathbf{a} \cdot \nabla\phi$	(2.6.4)
$\nabla \times (\phi\mathbf{a}) = \phi(\nabla \times \mathbf{a}) + (\nabla\phi) \times \mathbf{a}$	(2.6.5)
$\nabla(\mathbf{a} \cdot \mathbf{b}) = (\mathbf{a} \cdot \nabla)\mathbf{b} + (\mathbf{b} \cdot \nabla)\mathbf{a} + \mathbf{a} \times (\nabla \times \mathbf{b}) + \mathbf{b} \times (\nabla \times \mathbf{a})$	(2.6.6)
$\nabla \cdot (\mathbf{a} \times \mathbf{b}) = \mathbf{b} \cdot (\nabla \times \mathbf{a}) - \mathbf{a} \cdot (\nabla \times \mathbf{b})$	(2.6.7)
$\nabla \times (\mathbf{a} \times \mathbf{b}) = \mathbf{a}(\nabla \cdot \mathbf{b}) - \mathbf{b}(\nabla \cdot \mathbf{a}) + (\mathbf{b} \cdot \nabla)\mathbf{a} - (\mathbf{a} \cdot \nabla)\mathbf{b}$	(2.6.8)
$\mathbf{a} \times (\mathbf{b} \times \mathbf{c}) = (\mathbf{a} \cdot \mathbf{c})\mathbf{b} - (\mathbf{a} \cdot \mathbf{b})\mathbf{c}$	(2.6.9)

Example 2.6.1. Prove the identity $\nabla \cdot (\mathbf{a} \times \mathbf{b}) = \mathbf{b} \cdot (\nabla \times \mathbf{a}) - \mathbf{a} \cdot (\nabla \times \mathbf{b})$ using Cartesian index notation.

SOLUTION. Rewriting the term on the left-hand side in index notation and using the chain rule give

$$\nabla \cdot (\mathbf{a} \times \mathbf{b}) = \frac{\partial}{\partial x_i}(\varepsilon_{ijk}a_j b_k) = \varepsilon_{ijk}b_k\frac{\partial a_j}{\partial x_i} + \varepsilon_{ijk}a_j\frac{\partial b_k}{\partial x_i}.$$

Now make a cyclic change of indices in ε_{ijk} in the first term on the right-hand side of the above equation ($\varepsilon_{ijk} = \varepsilon_{kij}$) and exchange i and j in ε_{ijk} in the second term ($\varepsilon_{ijk} = -\varepsilon_{jik}$). Rewriting the result in terms of direct (vector) notation yields the identity in question.

Example 2.6.2. Prove the identity $\nabla \cdot (\nabla \times \mathbf{a}) = 0$.

SOLUTION. Writing the left-hand of this identity in index notation gives

$$\nabla \cdot (\nabla \times \mathbf{a}) = \varepsilon_{ijk}\frac{\partial^2 a_k}{\partial x_i\, \partial x_j}.$$

Recalling that ε_{ijk} is skew-symmetric in any two of its indices and that $\partial^2 a_k/(\partial x_i\, \partial x_j)$ is symmetric in i and j, the product in the above equation must vanish from Theorem 2.5.1.

Example 2.6.3. Prove the identity $\mathbf{a} \times (\mathbf{b} \times \mathbf{c}) = (\mathbf{a} \cdot \mathbf{c})\mathbf{b} - (\mathbf{a} \cdot \mathbf{b})\mathbf{c}$.

SOLUTION. Writing the left-hand side in index notation gives

$$\mathbf{a} \times (\mathbf{b} \times \mathbf{c}) = \varepsilon_{ijk}\varepsilon_{k\ell m}a_j b_\ell c_m \mathbf{e}_i.$$

A cyclic variation of the indices of the first permutation symbol ε_{ijk} and use of the identity (2.2.5) give

$$\mathbf{a}\times(\mathbf{b}\times\mathbf{c}) = \varepsilon_{kij}\varepsilon_{k\ell m}a_j b_\ell c_m \mathbf{e}_i = (\delta_{i\ell}\delta_{jm}-\delta_{im}\delta_{\ell j})a_j b_\ell c_m \mathbf{e}_i = (b_i a_j c_j - c_i a_j b_j)\mathbf{e}_i.$$

Converting back into direct (vector) notation gives the identity in question.

Example 2.6.4. Show that if $\mathbf{n}$ denotes the unit-normal vector to a plane surface S, then any vector $\mathbf{a}$ can be decomposed as the sum of a component $(\mathbf{a}\cdot\mathbf{n})\mathbf{n}$ normal to S and a projection $(\mathbf{n}\times\mathbf{a})\times\mathbf{n}$ tangent to S.

SOLUTION. The proof proceeds by first assuming the validity of the decomposition

$$\mathbf{a} = (\mathbf{a}\cdot\mathbf{n})\mathbf{n} + (\mathbf{n}\times\mathbf{a})\times\mathbf{n}, \tag{2.6.10}$$

converting each term to index notation, using the identity (2.2.5), and showing that an identity is obtained. In index notation, (2.6.10) becomes

$$\begin{aligned} a_i &= a_j n_j n_i + \varepsilon_{ijk}(\varepsilon_{j\ell m} n_\ell a_m) n_k \\ &= a_j n_j n_i + \varepsilon_{jki}\varepsilon_{j\ell m} a_m n_\ell n_k \\ &= a_j n_j n_i + (\delta_{k\ell}\delta_{im} - \delta_{km}\delta_{i\ell}) a_m n_\ell n_k \\ &= a_j n_j n_i + a_i n_k n_k - a_k n_i n_k \\ &= a_i n_k n_k. \end{aligned}$$

Since $\mathbf{n}$ is a unit vector, $n_k n_k = 1$ and the identity (2.6.10) is proved.

2.7 INTEGRAL IDENTITIES

Integral identities are used frequently in fluid dynamics to convert volume integrals to surface integrals and surface integrals to contour integrals. Since the conditions for validity of these integral identities require the surfaces and volumes to possess certain topological characteristics, we first review some definitions related to the geometry of volumes and surfaces.

Let S be a surface in a three-dimensional space. Here, S is said to be *closed* if it partitions the surrounding space into interior and exterior regions such that a continuous path between any point interior to S and any point exterior to S must cross the surface S. Letting S be specified by a relation of the form $f(\mathbf{x}) = \text{const}$ at any time t, then the unit normal $\mathbf{n}$ of S is given by $\mathbf{n} = \nabla f/|\nabla f|$, where the sense of the unit normal may be adjusted by changing the sign of $\mathbf{n}$. The surface S is said to be *smooth* if $\mathbf{n}$ varies continuously over S and to be *piecewise smooth* if it is formed of some finite number of smooth sub-surfaces. Similar definitions of smoothness can be made for a space curve. The surface S is said to be *simply connected* if all closed curves on S can be continuously shrunk to a point. A piecewise smooth surface

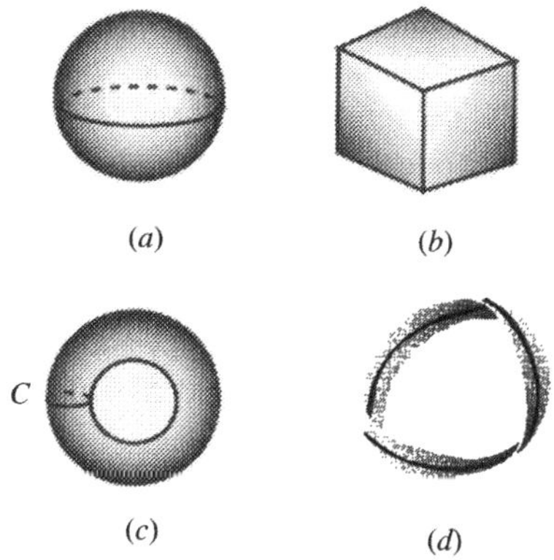

Figure 2.1 Examples of surfaces that have different topological characteristics: (*a*) sphere, (*b*) cube, (*c*) torus, and (*d*) Möbius strip.

S is said to be *orientable* if a positive direction can be assigned to the bounding curve about each smooth piece of S such that on the boundary C between any two adjacent smooth pieces S_1 and S_2 of S, the positive direction on C relative to S_1 is opposite to the positive direction on C relative to S_2. Examples of these definitions are shown for surfaces of several shapes in Figure 2.1. A sphere is closed, smooth, and simply connected. A cube is closed and simply connected but only piecewise smooth, since the unit normal changes direction discontinuously at the edges of the cube. A torus is closed and smooth, but not simply connected, since a circle such as that labeled C in Figure 2.1c cannot be continuously shrunk to a point while remaining on the surface S. Most commonly encountered surfaces are orientable; however, some strange surfaces (such as the Möbius strip shown in Figure 2.1d) can be found which do not possess this property.

Now suppose that a region V occupies the interior of a closed surface S. We say that V is an *elementary volume* if any straight line passing through V intersects the surface S in exactly two points. The interior of a sphere and a cube are elementary, but the interior of a torus is not. A region V is said to be a *composite volume* if it can be divided into some number of elementary volumes.

There are quite a variety of integral identities that are used in fluid dynamics; however, most of these identities can be derived from the following theorem.

Theorem 2.7.1 (Generalized Green's Theorem). If V is any composite volume with piecewise smooth bounding surface S and outward unit normal $\mathbf{n}$ and F is any continuous function in V whose gradient ∇F is also continuous in V, then

$$\int_V \nabla F dv = \int_S F\mathbf{n}\,da \qquad \text{or} \qquad \int_V \frac{\partial F}{\partial x_i} dv = \int_S F n_i\, da. \tag{2.7.1}$$

This theorem is first proved for an elementary volume V. The proof can then be extended to a composite volume by joining the elementary volumes from which the composite volume is formed and noting that since the outward unit normals on a surface separating two adjacent elementary volumes are in opposite directions, the contribution of this surface to the integral in the right-hand side of (2.7.1) vanishes. The proof is illustrated by considering differentiation in just one direction, say z. A line L which is tangent to the z-axis can be specified by constant values of x and y. Since the volume V is elementary, each such line L passing through V will intersect the bounding surface S of V exactly two times. Let us denote the value of z at the upper and lower of these intersections by $z^+(x, y)$ and $z^-(x, y)$, respectively. Now divide the surface S up into two halves S^+ and S^- such that $z = z^+$ on S^+ and $z = z^-$ on S^- (Figure 2.2). We may then write

$$\begin{aligned}\int_V \frac{\partial F}{\partial z}\, dv &= \int_A \left(\int_L \frac{\partial F}{\partial z}\, dz\right) dx\, dy \\ &= \int_A \left[F\left(x, y, z^+(x, y)\right) - F\left(x, y, z^-(x, y)\right)\right] dx\, dy, \qquad (2.7.2)\end{aligned}$$

where A denotes the projection of S on the x-y plane. We let n_3^+ and n_3^- denote the z-components of the outward unit normal on S^+ and S^-, respectively. The product $dx\, dy$ is related to the element of area da on S by $dx\, dy = n_3^+\, da^+$ on S^+ and by $dx\, dy = -n_3^-\, da^-$ on S^-. Equation (2.7.2) can then be written in terms of an integral over S as

$$\int_V \frac{\partial F}{\partial z}\, dv = \int_{S^+} F n_3^+\, da^+ + \int_{S^-} F n_3^-\, da^- = \int_S F n_3\, da, \qquad (2.7.3)$$

thus completing the proof.

The utility of (2.7.1) stems largely from the observation that the function F may be either a scalar or the components of a tensor quantity of any order. For instance, replacing the function F in (2.7.1b) with the components b_i of an arbitrary vector **b**

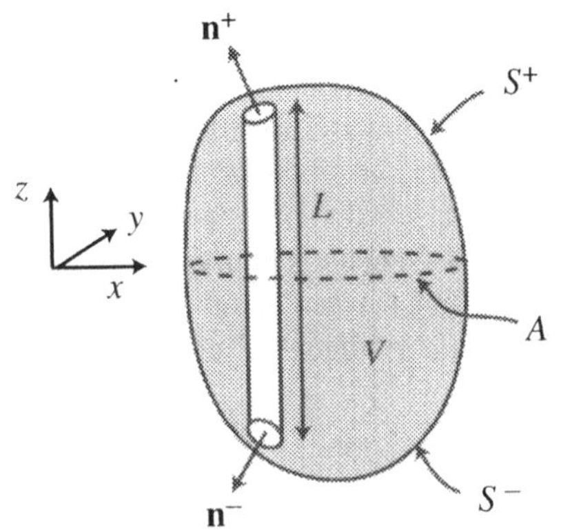

Figure 2.2 Division of the elementary surface $S = S^+ \cup S^-$ used in the proof of Theorem 2.7.1.

gives the usual *divergence theorem*

$$\int_V \frac{\partial b_i}{\partial x_i}\,dv = \int_S b_i n_i\,da \qquad \text{or} \qquad \int_V \nabla\cdot\mathbf{b}\,dv = \int_S \mathbf{n}\cdot\mathbf{b}\,da. \tag{2.7.4}$$

Replacing the function F in (2.7.1b) with the product $\varepsilon_{ijk}b_j$ gives the identity

$$\int_V \varepsilon_{ijk}\frac{\partial b_j}{\partial x_i}\,dv = \int_S \varepsilon_{ijk}n_i b_j\,da \qquad \text{or} \qquad \int_V \nabla\times\mathbf{b}\,dv = \int_S \mathbf{n}\times\mathbf{b}\,da. \tag{2.7.5}$$

Other standard forms of Green's theorem can be readily obtained from (2.7.4). For instance, letting ϕ and ψ denote two arbitrary scalar quantities, it follows from the chain rule that

$$\begin{aligned} \nabla\cdot(\phi\nabla\psi) &= \nabla\phi\cdot\nabla\psi + \phi\nabla^2\psi, \\ \nabla\cdot(\psi\nabla\phi) &= \nabla\psi\cdot\nabla\phi + \psi\nabla^2\phi. \end{aligned} \tag{2.7.6}$$

Subtracting these two equations gives

$$\nabla\cdot(\phi\nabla\psi - \psi\nabla\phi) = \phi\nabla^2\psi - \psi\nabla^2\phi. \tag{2.7.7}$$

Substituting this result into the divergence theorem (2.7.4) gives *Green's second identity*

$$\int_V (\phi\nabla^2\psi - \psi\nabla^2\phi)\,dv = \int_S \left(\phi\frac{\partial\psi}{\partial n} - \psi\frac{\partial\phi}{\partial n}\right)da. \tag{2.7.8}$$

A variety of useful identities relating integrals over a closed curve C to integrals over an open surface A bounded by C can be derived using the two-dimensional form of (2.7.1b), given by

$$\int_C G\hat{n}_j\,d\ell = \int_A \frac{\partial G}{\partial x_j}\,da, \tag{2.7.9}$$

where G is a continuous function defined on a plane open surface A with bounding curve C and $\hat{\mathbf{n}}$ is the outward unit normal to the curve C lying on a plane tangent to A. A result similar to (2.7.9) can be written when A is a curved surface, but this would involve introducing certain topological properties of the surface which are beyond the scope of the present text. We now replace the function G by $\varepsilon_{ijk}n_i f_k$, where $\mathbf{n}$ is the unit normal to the surface A and f_k are the components of a continuous vector function, giving

$$\int_C \varepsilon_{ijk}n_i\hat{n}_j f_k\,d\ell = \int_A \varepsilon_{ijk}n_i\frac{\partial f_k}{\partial x_j}\,da. \tag{2.7.10}$$

Since $\mathbf{n}$ and $\hat{\mathbf{n}}$ are normal both to C and to each other, the product $\varepsilon_{ijk}n_i\hat{n}_j$ in the integral on the left-hand side of (2.7.10) is equal to the components $\hat{t}_k$ of the unit tangent vector $\hat{\mathbf{t}} = \mathbf{n} \times \hat{\mathbf{n}}$ to C, where the direction of $\hat{\mathbf{t}}$ is set by the direction of $\mathbf{n}$ according to the right-hand rule. Equation (2.7.10) then reduces to the usual *Stokes theorem*

$$\int_C f_k \, dx_k = \int_A \varepsilon_{ijk} n_i \frac{\partial f_k}{\partial x_j} \, da, \tag{2.7.11a}$$

or in vector form

$$\int_C \mathbf{f} \cdot d\mathbf{x} = \int_A \mathbf{n} \cdot (\nabla \times \mathbf{f}) \, da, \tag{2.7.11b}$$

where $d\mathbf{x}$ is an oriented element of length tangent to C.

A similar integral theorem can be derived for the integral about C of the vector product of a vector $\mathbf{f}$ with the tangent vector $d\mathbf{x}$, where in this case the function G in (2.7.9) is replaced by $\varepsilon_{ijk}\varepsilon_{pqk} n_i f_p$, giving

$$\int_C \mathbf{f} \times d\mathbf{x} = \int_A [\mathbf{n}(\nabla \cdot \mathbf{f}) - (\mathbf{n} \cdot \nabla)\mathbf{f} - \mathbf{n} \times (\nabla \times \mathbf{f})] \, da. \tag{2.7.12}$$

Setting $\mathbf{f} = \phi \mathbf{a}$ in the Stokes theorem (2.7.11), where $\mathbf{a}$ is an arbitrary constant unit vector and ϕ is an arbitrary scalar, and using the property of invariance to cyclic permutations of the scalar triple product, an identity for the integral of a scalar around a closed curve C is obtained as

$$\int_C \phi d\mathbf{x} = \int_A \mathbf{n} \times \nabla \phi \, da. \tag{2.7.13}$$

Derivation of the integral identities (2.7.12) and (2.7.13) is left as exercises.

While the results (2.7.11)–(2.7.13) are derived in this section for cases where the surface A is planar, they in fact apply when A is any orientable surface bounded by a closed curve C. This property can be deduced by imagining the surface A to be approximated by a piecewise smooth surface A^* which is composed of some finite number of flat "panels" A_i that join at the edges and are each bounded by circuits C_i. The integral identities (2.7.11)–(2.7.13) apply to each of these panels individually, where A and C are replaced by A_i and C_i, respectively. Since A is orientable, the contour integrals on the left-hand side of (2.7.11)–(2.7.13) cancel out on all adjoining sides of these panels such that the integral over the whole surface A^* reduces to a contour integral over the circuit C^*, forming the boundary of A^*. As the number of panels approaches infinity, the surface and contour integrals over the piecewise smooth approximations A^* and C^* approach integrals over the curved surface A and contour C.

BIBLIOGRAPHY

Aris, R. (1962). *Vectors, Tensors and the Basic Equations of Fluid Mechanics*, Prentice-Hall, Englewood Cliffs, NJ (reprinted Dover Publications, New York, 1989).

Erickson, J.L. (1960). "Tensor fields," in *Handbuch der Physik* III/1, ed. S. Flugge, Springer-Verlag, Berlin, pp. 794–858.

Jeffreys, H. (1931). *Cartesian Tensors*, Cambridge University Press, Cambridge.

Sokolnikoff, I.S. (1951). *Tensor Analysis*, John Wiley & Sons, New York.

PROBLEMS

1. All of the following expressions use summation convention incorrectly. Explain why.
   (a) $A_{ij}\varepsilon_{ijk}u_k + v_k = c$
   (b) $u_i u_j A_{ij}\delta_{ij} = u_i u_i A_{jj}$
   (c) $u_i u_i + A_{ij} = 3$
   (d) $\varepsilon_{ijk}A_{ii} = r_k$
2. Write the following expressions in indicial form for a vector $\mathbf{u}$ and a second-order tensor $\mathbf{A}$:
   (a) $\nabla \times (\nabla \times \mathbf{u})$
   (b) $\nabla(\mathbf{u} \cdot \mathbf{A} \cdot \mathbf{u})$
   (c) $\mathbf{u} \cdot \nabla(\operatorname{tr}\mathbf{A})$
   (d) $\det(\mathbf{A})\nabla \times \mathbf{u}$
3. Prove that $\varepsilon_{ijk}\varepsilon_{ij\ell} = 2\delta_{k\ell}$.
4. Prove the identity $\mathbf{a} \times \mathbf{w} = \mathbf{B} \cdot \mathbf{a}$, where $\mathbf{w}$ is the axial vector of a skew-symmetric tensor $\mathbf{B}$ and $\mathbf{a}$ is an arbitrary vector.
5. Prove all of the identities in Table 2.1 by first rewriting in indicial notation and then using the theorems and identities in Sections 2.2–2.5.
6. Prove that for any second-order tensor $\mathbf{A}$ we can write $\det(\mathbf{A}) = \det(\mathbf{A}^{\mathrm{T}})$.
7. Prove that the determinant of a second-order tensor $\mathbf{A}$ can be written as

$$\det(\mathbf{A}) = \tfrac{1}{6}\varepsilon_{ijk}\varepsilon_{\ell mn}A_{\ell i}A_{mj}A_{nk}.$$

8. Derive the integral indentities (2.7.12) and (2.7.13).
9. Pressure exerts a force $-p\mathbf{n}$ on a surface S immersed in a fluid, where $\mathbf{n}$ is the unit normal of S. Use the integral identities in Section 2.7 to prove that there is no net pressure force acting on a closed body immersed in a uniform pressure field.
10. Show that for any closed circuit C and scalar ϕ, the following identity holds:

$$\int_C \nabla\phi \cdot d\mathbf{x} = 0.$$

CHAPTER 3

KINEMATICS OF FLUID MOTION

The subject of *kinematics* is concerned with the geometry of motion, without regard for the dynamical laws that govern motion. In the current chapter, we introduce some of the basic kinematical concepts and results that form the backbone of the study of fluid mechanics. Since kinematics is purely geometrical, the material introduced in this chapter applies equally well to all media, fluids and solids alike. Nevertheless, many of the kinematical results emphasized in this chapter, such as streamlines, vorticity, and decomposition of the velocity gradient, play a particularly important role in describing fluid motion but are not as important when describing deformation of most solids. Other kinematical quantities, which play an important role in describing solid deformation but are not as relevant to fluids, have been omitted.

3.1 DESCRIPTIONS OF FLUID MOTION

A fluid is assumed to occupy a region $\mathfrak{R}_0$ in space at some reference time t_0, where $\mathfrak{R}_0$ may be either bounded or unbounded. The position vector, with respect to a fixed coordinate frame, of a fluid particle P in $\mathfrak{R}_0$ is denoted by $\boldsymbol{\xi}$. The set consisting of the positions of all fluid particles in $\mathfrak{R}_0$ at time t_0 is called the *reference configuration* of the fluid. It is assumed that the fluid is a *continuum* such that for any particle P not on the boundary of $\mathfrak{R}_0$, any infinitesimal variation $\delta\boldsymbol{\xi}$ of $\boldsymbol{\xi}$ at time t_0 yields the position vector of another fluid particle in $\mathfrak{R}_0$. At some later time t, the fluid moves to some other space region $\mathfrak{R}$ such that the position vector to the particle P is now denoted by $\mathbf{x}$. If t denotes the current time, then we call the positions of all fluid particles in $\mathfrak{R}$ the *current configuration* of the fluid. The position vector $\mathbf{x}$ in the current configuration depends both on time and on the identity of the fluid particle

P to which the position vector points. The particle identity is most conveniently specified by the position vector $\boldsymbol{\xi}$ of the particle in the reference configuration, so that we write

$$\mathbf{x} = \mathbf{x}(\boldsymbol{\xi}, t). \tag{3.1.1}$$

The variation of the position vector $\mathbf{x}$ in time specifies the motion of the fluid, as is illustrated in Figure 3.1.

Several assumptions are commonly made with regard to the time variation of the vector $\mathbf{x}$. For instance, it is assumed that $\mathbf{x}(\boldsymbol{\xi}, t)$ is a *continuous* function of both t and $\boldsymbol{\xi}$. Continuity in t implies that a particle cannot instantaneously jump from one location to another but must trace a continuous path in space between any two locations of the particle. Continuity in $\boldsymbol{\xi}$ implies that particles which are initially in some infinitesimal neighborhood of a given particle remain in the neighborhood of the particle for all time. It is also assumed that the mapping $\boldsymbol{\xi} \to \mathbf{x}$ is *single valued*, such that a particle cannot occupy two positions at a given instant of time. Similarly, single-valuedness implies that two different particles cannot occupy the same position at a given instant of time. We shall also assume sufficient *smoothness* of the function $\mathbf{x}(\boldsymbol{\xi}, t)$, which implies that some number of derivatives of $\mathbf{x}(\boldsymbol{\xi}, t)$, typically through second order, with respect to t and $\boldsymbol{\xi}$ exist and are continuous in time and space. Exceptions to the smoothness assumption sometimes occur on surfaces, such as shock waves or interfaces between two different media, or on lines and points within the flow. Such cases where the derivatives of $\mathbf{x}(\boldsymbol{\xi}, t)$ become discontinuous must be treated with special care, as discussed in Chapter 5.

Another necessary assumption is that the equation (3.1.1) is invertible, so that we may write

$$\boldsymbol{\xi} = \boldsymbol{\xi}(\mathbf{x}, t). \tag{3.1.2}$$

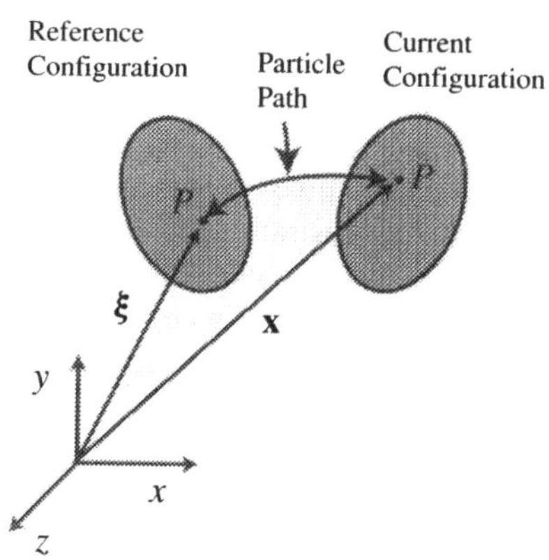

Figure 3.1 Motion of a fluid particle from the reference configuration to the current configuration.

Equation (3.1.2) implies that given knowledge of the current position of the particle at time t and of the motion of the particle for all times prior to t, one can deduce (at least in principle) the position of the particle at any previous time. While for most fluid dynamics applications it is most convenient to deal only with the current position of the fluid particles, there are a variety of cases where it is desirable to instead refer to the particles in terms of their location $\boldsymbol{\xi}$ in some reference configuration.

The variables describing a fluid, such as density, velocity, and so forth, are generally written as functions of t and the current spatial position $\mathbf{x}$. For instance, density, whose value is denoted by ρ, can be set equal to a function $\hat{\rho}$ of $\mathbf{x}$ and t by writing

$$\rho = \hat{\rho}(\mathbf{x}, t). \tag{3.1.3}$$

(While we generally use the same notation for a function, such as $\hat{\rho}$, and its value, ρ, for the current discussion it is necessary to make a distinction.) The expression (3.1.1) for $\mathbf{x}$ can be substituted into (3.1.3) to alternatively express density as a different function $\tilde{\rho}$ of $\boldsymbol{\xi}$ and t, or

$$\rho = \tilde{\rho}(\boldsymbol{\xi}, t). \tag{3.1.4}$$

The approach used in (3.1.3), in which the dependent variables are expressed as functions of the current position $\mathbf{x}$, is called the *Eulerian viewpoint*. The approach used in (3.1.4), in which dependent variables are functions of the particle positions $\boldsymbol{\xi}$ in some fixed reference configuration, is called the *Lagrangian viewpoint*. For instance, in constructing a computational approach for solution of a fluid flow using the Eulerian viewpoint, a fixed grid is placed over the region occupied by the fluid and the variation in time of the dependent variables describing the flow is examined at the nodes of the fixed grid. In solving the same flow using a Lagrangian viewpoint, a finite set of fluid particles is selected at some initial time and the motion of these particles is followed in time while examining the change of the variables describing the fluid on these particles. Examples of computational methods using both the Eulerian and Lagrangian viewpoints abound.

The dynamical laws governing fluid motion involve not only the positions of the fluid particles but also the time derivatives of the particle positions and of other variables that are functions of the particle positions. When using the Eulerian viewpoint, it is most convenient to evaluate these time derivatives at fixed points in space. When using the Lagrangian viewpoint, it is desirable to evaluate time derivatives while traveling with a given fluid particle. The time derivative evaluated at a fixed spatial point is called the *partial derivative* and is denoted by $\partial/\partial t$. The time derivative evaluated on a given fluid particle is called the *material derivative* and is denoted in the current text by D/Dt, or sometimes by a dot placed over a variable. In other literature, the material derivative is sometimes called the *total*, *convected*, or *substantial* derivative. A *steady flow* is one for which all variables describing the fluid are constant when evaluated at fixed points in space. Hence, the partial derivative with respect to time of all fluid variables vanishes in a steady flow, but the material derivative may not vanish. For instance, in the example of steady flow through a nozzle shown in Fig-

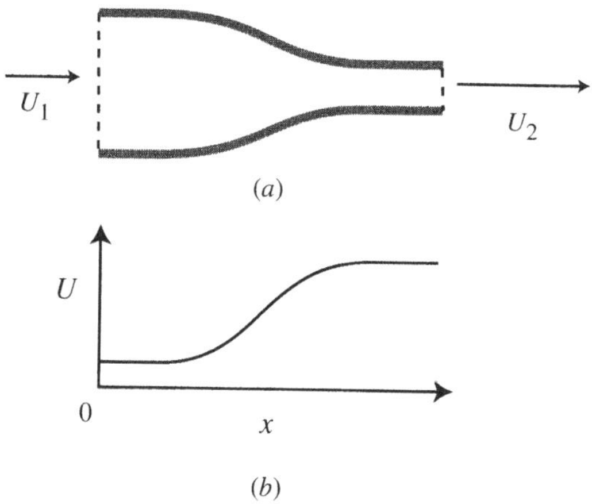

Figure 3.2 Steady flow through a converging nozzle: (*a*) schematic of nozzle; (*b*) variation of centerline velocity with distance along nozzle axis.

ure 3.2, the velocity component u in the x-direction is constant at all fixed points in space, but u increases with time when evaluated on a given fluid particle as the particle moves through the nozzle.

When taking the material derivative of a variable that is written as a function of the current position **x**, such as in (3.1.3), it is important to remember that the value of **x** for a given fluid particle varies in time. Using the standard chain rule of calculus, we can write

$$\frac{D\hat{\rho}}{Dt} = \frac{\partial \hat{\rho}}{\partial t} + \frac{dx}{dt}\frac{\partial \hat{\rho}}{\partial x} + \frac{dy}{dt}\frac{\partial \hat{\rho}}{\partial y} + \frac{dz}{dt}\frac{\partial \hat{\rho}}{\partial z} \qquad (3.1.5)$$

Equation (3.1.5) can be written more concisely using vector notation and by noting that the material derivative of the position vector **x** is simply the velocity **u**, such that

$$\frac{D\hat{\rho}}{Dt} = \frac{\partial \hat{\rho}}{\partial t} + \mathbf{u} \cdot \nabla \hat{\rho}. \qquad (3.1.6)$$

Similarly, expressing the velocity **u** as a function $\hat{\mathbf{u}}$ of **x** and t and recalling that the acceleration **a** is the material derivative of the velocity, we can write

$$\mathbf{a} = \frac{D\hat{\mathbf{u}}}{Dt} = \frac{\partial \hat{\mathbf{u}}}{\partial t} + (\hat{\mathbf{u}} \cdot \nabla)\hat{\mathbf{u}}. \qquad (3.1.7)$$

It is common practice to drop the "hats" over the functions in expressions such as (3.1.6) and (3.1.7). It is clear that the material derivative is simply the usual total derivative of calculus, and in fact many fluid mechanics texts write equations such as (3.1.6) and (3.1.7) with the standard lowercase d notation often used for the total derivative in mathematics texts. The capital D notation is used for the material derivative in the current text (following Batchelor, 1967) to remind us that the function being differentiated depends on the current position $\mathbf{x}$ rather than the reference position $\boldsymbol{\xi}$ of the fluid particles.

A point in space that coincides for all times with the same fluid particle is called a *material* point. Similarly, curves, surfaces, or volumes in space that consist of the same fluid particles for all time are called material curves, surfaces, or volumes, respectively. If the equation $f(\mathbf{x}, t) = \text{const}$ specifies the position of a material surface for all time it, then it follows that the value of f does not change when evaluated on fluid particles on the surface, such that

$$\frac{Df}{Dt} = 0. \tag{3.1.8}$$

Equation (3.1.8) is frequently used in deriving boundary conditions for material surfaces.

3.2 DEFORMATION MEASURES

The differential $d\mathbf{x}$ of the position vector in the current configuration represents an infinitesimal line segment of fluid particles. The line segment occupied by these same fluid particles in the reference configuration is denoted by $d\boldsymbol{\xi}$, where our assumptions of continuity and smoothness imply that the curve occupied by these particles continues to be a line segment in the limit of infinitesimal length. The orientation and length of this material line segment in the current and reference configurations are linearly related to each other by the expression

$$d\mathbf{x} = \mathbf{F} \cdot d\boldsymbol{\xi} \qquad \text{or} \qquad dx_i = F_{iA} d\xi_A. \tag{3.2.1}$$

The quantity $\mathbf{F}$ in (3.2.1) is a second-order tensor called the *deformation gradient* and is defined by

$$\mathbf{F} = \frac{\partial \mathbf{x}}{\partial \boldsymbol{\xi}} \qquad \text{or} \qquad F_{iA} = \frac{\partial x_i}{\partial \xi_A}. \tag{3.2.2}$$

In keeping with usual convention, we use a lowercase index for quantities that are evaluated in the current configuration and an uppercase index for quantities evaluated in the reference configuration in writing the components of the vectors and tensors in (3.2.1) and (3.2.2).

We have previously stated that the transformation $\mathbf{x} = \mathbf{x}(\boldsymbol{\xi}, t)$ is assumed to be invertible, such that it can be written in the form (3.1.2). The indicial form of (3.2.1) is a matrix equation relating the components of $d\mathbf{x}$ with those of $d\boldsymbol{\xi}$. From standard results of matrix algebra, the necessary and sufficient condition for existence of the inverse of (3.2.1) is that the determinant of the matrix F_{iA} be nonvanishing. We define the *Jacobian of deformation* J by

$$J \equiv \det(\mathbf{F}) = \varepsilon_{ABC} \frac{\partial x_1}{\partial \xi_A} \frac{\partial x_2}{\partial \xi_B} \frac{\partial x_3}{\partial \xi_C}, \tag{3.2.3}$$

where the last part of (3.2.3) follows from the identity (2.3.8). In the reference configuration, the deformation gradient $\mathbf{F}$ is equal to the identity tensor $\mathbf{I}$, so that J is equal to unity. Since J must vary continuously and cannot equal zero, we deduce that $0 < J < \infty$ for all time.

In addition to the deformation gradient, the *velocity gradient* $\nabla\mathbf{u}$ plays a key role in fluid mechanics. Since the velocity gradient is a second-order tensor, according to the tensor representation in Theorem 2.5.2, we can decompose $\nabla\mathbf{u}$ as the sum of symmetric and skew-symmetric tensors, such that

$$\nabla\mathbf{u} = \mathbf{D} + \mathbf{W} \qquad \text{or} \qquad \frac{\partial u_i}{\partial x_j} = D_{ij} + W_{ij}, \tag{3.2.4}$$

where

$$D_{ij} = \frac{1}{2}\left(\frac{\partial u_i}{\partial x_j} + \frac{\partial u_j}{\partial x_i}\right) \qquad \text{and} \qquad W_{ij} = \frac{1}{2}\left(\frac{\partial u_i}{\partial x_j} - \frac{\partial u_j}{\partial x_i}\right). \tag{3.2.5}$$

The symmetric part $\mathbf{D}$ of the velocity gradient is called the *rate of deformation tensor* (or sometimes the *rate of strain tensor*) and the skew-symmetric part $\mathbf{W}$ is called the *vorticity tensor*. We note that from Theorem 2.5.1 the scalar product $D_{ij}W_{ij}$ must vanish identically.

A useful relationship exists between the velocity gradient and the material derivative of the deformation gradient $\mathbf{F}$. Taking the material derivative of (3.2.2) and using the raised-dot notation for the material derivative give

$$\dot{F}_{iA} = \frac{D}{Dt}\frac{\partial x_i}{\partial \xi_A} = \frac{\partial}{\partial \xi_A}\frac{Dx_i}{Dt} = \frac{\partial u_i}{\partial \xi_A}, \tag{3.2.6}$$

where the fact that $\boldsymbol{\xi}$ is evaluated at a fixed reference time, and is therefore independent of t, is used to commute the material derivative and the partial derivative with respect to ξ_A in (3.2.6). Using the chain rule of calculus, equation (3.2.6) can alternatively be written as

$$\dot{F}_{iA} = \frac{\partial u_i}{\partial x_j}\frac{\partial x_j}{\partial \xi_A}. \tag{3.2.7}$$

3.3 STRETCH RATE AND VORTICITY

Let us decompose an infinitesimal line segment $d\mathbf{x}$ as the product of its magnitude dx and a unit vector $\mathbf{a}$ that indicates its direction, so that

$$d\mathbf{x} = dx\,\mathbf{a}. \tag{3.3.1a}$$

If $d\boldsymbol{\xi}$ denotes the line segment occupied by the fluid particles in the reference configuration, then a similar decomposition of $d\boldsymbol{\xi}$ as the product of its magnitude $d\xi$ and a unit vector **c** gives

$$d\boldsymbol{\xi} = d\xi\,\mathbf{c}. \tag{3.3.1b}$$

The *stretch* Λ of an infinitesimal material line segment is defined as the ratio of the length of the line segment in the current configuration to its length in the reference configuration, or

$$\Lambda \equiv \frac{dx}{d\xi}. \tag{3.3.2}$$

Dividing (3.2.1) by $d\xi$ and using the definitions (3.3.1) and (3.3.2) yield

$$\Lambda\mathbf{a} = \mathbf{F}\cdot\mathbf{c}. \tag{3.3.3}$$

Taking the material derivative of (3.3.3) and using the result (3.2.7) for $\dot{\mathbf{F}}$ give

$$\begin{aligned}\dot{\Lambda}a_i + \Lambda\dot{a}_i &= \dot{F}_{iB}c_B \\ &= \frac{\partial u_i}{\partial x_j}F_{jB}c_B \\ &= \frac{\partial u_i}{\partial x_j}\Lambda a_j.\end{aligned} \tag{3.3.4}$$

The unit vector **c** passes through the time derivative in (3.3.4) since it is evaluated in the reference configuration and is therefore independent of time. Taking the scalar product of (3.3.4) with a_i and noting that $D(\mathbf{a}\cdot\mathbf{a})/Dt = 2\mathbf{a}\cdot\dot{\mathbf{a}} = 0$ since **a** is a unit vector give

$$\frac{\dot{\Lambda}}{\Lambda} = \frac{\partial u_i}{\partial x_j}a_i a_j. \tag{3.3.5}$$

The ratio on the left-hand side of (3.3.5) is called the *logarithmic rate of stretching*, since $D(\ln|\Lambda|)/Dt = \dot{\Lambda}/\Lambda$. Substituting the decomposition (3.2.4) of the velocity gradient into (3.3.5) and noting that $W_{ij}a_i a_j$ vanishes identically since $a_i a_j$ is symmetric and W_{ij} is skew-symmetric give

$$\frac{\dot{\Lambda}}{\Lambda} = D_{ij}a_i a_j. \tag{3.3.6}$$

The result (3.3.6) indicates that the change in length of an infinitesimal material line segment is controlled only by the symmetric part D_{ij} of the velocity gradient. Since the unit vector **a** can be oriented in any direction, we might consider the special case in which **a** is oriented along one of the coordinate directions. For instance, if **a** is tangent to the base vector $\mathbf{e}_1$ in the x-direction, (3.3.6) reduces to

$$D_{11} = \frac{\dot{\Lambda}}{\Lambda}. \tag{3.3.7}$$

The other two diagonal components of D_{ij} can similarly be set equal to the logarithmic rate of stretching of line segments oriented in the y- and z-directions.

While the result (3.3.6) can be used to determine the stretching rate of material line segments, it indicates nothing about the rotation rate of the line segments. To determine rotation rate, we divide (3.3.3) by the stretch Λ and then take the material derivative of the resulting equation, giving

$$\dot{\mathbf{a}} = \frac{1}{\Lambda}\dot{\mathbf{F}} \cdot \mathbf{c} - \frac{\dot{\Lambda}}{\Lambda^2}\mathbf{F} \cdot \mathbf{c}. \tag{3.3.8}$$

Using (3.2.7) for $\dot{\mathbf{F}}$ and (3.3.3) for $\mathbf{F} \cdot \mathbf{c}$, this result becomes (in index notation)

$$\dot{a}_i = \frac{\partial u_i}{\partial x_j} a_j - \frac{\dot{\Lambda}}{\Lambda} a_i. \tag{3.3.9}$$

Substituting the velocity gradient decomposition (3.2.4) and the result (3.3.6) for the stretch rate into (3.3.9) yields

$$\dot{a}_i = D_{ij} a_j + W_{ij} a_j - D_{jk} a_j a_k a_i. \tag{3.3.10}$$

Equation (3.3.10) indicates that, in general, both the rate of deformation tensor $\mathbf{D}$ and the vorticity tensor $\mathbf{W}$ contribute to the rotation of material line segments.

We now consider the special case in which the material line segment $\mathbf{a}$ is coincident with one of the three principal directions $\mathbf{d}^{(k)}$ of the rate of deformation tensor $\mathbf{D}$, defined by

$$\mathbf{D} \cdot \mathbf{d}^{(k)} = \lambda^{(k)} \mathbf{d}^{(k)}, \tag{3.3.11}$$

where $\lambda^{(k)}$ are the eigenvalues of $\mathbf{D}$ and the index $k \in \{1, 2, 3\}$ indicates the principal direction to which the quantity is referred. In this case, (3.3.6) becomes

$$\left. \frac{\dot{\Lambda}}{\Lambda} \right|_{\mathbf{a}=\mathbf{d}^{(k)}} = \lambda^{(k)}, \tag{3.3.12}$$

which indicates that the eigenvalues of the rate of deformation tensor $\mathbf{D}$ are equal to the logarithmic rate of stretching along the respective principal directions.

Differentiation of (3.3.6) with respect to a_k gives

$$\frac{\partial}{\partial a_k} \frac{\dot{\Lambda}}{\Lambda} = D_{ij} \frac{\partial a_i}{\partial a_k} a_j + D_{ij} a_i \frac{\partial a_j}{\partial a_k}. \tag{3.3.13}$$

When $\mathbf{a}$ is coincident with one of the principal directions $\mathbf{d}^{(m)}$, we can use (3.3.11) to write (3.3.13) as

$$\left. \frac{\partial}{\partial a_k} \frac{\dot{\Lambda}}{\Lambda} \right|_{\mathbf{a}=\mathbf{d}^{(m)}} = 2\lambda^{(m)} a_i \frac{\partial a_i}{\partial a_k}. \tag{3.3.14}$$

However, since **a** is a unit vector, so that $a_i a_i = 1$, the product $a_i(\partial a_i/\partial a_k)$ on the right-hand side of (3.3.14) vanishes identically. This result indicates that the logarithmic stretching rate attains a local extremum or point of inflection along each of the principal directions of **D**. When the three eigenvalues of **D** are distinct, the maximum stretching rate occurs along one principal direction (say, $k = 1$) and the minimum stretching rate occurs along another principal direction (say, $k = 3$). The third principal direction ($k = 2$) has the form of a saddle point, where the stretching rate increases when the line segment is rotated toward the $\mathbf{d}^{(1)}$ direction and decreases when it is rotated toward the $\mathbf{d}^{(3)}$ direction. This situation can be illustrated by an ellipsoid, as shown in Figure 3.3, where the distance from the center of the ellipsoid to its surface represents the stretching rate in the corresponding direction.

Another result obtained by letting the unit vector **a** coincide with a principal direction of **D** is that the first and third terms on the right-hand side of (3.3.10) cancel, giving

$$\dot{a}_i|_{\mathbf{a}=\mathbf{d}^{(m)}} = W_{ij} d_j^{(m)}. \tag{3.3.15}$$

Equation (3.3.15) indicates that the rotation rate of a material line segment oriented tangent to one of the principal directions of **D** at time t is controlled solely by the vorticity tensor **W**. This result can be written in an alternative form with the introduction of the *vorticity vector* $\boldsymbol{\omega}$, defined by

$$\boldsymbol{\omega} \equiv \nabla \times \mathbf{u}. \tag{3.3.16}$$

Using index notation and the decomposition (3.2.4), this definition can be rewritten as

$$\omega_i = \varepsilon_{ijk} \frac{\partial u_k}{\partial x_j} = \varepsilon_{ijk} W_{jk}, \tag{3.3.17}$$

where we note that $\varepsilon_{ijk} D_{jk}$ vanishes identically since D_{jk} is symmetric and ε_{ijk} is skew-symmetric in j and k. It is evident from the form of (3.3.17) that the vorticity vector is equal to twice the axial vector associated with the skew-symmetric tensor **W**, defined in (2.5.3). Using (2.5.2), the result (3.3.17) can be inverted to yield

$$W_{ij} = \tfrac{1}{2} \varepsilon_{ijk} \omega_k. \tag{3.3.18}$$

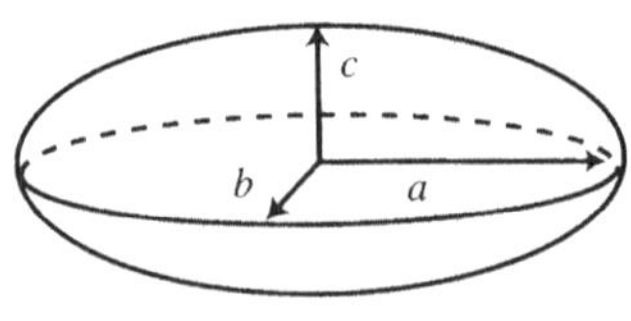

Figure 3.3 Ellipsoid with semiaxes a, b, and c, which represent the principal values of the rate of deformation tensor corresponding to a maximum, minimum, and saddle point of the logarithmic rate of stretching, respectively.

Substituting (3.3.18) into (3.3.15) gives

$$\dot{a}_i \big|_{\mathbf{a}=\mathbf{d}^{(k)}} = \tfrac{1}{2}\varepsilon_{ij\ell} d_j^{(k)} \omega_\ell \qquad \text{or} \qquad \dot{\mathbf{a}} \big|_{\mathbf{a}=\mathbf{d}^{(k)}} = \tfrac{1}{2}\mathbf{d}^{(k)} \times \boldsymbol{\omega}. \tag{3.3.19}$$

The result (3.3.19) has the form of a rigid-body rotation of the material line segment **a** with rate $|\mathbf{a} \times \boldsymbol{\omega}|/2$.

3.4 DILATATION

This section is concerned with the change in volume of a material region during the fluid motion. It is recalled that a volume element dv can be constructed from the scalar triple product of three noncoplanar line segments $d\mathbf{x}^{(1)}$, $d\mathbf{x}^{(2)}$, and $d\mathbf{x}^{(3)}$, or

$$dv = d\mathbf{x}^{(1)} \cdot (d\mathbf{x}^{(2)} \times d\mathbf{x}^{(3)}) = \varepsilon_{ijk}\, dx_i^{(1)}\, dx_j^{(2)}\, dx_k^{(3)}. \tag{3.4.1}$$

The quantity dv is the volume contained within a parallelogram whose sides are the line segments $d\mathbf{x}^{(1)}$, $d\mathbf{x}^{(2)}$, and $d\mathbf{x}^{(3)}$, as shown in Figure 3.4. Similarly, the volume dV of this same material region in the reference configuration can be expressed in terms of the material line segments $d\boldsymbol{\xi}^{(1)}$, $d\boldsymbol{\xi}^{(2)}$, and $d\boldsymbol{\xi}^{(3)}$ as

$$dV = d\boldsymbol{\xi}^{(1)} \cdot (d\boldsymbol{\xi}^{(2)} \times d\boldsymbol{\xi}^{(3)}) = \varepsilon_{ABC}\, d\xi_A^{(1)}\, d\xi_B^{(2)}\, d\xi_C^{(3)}, \tag{3.4.2}$$

where (3.2.1) relates the material line segments in the current and reference configurations by

$$dx_i^{(1)} = F_{iA}\, d\xi_A^{(1)}. \tag{3.4.3}$$

Substituting (3.4.3) and similar relationships for $d\mathbf{x}^{(2)}$ and $d\mathbf{x}^{(3)}$ into (3.4.1) gives

$$dv = (\varepsilon_{ijk} F_{iA} F_{jB} F_{kC})\, d\xi_A^{(1)}\, d\xi_B^{(2}\, d\xi_C^{(3)}. \tag{3.4.4}$$

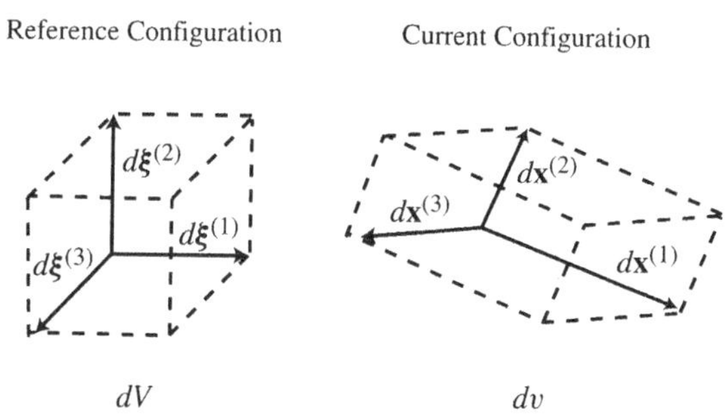

Figure 3.4 The volume elements dV and dv in the reference and current configurations are the volumes contained within the parallelograms whose sides are the line segments $\{d\boldsymbol{\xi}^{(1)}, d\boldsymbol{\xi}^{(2)}, d\boldsymbol{\xi}^{(3)}\}$ and $\{d\mathbf{x}^{(1)}, d\mathbf{x}^{(2)}, d\mathbf{x}^{(3)}\}$, respectively.

Recalling the identity (2.3.9), we can replace the terms in parentheses in (3.4.4) by $\varepsilon_{ABC}\det(\mathbf{F})$, which after using (3.4.2) and noting that $J = \det(\mathbf{F})$ yields

$$dv = J\,dV. \tag{3.4.5}$$

This relationship shows that the Jacobian of deformation J can be interpreted as the ratio of volume dv of a material region in the current configuration to the volume dV of this same region in the reference configuration. To emphasize this physical interpretation, the quantity J is often referred to as the *dilatation*.

Just as the relationship between the rates of stretching and rotation and the velocity gradient is examined in Section 3.3, it is also of interest to examine the relationship between the rate of dilatation and the velocity gradient. The rate of J can be obtained by using the following identity, which is valid for any second-order tensor $\mathbf{A}$ with components A_{ij}:

$$\frac{D}{Dt}(\det\mathbf{A}) = \mathrm{tr}\left(\frac{D\mathbf{A}}{Dt}\cdot\mathbf{A}^{-1}\right)\det(\mathbf{A}) = \frac{DA_{ij}}{Dt}A_{ji}^{-1}\det(\mathbf{A}). \tag{3.4.6}$$

A proof of the identity (3.4.6) is given at the end of this section. Using (3.4.6) with the arbitrary tensor $\mathbf{A}$ identified with the deformation gradient $\mathbf{F}$ and noting that $\dot{F}_{iB} = \partial u_i/\partial\xi_B$ from (3.2.7) and $F_{iB}^{-1} = \partial\xi_B/\partial x_i$ from the definition (3.2.2), we can immediately write

$$\frac{DJ}{Dt} = \frac{\partial u_i}{\partial\xi_B}\frac{\partial\xi_B}{\partial x_i}J = \frac{\partial u_i}{\partial x_i}J,$$

or in direct notation

$$\frac{DJ}{Dt} = (\nabla\cdot\mathbf{u})J. \tag{3.4.7}$$

From (3.4.7), the rate of dilatation, $D(\ln J)/Dt$, is equal to the divergence of velocity.

A special case of particular interest is that of an *incompressible* fluid, which is defined as a fluid in which the volume of any material region is constant in time. While all real fluids are somewhat compressible, incompressibility is frequently a useful idealization that simplifies the complexity of the equations governing the fluid flow. The incompressibility assumption applies to nearly all liquid flows and to many gas flows. For external flows and internal flows with fixed boundaries, a common rule-of-thumb is that gas flows can be treated as incompressible if the fluid speed is less than about 30% of the speed of sound. For an incompressible flow, $dv = dV$, and from (3.4.5) it follows that $J = 1$ for all time (since $J = 1$ in the reference configuration). From (3.4.7), a necessary and sufficient condition for a flow to be incompressible is that

$$\nabla\cdot\mathbf{u} = 0 \tag{3.4.8}$$

be everywhere satisfied for all time. Equation (3.4.8) is most commonly called the *continuity equation* in the fluid dynamics literature, on the basis of an alternative derivation in which this equation is obtained by setting density equal to a constant at material points in the mass balance law (Section 4.1). However, this equation is purely kinematical and, as we have shown, does not require the assumption of mass conservation for its derivation. Alternative names for equation (3.4.8) are the *incompressibility condition* or the *incompressibility constraint.*

Appendix 3.4.1 Proof of the Identity (3.4.6)

Let **A** be an arbitrary second-order tensor with components A_{ij}. Two other identities must be proved prior to obtaining the result (3.4.6). We begin by defining the components T_{ijk} of some third-order tensor by

$$T_{ijk} \equiv \varepsilon_{\ell mn} A_{\ell i} A_{mj} A_{nk}. \tag{3.4.9}$$

Using the identity (2.3.8), it is clear that $T_{123} = \det(\mathbf{A})$. Making use of the fact that ε_{ijk} is skew-symmetric in any two of its indices, it follows that T_{ijk} is equal to $\det(\mathbf{A})$ for any cyclic ordering of its indices, $-\det(\mathbf{A})$ for any anticyclic ordering of its indices and 0 whenever any two of its indices are the same. Recalling the definition of ε_{ijk} in Section 2.2, we find that

$$T_{ijk} = \varepsilon_{ijk} \det(\mathbf{A}) = \varepsilon_{\ell mn} A_{\ell i} A_{mj} A_{nk}. \tag{3.4.10}$$

Multiplying this result by $\varepsilon_{ijk}/6$ and recalling the result of Example 2.2.1, we find that the determinant of an arbitrary second-order tensor may be written in the form

$$\det(\mathbf{A}) = \tfrac{1}{6}\varepsilon_{ijk}\varepsilon_{\ell mn} A_{\ell i} A_{mj} A_{nk}. \tag{3.4.11}$$

The second identity is obtained by differentiating (3.4.11) with respect to **A**. Recall that if the components of **A** are independent, then

$$\frac{\partial A_{ij}}{\partial A_{rs}} = \delta_{ir}\delta_{js}. \tag{3.4.12}$$

Differentiating (3.4.11) by A_{rs} and making use of (3.4.12) yield, after a few lines of straightforward algebra,

$$\frac{\partial(\det \mathbf{A})}{\partial A_{rs}} = \frac{1}{2}\varepsilon_{sjk}\varepsilon_{rmn} A_{mj} A_{nk}. \tag{3.4.13}$$

Supposing now that **A** is nonsingular, so that its determinant is nonzero; there will exist a tensor **B** with components B_{ij}, called the *inverse* of **A**, such that

$$A_{\ell i} B_{ir} = \delta_{\ell r}. \tag{3.4.14}$$

Writing $\varepsilon_{rmn} = \varepsilon_{\ell mn}\delta_{r\ell} = \varepsilon_{\ell mn} A_{\ell i} B_{ir}$ in (3.4.13) and then using the identity (2.3.9) yield

$$\begin{aligned}\frac{\partial(\det \mathbf{A})}{\partial A_{rs}} &= \frac{1}{2}\varepsilon_{sjk}(\varepsilon_{\ell mn} A_{\ell i} A_{mj} A_{nk}) B_{ir} \\ &= \frac{1}{2}\varepsilon_{sjk}\varepsilon_{ijk} \det(\mathbf{A}) B_{ir} \\ &= \det(\mathbf{A}) B_{sr}. \end{aligned} \tag{3.4.15}$$

The identity $\varepsilon_{sjk}\varepsilon_{ijk} = 2\delta_{si}$, which follows directly from (2.2.5), was used in the last step of this calculation. In direct notation, this identity becomes

$$\frac{\partial(\det \mathbf{A})}{\partial \mathbf{A}} = (\det \mathbf{A})\mathbf{A}^{-1}. \tag{3.4.16}$$

The final step in obtaining (3.4.6) is to use the chain rule to write

$$\frac{D}{Dt}(\det \mathbf{A}) = \frac{DA_{rs}}{Dt}\frac{\partial(\det \mathbf{A})}{\partial A_{rs}}, \tag{3.4.17}$$

which after substituting (3.4.16) yields (3.4.6).

3.5 PATHLINES, STREAKLINES, STREAMLINES, AND VORTEX LINES

Several different types of curves are commonly constructed to provide a geometrical representation of fluid flow. Each of these curve types indicates different aspects of a flow, and taken together they can considerably enhance understanding of complex fluid motions.

One such curve is a *pathline*, which is the path that a single fluid particle traces by its motion during some time interval. A pathline is constructed by plotting the particle position $\mathbf{x} = \mathbf{x}(\boldsymbol{\xi}, \tau)$ over a time interval $0 \le \tau \le t$ for a fixed value of $\boldsymbol{\xi}$ (i.e., a fixed fluid particle). The assumption of continuity of $\mathbf{x} = \mathbf{x}(\boldsymbol{\xi}, t)$ in t implies that pathlines are continuous. Pathlines may cross each other, or a single pathline may cross itself, since two fluid particles may occupy the same position at different times. A pathline can be produced experimentally, for instance, by taking a photograph, with very long time exposure, of a small reflective or fluorescent seed particle immersed in a flow. The photograph would contain an illuminated curve indicating the particle position at any time during the interval over which the film was exposed.

A *streakline* is defined as the location, evaluated at a given instant of time t, of all fluid particles that have passed through a fixed point $\mathbf{x}_0$ in space at some previous time, say τ. Mathematically, a streakline may be expressed as the location $\mathbf{x} = \mathbf{x}(\boldsymbol{\xi}, t)$, at a fixed value of t, for all values of $\boldsymbol{\xi}$ for which $\boldsymbol{\xi} = \boldsymbol{\xi}(\mathbf{x}_0, \tau)$ for some $\tau \le t$. The assumption of continuity of $\mathbf{x} = \mathbf{x}(\boldsymbol{\xi}, t)$ in $\boldsymbol{\xi}$ implies that streaklines are continuous. Streaklines are easily produced experimentally by releasing some dye into the flow at a fixed location over a period of time. The curve marked by the dye at any instant of time (neglecting diffusion of the dye in the fluid) is the streakline associated with the point of dye release.

A *streamline* is defined as a curve that is everywhere tangent to the velocity vector at a given instant of time. If $d\mathbf{x}$ is an increment of length along a streamline, this definition implies that we can write

$$d\mathbf{x} = \mathbf{u}\, ds, \tag{3.5.1}$$

where s is a parameter that varies along the streamlines. In component form, using the usual convention of writing the components of velocity $\mathbf{u}$ in the (x, y, z) directions as (u, v, w), respectively, (3.5.1) becomes

$$\frac{dx}{u} = \frac{dy}{v} = \frac{dz}{w} = ds. \tag{3.5.2}$$

Streamlines are most easily obtained by solution of the parametric equation

$$\frac{d\mathbf{x}}{ds} = \mathbf{u}(\mathbf{x}, t) \tag{3.5.3}$$

for a fixed value of t. The assumption of single-valuedness of the mapping $\boldsymbol{\xi} \to \mathbf{x}$ implies that streamlines cannot cross; however, streamlines can bifurcate at a stagnation point. For example, in the problem of uniform flow past a fixed cylinder, the streamline C that passes into the stagnation point at the front of the body bifurcates and travels about the body surface on both sides of the body.

In a steady flow, pathlines are fixed in time and are therefore identical to streaklines. Since the velocity $\mathbf{u}$ is independent of t in a steady flow, the parameter s may be replaced by time t in (3.5.3), yielding the equation for variation of $\mathbf{x}$ on a pathline. Hence, all three types of curves are identical in steady flows. In unsteady flows, on the other hand, these three types of curves may appear very different. For example, we consider the case of a two-dimensional vortex flow with an oscillatory point source placed at the vortex center. The velocity components in the r- and θ-directions are given by

$$u_r = \frac{m(t)}{2\pi r}, \qquad u_\theta = \frac{C}{2\pi r}, \tag{3.5.4}$$

where C is a positive constant. The source strength m varies periodically in time according to $m(t) = A\sin(2\pi t/P)$, where A and P are adjustable constants. The streamlines associated with this flow are outward spirals whenever m is positive and inward spirals whenever m is negative (Figure 3.5*a*). The pathlines are wavy curves (dashed line in Figure 3.5*b*) and the streaklines form a sawtooth pattern (solid line in Figure 3.5*b*), both of which wrap around the vortex center.

A *stream tube* is a surface (or tube) formed of all streamlines passing through a closed circuit C at some instant of time. An illustration of a stream tube is shown

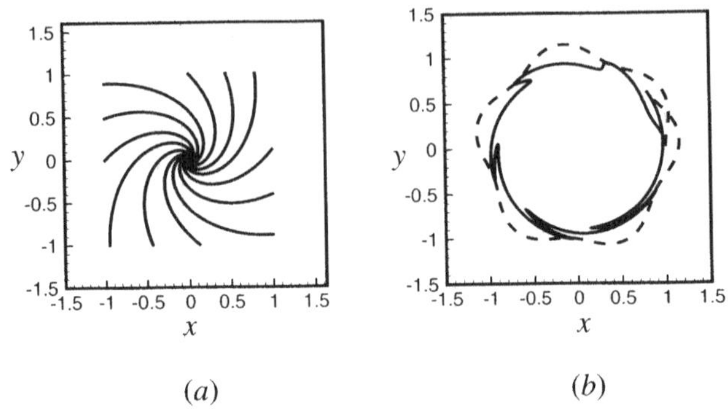

Figure 3.5 Flow lines for oscillating vortex flow: (*a*) streamlines and (*b*) pathline (dashed line) and streakline (solid line) for a point originating at $(x, y) = (1, 0)$.

in Figure 3.6. The *strength* Q of a stream tube is the volumetric flow rate across the open surface A bounded by the curve C, or

$$Q \equiv \int_A \mathbf{u} \cdot \mathbf{n}\, da, \tag{3.5.5}$$

where the unit vector $\mathbf{n}$ is normal to A and points in the same sense for each cross section of the stream tube. The incompressibility condition (3.4.8) can be used to prove a useful theorem regarding the variation of the stream tube strength.

Theorem 3.5.1. The strength of a stream tube is uniform along the stream tube in any incompressible fluid flow.

This theorem can be proved by integrating the dilatation rate $\nabla \cdot \mathbf{u}$ over the region V shown in Figure 3.6. The region V is bounded at the ends by two open surfaces A_1 and A_2, whose bounding curves C_1 and C_2 lie on the same stream tube, and on the sides by the stream tube surface S. If $\hat{\mathbf{n}}$ is the outward unit normal to the bounding surface of the region V, we can use the divergence theorem to write

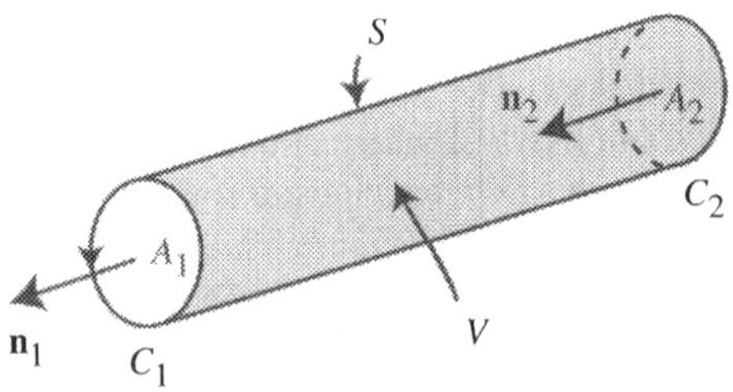

Figure 3.6 Illustration of a stream tube.

$$\int_V \nabla \cdot \mathbf{u}\, dv = \int_S \mathbf{u} \cdot \hat{\mathbf{n}}\, da + \int_{A_1} \mathbf{u} \cdot \hat{\mathbf{n}}\, da + \int_{A_2} \mathbf{u} \cdot \hat{\mathbf{n}}\, da = 0. \tag{3.5.6}$$

The above integral equals zero since $\nabla \cdot \mathbf{u}$ vanishes everywhere in an incompressible flow. The integral over S also vanishes identically, since the velocity is everywhere tangent to the stream tube S (so that $\mathbf{u} \cdot \hat{\mathbf{n}} = 0$ on S). We note that if the sense of the unit vector $\mathbf{n}$ in the definition (3.5.5) of the stream tube strength is chosen such that $\mathbf{n}$ coincides with the outward unit normal of V on one end, say $\mathbf{n} = \hat{\mathbf{n}}_1$, then $\mathbf{n}$ will be oriented in the opposite direction to the unit normal on the other end, or $\mathbf{n} = -\hat{\mathbf{n}}_2$. Writing the outward unit normal $\hat{\mathbf{n}}$ in the integrals over the end regions in the above equation in terms of the vector $\mathbf{n}$, we obtain

$$\int_{A_1} \mathbf{u} \cdot \mathbf{n}\, da = \int_{A_2} \mathbf{u} \cdot \mathbf{n}\, da. \tag{3.5.7}$$

Since the end regions, A_1 and A_2, of V were chosen arbitrarily, this result confirms that the stream tube strength must be the same when evaluated over any cross section of a stream tube at a given instant of time.

One final type of curve, which can provide valuable information for flows with nonvanishing vorticity, is the *vortex line*. Vortex lines are defined as curves that are everywhere tangent to the vorticity field at a given instant of time. Vortex lines thus play the same role for the vorticity field as streamlines play for the velocity field. The equations governing vortex lines are hence the same as (3.5.1)–(3.5.3), with the velocity replaced by the vorticity. A *vortex tube* is defined as the surface formed by the set of all vortex lines passing through a closed circuit C at some instant of time, which parallels our previous definition of a stream tube. The *strength* of a vortex tube, denoted by K, can be defined, in analogy to (3.5.5), by

$$K \equiv \int_A \boldsymbol{\omega} \cdot \mathbf{n}\, da, \tag{3.5.8}$$

where A is an open surface bounded by the curve C.

Another useful kinematical quantity is the circulation Γ about a closed circuit C, defined by

$$\Gamma \equiv \int_C \mathbf{u} \cdot d\mathbf{x}, \tag{3.5.9}$$

where $d\mathbf{x}$ is a directed element of length on C and by convention the integration in (3.5.9) is taken in the counterclockwise direction. Using Stokes's theorem (2.7.11), the circulation can be rewritten as

$$\Gamma = \int_A \mathbf{n} \cdot (\nabla \times \mathbf{u})\, da. \tag{3.5.10}$$

Comparing (3.5.8) and (3.5.10) and recalling the vorticity definition $\boldsymbol{\omega} \equiv \nabla \times \mathbf{u}$, we conclude that the strength of a vortex tube is equal to the circulation over any nonreducible circuit C lying on the tube.

Since the vorticity is defined as the curl of the velocity, the vector identity (2.6.2) requires that the vorticity be everywhere divergence free, or

$$\nabla \cdot \boldsymbol{\omega} = 0. \tag{3.5.11}$$

We can then utilize the same argument as that used to prove Theorem 3.5.1, with velocity replaced by vorticity, to prove the following.

Theorem 3.5.2. The strength of a vortex tube must be uniform along the vortex tube at any fixed instant of time.

This theorem, which is the first of three "vortex laws" attributed to Helmholtz, is valid for both incompressible and compressible flows, since (3.5.11) follows by vector identity. One consequence of Theorems 3.5.1 and 3.5.2 is that neither stream tubes nor vortex tubes can end in the middle of a flow. If the boundary conditions allow, stream tubes and vortex tubes may end on an interface with some other media. A stream tube may end on a solid surface, for instance, if the surface is in motion or if fluid is ejected through small pores in the wall. A vortex tube may end on a solid surface in an inviscid flow, but the no-slip boundary condition does not allow this to occur in a viscous flow unless the solid body is rotating. Alternatively, stream tubes and vortex tubes can form closed surfaces that have no ends. Common examples include the ringlike stream tubes that wrap about a line vortex and the ringlike vortex tubes present in the core of a vortex ring.

3.6 TRANSPORT THEOREM

The conservation equations, in their most basic form, deal with the rate of change of an integral over a material region $\wp$ of the fluid of some specific quantity f, such as density, specific momentum, and specific energy. The conservation equations thus have the form

$$\frac{d}{dt}\int_{\wp} f\, dv = R, \tag{3.6.1}$$

where R is an integral over $\wp$ or its bounding surface $\partial\wp$. The total derivative notation d/dt is used in (3.6.1) since the integral over $\wp$ is a function only of time. In this form, the conservation equations are not very useful for solving fluid flow problems. We discuss the conservation laws in detail in the next chapter and show how they can be converted either into local differential equations or into equations involving integrals over fixed regions in space.

The first step in converting the conservation laws into more usable forms is to take the time derivative in (3.6.1) inside of the integral over the material region $\wp$. One cannot simply exchange the derivative and the integration, since the material region $\wp$ in general deforms in time. The trick to dealing with this problem is to note that one can rewrite the integral in (3.6.1) as an integral over the region $\wp_0$ that is occupied by the fluid particles in $\wp$ in the reference configuration. This transformation can be accomplished by using the relationship (3.4.5) between the volume dv of a fluid element in the current configuration and the volume dV occupied by the same fluid particles in the reference configuration. Once the integral is expressed over the region $\wp_0$, the material derivative and the integration can be exchanged, since $\wp_0$ is evaluated at a fixed instant of time. We thus write

$$\begin{aligned}\frac{d}{dt}\int_{\wp} f\,dv &= \frac{d}{dt}\int_{\wp_0} fJ\,dV \\ &= \int_{\wp_0}\frac{D}{Dt}(fJ)\,dV \\ &= \int_{\wp_0}\left(\frac{Df}{Dt}J + f\frac{DJ}{Dt}\right)dV. \end{aligned} \tag{3.6.2}$$

The material derivative of J is given by (3.4.7), which when substituted into the above equation yields

$$\frac{d}{dt}\int_{\wp} f\,dv = \int_{\wp_0}\left(\frac{Df}{Dt} + f\nabla\cdot\mathbf{u}\right)J\,dV. \tag{3.6.3}$$

Using (3.4.5) to convert the region of integration back to the material region in the current configuration, we obtain the *Reynolds transport theorem*

$$\frac{d}{dt}\int_{\wp} f\,dv = \int_{\wp}\left(\frac{Df}{Dt} + f\nabla\cdot\mathbf{u}\right)dv. \tag{3.6.4}$$

There are several alternative forms in which the transport theorem is frequently written. For instance, if the material derivative is expanded as in (3.1.6) and the convective term is grouped together with the last term in (3.6.4), the transport theorem becomes

$$\frac{d}{dt}\int_{\wp} f\,dv = \int_{\wp}\left[\frac{\partial f}{\partial t} + \nabla\cdot(f\mathbf{u})\right]dv. \tag{3.6.5}$$

Using the divergence theorem to convert the last integral in (3.6.5) to a surface integral gives

$$\frac{d}{dt}\int_{\wp} f\,dv = \int_{\wp}\frac{\partial f}{\partial t}\,dv + \int_{\partial\wp} f\mathbf{u}\cdot\mathbf{n}\,da. \tag{3.6.6}$$

If we consider a fixed *control volume* V which at the instant of time t coincides with the material region, as shown in Figure 3.7, the integral on the right-hand side

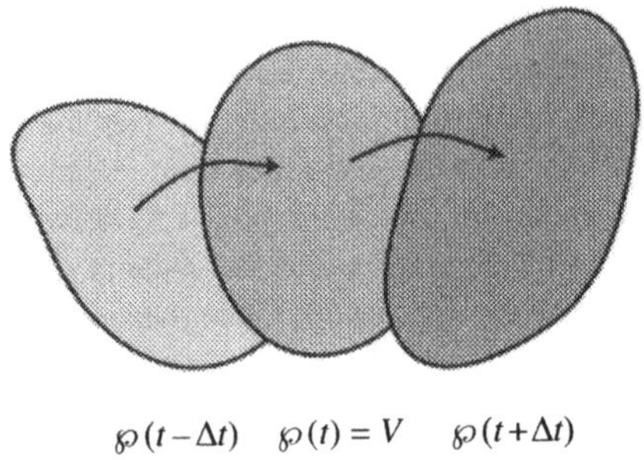

Figure 3.7 Illustration showing a fixed volume V that coincides with a material region $\wp$ at a given instant of time.

of (3.6.5) can be written in terms of an integral over the region V instead of $\wp$. Since V is fixed, we may bring the partial time derivative outside of the integral over V in the first term. We thus obtain the control-volume form of the transport theorem as

$$\frac{d}{dt}\int_{\wp} f\,dv = \frac{d}{dt}\int_{V} f\,dv + \int_{S} f\mathbf{u}\cdot\mathbf{n}\,da. \tag{3.6.7}$$

The first term on the right-hand side of (3.6.7) represents the rate of change of the integral quantity associated with f (e.g., mass or momentum) in the fixed region V, and the second term on the right-hand side of (3.6.7) represents the rate of flow of this quantity across the bounding surface S of V. The region $\wp$ cannot be replaced by V within the time derivative in the integral on the left-hand side of (3.6.7) since $\wp$ varies with time.

BIBLIOGRAPHY

Batchelor, G.K. (1967). *An Introduction to Fluid Dynamics*, Cambridge University Press, Cambridge.

Ottino, J.M. (1989). *The Kinematics of Mixing: Stretching, Chaos and Transport*, Cambridge University Press, Cambridge.

Truesdell, C.A., and R. Toupin (1960). "The classical field theories," in *Handbuch der Physik* III/1, ed. S. Flugge, Springer-Verlag, Berlin.

PROBLEMS

1. An elliptical curve C in the x-y plane is specified by

$$f(x, y, t) = \frac{(x\cos\theta + y\sin\theta)^2}{a^2} + \frac{(-x\sin\theta + y\cos\theta)^2}{b^2} - 1 = 0,$$

where $a(t)$ and $b(t)$ denote the major and minor axes of the ellipse and $\theta(t)$ is the ellipse orientation. The ellipse is immersed in a plane straining flow with velocity components $u = ex$ and $v = -ey$, where e is the straining rate. Derive expressions for the rate of change of a, b, and θ if the curve C moves as a material circuit. Is it possible for C to maintain an elliptical shape for all time t?

2. Determine the extrema values of the logarithmic rate of stretching and the direction along which these values occur, for the flow fields below.

   (a) Flow down an inclined plane [in Cartesian coordinates (x, y, z)]:

$$\mathbf{u} = C(y^2 - 2hy)\mathbf{e}_x$$

   (b) Burger's vortex with $2\pi\sigma^2 c/\Gamma = 0.1$ [in cylindrical polar coordinates (R, α, z)]:

$$\mathbf{u} = -cR\mathbf{e}_R + \frac{\Gamma}{2\pi R}\left[1 - e^{-R^2/\sigma^2}\right]\mathbf{e}_\alpha + 2cz\mathbf{e}_z.$$

3. A polymeric fluid is composed of long-chain molecules, the stretching of which alters the rheological properties of the flow. Consider a polymeric fluid in a uniform shear flow $\mathbf{u} = Sy\mathbf{e}_x$, where S is the constant shearing rate. Assuming that the long-chain molecules move independently of each other and neglecting Brownian motion, derive expressions for the rate of stretching and rate of rotation of a molecule in the x-y plane. Solve these expressions to obtain the stretch and orientation angle θ of the molecule as functions of time.

4. A one-dimensional sound wave oscillates with velocity $u = A\sin(kx - \sigma t)$, where A, k, and σ are known constants. Determine the maximum dilatation J of the fluid, assuming that $kA/\sigma \ll 1$.

5. Compute and plot the streamlines, streaklines, and pathlines at time $t = 10$ for a flow with velocity field $\mathbf{u} = m/2\pi r\mathbf{e}_r + C(t)/2\pi r\mathbf{e}_\theta$, where $m = -1$ is a constant and $C(t) = 1 + 0.5\sin(t)$. For streaklines, dye is released at $(r_0, \theta_0) = (1, 0)$ over the time interval $0 \le t \le 10$. For pathlines, a particle is released at $(r_0, \theta_0) = (1, 0)$ at $t = 0$.

CHAPTER 4

LAWS OF FLUID DYNAMICS

Fluid motion is constrained by a variety of dynamical laws, representing conservation of mass, momentum, and moment of momentum. In the current chapter, these laws are first presented in their most basic form, which follow directly from Newton's laws of motion, and then alternative forms of these equations are derived that are more useful in solving fluid flow problems. Since we do not consider heat transfer in the current text, our development proceeds in terms of a purely mechanical theory. The assumption of smoothness of the variables describing fluid motion (such that the dependent variables and a sufficient number of their derivatives exist and are continuous everywhere) will be made throughout the chapter.

4.1 MASS CONSERVATION

The law of mass conservation requires that the mass of a given set of fluid particles (i.e., a material region of the fluid) does not change in time. If $\rho = \rho(\mathbf{x}, t)$ denotes the density of the fluid at time t and position $\mathbf{x}$ in the current configuration, then mass conservation may be expressed as

$$\frac{d}{dt}\int_{\wp} \rho \, dv = 0, \tag{4.1.1}$$

where $\wp$ is a material region of the fluid. Using the transport theorem (3.6.4) to take the time derivative inside of the integral, (4.1.1) becomes

$$\int_{\wp} \left(\frac{D\rho}{Dt} + \rho \nabla \cdot \mathbf{u} \right) dv = 0. \tag{4.1.2}$$

Since (4.1.2) holds for an arbitrary material region $\wp$, the integrand must vanish at any point of the fluid, or

$$\frac{D\rho}{Dt} + \rho\nabla \cdot \mathbf{u} = 0. \tag{4.1.3}$$

The differential equation (4.1.3) is the *local* form of the mass conservation equation, which applies at every point within the fluid for which the smoothness assumptions on ρ and $\mathbf{u}$ are satisfied. Using the expansion (3.1.6) of the material derivative, (4.1.3) can be written in the alternative form

$$\frac{\partial\rho}{\partial t} + \nabla \cdot (\rho\mathbf{u}) = 0. \tag{4.1.4}$$

For an incompressible flow, the requirement that volume of a material region be conserved in time results in the restriction that the velocity be divergence free, as derived in Section 3.4. The mass conservation law for an incompressible flow thus reduces to

$$\frac{D\rho}{Dt} = 0, \tag{4.1.5}$$

which requires that the density evaluated at a given fluid particle be constant with time. It is noted that (4.1.5) does not require that density be uniform in space, and in fact density-stratified incompressible flows are important in the study of large water bodies, such as oceans and lakes, and can also be used as a reasonable approximation of many atmospheric motions. For a steady flow, the fluid properties are constant at any fixed spatial point, so that the partial time derivative of ρ must vanish in (4.1.4), yielding

$$\nabla \cdot (\rho\mathbf{u}) = 0. \tag{4.1.6}$$

In a steady flow that is also incompressible, the density gradient must vanish along streamlines.

Integration of (4.1.4) over a fixed region V with bounding surface S and use of the divergence theorem (2.7.4) yield the control-volume form of the mass conservation law as

$$\frac{d}{dt}\int_V \rho\, dv + \int_S \rho\mathbf{n} \cdot \mathbf{u}\, da = 0, \tag{4.1.7}$$

where $\mathbf{n}$ is the outward unit normal of S. The first term in (4.1.7) represents the rate of change of mass within the control volume V, and the second term represents the rate at which mass is carried across the surface S by the flow.

With use of the local form of mass conservation, the transport theorem developed in Section 3.6 can be cast in an alternative form. Suppose that f is some specific quantity, measured per unit mass, and that we are concerned with taking the time derivative inside a volume integral of ρf over a material region $\wp$. Using the transport theorem (3.6.4), we can write

$$\frac{d}{dt}\int_{\wp} \rho f\, dv = \int_{\wp}\left[\frac{D}{Dt}(\rho f) + \rho f\, \nabla\cdot\mathbf{u}\right] dv$$
$$= \int_{\wp}\left[\rho\frac{Df}{Dt} + f\left(\frac{D\rho}{Dt} + \rho\,\nabla\cdot\mathbf{u}\right)\right] dv. \tag{4.1.8}$$

The last term on the right-hand side of (4.1.8) vanishes due to the local mass conservation equation (4.1.3), so that the transport theorem reduces to

$$\frac{d}{dt}\int_{\wp} \rho f\, dv = \int_{\wp} \rho\frac{Df}{Dt}\, dv. \tag{4.1.9}$$

4.2 MOMENTUM CONSERVATION

The law of conservation of momentum, as stated by Newton, requires that the rate of change of momentum of a set of fluid particles is equal to the sum of the forces acting on the particles. Applying the momentum conservation law to the fluid occupying some material region $\wp$ gives

$$\frac{d}{dt}\int_{\wp} \rho\mathbf{u}\, dv = \sum_{\wp} \text{forces}. \tag{4.2.1}$$

The forces acting on the fluid are of two types, which are illustrated in Figure 4.1. For some types of forces, called *contact forces*, we need to account only for the action of the forces on the bounding surface $\partial\wp$ of $\wp$. Common examples of contact forces

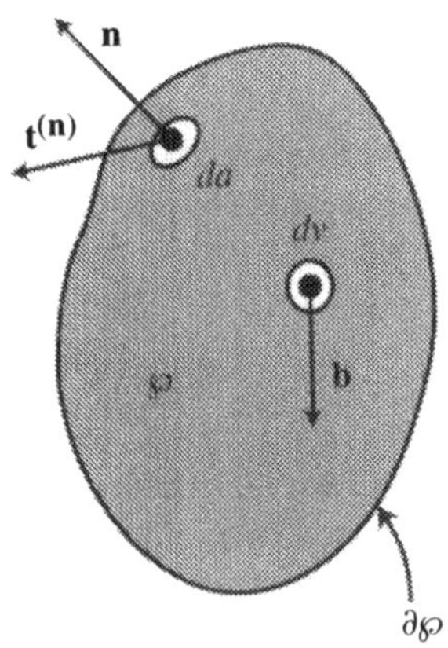

Figure 4.1 Illustration of body force $\mathbf{b}$ acting on each point of a material region $\wp$ and contact force $\mathbf{t}^{(\mathbf{n})}$ acting on the surface $\partial\wp$ of $\wp$.

include pressure force and shearing stress. While pressure and shearing stress also exist at points interior to $\wp$, the effect of these forces at interior points cancel out and produce no net force on the fluid in $\wp$. Other types of forces, called *body forces*, must be evaluated as acting on each point of the fluid within $\wp$. Common examples of body forces are gravity and electromagnetic force. The body force acting per unit mass of the fluid is denoted by the vector $\mathbf{b}$, which is in general a prescribed function of $\mathbf{x}$ and t. The contact force acting per unit area of the surface $\partial\wp$ is denoted by the vector $\mathbf{t}^{(\mathbf{n})}$, which is called the *stress vector*. The subscript $(\mathbf{n})$ attached to the stress vector is used to emphasize that this vector depends not only on time and location in space but also on the orientation of the surface $\partial\wp$ on which the stress acts. At a given $\mathbf{x}$ and t, the stress vector is in general different for two surfaces S_1 and S_2 passing through this point. As shown later in this section, the stress vector depends only on the orientation of the surface unit normal $\mathbf{n}$ and not on other geometrical characteristics of the surface.

With this notation, the conservation-of-momentum law for a material region can be written as

$$\frac{d}{dt}\int_{\wp} \rho\mathbf{u}\,dv = \int_{\wp} \rho\mathbf{b}\,dv + \int_{\partial\wp} \mathbf{t}^{(\mathbf{n})}\,da. \tag{4.2.2}$$

In order to express the momentum conservation law in the form of a local differential equation, it is necessary to both take the time derivative inside of the integration over $\wp$ on the left-hand side of (4.2.2) and express the contact force term as a volume integral over $\wp$. The first of these tasks is easily accomplished by using the form (4.1.9) of the transport theorem to write

$$\frac{d}{dt}\int_{\wp} \rho\mathbf{u}\,dv = \int_{\wp} \rho\frac{D\mathbf{u}}{Dt}\,dv. \tag{4.2.3}$$

To perform the second of these tasks, it is necessary to develop an expression attributed to Cauchy that relates the stress vector to the unit normal of the surface $\partial\wp$. Consider a small region $\mathfrak{R}$ of fluid which has characteristic dimension d in all directions (i.e., is not highly elongated). This region is made progressively smaller without changing its shape. As $d \to 0$, the volume of $\mathfrak{R}$ approaches zero in proportion to d^3 and the surface area of approaches zero in proportion to d^2. Turning now to the momentum conservation law (4.2.2), we observe that the rate of momentum and body force terms are volume integrals which approach zero as d^3, but the contact force term is a surface integral which approaches zero only as d^2. It follows that in the limit of an infinitesimally small region, the contract stresses are in equilibrium, such that

$$\lim_{d\to 0}\frac{1}{d^2}\int_{\partial\mathfrak{R}} \mathbf{t}^{(\mathbf{n})}\,da = 0. \tag{4.2.4}$$

Consider now a special case of this argument where the region $\mathfrak{R}$ has the form of a small tetrahedron, as shown in Figure 4.2, with three of its faces (labeled 1, 2, 3)

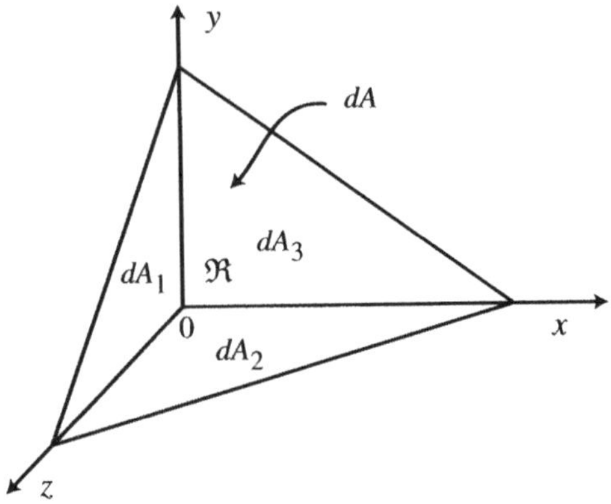

Figure 4.2 Small tetrahedron used to illustrate the Cauchy stress formula (4.2.6).

oriented orthogonal to the (x, y, z) coordinate directions and one face inclined with outward unit normal $\mathbf{n}$. The condition of equilibrium of the contact forces implies that the resultant force $\mathbf{t}^{(\mathbf{n})}\,dA$ on the inclined surface be balanced by the sum of the contact forces, e.g., $\mathbf{t}^{(1)}\,dA_1$, on the other three faces, or

$$\mathbf{t}^{(\mathbf{n})}\,dA = \mathbf{t}^{(1)}\,dA_1 + \mathbf{t}^{(2)}\,dA_2 + \mathbf{t}^{(3)}\,dA_3. \tag{4.2.5}$$

We let T_{ji} denote the ith component of the stress vector $\mathbf{t}^{(j)}$ with respect to a coordinate axis in the x_j-direction and $t_i^{(\mathbf{n})}$ denote the ith component of $\mathbf{t}^{(\mathbf{n})}$. Making use of this notation and the fact that the areas of the three orthogonal faces of the tetrahedron are given by $dA_i = n_i\,dA$, (4.2.5) can be rewritten, after dividing by the area dA of the inclined surface of the tetrahedron, as

$$t_i^{(\mathbf{n})} = n_j T_{ji} \qquad \text{or} \qquad \mathbf{t}^{(\mathbf{n})} = \mathbf{n} \cdot \mathbf{T}. \tag{4.2.6}$$

Equation (4.2.6) is called the *Cauchy stress formula*, and it states that the stress vector $\mathbf{t}^{(\mathbf{n})}$ acting on a surface $\partial\wp$ is linearly related to the outward unit normal $\mathbf{n}$ of $\partial\wp$. The symbols T_{ij} are the components of the second-order *stress tensor* T. From the indicial form of (4.2.6), it is clear that the first index of the stress tensor determines the direction of the unit normal of the surface on which the force acts and the second index of the stress tensor determines the direction of the force acting on the surface.

Substituting (4.2.6) into the last term in (4.2.2), the integral identity (2.7.1) can be used to write

$$\int_{\partial\wp} t_i^{(\mathbf{n})}\,da = \int_{\partial\wp} T_{ji} n_j\,da = \int_{\wp} \frac{\partial T_{ji}}{\partial x_j}\,dv. \tag{4.2.7}$$

Substituting (4.2.3) and (4.2.7) into (4.2.2), the momentum conservation law, in indicial form, becomes

$$\int_{\wp} \left(\rho \frac{Du_i}{Dt} - \rho b_i - \frac{\partial T_{ji}}{\partial x_j} \right) dv = 0. \tag{4.2.8}$$

Since this result is valid for arbitrary fluid regions $\wp$, the integrand must vanish at every point within the fluid. The local form of the momentum conservation law is thus obtained as

$$\rho \frac{Du_i}{Dt} = \rho b_i + \frac{\partial T_{ji}}{\partial x_j}, \tag{4.2.9a}$$

or in direct notation

$$\rho \frac{D\mathbf{u}}{Dt} = \rho \mathbf{b} + \nabla \cdot \mathbf{T}. \tag{4.2.9b}$$

Expanding the material derivative in (4.2.9) gives

$$\rho \left[\frac{\partial u_i}{\partial t} + u_j \frac{\partial u_i}{\partial x_j} \right] = \rho b_i + \frac{\partial T_{ji}}{\partial x_j}, \tag{4.2.10a}$$

or

$$\rho \left[\frac{\partial \mathbf{u}}{\partial t} + (\mathbf{u} \cdot \nabla)\mathbf{u} \right] = \rho \mathbf{b} + \nabla \cdot \mathbf{T}. \tag{4.2.10b}$$

The local mass conservation equation (4.1.4) can be used to rewrite (4.2.10) in the alternative form

$$\frac{\partial(\rho u_i)}{\partial t} + \frac{\partial(\rho u_j u_i)}{\partial x_j} = \rho b_i + \frac{\partial T_{ji}}{\partial x_j}. \tag{4.2.11}$$

Integration of (4.2.11) over a fixed region V with bounding surface S and use of the divergence theorem give the control-volume form of the momentum conservation equation as

$$\frac{d}{dt} \int_V \rho u_i \, dv + \int_S (u_j n_j) \rho u_i \, da = \int_V \rho b_i \, dv + \int_S T_{ji} n_j \, da. \tag{4.2.12}$$

The first term on the left-hand side of (4.2.12) is the rate of change of momentum within V and the second term is the rate at which momentum is carried across the bounding surface of V by the flow.

4.3 MOMENT OF MOMENTUM CONSERVATION

The final conservation law that we consider is conservation of moment of momentum (or angular momentum), which requires that the rate of change of the moment of momentum of a set of fluid particles is equal to the sum of the torques exerted on the particles. The moment of momentum is given by the cross product of the position vector **x** with the momentum, and the torques exerted on the fluid are given by the cross product of **x** with the body and contact forces, such that the law of conservation of moment of momentum can be expressed as

$$\frac{d}{dt}\int_{\wp} \mathbf{x} \times \rho \mathbf{u}\, dv = \int_{\wp} \mathbf{x} \times \rho \mathbf{b}\, dv + \int_{\partial\wp} \mathbf{x} \times \mathbf{t}^{(\mathbf{n})}\, da. \tag{4.3.1}$$

In order to derive a local form of this conservation law, we follow an analogous procedure to that used to obtain the local form (4.2.9) of momentum conservation.

The time differentiation on the left-hand side of (4.3.1) can be taken inside the integral with use of the transport theorem in the form (4.1.9), giving

$$\frac{d}{dt}\int_{\wp} \mathbf{x} \times \rho \mathbf{u}\, dv = \int_{\wp} \rho \frac{D}{Dt}(\mathbf{x} \times \mathbf{u})\, dv = \int_{\wp} \rho \left(\mathbf{x} \times \frac{D\mathbf{u}}{Dt}\right) dv, \tag{4.3.2}$$

where we note that $\dot{\mathbf{x}} \times \mathbf{u} = \mathbf{u} \times \mathbf{u} = 0$. From (4.3.1) and (4.3.2), the moment of momentum equation can be rearranged as

$$\int_{\wp} \mathbf{x} \times \left(\rho \frac{D\mathbf{u}}{Dt} - \rho \mathbf{b} - \nabla \cdot \mathbf{T}\right) dv = \int_{\partial\wp} \mathbf{x} \times \mathbf{t}^{(\mathbf{n})}\, da - \int_{\wp} \mathbf{x} \times (\nabla \cdot \mathbf{T})\, dv, \tag{4.3.3}$$

where the last integral in (4.3.3) has been subtracted from both sides.

The left-hand side of (4.3.3) vanishes due to satisfaction of the local momentum equation (4.2.9). In order to determine what sort of restriction is placed on the stress tensor from the requirement that the right-hand side of (4.3.3) also vanish, it is useful to rewrite the right-hand side using indicial notation and to express the stress vector in terms of the stress tensor by using the Cauchy stress formula (4.2.6), giving

$$\int_{\partial\wp} \varepsilon_{ijk} x_j n_\ell T_{\ell k}\, da - \int_{\wp} \varepsilon_{ijk} x_j \frac{\partial T_{\ell k}}{\partial x_\ell}\, v = 0. \tag{4.3.4}$$

Using the integral identity (2.7.1), the first integral in (4.3.4) can be converted to the volume integral

$$\int_{\wp} \varepsilon_{ijk} \frac{\partial}{\partial x_\ell}(x_j T_{\ell k})\, dv.$$

Using the fact that $\partial x_j / \partial x_\ell = \delta_{j\ell}$, (4.3.4) reduces to

$$\int_{\wp} \varepsilon_{ijk} T_{jk}\, dv = 0. \tag{4.3.5}$$

Since (4.3.5) is satisfied for arbitrary material parts $\wp$ of the fluid, the integrand must vanish at every point, or

$$\varepsilon_{ijk} T_{jk} = 0. \tag{4.3.6}$$

We recall that the permutation symbol ε_{ijk} is skew-symmetric in any two of its indices. The restriction (4.3.6) thus requires that the stress tensor $\mathbf{T}$ be symmetric, or that

$$T_{ij} = T_{ji}, \tag{4.3.7}$$

where we recall the result of Theorem 2.5.1 stating that the scalar product of any skew-symmetric tensor with a symmetric tensor must vanish. The law of conservation of moment of momentum is thus satisfied identically if the local momentum conservation equation (4.2.9) is satisfied and if the stress tensor is symmetric.

4.4 KINETIC ENERGY TRANSPORT THEOREM

The kinetic energy per unit mass of the fluid, denoted by κ, is defined by

$$\kappa \equiv \tfrac{1}{2}\mathbf{u} \cdot \mathbf{u}. \tag{4.4.1}$$

An evolution equation for kinetic energy can be derived by taking the scalar product of the local momentum equation (4.2.9) with the velocity vector, giving

$$\rho \frac{D\kappa}{Dt} = \rho u_i b_i + \frac{\partial}{\partial x_j}(u_i T_{ij}) - T_{ij}\frac{\partial u_i}{\partial x_j}. \tag{4.4.2}$$

The physical significance of the terms in the kinetic energy transport equation can be understood by integrating (4.4.2) over a material volume $\wp$ with bounding surface $\partial\wp$, to obtain

$$\frac{d}{dt}\int_{\wp} \rho\kappa \, dv = \int_{\wp} \rho \mathbf{b} \cdot \mathbf{u} \, dv + \int_{\partial\wp} \mathbf{u} \cdot \mathbf{t}^{(\mathbf{n})} \, da - \int_{\wp} P \, dv. \tag{4.4.3}$$

In deriving (4.4.3), use has been made of the form (4.1.9) of the transport theorem and the Cauchy stress theorem (4.2.6). The variable P in the last term in (4.4.3) is called the *mechanical power* and is defined by

$$P \equiv T_{ij}\frac{\partial u_i}{\partial x_j}. \tag{4.4.4}$$

Since T_{ij} must be symmetric, from conservation of moment of momentum, the scalar product of T_{ij} with the vorticity tensor vanishes identically, so that (4.4.4) can be rewritten as

$$P \equiv T_{ij} D_{ij}. \tag{4.4.5}$$

Equation (4.4.3) states that the change of the kinetic energy of a set of fluid particles in a region occurs due to (i) work performed by body forces at all points within $\wp$,

(ii) work performed by contact forces on the surface of $\wp$, and (iii) internal loss or production of kinetic energy within $\wp$ due to the mechanical power.

In order to understand the physical significance of the mechanical power P in fluid flows, it is useful to temporarily broaden our scope to consider compressible, viscous flows in general. Up to this point, all of the expressions developed in the text are valid for any media, including both fluids and solids. The differences in the governing equations for different classes of continua become apparent, however, in the relationship between the stress tensor $\mathbf{T}$ and other variables describing the medium (such as velocity, density, and their gradients). Such an equation is called a *constitutive equation* because its form varies depending on the material characteristics of the medium. For instance, in an elastic solid, the shear stress depends on the gradient of the displacement of the media from some equilibrium configuration. In most gases or liquids, it is commonly known that shear stress depends on the gradient of the fluid velocity. While constitutive equations have historically been obtained by a combination of experiments and modeling at the molecular level, there are several restrictions on the form of the constitutive equations that must also be satisfied in order for the continuum theory to be consistent with itself and with basic physical principles. One of these restrictions, of course, is that the stress tensor be symmetric in order to satisfy conservation of moment of momentum. Other restrictions follow from thermodynamic inequalities, such as the restriction on internal entropy production expressed by the second law of thermodynamics, and from invariance properties which the equations of motion are required to satisfy. The interested reader might consult Coleman and Noll (1963) or Truesdell and Toupin (1960) for an account of these restrictions and the way in which they are applied to the stress constitutive equation.

For any compressible, Newtonian viscous fluid, which includes most common gases and liquids, the constitutive equation for stress is given by

$$T_{ij} = -p\delta_{ij} + \lambda D_{kk}\delta_{ij} + 2\mu D_{ij}, \tag{4.4.6}$$

where p, λ, and μ are material coefficients that are functions of density and temperature (assuming that the fluid has uniform chemical composition). A well-known expression for pressure p that applies to many gases, for instance, is the ideal gas equation $p = \rho RT$, where T is temperature and R is the gas constant. The coefficients λ and μ are the bulk and dynamic viscosities, respectively. The second law of thermodynamics requires that the viscosity coefficients satisfy the inequalities

$$\mu \geq 0, \qquad \lambda + \tfrac{2}{3}\mu \geq 0. \tag{4.4.7}$$

Substituting the stress constitutive equation (4.4.6) into (4.4.5), the mechanical power becomes

$$P = -pD_{kk} + \lambda D_{kk}D_{ii} + 2\mu D_{ij}D_{ij}. \tag{4.4.8}$$

The first term on the right-hand side of (4.4.8) represents storage of energy by fluid dilatation. We recall from Section 3.4 that the dilatation rate $D_{ii} = \nabla \cdot \mathbf{u}$ vanishes in an incompressible flow. Since the absolute pressure p is a positive quantity, it follows from (4.4.8) and (4.4.3) that an expansion (for which $\dot{J}/J = D_{ii} > 0$) results in an increase in kinetic energy in the fluid, and consequently a decrease in stored energy, since $pD_{kk} > 0$. By contrast, a compression (for which $\dot{J}/J = D_{ii} < 0$) results in a decrease in kinetic energy, and consequently an increase in stored energy, in the medium since $pD_{kk} < 0$.

The second and third terms on the right-hand side of (4.4.8) are positive for both an expansion and a compression and always result in a decrease in kinetic energy. Both of these terms cause a drain of kinetic energy from the fluid, which is subsequently converted into heat. The term in (4.4.8) proportional to λ arises from the dilatation of the fluid and vanishes for an incompressible flow. The term proportional to μ arises from both shear and normal stresses, although it is more typically dominated by shear stresses. This term is commonly referred to as the *energy dissipation rate* when dealing with an incompressible flow. In an inviscid flow, both λ and μ are set to zero, so that there is no conversion of kinetic energy into heat. For a flow that is both incompressible and inviscid, the mechanical power P is everywhere zero.

Although the kinetic energy equation (4.4.2) contains terms with the dimensions of energy, it is not a statement of the conservation of energy and is in fact derived from the momentum conservation equation. By comparison, the energy conservation equation also has terms related to change in internal energy and heat transfer within the fluid. While it is true that in the limit of an isothermal flow, heat transfer terms are negligible and the energy conservation equation has a form similar to the kinetic energy transport equation (4.4.2), the derivation of (4.4.2) from momentum conservation does not require the flow to be isothermal.

4.5 CONSTRAINT OF INCOMPRESSIBILITY

The equation for the stress tensor in an incompressible, inviscid fluid can be derived using the physical argument, noted in the last section, that because such a flow admits neither energy storage through dilatation nor conversion of kinetic energy into heat through viscous effects, the mechanical power P must vanish. We therefore search for the most general expression for the stress tensor T_{ij} for which

$$P = T_{ij} D_{ij} = 0 \tag{4.5.1}$$

under the condition that the rate of deformation tensor satisfy the incompressibility condition, written in the form

$$\delta_{ij} D_{ij} = 0. \tag{4.5.2}$$

If we were to require (4.5.1) to be satisfied for all values of D_{ij}, then the corresponding components of the stress tensor would have to either be zero or be such that the determinant of the stress tensor vanishes. Neither of these results would be acceptable. Fortunately, (4.5.1) does not need to be satisfied for all D_{ij}, but only for those values of D_{ij} that satisfy (4.5.2). The corresponding expression for stress tensor can be obtained by multiplying (4.5.2) by some function $p(\mathbf{x}, t)$ and then adding the resulting equation to (4.5.1), giving

$$(T_{ij} + p\delta_{ij})D_{ij} = 0. \tag{4.5.3}$$

Requiring that (4.5.3) be satisfied for all D_{ij} yields an expression for the stress tensor as

$$T_{ij} = -p(\mathbf{x}, t)\delta_{ij}. \tag{4.5.4}$$

The result (4.5.4) appears at first glance to be the same as the stress constitutive equation (4.4.6) in the limit of incompressible flow ($D_{ii} = 0$) and vanishing viscosity. There is, however, an important difference. In the constitutive equation (4.4.6), the pressure is a function of density and temperature (as in the ideal gas law, for instance). In the limit of an isothermal flow with constant density, the limiting form of (4.4.6) would require that pressure is constant. This result, however, would make the governing equations improperly posed. For instance, if there were no body force, the momentum conservation equation (4.2.9) would require that the acceleration vanishes everywhere, which might be inconsistent with the boundary and initial conditions. By contrast in the result (4.5.4), the quantity p, which is also called the pressure, is an arbitrary function of $\mathbf{x}$ and t. A similar argument can be made in a viscous incompressible flow but for this case there exists an additional term added to the right-hand side of (4.5.4), corresponding to the last term in (4.4.6), which is related to energy dissipation by viscosity.

In a compressible flow, the *primitive variables* are density and the three components of velocity, whose solutions are obtained from the momentum and mass conservation equations as functions of $\mathbf{x}$ and t. Other quantities, such as pressure or shear stress, are written as functions of these primitive variables or their derivatives. By contrast, the primitive variables in an incompressible flow are pressure and the three velocity components, whose solutions are obtained from the governing equations as functions of $\mathbf{x}$ and t.

We thus observe that the governing equations for an incompressible flow cannot be obtained as a special case of those for a compressible flow but must be derived in terms of a special "constraint theory." A *constraint* is a restriction on the values that the primitive variables may attain. For incompressible flow, the constraint is given by the expression (4.5.2), requiring that the rate of fluid dilatation vanish. Examples of other types of constraint theories can be found in classical textbooks on dynamics, continuum mechanics, and calculus of variations (Segal, 1977). A constraint theory typically requires some additional *constraint force* be admitted to maintain the constraint. For the case of incompressible flow, the constraint force is given by (4.5.4), which is added to the constitutive equation (4.4.6). Because the first term on the right-hand side of (4.4.6) and the expression (4.5.4) have the same form, these two terms can be combined, with the resulting multiplier of δ_{ij} being again called simply "pressure" and denoted by the symbol p.

4.6 NAVIER-STOKES EQUATION AND SOME SIMPLE SOLUTIONS

In this section we diverge from the our principal theme of inviscid flows to consider the Navier-Stokes equation, which governs the motion of viscous fluid flows, and a few simple solutions of this equation. The main motivation for this discussion is that some understanding of the modifications brought about by introduction of viscosity is necessary in order to appreciate the relationship between inviscid flow theory and the motion of real fluids. In addition, there is a class of simple exact solutions of the viscous flow equations for which the velocity field can also be regarded as a solution of the inviscid flow equations. Several of these flows are used in Chapter 16 as base flows for inviscid stability analysis.

The Navier-Stokes equation is obtained by substituting the constitutive equation (4.4.6) for stress into the momentum equation (4.2.9). For an incompressible fluid, the stress constitutive equation reduces to

$$T_{ij} = -p\delta_{ij} + 2\mu D_{ij}, \tag{4.6.1}$$

where p is an arbitrary function of $\mathbf{x}$ and t. Taking the divergence of $\mathbf{T}$ and making the usual assumption that the viscosity μ is constant give

$$\nabla \cdot \mathbf{T} = -\nabla p + 2\mu \nabla \cdot \mathbf{D}. \tag{4.6.2}$$

Written in index notation and using the definition of $\mathbf{D}$, $\nabla \cdot \mathbf{D}$ reduces to

$$\frac{\partial D_{ij}}{\partial x_j} = \frac{1}{2}\left(\frac{\partial^2 u_i}{\partial x_j^2} + \frac{\partial^2 u_j}{\partial x_j \partial x_i}\right). \tag{4.6.3}$$

The last term in (4.6.3) vanishes because of the assumption of incompressibility, which requires that $\partial u_j/\partial x_j = 0$. The first term on the right-hand side of (4.6.3) is simply the Laplacian of the velocity vector. The momentum equation (4.2.9) therefore reduces to the *Navier-Stokes equation*

$$\rho\frac{D\mathbf{u}}{Dt} = \rho\mathbf{b} - \nabla p + \mu \nabla^2\mathbf{u}. \tag{4.6.4}$$

In the absence of viscosity, (4.6.4) reduces to the *Euler equation*

$$\rho\frac{D\mathbf{u}}{Dt} = \rho\mathbf{b} - \nabla p. \tag{4.6.5}$$

It is important to note that the viscous term in (4.6.4) involves second-order spatial derivatives, whereas equation (4.6.5) involves only first-order spatial derivatives. A mathematical consequence of this difference is that the Navier-Stokes equation must have one additional boundary condition in each spatial direction than the Euler equation. Physically, the result of this observation is that viscous flow solutions are required to satisfy both the no-penetration and no-slip conditions on the surface of an immersed solid body, whereas inviscid flow solutions are in general only required to satisfy the no-penetration condition. It is always possible to construct an inviscid flow solution that satisfies the no-slip condition as well by appropriate specification of the vorticity field (indeed, this is the basis of the vortex panel method described in Chapter 12), but this inviscid solution might not much resemble the viscous flow solution in the region near the boundaries where viscous shear is important.

There is a special set of solutions of the Navier-Stokes equation for which the inertia term $D\mathbf{u}/Dt$ vanishes. The velocity field of such flows may be regarded as a solution of either the Navier-Stokes or Euler equations after suitable modification of the pressure field. In the current section, we consider a special class of flows, called *parallel shear flows*, which have the property that the velocity is everywhere oriented along one coordinate direction and it varies as a function only of a coordinate in an orthogonal direction. For instance, in Cartesian coordinates an example of a parallel shear flow is a velocity field of the form

$$\mathbf{u} = u(y)\mathbf{e}_x. \tag{4.6.6}$$

In cylindrical polar coordinates (R, α, z), examples of parallel shear flows include axisymmetric flow

$$\mathbf{u} = w(R)\mathbf{e}_z \tag{4.6.7}$$

and swirling flow

$$\mathbf{u} = v(R)\mathbf{e}_\alpha. \tag{4.6.8}$$

That the inertia term vanishes for velocity fields of the form (4.6.6) and (4.6.7) can be verified by direct substitution. For swirling flows of the form (4.6.8), the inertia term reduces to the centrifugal force $D\mathbf{u}/Dt = -(v^2/R)\mathbf{e}_R$, which can be eliminated by combining with the radial pressure gradient. Several examples of solutions for parallel shear flows are given below.

Example 4.6.1 Plane Couette Flow. Couette flow refers to a fluid flow that is driven by the relative motion of parallel surfaces. In plane Couette flow, the fluid is confined between two infinite parallel plates separated by a distance H that are translated at different velocities in a direction parallel to the plate surface (Figure 4.3a). The flow is assumed to be fully developed, so that velocity changes neither in time nor along the length of the plate. Writing the velocity field in the form (4.6.6) and assuming that the pressure gradient and the body force vanish everywhere, the Euler equation (4.6.5) is identically satisfied for any velocity profile $u(y)$. The Navier-Stokes equation (for $\mu \neq 0$) reduces to $d^2u/dy^2 = 0$, which requires that u vary linearly with y. Furthermore, the viscous theory requires that the no-slip condition be satisfied on the plates at $y = 0$ and H. If the reference frame is selected such that the lower plate is fixed and the upper plate is translated with velocity U, then the viscous flow solution for the velocity profile is

$$u = \frac{yU}{H}. \tag{4.6.9}$$

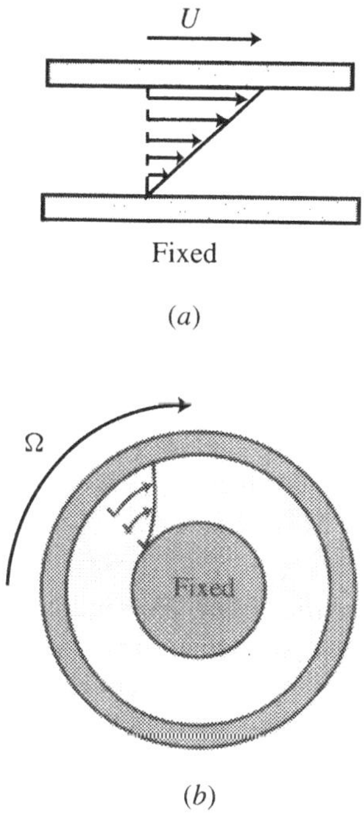

Figure 4.3 Velocity profiles for (*a*) plane and (*b*) rotating Couette flow.

Plane Couette flow exhibits uniform shear stress throughout the flow, given by $T_{12} = \mu U/H$.

Example 4.6.2 Rotating Couette Flow. In rotating Couette flow, the fluid is confined between two circular cylinders with radii R_1 and R_2 that are rotating at different angular rotation rates Ω_1 and Ω_2, as indicated in Figure 4.3*b*. The cylinders are assumed to be of infinite length in the axial direction. The velocity field has the form given in (4.6.8), for which the inertia term reduces to $D\mathbf{u}/Dt = -(v^2/R)\mathbf{e}_R$. Since both the azimuthal pressure gradient and the azimuthal component of inertia vanish, the Euler equation is satisfied for any azimuthal velocity profile $v(R)$. The azimuthal component of the Navier-Stokes equation reduces to

$$\frac{1}{R}\frac{d}{dR}\left(R\frac{dv}{dR}\right) - \frac{v}{R^2} = 0. \tag{4.6.10}$$

Equation (4.6.10) can be rearraged in the form of an Euler-Cauchy equation, which is known to have solutions proportional to R^m, where m is an undetermined constant. Substituting this solution into (4.6.10) yields $m = 1$ and $m - 1$, indicating a general solution for v of the form

$$v(R) = AR + \frac{B}{R}. \tag{4.6.11}$$

The no-slip condition at the two cylinders requires that $v(R_1) = \Omega_1 R_1$ and $v(R_2) = \Omega_2 R_2$, which yield expressions for the constants A and B as

$$A = \frac{\Omega_2 R_2^2 - \Omega_1 R_1^2}{R_2^2 - R_1^2}, \qquad B = -R_1^2 R_2^2 \left(\frac{\Omega_2 - \Omega_1}{R_2^2 - R_1^2} \right). \tag{4.6.12}$$

The solution for the problem of swirling flow external to a single rotating cylinder can be obtained as a special case of the above solution by letting $R_2 \to \infty$ and $\Omega_2 \to 0$, which gives $A = 0$ and $B = \Omega_1 R_1^2$. Similarly, the problem of flow internal to a single rotating cylinder can be obtained by letting $R_1 \to 0$ and $\Omega_1 \to 0$, yielding solid-body rotation with $A = \Omega_2$ and $B = 0$.

Example 4.6.3 Plane Poiseuille Flow. Poiseuille flow refers to a fluid flow that is driven in a conduit between parallel surfaces by a uniform pressure gradient. In plane Poiseuille flow, the fluid is confined between two infinite parallel plates separated by a distance H, and the velocity field has the form (4.6.6). The inertia term $\mathbf{Du}/Dt$ vanishes for this velocity profile, so in the absence of a body force the Euler equation in the direction of flow becomes $dp/dx = 0$. Any velocity profile $u(y)$ thus corresponds to a solution of the Euler equation with pressure everywhere constant. The Navier-Stokes equation for this flow reduces to

$$0 = -\frac{dp}{dx} + \mu \frac{d^2 u}{dx^2}. \tag{4.6.13}$$

Since the pressure gradient is constant, (4.6.13) can be integrated to yield a quadratic expression for the velocity profile (Figure 4.4). After applying the no-slip condition at $y = 0$ and H, the velocity profile becomes

$$u(y) = -\frac{1}{2\mu}\frac{dp}{dx} y(H - y). \tag{4.6.14}$$

The shear stress for this flow varies linearly over the channel as

$$T_{12} = -\frac{dp}{dx}\left(\frac{H}{2} - y \right). \tag{4.6.15}$$

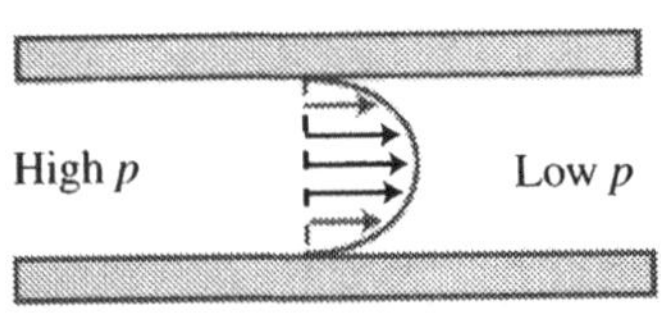

Figure 4.4 Velocity profile for Poiseuille flow.

The flow rate per unit width of the channel is given by

$$Q = -\frac{H^2}{12\mu}\frac{dp}{dx}. \tag{4.6.16}$$

Example 4.6.4 Axisymmetric Poiseuille Flow. In axisymmetric Poiseuille flow, the fluid is confined in a circular pipe or annulus and the velocity profile is of the form (4.6.7). In the absence of a body force, the Euler equation is satisfied for any velocity profile $w(R)$ if the pressure is maintained uniform. The Navier-Stokes equation reduces to

$$0 = -\frac{dp}{dz} + \frac{\mu}{R}\frac{d}{dR}\left(R\frac{dw}{dR}\right), \tag{4.6.17}$$

which for a uniform pressure gradient yields a parabolic solution for $w(R)$. For the case of flow in a circular pipe of radius a, application of the symmetry condition $dw/dR = 0$ at the pipe center ($R = 0$) and the no-slip condition $w = 0$ at the pipe wall ($R = a$) yields the velocity profile

$$w(R) = -\frac{1}{4\mu}\frac{dp}{dz}(a^2 - R^2). \tag{4.6.18}$$

The discharge through the pipe is obtained by integrating (4.6.18) over the pipe cross section as

$$Q = 2\pi\int_0^a w(R)R\,dR = -\frac{\pi a^4}{8\mu}\frac{dp}{dz}. \tag{4.6.19}$$

BIBLIOGRAPHY

Coleman, B.D., and W. Noll (1963). "The thermodynamics of elastic materials with heat conduction and viscosity," *Archive for Rational Mechanics and Analysis* **13**(3), 167–178.

Segal, L.A. (1977). *Mathematics Applied to Continuum Mechanics*, Macmillan Publishing Company, New York (reprinted Dover Publications, New York, 1987).

Serrin, J. (1959). "Mathematical principals of classical fluid mechanics," in *Handbuch der Physik* VIII/1, ed. S. Flugge and C. Truesdell, Springer-Verlag, Berlin.

Truesdell, C.A., and R. Toupin (1960). "The classical field theories," in *Handbuch der Physik* III/1, ed. S. Flugge, Springer-Verlag, Berlin.

PROBLEMS

1. The law of conservation of energy can be stated with respect to a material part $\wp$ of a fluid as

$$\frac{d}{dt}\int_{\wp} \rho(\varepsilon + \kappa)\, dv = \int_{\wp} \rho\,(r + \mathbf{b}\cdot\mathbf{u})\; dv + \int_{\partial\wp} (\mathbf{t}^{(\mathbf{n})}\cdot\mathbf{u} - \mathbf{q}\cdot\mathbf{n})\, da,$$

where ε is the internal energy per unit mass, r is the rate of external heat supply per unit mass, $\mathbf{q}$ is the heat flux vector, and the other variables are defined in the text.

(a) Derive the local form of the energy conservation equation.

(b) Use the kinetic energy transport theorem to write a differential equation for only.

(c) Suppose that $c \equiv \partial\varepsilon/\partial T$ and $\mathbf{q} = -k\nabla T$, where T is an absolute temperature, c is a specific heat, and $k > 0$ is a constant thermal conductivity. If $\varepsilon = \varepsilon(T)$, write an equation for T using the constitutive equation for the stress vector $\mathbf{t}^{(\mathbf{n})}$ for an incompressible, Newtonian viscous fluid.

2. Let c denote the mass of a CFC (chlorofluorocarbon) in the atmosphere per unit mass of atmosphere. The conservation law for c states that the rate of change of the mass of the CFC in any material region $\wp$ is equal to the rate at which the CFC is produced within $\wp$ plus the rate at which the CFC diffuses over the bounding surface $\partial\wp$ of $\wp$, or

$$\frac{d}{dt}\int_{\wp} \rho c\, dv = \int_{\wp} r\, dv - \int_{\partial\wp} \mathbf{q}\cdot\mathbf{n}\, da,$$

where $\rho(\mathbf{x}, t)$ is the mass density of the atmosphere (not necessarily constant), $r(\mathbf{x}, t)$ is the rate at which the CFC is produced per unit volume, and $\mathbf{q}$ is the diffusive flux of the CFC into $\wp$.

(a) Derive a local differential equation for c. Assume that the diffusive flux is given by Fick's law, $\mathbf{q} = -D\nabla c$, where D is a constant diffusion coefficient. Clearly state any assumptions that you make.

(b) If V is a fixed control volume with bounding surface S, write the control-volume form of the CFC conservation equation. For steady flow with no source term ($r = 0$), write this equation entirely in terms of integrals over S.

3. A common assumption in water wave theory is that the number of wave crests contained within a material region is conserved. Consider wave propagation in the x-direction in the presence of a current with velocity $\mathbf{u} = U(x, y)\mathbf{e}_x$, where x and y are the horizontal directions and z denotes depth. Let L denote the length in the x-direction of a rectangular material region, as shown in Figure 4.5. The assumption of conservation of waves can be expressed as

$$\frac{d}{dt}\int_{L} k\, dx = 0,$$

where the wavenumber k is equal to 2π times the number of wave crests per unit length.

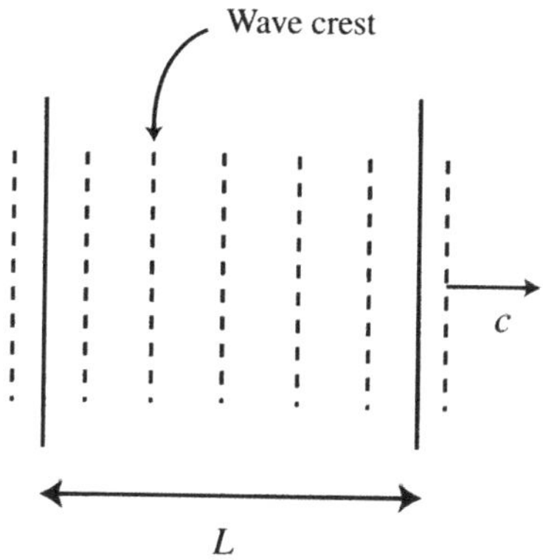

Figure 4.5 Control volume of length L covering a series of wave crests propagating with speed c.

(a) Derive a transport theorem that can be used to take the time derivative inside the integral in the above equation.

(b) Derive the local form of the conservation of waves equation.

4. Determine if each of the following constitutive equations for the stress tensor **T** satisfies the restriction imposed by conservation of moment of momentum and explain why or why not.

   (a) $T_{ij} = -p\delta_{ij} + \mu \dfrac{\partial u_i}{\partial x_j}$

   (b) $T_{ij} = -p\delta_{ij} + \mu_1 W_{ik} W_{kj} + \mu_2 D_{ij}$

   (c) $T_{ij} = \mu_1 \dot{D}_{ij} + 2\mu_2 D_{ij}$

   (d) $T_{ij} = -p\delta_{ij} + \mu_1 D_{ij} + \mu_2 u_i \omega_j$

5. Derive an integral form for the kinetic energy transport theorem applied to a fixed control volume V with bounding surface S. In the special case of a steady, inviscid, incompressible flow with no body force, show that this equation reduces to an integral over S.

6. For the Poiseuille pipe flow solution given in Example 4.6.4, determine the rate of viscous dissipation per unit length of the pipe.

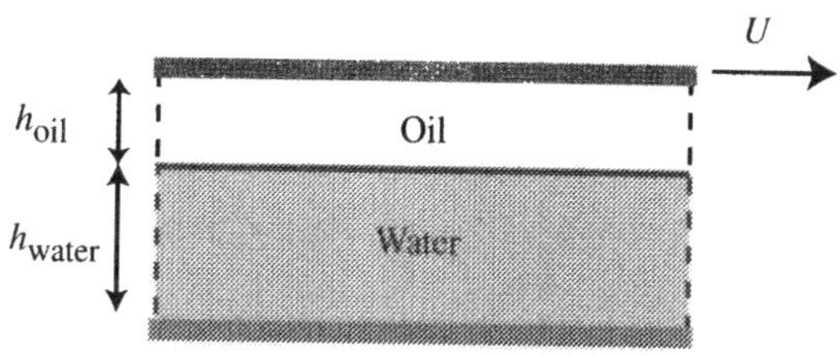

Figure 4.6 Couette flow of oil and water layers confined between two parallel plates.

7. Solve for the velocity profile and volumetric flow rate for axisymmetric Poiseuille flow in an annular region $R_1 \leq R \leq R_2$, with the no-slip boundary condition applied at R_1 and R_2.
8. Solve for the velocity profile and energy dissipation rate per unit length for a Couette flow in a system consisting of water and oil layers confined between two horizontal plates (Figure 4.6). The bottom plate is fixed and the top plate translates to the right at speed U.
9. Show that the velocity profile for rotating Couette flow approaches that for plane Couette flow as the gap width $d = R_2 - R_1$ becomes much smaller than R_1.

CHAPTER 5

DYNAMICS OF DISCONTINUITY SURFACES

In deriving the local forms of the conservation laws in Chapter 4, we rely on the assumption of smoothness of the function $\mathbf{x} = \mathbf{x}(\boldsymbol{\xi}, t)$ in order to ensure existence of first and second derivatives of the velocity and density in space and time. There are several cases, however, where the smoothness assumption is violated on specific surfaces in various fluid flows, such that on these surfaces the derivatives of the variables describing the fluid flow may not exist. Two examples of such situations include the interface between two immiscible fluids, such as an air-water or an oil-water interface, and shock waves in a compressible flow. We refer to surfaces across which either the density or velocity or their derivatives are discontinuous as *surfaces of discontinuity*. While in such cases the local conservations equations, such as (4.1.3) and (4.2.9), are valid on either side of the discontinuity surface, we must obtain additional *jump conditions* to specify the change in the variables describing the flow across the discontinuity surface.

5.1 EXTENDED TRANSPORT THEOREM FOR A REGION CONTAINING A DISCONTINUITY SURFACE

The jump conditions over a surface of discontinuity are derived from the global conservations laws, such as (4.1.1) and (4.2.2), using an extension of the argument used previously in deriving the local forms of the conservation laws. We consider a material region $\wp$, with bounding surface $\partial\wp$, which contains a section of a discontinuity surface, as shown in Figure 5.1. The discontinuity surface as a whole is denoted by Σ and the section of the discontinuity surface lying within $\wp$ is denoted by σ. The two parts of the region $\wp$ on either side of σ are denoted by $\wp_1$ and $\wp_2$, and the

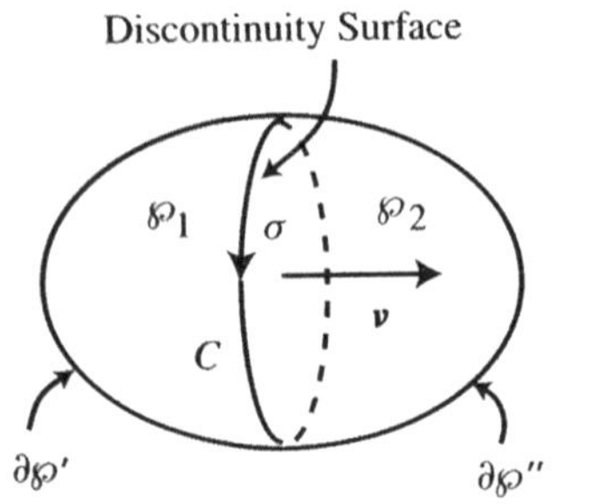

Figure 5.1 A material region $\wp$ with bounding surface $\partial\wp$, which is bisected by a surface of discontinuity σ and divided into two parts $\wp_1$ and $\wp_2$ on either side of the discontinuity surface.

parts of the surface $\partial\wp$ on either side of σ are denoted by $\partial\wp'$ and $\partial\wp''$. The local conservation laws (4.1.3) and (4.2.9) apply within the interior of each of the regions $\wp_1$ and $\wp_2$. The bounding surfaces of $\wp_1$ and $\wp_2$, denoted by $\partial\wp_1$ and $\partial\wp_2$, consist of the union of the open surfaces $\partial\wp'$ and $\partial\wp''$ and the discontinuity surface σ, or

$$\partial\wp_1 = \partial\wp' + \sigma, \qquad \partial\wp_2 = \partial\wp'' + \sigma. \tag{5.1.1}$$

The discontinuity surface σ is assumed to move with speed U_n along the direction of the unit normal $\boldsymbol{\nu}$ of σ, which by convention is chosen to point out of region $\wp_1$ and into region $\wp_2$. Since σ is simply a space surface in the region $\wp$, the normal velocity of σ may or may not coincide with the velocity of fluid particles that lie on σ at some instant of time. For instance, fluid particles pass through a shock wave, and hence travel at a velocity different from that of the shock, while an air-water interface is a material surface (in the absence of effects such as evaporation and condensation). The relative normal velocity W of the fluid at the surface σ is defined by the difference of the normal component of the velocity of fluid particles on the surface and the normal velocity U_n of the surface itself, or

$$W = \mathbf{u} \cdot \boldsymbol{\nu} - U_n. \tag{5.1.2}$$

When the discontinuity is a material surface, W vanishes everywhere on σ.

Before addressing the derivation of jump conditions from the global conservation laws, it is convenient to introduce an extension of the transport theorem of Section 3.6 for the case in which a discontinuity surface divides the region $\wp$. We have previously derived the transport theorem, in the forms (3.6.4)–(3.6.6), for a region $\wp$ that is material with respect to the fluid velocity $\mathbf{u}$ and within which the variables describing the fluid are continuous and sufficiently smooth. When we say that $\wp$ is *material with respect to a vector field* $\mathbf{v}$, we mean that the normal velocity of the bounding surface $\partial\wp$ of at any point $\mathbf{x}$ on $\partial\wp$ is the same as the component of the velocity field $\mathbf{v}$ normal to $\partial\wp$ at $\mathbf{x}$. Usually we refer simply to a "material region," implying that the region is material with respect to the velocity field of the fluid particles. However, the usual transport theorem, in the form (3.6.6), can be readily extended to regions that are material with respect to any vector field $\mathbf{v}$, having dimension of velocity, by writing

$$\frac{d}{dt}\int_{\wp} f\,dv = \int_{\wp}\frac{\partial f}{\partial t}\,dv + \int_{\partial\wp} f\mathbf{v}\cdot\mathbf{n}\,da, \tag{5.1.3}$$

where $\mathbf{v}\cdot\mathbf{n}$ in the last integral is the normal velocity of the surface $\partial\wp$.

We now introduce two vector fields $\mathbf{v}^-$ and $\mathbf{v}^+$, defined by

$$\mathbf{v}^- \equiv \begin{cases}\dot{\mathbf{x}} \text{ on } \partial\wp', \\ U_n\boldsymbol{\nu} \text{ on } \sigma,\end{cases} \qquad \text{and} \qquad \mathbf{v}^+ \equiv \begin{cases}\dot{\mathbf{x}} \text{ on } \partial\wp'', \\ U_n\boldsymbol{\nu} \text{ on } \sigma,\end{cases} \tag{5.1.4}$$

where $\mathbf{u} = \dot{\mathbf{x}}$ is the velocity of fluid particles. The values of $\mathbf{v}^-$ and $\mathbf{v}^+$ at positions other than the boundaries noted above are not relevant and are left arbitrary. Although the two parts $\wp_1$ and $\wp_2$ of $\wp$ may not be material with respect to the velocity of fluid particles, due to the possibility of fluid transport across the discontinuity surface, they are material with respect to the vector fields $\mathbf{v}^-$ and $\mathbf{v}^+$ defined in (5.1.4). Since $\wp$ consists of the sum of $\wp_1$ and $\wp_2$, we can write

$$\frac{d}{dt}\int_{\wp} f\,dv = \frac{d}{dt}\int_{\wp_1} f\,dv + \frac{d}{dt}\int_{\wp_2} f\,dv \tag{5.1.5}$$

The function f is continuous within both $\wp_1$ and $\wp_2$, so the form (5.1.3) of the transport theorem may be used within each of these regions to write

$$\begin{aligned}\frac{d}{dt}\int_{\wp_1} f\,dv &= \int_{\wp_1}\frac{\partial f}{\partial t}\,dv + \int_{\partial\wp_1} f\mathbf{v}^-\cdot\mathbf{n}\,da \\ &= \int_{\wp_1}\frac{\partial f}{\partial t}\,dv + \int_{\partial\wp'} f\mathbf{u}\cdot\mathbf{n}\,da + \int_{\sigma} f_1 U_n\,da\end{aligned} \tag{5.1.6a}$$

and

$$\begin{aligned}\frac{d}{dt}\int_{\wp_2} f\,dv &= \int_{\wp_2}\frac{\partial f}{\partial t}\,dv + \int_{\partial\wp_2} f\mathbf{v}^+\cdot\mathbf{n}\,da \\ &= \int_{\wp_2}\frac{\partial f}{\partial t}\,dv + \int_{\partial\wp''} f\mathbf{u}\cdot\mathbf{n}\,da - \int_{\sigma} f_2 U_n\,da.\end{aligned} \tag{5.1.6b}$$

A negative sign precedes the last term in (5.1.6b) because $\boldsymbol{\nu}$ points inward relative to region $\wp_2$ and is hence in the opposite direction to the unit normal $\mathbf{n}$ of $\partial\wp_2$. Here f_1 and f_2 denote the values of f on the respective sides of the discontinuity surface. The jump in f across the discontinuity surface is defined by $\langle f\rangle \equiv f_2 - f_1$.

Substituting (5.1.6) into (5.1.5) and collecting terms gives the extended form of the transport theorem for a region containing a discontinuity surface σ as

$$\frac{d}{dt}\int_{\wp} f\,dv = \int_{\wp_1+\wp_2} \frac{\partial f}{\partial t}\,dv + \int_{\partial\wp'+\partial\wp''} f\mathbf{u}\cdot\mathbf{n}\,da - \int_{\sigma} \langle fU_n\rangle\,da. \tag{5.1.7}$$

Adding and subtracting the integral $\int_\sigma f\mathbf{u}\cdot\mathbf{n}\,da$ on each side of σ and making use of the divergence theorem, the result (5.1.7) becomes

$$\frac{d}{dt}\int_{\wp} f\,dv = \int_{\wp_1}\left(\frac{Df}{Dt} + f\nabla\cdot\mathbf{u}\right)dv + \int_{\wp_2}\left(\frac{Df}{Dt} + f\nabla\cdot\mathbf{u}\right)dv + \int_{\sigma}\langle fW\rangle\,da. \tag{5.1.8}$$

If the region $\wp$ has no discontinuity surface, the last term in (5.1.8) vanishes and the above result reduces to the standard transport theorem for a continuous media. Similar to the development of the alternative form (4.1.9) of the transport theorem, the local mass conservation equation can be used to write (5.1.8) in the alternative form

$$\frac{d}{dt}\int_{\wp} \rho f\,dv = \int_{\wp_1} \rho\frac{Df}{Dt}\,dv + \int_{\wp_2} \rho\frac{Df}{Dt}\,dv + \int_{\sigma}\langle\rho fW\rangle\,da \tag{5.1.9}$$

where f represents a specific quantity defined per unit mass.

5.2 JUMP CONDITIONS ACROSS A SURFACE OF DISCONTINUITY

With use of the extended transport theorem (5.1.8), it is a straightforward task to derive the jump conditions over a discontinuity surface. We start with mass conservation, which is expressed in global form for a material region $\wp$ as

$$\frac{d}{dt}\int_{\wp} \rho\,dv = 0. \tag{5.2.1}$$

In writing (5.2.1), it is assumed that there is no mass source on the discontinuity surface. The extended transport theorem (5.1.8) can be used to write the mass conservation law as

$$\int_{\wp_1}\left(\frac{D\rho}{Dt} + \rho\nabla\cdot\mathbf{u}\right)dv + \int_{\wp_2}\left(\frac{D\rho}{Dt} + \rho\nabla\cdot\mathbf{u}\right)dv + \int_{\sigma}\langle\rho W\rangle\,da = 0, \tag{5.2.2}$$

where the two parts $\wp_1$ and $\wp_2$ of the region are separated by the discontinuity surface, as indicated in Figure 5.1. Since all variables are assumed to be continuous and smooth within the regions $\wp_1$ and $\wp_2$, excluding their common boundary σ, the local mass conservation equation (4.1.3) applies in these regions. The first two integrals in (5.2.2) therefore vanish, leaving only

$$\int_{\sigma}\langle\rho W\rangle\,da = 0. \tag{5.2.3}$$

Since σ is an arbitrary part of the discontinuity surface Σ, the integrand in (5.2.3) must vanish, giving the mass jump condition as

$$\langle \rho W \rangle = 0. \tag{5.2.4}$$

This jump condition requires the product of the density and the normal component of the fluid velocity measured relative to the velocity of the discontinuity surface, which is simply the mass flux across Σ, to be continuous over the discontinuity surface. When the discontinuity is a material surface, W vanishes everywhere on Σ and (5.2.4) is identically satisfied.

The global momentum conservation law for the material region $\wp$ is given by

$$\frac{d}{dt}\int_{\wp} \rho\mathbf{u}\,dv = \int_{\wp} \rho\mathbf{b}\,dv + \int_{\partial\wp} \mathbf{t}^{(\mathbf{n})}\,da. \tag{5.2.5}$$

In writing (5.2.5), we have not included the possibility that the discontinuity surface itself can exert a force on the region $\wp$. This assumption must be relaxed to account for the "surface tension" effect that occurs at interfaces between immiscible fluids, which is discussed in the next section. Using the form (5.1.9) of the extended transport theorem, the left-hand side of the momentum balance is written as

$$\frac{d}{dt}\int_{\wp} \rho\mathbf{u}\,dv = \int_{\wp_1} \rho\frac{D\mathbf{u}}{Dt}\,dv + \int_{\wp_2} \rho\frac{D\mathbf{u}}{Dt}\,dv + \int_{\sigma} \langle \rho\mathbf{u}W \rangle\,da. \tag{5.2.6}$$

The Cauchy stress formula (4.2.6) is used to write the contact force acting on the bounding surface $\partial\wp = \partial\wp' + \partial\wp''$ of $\wp$ as

$$\int_{\partial\wp} \mathbf{t}^{(\mathbf{n})}\,da = \int_{\partial\wp'} \mathbf{n}\cdot\mathbf{T}\,da + \int_{\partial\wp''} \mathbf{n}\cdot\mathbf{T}\,da. \tag{5.2.7}$$

While the divergence theorem cannot be applied to the region $\wp$ as a whole, since the stress tensor is not necessarily continuous on σ, it can be applied separately to the regions $\wp_1$ and $\wp_2$ on either side of the discontinuity surface. If $\boldsymbol{\nu}$ denotes the unit normal of the discontinuity surface σ, pointing from region $\wp_1$ into region $\wp_2$, $\mathbf{n}$ denotes the outward unit normal of $\partial\wp$, and $\mathbf{n}_1$ and $\mathbf{n}_2$ denote the outward unit normals of $\partial\wp_1$ and $\partial\wp_2$, respectively, we can write

$$\int_{\partial\wp_1} \mathbf{n}_1\cdot\mathbf{T}\,da = \int_{\partial\wp'} \mathbf{n}\cdot\mathbf{T}\,da + \int_{\sigma^-} \boldsymbol{\nu}\cdot\mathbf{T}\,da, \tag{5.2.8a}$$

$$\int_{\partial\wp_2} \mathbf{n}_2\cdot\mathbf{T}\,da = \int_{\partial\wp''} \mathbf{n}\cdot\mathbf{T}\,da - \int_{\sigma^+} \boldsymbol{\nu}\cdot\mathbf{T}\,da. \tag{5.2.8b}$$

The negative sign in front of the last term in (5.2.8b) is due to the fact that $\boldsymbol{\nu}$ points inward to the region $\wp_2$, so that the appropriate outward unit normal is $-\boldsymbol{\nu}$. The notations σ^- and σ^+ indicate evaluation of the integral on the $\wp_1$ and $\wp_2$ sides of σ, respectively. Substituting (5.2.8) into the right-hand side of (5.2.7) and using the divergence theorem for the surface integrals over $\partial\wp_1$ and $\partial\wp_2$ give

$$\int_{\partial\wp} \mathbf{t}^{(\mathbf{n})}\,da = \int_{\wp_1} \nabla\cdot\mathbf{T}\,dv + \int_{\wp_2} \nabla\cdot\mathbf{T}\,dv + \int_{\sigma} \langle \boldsymbol{\nu}\cdot\mathbf{T}\rangle\,da. \tag{5.2.9}$$

Substituting (5.2.6) and (5.2.9) into (5.2.5) gives the momentum balance for $\wp$ as

$$\int_{\wp_1}\left(\rho\frac{D\mathbf{u}}{Dt} - \rho\mathbf{b} - \nabla\cdot\mathbf{T}\right)dv + \int_{\wp_2}\left(\rho\frac{D\mathbf{u}}{Dt} - \rho\mathbf{b} - \nabla\cdot\mathbf{T}\right)dv + \int_{\sigma}\langle \rho W\mathbf{u} - \boldsymbol{\nu}\cdot\mathbf{T}\rangle\,da = 0. \tag{5.2.10}$$

Since the local momentum equation (4.2.9) applies within the regions $\wp_1$ and $\wp_2$, the first two integrals in (5.2.10) vanish, leaving

$$\int_{\sigma}\langle \rho W\mathbf{u} - \boldsymbol{\nu}\cdot\mathbf{T}\rangle\,da = 0. \tag{5.2.11}$$

The result (5.2.11) holds for an arbitrary part of the discontinuity surface, which implies that the integrand of (5.2.11) vanishes at any point on Σ. The resulting momentum jump condition is given by

$$\langle \rho W\mathbf{u} - \boldsymbol{\nu}\cdot\mathbf{T}\rangle = 0. \tag{5.2.12}$$

The result (5.2.12) requires that the jump in the momentum flux $\rho W\mathbf{u}$ across the surface be balanced by a jump in stress. For the case in which the discontinuity is a material surface, the momentum flux across the surface vanishes. Using the expression (4.5.4) for the stress tensor in an inviscid fluid, the momentum jump condition reduces to a requirement that pressure is continuous across a material discontinuity surface, or

$$\langle p\rangle = 0. \tag{5.2.13}$$

5.3 SURFACE TENSION

Within a region that is filled by a gas or a liquid of uniform chemical composition, the attractive force exerted by neighboring molecules on a given fluid molecule is approximately the same in all directions when averaged over time. However, at the interface between materials of different phases, such as that separating a liquid and a gas, or between two liquids that do not mix with each other, the average molecular attraction shows substantial directional variation. For instance, at a liquid-gas interface, molecules of the liquid within a region with a thickness on the order of the liquid mean free path (typically about 10^{-7} cm) experience a much greater attraction from the surrounding liquid molecules, which lie primarily in one direction, than they do from the gas molecules lying primarily in the opposite direction. The effect of this strong directional dependence of the molecular force is to compress the molecules in the thin region along the interface, or rather to make the number density of liquid molecules larger in this interfacial region than in the nearby internal parts of the liquid. The individual liquid molecules in this interfacial region still have about the same kinetic energy as molecules in the other parts of the liquid, but they are spaced closer together because the molecular attraction is stronger in one direction than in the opposite direction.

The increased number density (and decreased spacing) of liquid molecules in the interfacial region causes an excess energy of the fluid along the interface. On a continuum scale, the material interface separating two immiscible fluids behaves in a manner analogous to a thin elastic membrane. Increase in the interface area requires an addition of energy, by an amount proportional to the change in area, to provide for the excess energy required for the new interface area. This energy is obtained from the surrounding fluid by work performed by a force which, like the tensile force of an elastic membrane, acts tangent to the interface. The energy supplied upon stretching of the interface is reversible and is returned to the fluid when the interface returns to its original shape.

The *surface tension*, γ, is defined as the excess interface energy per unit surface area. The surface tension has the dimensions of force over length and is a property of temperature and of the chemical composition of the materials on either side of the interface. The surface tension of a material interface is generally positive, which implies that molecules of a given type are attracted (on average) back to the mass of molecules of that same type. For instance, in a liquid-gas interface, the liquid molecules near the interface are attracted more to the liquid than toward the gas. If two liquid regions have positive surface tension, the molecules of one liquid are drawn back toward molecules of a similar type, such that the liquid regions do not mix. The liquids in this case are said to be *immiscible*. If two immiscible liquids, such as oil and water, are vigorously stirred together, the interface will be broken up and small droplets of one liquid will form within the other liquid. After sufficient time, however, the tiny droplets will coalesce (e.g., through gravitational drift or random collisions) and two disjoint liquid regions will again form with a sharp interface in-between. On the other hand, if a system with two liquids having negative surface tension are initially separated by an interface, the molecules of both liquids will be more attracted to those of the opposing type of liquid than to molecules of their own type. In this case, the interface soon disintegrates and the two liquids permanently mix together. Even if the mixture is left sitting for a long time, the system never re-forms a state in which the two different liquids are separated by a sharp interface.

In order to derive the effect of surface tension on the momentum jump conditions, we reconsider the case of a material region $\wp$ containing a part σ of a material interface, as sketched in Figure 5.1. The global momentum conservation equation for this region in the absence of surface tension is given in (5.2.5). When surface tension is present, an additional contact force must be added to account for the force exerted on $\wp$ by the tension along the material interface σ. This additional contact force acts along the curve C at the intersection of the bounding surface $\partial\wp$ of $\wp$ and the discontinuity surface Σ and is oriented tangent to Σ and normal to C. The magnitude of this interfacial force per unit length of C is equal to the surface tension γ. The interfacial force can thus be written as

$$-\int_C \gamma \boldsymbol{\nu} \times d\mathbf{x},$$

where $d\mathbf{x}$ is a line segment of infinitesimal length tangent to C. This term is added to the right-hand side of the global momentum equation (4.2.2) for regions $\wp$ that intersect a material interface.

If the region $\wp$ is selected to have the form of a thin wafer with small thickness of order d, the volume integrals in the momentum balance approach zero in proportion to d while the surface integrals remain finite. Making an argument similar to that used in Section 4.2 in the derivation of the Cauchy stress formula (4.2.6), we find that the contact forces acting on $\partial\wp$ must be in equilibrium as the width of the region $\wp$ becomes small. Using the form (4.5.4) for the stress tensor in an inviscid fluid and balancing the contact forces on $\partial\wp$ gives

$$\int_{\partial\wp'} p\mathbf{n}\, da + \int_{\partial\wp''} p\mathbf{n}\, da + \int_C \gamma \boldsymbol{\nu} \times d\mathbf{x} = 0. \tag{5.3.1}$$

In the limit as $\wp$ approaches a thin wafer enclosing the part σ of the interface, as shown in Figure 5.2, the outward unit normals to $\partial\wp'$ and $\partial\wp''$ approach $-\boldsymbol{\nu}$ and $\boldsymbol{\nu}$, respectively, such that (5.3.1) becomes

$$\int_\sigma \langle p \rangle \boldsymbol{\nu}\, da + \int_C \gamma \boldsymbol{\nu} \times d\mathbf{x} = 0. \tag{5.3.2}$$

In order to make further progress, it is necessary to convert the contour integral in (5.3.2) into a surface integral over σ. This can be accomplished using the vector identity (2.7.12), which for this case can be expressed as

$$\int_C \mathbf{f} \times d\mathbf{x} = \int_\sigma [\boldsymbol{\nu}(\nabla \cdot \mathbf{f}) - (\boldsymbol{\nu} \cdot \nabla)\mathbf{f} - \boldsymbol{\nu} \times (\nabla \times \mathbf{f})]\, da. \tag{5.3.3}$$

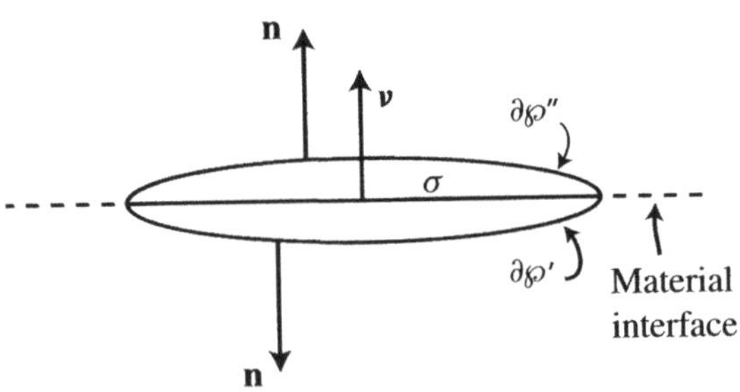

Figure 5.2 A thin waferlike material region $\wp$ enclosing a part σ of the discontinuity surface, such that the unit normal $\mathbf{n}$ to $\partial\wp$ approaches $\boldsymbol{\nu}$ on $\partial\wp''$ and $-\boldsymbol{\nu}$ on $\partial\wp'$, where $\boldsymbol{\nu}$ is the unit normal of σ.

Letting $\mathbf{f} = \gamma\boldsymbol{\nu}$ and noting that the gradient of the surface tension must be tangent to the interface σ, we have

$$\int_C \gamma\boldsymbol{\nu}\times d\mathbf{x} = \int_\sigma [\gamma\boldsymbol{\nu}(\nabla\cdot\boldsymbol{\nu}) - \gamma(\boldsymbol{\nu}\cdot\nabla)\boldsymbol{\nu} - \boldsymbol{\nu}\times(\nabla\gamma\times\boldsymbol{\nu}) - \gamma\boldsymbol{\nu}\times(\nabla\times\boldsymbol{\nu})]\,da. \quad (5.3.4)$$

Recalling the result of Example 2.6.4, we know that $\boldsymbol{\nu}\times(\nabla\gamma\times\boldsymbol{\nu})$ is the projection of $\nabla\gamma$ onto the surface with unit normal $\boldsymbol{\nu}$, or simply $\nabla\gamma$ itself in this case. Also, since $\boldsymbol{\nu}$ is a unit vector, the sum $(\boldsymbol{\nu}\cdot\nabla)\boldsymbol{\nu} + \boldsymbol{\nu}\times(\nabla\times\boldsymbol{\nu}) = (\frac{1}{2})\nabla(\boldsymbol{\nu}\cdot\boldsymbol{\nu})$ must vanish. The right-hand side of (5.3.4) thus reduces to

$$\int_C \gamma\boldsymbol{\nu}\times d\mathbf{x} = \int_\sigma [\gamma\boldsymbol{\nu}(\nabla\cdot\boldsymbol{\nu}) - \nabla\gamma]\,da, \quad (5.3.5)$$

which when substituted into (5.3.2) yields

$$\int_\sigma [\langle p\rangle\boldsymbol{\nu} + \gamma\boldsymbol{\nu}(\nabla\cdot\boldsymbol{\nu}) - \nabla\gamma]\,da = 0. \quad (5.3.6)$$

Since (5.3.6) holds for arbitrary parts σ of the material interface, the integrand must vanish, yielding

$$\langle p\rangle\boldsymbol{\nu} = -\gamma\boldsymbol{\nu}(\nabla\cdot\boldsymbol{\nu}) + \nabla\gamma. \quad (5.3.7)$$

The component of (5.3.7) normal to σ is

$$\langle p\rangle = -\gamma(\nabla\cdot\boldsymbol{\nu}), \quad (5.3.8)$$

since the gradient of γ is oriented tangent to the interface. The tangential component of (5.3.7) yields the restriction that the gradient of γ must vanish, such that the surface tension must be uniform over a material interface in an inviscid fluid. In a real fluid, variation in surface tension can occur due to a temperature gradient or variation in surfactant concentration. The contact force due to a nonuniform surface tension in a real fluid is balanced by a discontinuity in viscous shear force at the interface (Sarpkaya, 1996).

The divergence of the unit normal vector can be related to the surface curvature using the formula of Rodrigues (Struik, 1961). Rodrigues's formula states that a change $d\mathbf{n}$ in the surface unit normal $\mathbf{n}$ corresponding to a displacement $d\mathbf{x}$ along a line of curvature on the surface is related to the curvature κ in this direction by

$$d\mathbf{n} + \kappa\,d\mathbf{x} = 0. \quad (5.3.9)$$

The divergence of the unit normal is then obtained as

$$\nabla\cdot\mathbf{n} = -(\kappa_1 + \kappa_2), \quad (5.3.10)$$

where κ_1 and κ_2 are the curvatures evaluated in two orthogonal directions on the surface. It can be shown that the sum $\kappa_1 + \kappa_2$ is independent of orientation of these

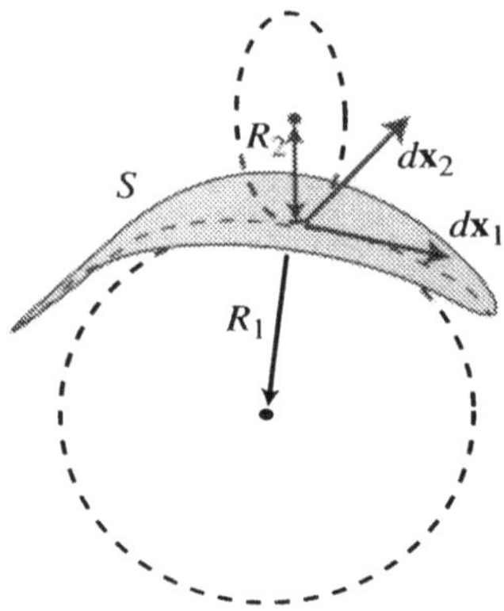

Figure 5.3 Sketch showing tangent circles of a surface S along the orthogonal directions $d\mathbf{x}_1$ and $d\mathbf{x}_2$ with radii of curvature R_1 and R_2.

two orthogonal directions. The magnitude of the curvatures κ_1 and κ_2 is simply the inverse of the radii R_1 and R_2 of tangent circles at a given point on the surface, where the circles are oriented in two orthogonal directions $d\mathbf{x}_1$ or $d\mathbf{x}_2$, as shown in Figure 5.3. The sign of the curvature is positive when the unit normal $\mathbf{n}$ points toward the center of the tangent circle and negative when the normal points away from the center of the tangent circle. Using (5.3.10), the momentum jump condition (5.3.8) yields the pressure jump across a material interface in an inviscid fluid as

$$\langle p \rangle = \gamma(\kappa_1 + \kappa_2), \tag{5.3.11}$$

which is the well-known *Laplace formula*.

Example 5.3.1. Determine the difference in pressure across the boundary of a spherical gas bubble of radius R immersed in a liquid.

SOLUTION. The pressure of the gas inside the bubble is denoted by p_g and that of the surrounding liquid is denoted by p_ℓ. We select the liquid side of the interface as region $\wp_1$ and the gas side as region $\wp_2$, such that the interface normal $\boldsymbol{\nu}$ points into the gas. The bubble is spherical and has radius R, such that $\kappa_1 = \kappa_2 = 1/R$. The result (5.3.11) then yields

$$p_g - p_\ell = \frac{2\gamma}{R}. \tag{5.3.12}$$

Had we selected the sides $\wp_1$ and $\wp_2$ in the opposite manner, the interface normal would have pointed into the liquid and the curvatures would be negative, giving the same result.

5.4 BOUNDARY CONDITIONS FOR FLUID FLOW

The Euler equation, which governs the motion of inviscid fluids, can be expressed by expanding the material derivative in (4.6.5) as

$$\rho\left[\frac{\partial \mathbf{u}}{\partial t} + (\mathbf{u}\cdot\nabla)\,\mathbf{u}\right] = \rho\mathbf{b} - \nabla p, \tag{5.4.1}$$

where pressure p is determined in a compressible flow as a function of density and temperature by an equation of state (such as the ideal gas law) and in an incompressible flow it is solved for as part of the solution (as a function of position and time).

The boundary conditions for an incompressible flow are obtained by applying the discontinuity jump conditions derived in the previous two sections to different types of boundaries. For instance, at the boundary S of an inviscid fluid and a rigid solid, the jump condition (5.2.4) in mass requires that the normal mass flux at the boundary is continuous. If the solid is impermeable, then the boundary is a material surface and the relative normal velocity W vanishes. Assuming that the normal velocity $U_n(\mathbf{x}, t)$ on the surface of the rigid body is known, the boundary condition reduces to the so-called *no-penetration condition*

$$\mathbf{u}\cdot\mathbf{n} = U_n(\mathbf{x}, t), \tag{5.4.2}$$

where $\mathbf{n}$ is the normal to the body surface S. The momentum jump condition (5.2.12) determines the pressure on the solid surface and does not affect the fluid flow, except perhaps indirectly through motion of the solid body.

Another common type of boundary is an interface separating two fluids, such as water and air. For such a case, the relative normal velocity W vanishes and the no-penetration condition (5.4.2) again applies. A second boundary condition is necessary for a moving interface since the location of the interface is unknown and is solved for as part of the problem solution. This second boundary condition is supplied by momentum conservation, resulting in the expression (5.3.11) for pressure jump across the interface. It is customary to refer to (5.4.2) as the *kinematic interface condition* and to (5.3.11) as the *dynamic interface condition.*

Other types of discontinuities are common for inviscid flows in which the velocity is continuous across a surface but the velocity gradient is discontinuous on the surface. A common example of this type of discontinuity is found at the lateral boundary of a Rankine vortex core, where the vorticity is uniform everywhere within a circular core region and zero outside of this region. Since for such a discontinuity both the normal and tangential components of velocity are continuous at the discontinuity surface, the kinematic boundary condition is automatically satisfied. The remaining matching condition at the surface for an inviscid fluid is continuity in pressure, as required by the momentum jump condition.

The momentum equation for a compressible viscous fluid with constant viscosity coefficients is given by

$$\rho\left[\frac{\partial \mathbf{u}}{\partial t}+(\mathbf{u}\cdot\nabla)\,\mathbf{u}\right]=\rho\mathbf{b}-\nabla p+\mu\nabla^2\mathbf{u}+(\lambda+\mu)\nabla(\nabla\cdot\mathbf{u}), \tag{5.4.3}$$

where the last term in (5.4.3) vanishes for incompressible flows. The viscous terms in (5.4.3) contain second derivatives in the velocity, whereas the momentum equation (5.4.1) for an inviscid flow contains only first derivatives in velocity (in the advection term). As a consequence, the viscous equations require additional boundary conditions on velocity at each boundary. One additional boundary condition follows from the tangential component of the momentum jump condition (5.2.12), which requires that the tangential component of the stress must be continuous across a material interface. This boundary condition is used at the interface between two immiscible fluids.

An additional boundary condition that is used in viscous flows both for fluid-fluid interfaces and at the surface of a rigid body is that of *no slip*, which requires that the tangential component of the velocity is continuous across the interface. The no-slip condition does not follow from the mass or momentum jump conditions, although the jump in energy can be used to show that the no-slip condition is consistent with the assumption of vanishing heat source at the interface. Because it does not follow directly from the jump conditions but must be viewed as an additional assumption, there was considerable debate about the validity of the no-slip condition near the end of the nineteenth century (Goldstein, 1938). There are also several instances where the no-slip condition leads to inconsistencies in the general theory of fluid flow. For instance, at the contact line separating two fluids and a solid, such as at the intersection of a water-air interface with the side of a tank, enforcement of the no-slip condition together with the Navier-Stokes equation leads to a condition in which it is impossible for the contact line to move without the exertion of an infinite force! This situation is clearly not in agreement with common experience. One common way of fixing this problem is to allow slip within a small region around the contact line (Dussan, 1979). In other cases, such as for rarefied gases or certain viscoelastic fluids, considerable slip of the fluid is observed at the surface with a solid body and the no-slip assumption is clearly not applicable.

In an inviscid theory, it is not necessary to satisfy these additional viscous boundary conditions, such as no slip. However, there do exist certain special flows that are solutions of both the Euler and Navier-Stokes equations in which the inviscid solutions satisfy both the no-penetration and no-slip conditions simultaneously. Several examples of such flow fields are given in Section 4.6.

BIBLIOGRAPHY

Dussan, E.B. (1979). "On the spreading of liquids on solid surfaces: static and dynamic contact lines," *Annual Review of Fluid Mechanics* 11, 371–400.

Goldstein, S. (1938). *Modern Developments in Fluid Dynamics*, Vol. II, Claredon Press, Oxford (reprinted Dover Publications, New York, 1965).

Sarpkaya, T. (1996). "Vorticity, free surface and surfactants," *Annual Review of Fluid Mechanics* **28**, 83–128.

Struik, D.J. (1961). *Lectures on Classical Differential Geometry*, 2nd ed., Addison-Wesley Publishing Company, Reading, MA (reprinted Dover Publications, New York, 1988).

PROBLEMS

1. Derive an expression for the jump in density and the normal and tangential components of velocity for a stationary normal shock in an inviscid, barotropic fluid [i.e., a fluid in which $p = p(\rho)$]. In the more general case that $p = p(\rho, T)$, where T denotes temperature, what additional jump condition is needed to find the density jump?
2. Use the global energy conservation law given in Problem 4.1 to derive the jump condition for energy over a surface of discontinuity Σ.
3. Consider the special case in which the discontinuity Σ is a material surface and there is no thermal conduction or temperature difference across Σ. If there is no momentum or energy supply at the surface, use the jump condition in energy (Problem 5.2), momentum, and mass to show that the velocity vector $\mathbf{u}$ must be continuous across Σ for a Newtonian viscous fluid.
4. Consider two-dimensional water wave propagation in an inviscid fluid with interface displacement $y = \eta(x, t)$. Assume that the interface displacement can be expressed as $\eta = A\sin(kx - \omega t)$, where A, k, and ω are constants. Show that for small-amplitude waves, such that $\varepsilon \equiv Ak \ll 1$, the jump in pressure can be approximated to leading order in ε as

$$\langle p \rangle = -\gamma(\partial^2\eta/\partial x^2).$$

CHAPTER 6

VELOCITY REPRESENTATIONS AND ASSOCIATED THEOREMS

In solving for the velocity and pressure fields for specific fluid flows, it is quite common, particularly for inviscid flows, to recast both the velocity field and the momentum equation in one of a variety of different forms. For instance, instead of solving directly for velocity, one might choose to solve instead for the vorticity field and then use the vorticity solution to obtain the velocity and pressure. In certain classes of flows, it is possible to solve for a single scalar quantity (e.g., the velocity potential or the stream function) and then obtain the velocity vector by differentiation of this scalar quantity. The basis for all such alternative methods for solving the momentum equation lies in a set of vector representations that can be applied to the velocity vector. The current chapter explores these representations and certain fundamental theorems that result from use of the representations for the velocity vector.

6.1 IRROTATIONAL (LAMELLAR) AND RELATED FLOWS

A flow is said to be *irrotational* in a region $\mathfrak{R}$ if the vorticity vanishes everywhere in $\mathfrak{R}$, or

$$\nabla \times \mathbf{u} = 0. \tag{6.1.1}$$

Vectors satisfying (6.1.1) are sometimes called *lamellar* in texts on vector analysis. Most fluid flows possess regions in which the flow is irrotational, and for certain flows, including many free-surface wave problems and two-dimensional airfoil problems, it is often a reasonable assumption to suppose that the flow is everywhere irrotational. In particular, for inviscid flows it is found that if the vorticity initially vanishes on a fluid particle, then it will always vanish on that particle (Chapter 7).

It is clear from the vector identity (2.6.1), which states that the curl of the gradient of any scalar is zero, that a velocity field that can be expressed as the gradient of some scalar, or

$$\mathbf{u} = \nabla\phi, \tag{6.1.2}$$

satisfies (6.1.1) for any choice of ϕ. The scalar ϕ is called the *velocity potential*, or sometimes the *scalar potential* to distinguish it from another vector-valued quantity introduced in the next section. However, we have not yet shown that all irrotational vectors must have a representation of the form (6.1.2).

The necessity of the representation (6.1.2) can be addressed by applying Stokes's theorem (2.7.11) to any closed circuit C bounding an open surface A to write

$$\int_C \mathbf{u} \cdot d\mathbf{x} = \int_A \mathbf{n} \cdot (\nabla \times \mathbf{u})\, da = 0 \tag{6.1.3}$$

for any irrotational vector $\mathbf{u}$. Consider now two closed circuits, as shown in Figure 6.1. The first circuit is composed of the curve C_1 plus the common section C_2, and the other circuit is composed of the curve C_3 plus C_2, where the curves C_1 and C_3 join at the two points P and Q. Applying the result (6.1.3) to both of these circuits individually gives

$$\int_{C_1} \mathbf{u} \cdot d\mathbf{x} + \int_{C_2} \mathbf{u} \cdot d\mathbf{x} = 0, \qquad \int_{C_3} \mathbf{u} \cdot d\mathbf{x} + \int_{C_2} \mathbf{u} \cdot d\mathbf{x} = 0,$$

so that

$$\int_{C_1} \mathbf{u} \cdot d\mathbf{x} = \int_{C_3} \mathbf{u} \cdot d\mathbf{x}.$$

More generally, we conclude that the integral $\int_Q^P \mathbf{u} \cdot d\mathbf{x}$ between any two points P and Q must be independent of path and depend only on the location of the two end points for an irrotational flow. Selecting some fixed origin O as a reference point and

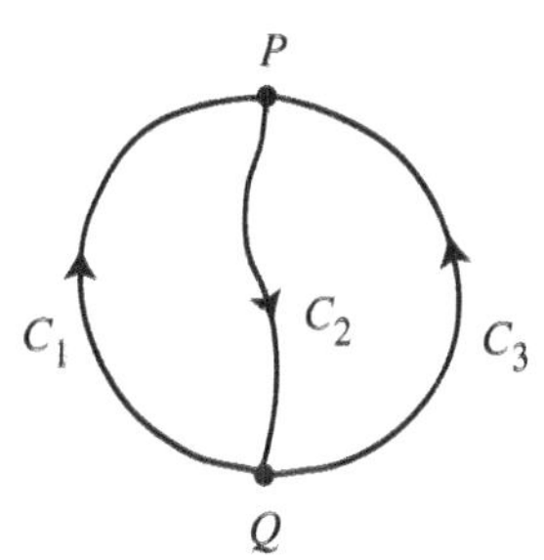

Figure 6.1 Schematic showing three paths connecting two points, P and Q.

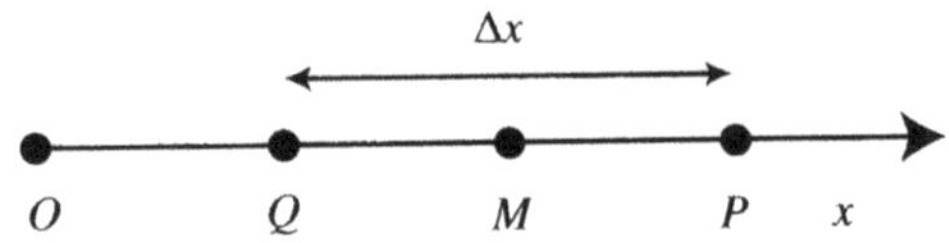

Figure 6.2 Short section lying parallel to the x-axis, with end points Q and P, midpoint M, and length Δx.

defining a quantity ϕ by

$$\phi \equiv \int_O^P \mathbf{u} \cdot d\mathbf{x}, \tag{6.1.4}$$

we conclude that since the integral on the right-hand side of (6.1.4) is independent of path, the quantity ϕ is a function, depending only on the position $\mathbf{x}$ of the point P and time.

We consider a special case where the two end points P and Q are separated only by a small distance Δx in the x-direction of a Cartesian coordinate system, as shown in Figure 6.2. Applying the definition (6.1.4) to the path shown in Figure 6.2, which joins points O and Q and points Q and P, we can write

$$\phi(x_0 + \Delta x, y_0, z_0, t) - \phi(x_0, y_0, z_0, t) = \int_Q^P \mathbf{u} \cdot d\mathbf{x} \cong u(M)\Delta x, \tag{6.1.5}$$

where M is the midpoint of the line segment PQ and (x_0, y_0, z_0) are the coordinates of point Q. Dividing this result by Δx and letting Δx approach zero yields

$$u(x_0, y_0, z_0, t) = \lim_{\Delta x \to 0} \frac{\phi(x_0 + \Delta x, y_0, z_0, t) - \phi(x_0, y_0, z_0, t)}{\Delta x}, \tag{6.1.6}$$

or by definition of the partial derivative $u = \partial\phi/\partial x$. Repeating this argument for two points separated by a small distance Δy in the y-direction or Δz in the z-direction yields the representation (6.1.2).

One property of irrotational flows that follows immediately from (6.1.2) is that the projection of the velocity vector must vanish along any surface over which $\phi =$ const. Such surfaces are referred to as *equipotential surfaces* and curves on such a surface are called *equipotential lines*. Since $\mathbf{u} \cdot \nabla\phi = 0$ at points where $\mathbf{u}$ is tangent to a surface on which ϕ is constant, we conclude that *streamlines are normal to equipotential surfaces*.

There are other types of flow fields that also exhibit the property that streamlines are normal to surfaces over which a scalar function is constant. An example is any flow in which the streamlines and the vortex lines are everywhere orthogonal, such that

$$\mathbf{u} \cdot (\nabla \times \mathbf{u}) = 0. \tag{6.1.7}$$

Vectors satisfying (6.1.7) are called *complex lamellar* in the vector analysis literature, and any flow field satisfying (6.1.7) is called a complex lamellar flow. Common examples of flows satisfying (6.1.7) are any two-dimensional flow or any axisymmetric flow without swirl. Complex lamellar vector fields admit a representation of the form

$$\mathbf{u} = \lambda \nabla \eta, \tag{6.1.8}$$

where γ and η are scalars. Taking the curl of (6.1.8) gives

$$\nabla \times \mathbf{u} = \nabla \lambda \times \nabla \eta, \tag{6.1.9}$$

which is orthogonal to $\lambda \nabla \eta$, thus satisfying (6.1.7).

6.2 INCOMPRESSIBLE (SOLENOIDAL) FLOWS

It is proved in Section 3.4 that for any incompressible flow the divergence of the velocity everywhere vanishes. Vectors satisfying this condition are called *solenoidal* (or sometimes *divergence free*) in the vector analysis literature. Integrating the restriction $\nabla \cdot \mathbf{u} = 0$ over a volume V with bounding surface S and using the divergence theorem yields

$$\int_S \mathbf{u} \cdot \mathbf{n}\, da = 0, \tag{6.2.1}$$

where $\mathbf{n}$ is the outward unit normal of S. Equation (6.2.1) states that for any solenoidal velocity field, the net volumetric flow rate over a closed surface must vanish.

Any solenoidal vector can be written as the curl of another vector, or

$$\mathbf{u} = \nabla \times \boldsymbol{\beta}, \tag{6.2.2}$$

where $\boldsymbol{\beta}$ is called the *vector potential*. That the form (6.2.2) satisfies the incompressibility condition (3.4.8) follows immediately from the vector identity (2.6.2), which states that the divergence of the curl of any vector must vanish identically. The vector potential $\boldsymbol{\beta}$ is not determined uniquely by (6.2.2) but is arbitrary to within a scalar gradient. To see this, define some other vector $\boldsymbol{\beta}^* \equiv \boldsymbol{\beta} + \nabla \alpha$, where α is an arbitrary scalar quantity. Since $\nabla \times \nabla \alpha = 0$ by the vector identity (2.6.1), we have $\nabla \times \boldsymbol{\beta}^* = \nabla \times \boldsymbol{\beta}$, so that $\boldsymbol{\beta}^*$ would make an equally suitable choice for the vector potential in (6.2.2).

It is often desirable to chose $\boldsymbol{\beta}$ as a vector that is itself solenoidal, such that $\nabla \cdot \boldsymbol{\beta} = 0$. That this is always possible follows from a more general vector representation attributed to Helmholtz, which is derived in Section 6.4. Helmholtz's representation states that any vector can be written as the sum of the gradient of a scalar and the curl of another vector [see equation (6.4.1)]. Hence, we can always write $\boldsymbol{\beta} = \nabla\gamma + \nabla \times \boldsymbol{\chi}$, such that using the vector identity (2.6.2), $\nabla \cdot \boldsymbol{\beta} = \nabla^2\gamma$. The vector potential can therefore be made solenoidal by selecting the scalar to be a solution of the Laplace equation $\nabla^2\gamma = 0$.

We note in passing that it is also always possible (although infrequently employed in the study of fluid flows) to choose the vector potential $\boldsymbol{\beta}$ as a complex lamellar vector of the form $\boldsymbol{\beta} = \lambda\nabla\eta$, such that (6.2.2) becomes

$$\mathbf{u} = \nabla \times (\lambda\nabla\eta) = \nabla\lambda \times \nabla\eta. \tag{6.2.3}$$

The streamlines of $\mathbf{u}$ are therefore always tangent to the iso-surfaces of the potentials γ and η. A proof that such a representation for $\mathbf{u}$ always exists can be found in Aris (1962, pp. 67–69).

For a two-dimensional flow in a plane A, the vector potential $\boldsymbol{\beta}$ is oriented normal to A. For instance, for a two-dimensional flow in the x-y plane, $\boldsymbol{\beta}$ is given by $\boldsymbol{\beta} = \psi\mathbf{e}_z$. The function ψ is called the *stream function*, for reasons that will become clear presently. Substituting this expression for $\boldsymbol{\beta}$ into (6.2.2) gives the velocity components u and v in the x- and y-directions, respectively, as

$$u = \frac{\partial\psi}{\partial y}, \qquad v = -\frac{\partial\psi}{\partial x}. \tag{6.2.4}$$

The gradient of ψ can thus be written as

$$\nabla\psi = \frac{\partial\psi}{\partial x}\mathbf{e}_x + \frac{\partial\psi}{\partial y}\mathbf{e}_y = -v\mathbf{e}_x + u\mathbf{e}_y. \tag{6.2.5}$$

Taking the scalar product of (6.2.5) with $\mathbf{u}$ gives

$$\mathbf{u} \cdot \nabla\psi = 0, \tag{6.2.6}$$

so the stream function ψ is constant along streamlines.

The strength Q of a stream tube, defined in general by (3.5.5), can be written for an incompressible flow as

$$Q \equiv \int_A \mathbf{u} \cdot \mathbf{n}\, da = \int_A \mathbf{n} \cdot (\nabla \times \boldsymbol{\beta})\, da = \int_C \boldsymbol{\beta} \cdot d\mathbf{x}, \tag{6.2.7}$$

where A is any open surface spanning the cross section of the stream tube with bounding curve C and unit normal $\mathbf{n}$. The last form of the above integral follows from Stokes's theorem (2.7.11). For the case of a two-dimensional flow in the x-y plane, let us consider the rectangular surface A shown in Figure 6.3, where one edge lies on the x-y plane and has length L and the other edge is oriented in the z-direction and has length b. The edge lying in the x-y plane ends at points P_1 and P_2. The value of the stream function on the two streamlines passing through these two points is ψ_1 and ψ_2, respectively. Since $\boldsymbol{\beta} = \psi\mathbf{e}_z$, the edges of the surface A lying tangent to

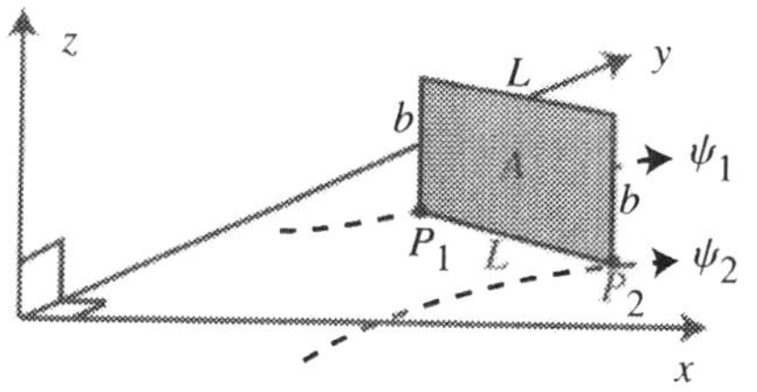

Figure 6.3 Schematic showing a rectangular surface A oriented normal to the x-y plane. One edge of the rectangle lies on the x-y plane, and the end points P_1 and P_2 of this edge lie on streamlines on which the stream function has the values ψ_1 and ψ_2, respectively.

the x-y plane make no contribution to the integral in (6.2.7). Taking into account the direction of integration in (6.2.7), which is in the positive z-direction on one side of A and in the negative z-direction on the opposite side, we obtain

$$Q = b(\psi_2 - \psi_1). \tag{6.2.8}$$

The difference in value of stream function between any two streamlines is equal to the volumetric flow rate passing between the two streamlines per unit width normal to the plane of motion.

6.3 FLOWS THAT ARE BOTH INCOMPRESSIBLE AND IRROTATIONAL

A flow that is both irrotational and incompressible must satisfy both the condition (6.1.1) of zero curl and the condition (3.4.8) of zero divergence. Since the flow is irrotational, it must admit a velocity potential such that the representation (6.1.2) applies. Substituting (6.1.2) into the condition of zero divergence gives an equation for the velocity potential as

$$\nabla^2 \phi = 0. \tag{6.3.1}$$

Alternatively, applying the vector identity (2.6.3) to $\mathbf{u}$ gives

$$\nabla^2 \mathbf{u} = \nabla(\nabla \cdot \mathbf{u}) - \nabla \times (\nabla \times \mathbf{u}), \tag{6.3.2}$$

such that if $\mathbf{u}$ is both solenoidal and irrotational, we can write

$$\nabla^2 \mathbf{u} = 0. \tag{6.3.3}$$

Hence, for a flow that is both irrotational and incompressible, both the velocity field and its scalar potential must satisfy Laplace's equation. Any function that is a solution of Laplace's equation is said to be a *harmonic function*. There exists a substantial literature on harmonic functions that can be employed to solve particular flow fields, some of which are explored in Chapter 9. In the present section, we consider only some very general properties of harmonic functions.

For a flow that is both irrotational and incompressible, the velocity field is completely determined by (6.3.3), aside from boundary conditions, without use of the momentum equation. In order for this to be possible without violation of conservation of momentum, the three components of the momentum equation must reduce to a single equation for the pressure. We examine conditions under which this reduction may occur in Chapters 7 and 8.

One of the more important properties of harmonic functions is that the value of the function within any domain V can be generated by integration of the function and its normal derivative over the boundary S of the domain. To see this, we apply Green's second identity (2.7.8) to two functions, $\phi(\mathbf{x})$ and $G(\mathbf{x}-\mathbf{x}')$, where ϕ is the scalar velocity potential and G is a Green's function, giving

$$\int_V (\phi\nabla^2 G - G\nabla^2\phi)\,dv = \int_S \left(\phi\frac{\partial G}{\partial n} - G\frac{\partial \phi}{\partial n}\right) da. \tag{6.3.4}$$

The Green's function $G(\mathbf{x}-\mathbf{x}')$ for the Laplace equation is defined to be a particular solution of the fundamental equation

$$\nabla^2 G = \delta(\mathbf{x}-\mathbf{x}'), \tag{6.3.5}$$

where $\delta(\mathbf{x}-\mathbf{x}')$ is the Dirac delta (which has the value of infinity at $\mathbf{x}=\mathbf{x}'$ and zero elsewhere). Substituting (6.3.5) into (6.3.4), recalling that

$$\int_V \phi(\mathbf{x}')\delta(\mathbf{x}-\mathbf{x}')\,dv' = \begin{cases} \phi(\mathbf{x}) & \text{if } \mathbf{x}\in V, \\ 0 & \text{if } \mathbf{x}\notin V, \end{cases} \tag{6.3.6}$$

and making use of the fact that ϕ satisfies the Laplace equation, we obtain

$$\phi(\mathbf{x},t) = \int_S \left(\phi\frac{\partial G}{\partial n} - G\frac{\partial \phi}{\partial n}\right) da \tag{6.3.7}$$

for any $\mathbf{x}\in V$. A similar equation can be written for the velocity vector $\mathbf{u}$. Hence, if the value of ϕ and its normal derivative are known on the bounding surface S of V, then ϕ can be obtained at any other point in V simply by integration over S.

Another property of flows that are both irrotational and incompressible is stated in the following theorem.

Theorem 6.3.1 (Velocity Maximum Theorem). In an irrotational flow of an incompressible fluid in a bounded region V, the maximum velocity magnitude must occur along the boundary S of V.

This theorem is a consequence of a more general result for harmonic functions, but we confine our attention to the fluid mechanics application stated above. The theorem also applies to the more general case of flow in an unbounded region provided that the velocity is finite at infinity.

In order to prove the velocity maximum theorem, it is convenient to recall the kinetic energy κ per unit mass, defined as $\kappa \equiv \mathbf{u} \cdot \mathbf{u}/2$. Our task is equivalent to proving that the maximum value of κ must occur on the boundary S of V. If the velocity vector is harmonic, the Laplacian of κ is positive definite, for upon expanding we have

$$\nabla^2 \kappa = \frac{\partial u_i}{\partial x_j} \frac{\partial u_i}{\partial x_j} \geq 0. \tag{6.3.8}$$

This result can alternatively be written using the velocity decomposition (3.2.4) as

$$\nabla^2 \kappa = D_{ij} D_{ij} \geq 0 \tag{6.3.9}$$

since the vorticity vector is everywhere zero.

The proof of the velocity maximum theorem follows by contradiction. That is, we suppose that there exists some point P in the interior of V at which κ attains a local maximum. If $\mathfrak{R}$ is an infinitesimally small region surrounding the point P, with bounding surface $\partial\mathfrak{R}$, then from the assumption that κ attains a maximum value at P it follows that κ decreases as we progress away from P, or

$$\int_{\partial\mathfrak{R}} \mathbf{n} \cdot \nabla\kappa \, da < 0, \tag{6.3.10}$$

where $\mathbf{n}$ is the outward unit normal of $\partial\mathfrak{R}$. However, using the divergence theorem, we can write

$$\int_{\partial\mathfrak{R}} \mathbf{n} \cdot \nabla\kappa \, da = \int_{\mathfrak{R}} \nabla^2\kappa \, dv. \tag{6.3.11}$$

We have already shown that $\nabla^2\kappa$ is positive definite, so it is not possible for (6.3.10) to be satisfied. Thus, no point in the interior of V exists at which the velocity magnitude attains a local maximum. Of course, if the point P were on the boundary S of V, then no region $\mathfrak{R}$ exists that surrounds P but also lies entirely in V, and so this contradiction does not apply.

6.4 HELMHOLTZ REPRESENTATION THEOREM

We now turn to representations that apply to general vector fields. One of the most useful general representation theorems, usually attributed to Helmholtz, states that any vector field may be written as the sum of a solenoidal vector and an irrotational (or lamellar) vector. Using the representation theorems (6.1.2) and (6.2.2) for irrotational and solenoidal vectors, respectively, the Helmholtz representation for an arbitrary velocity field $\mathbf{u}$ can be expressed as

$$\mathbf{u} = \mathbf{v} + \nabla\alpha + \nabla \times \boldsymbol{\beta}, \tag{6.4.1}$$

where $\nabla \cdot \mathbf{v} = 0$ and $\nabla \times \mathbf{v} = 0$. It is common in the vector analysis literature not to include the harmonic vector field $\mathbf{v}$ in the Helmholtz representation (6.4.1), since it can be combined with either of the other two terms on the right-hand side of (6.4.1); however, in applying this representation to fluid flows, it is somewhat clearer to write out this term explicitly. As explained previously in Section 6.2, the vector potential $\boldsymbol{\beta}$ is not uniquely determined by (6.4.1) and can be chosen to be a solenoidal vector without loss in generality. While the decomposition (6.4.1) can be applied to any vector, such as the vorticity, velocity, or acceleration, it is most common in fluid mechanics to apply it to the velocity vector. This velocity representation is the basis for nearly all of the solution methods used in the remainder of the text.

To prove the representation (6.4.1), we apply Green's second identity (2.7.8), for a region V with bounding surface S, to the velocity vector $\mathbf{u}$ and a Green's function G to write

$$\int_V (G\nabla^2\mathbf{u} - \mathbf{u}\nabla^2 G)\, dv = \int_S \left(G\frac{\partial \mathbf{u}}{\partial n} - \mathbf{u}\frac{\partial G}{\partial n} \right) da. \tag{6.4.2}$$

The Green's function G is defined as a solution of the fundamental equation (6.3.5), so

$$\int_V \mathbf{u}(\mathbf{x}', t)\nabla'^2 G(\mathbf{x} - \mathbf{x}')\, dv' = \int_V \mathbf{u}(\mathbf{x}', t)\delta(\mathbf{x} - \mathbf{x}')dv' = \mathbf{u}(\mathbf{x}, t). \tag{6.4.3}$$

The integration and differentiation in (6.4.2) and (6.4.3) are taken over the primed variable $\mathbf{x}'$, as implied by the primes attached to dv and ∇ in (6.4.3).

The surface integral on the right-hand side of (6.4.2) generates a harmonic vector, say $-\mathbf{v}$, for any arbitrary vector $\mathbf{u}$. This can easily be verified by taking the Laplacian of this term with respect to the unprimed variable $\mathbf{x}$ and using (6.3.5), which yields

$$\int_S \left[\delta(\mathbf{x} - \mathbf{x}')\frac{\partial \mathbf{u}}{\partial n}(\mathbf{x}', t) - \mathbf{u}(\mathbf{x}', t)\frac{\partial \delta}{\partial n}(\mathbf{x} - \mathbf{x}') \right] da'. \tag{6.4.4}$$

This integral vanishes identically for any $\mathbf{x}$ that does not lie on the boundary S, since the delta function $\delta(\mathbf{x} - \mathbf{x}')$ is everywhere zero except at $\mathbf{x} = \mathbf{x}'$. Substituting (6.4.3) into (6.4.2) and solving for $\mathbf{u}$ yield

$$\mathbf{u}(\mathbf{x}, t) = \mathbf{v}(\mathbf{x}, t) + \int_V G(\mathbf{x} - \mathbf{x}')\nabla'^2\mathbf{u}(\mathbf{x}', t)\, dv'. \tag{6.4.5}$$

Substituting the vector identity (2.6.3) for $\nabla^2\mathbf{u}$ into the volume integral in (6.4.5) gives

$$\mathbf{u}(\mathbf{x}, t) = \mathbf{v}(\mathbf{x}, t) + \int_V G(\mathbf{x} - \mathbf{x}')\nabla'\Delta(\mathbf{x}', t)\, dv' - \int_V G(\mathbf{x} - \mathbf{x}')\nabla' \times \boldsymbol{\omega}(\mathbf{x}', t)\, dv', \tag{6.4.6}$$

where the dilatation rate and the vorticity are defined as $\Delta \equiv \nabla \cdot \mathbf{u}$ and $\boldsymbol{\omega} \equiv \nabla \times \mathbf{u}$.

Using the vector identities (2.6.4) and (2.6.5), the last two terms in (6.4.6) become

$$\int_V G(\mathbf{x}-\mathbf{x}')\nabla'\Delta(\mathbf{x}',t)\,dv' = \int_V \nabla'[G(\mathbf{x}-\mathbf{x}')\Delta(\mathbf{x}',t)]\,dv' - \int_V [\nabla' G(\mathbf{x}-\mathbf{x}')]\Delta(\mathbf{x}',t)dv', \tag{6.4.7a}$$

$$\int_V G(\mathbf{x}-\mathbf{x}')\nabla'\times\boldsymbol{\omega}(\mathbf{x}',t)dv' = \int_V \nabla'\times[G(\mathbf{x}-\mathbf{x}')\boldsymbol{\omega}(\mathbf{x}',t)]\,dv' - \int_V [\nabla' G(\mathbf{x}-\mathbf{x}')]\times\boldsymbol{\omega}(\mathbf{x}',t)dv'. \tag{6.4.7b}$$

The first integral on the right-hand side of both (6.4.7a) and (6.4.7b) can be converted to integrals over the bounding surface S with use of the integral identities (2.7.1) and (2.7.5), giving

$$\int_V \nabla'[G(\mathbf{x}-\mathbf{x}')\Delta(\mathbf{x}',t)]\,dv' = \int_S G(\mathbf{x}-\mathbf{x}')\Delta(\mathbf{x}',t)\mathbf{n}'\,da', \tag{6.4.8a}$$

$$\int_V \nabla'\times[G(\mathbf{x}-\mathbf{x}')\boldsymbol{\omega}(\mathbf{x}',t)]\,dv' = \int_S G(\mathbf{x}-\mathbf{x}')\mathbf{n}'\times\boldsymbol{\omega}(\mathbf{x}',t)\,da'. \tag{6.4.8b}$$

The integrals in (6.4.8) generate harmonic flows within the region V, since the Laplacian of the surface integrals vanish identically for any $\mathbf{x}$ not on S. In the following, these terms are combined with the arbitrary harmonic vector $\mathbf{v}$ in (6.4.5).

Substitution of (6.4.7) into (6.4.6) yields

$$\mathbf{u}(\mathbf{x},t) = \mathbf{v}(\mathbf{x},t) - \int_V [\nabla' G(\mathbf{x}-\mathbf{x}')]\Delta(\mathbf{x}',t)\,dv' + \int_V [\nabla' G(\mathbf{x}-\mathbf{x}')]\times\boldsymbol{\omega}(\mathbf{x}',t)\,dv'. \tag{6.4.9}$$

Because G is not a function of $\mathbf{x}$ and $\mathbf{x}'$ independently, but rather of the combination $\mathbf{x}-\mathbf{x}'$, we can write

$$\nabla' G(\mathbf{x}-\mathbf{x}') = -\nabla G(\mathbf{x}-\mathbf{x}'), \tag{6.4.10}$$

where the differentiation on the right-hand side of (6.4.10) is performed with respect to the unprimed variable $\mathbf{x}$. Substituting (6.4.10) into (6.4.9) and allowing the gradient with respect to $\mathbf{x}$ to pass through the integral, since the integration is with respect to $\mathbf{x}'$, give

$$\mathbf{u}(\mathbf{x},t) = \mathbf{v}(\mathbf{x},t) + \nabla\left[\int_V G(\mathbf{x}-\mathbf{x}')\Delta(\mathbf{x}',t)\,dv'\right] + \nabla\times\left[-\int_V G(\mathbf{x}-\mathbf{x}')\boldsymbol{\omega}(\mathbf{x}',t)\,dv'\right]. \tag{6.4.11}$$

The result (6.4.11) has the same form as the representation (6.4.1), where the scalar potential α is identified with the integral over the dilatation rate and the vector potential $\boldsymbol{\beta}$ is identified with the integral over the vorticity.

While every vector field admits a representation of the form (6.4.1), other general representations also exist. For instance, any vector field can also be written as the sum of a scalar gradient and a complex lamellar vector, such that using the representations (6.1.2) and (6.1.8), we can write

$$\mathbf{u} = \nabla\phi + \lambda\nabla\eta. \tag{6.4.12}$$

The scalars ϕ, γ, and η are called *Monge potentials* and have the property that

$$\mathbf{u}\cdot(\nabla\times\mathbf{u}) = \nabla\phi\cdot(\nabla\lambda\times\nabla\eta). \tag{6.4.13}$$

The representation (6.4.12) can be proved by noting that because the divergence of the curl of any vector vanishes, the vector $\nabla\times\mathbf{u}$ is solenoidal and admits a representation of the form (6.2.3). We can thus write

$$\nabla\times(\mathbf{u}-\gamma\nabla\eta) = 0. \tag{6.4.14}$$

The combination $\mathbf{u}-\gamma\nabla\eta$ must therefore be irrotational and hence can be expressed as the gradient of a potential, which results in the representation (6.4.12).

6.5 BIOT-SAVART LAW

Substituting the Helmholtz representation (6.4.1) for the velocity into the definition of the dilatation rate, $\Delta \equiv \nabla\cdot\mathbf{u}$, and recalling that both the harmonic vector $\mathbf{v}$ and $\nabla\times\boldsymbol{\beta}$ have zero divergence, yield a Poisson equation for the scalar potential α in (6.4.1) as

$$\nabla^2\alpha = \Delta. \tag{6.5.1}$$

Similarly, substituting (6.4.1) into the definition of vorticity, $\boldsymbol{\omega}\equiv\nabla\times\mathbf{u}$, and recalling that both $\mathbf{v}$ and $\nabla\alpha$ have zero curl, yield a Poisson equation for the vector potential $\boldsymbol{\beta}$ of the form

$$\nabla^2\boldsymbol{\beta} = -\boldsymbol{\omega}. \tag{6.5.2}$$

The vector identity (2.6.3) is used in obtaining (6.5.2) as well as the fact that $\boldsymbol{\beta}$ can in general be chosen to be solenoidal.

Although neither α nor $\boldsymbol{\beta}$ are determined uniquely by (6.5.1) and (6.5.2), it is common practice to set α and $\boldsymbol{\beta}$ equal to the particular solutions of the Poisson equation, the general solution being combined with the harmonic vector $\mathbf{v}$. These particular solutions for α and $\boldsymbol{\beta}$ can be expressed in terms of the Green's function using the result (6.4.11) of the previous section, giving

$$\alpha(\mathbf{x}, t) = \int_V G(\mathbf{x} - \mathbf{x}')\Delta(\mathbf{x}', t)\, dv', \tag{6.5.3}$$

$$\boldsymbol{\beta}(\mathbf{x}, t) = -\int_V G(\mathbf{x} - \mathbf{x}')\boldsymbol{\omega}(\mathbf{x}', t)\, dv'. \tag{6.5.4}$$

The Green's function for the Poisson equation is

$$G(\mathbf{r}) = \begin{cases} \dfrac{1}{2\pi}\ln(r) & \text{in two dimensions,} \\ -\dfrac{1}{4\pi r} & \text{in three dimensions,} \end{cases} \tag{6.5.5}$$

where $\mathbf{r} = \mathbf{x} - \mathbf{x}'$ and $r = |\mathbf{r}|$.

It is sometimes useful to decompose the velocity $\mathbf{u}$ as the sum of a harmonic part $\mathbf{v}$, a part $\mathbf{u}_D$ generated by the dilatation rate, and a part $\mathbf{u}_V$ generated by the vorticity. These three parts can, of course, be identified with the three terms in the Helmholtz representation (6.4.1), respectively. Taking the gradient of the particular solution (6.5.3) for α gives an expression for $\mathbf{u}_D$ as

$$\mathbf{u}_D(\mathbf{x}, t) = \frac{1}{2\pi}\int_A \frac{\mathbf{r}}{r^2}\Delta(\mathbf{x}', t)\, da' \tag{6.5.6a}$$

for two-dimensional flows and as

$$\mathbf{u}_D(\mathbf{x}, t) = \frac{1}{4\pi}\int_V \frac{\mathbf{r}}{r^3}\Delta(\mathbf{x}', t)\, dv' \tag{6.5.6b}$$

for three-dimensional flows. Taking the curl of the solution (6.5.4) for $\boldsymbol{\beta}$ gives an expression for $\mathbf{u}_V$ as

$$\mathbf{u}_V(\mathbf{x}, t) = -\frac{1}{2\pi}\int_A \frac{\mathbf{r} \times \boldsymbol{\omega}(\mathbf{x}', t)}{r^2}\, da' \tag{6.5.7a}$$

for two-dimensional flows and as

$$\mathbf{u}_V(\mathbf{x}, t) = -\frac{1}{4\pi}\int_V \frac{\mathbf{r} \times \boldsymbol{\omega}(\mathbf{x}', t)}{r^3}\, dv' \tag{6.5.7b}$$

for three-dimensional flows. Equation (6.5.7), in either two or three dimensions, is called the *Biot-Savart equation*. There exists a similar relationship in electrodynamics between the magnetic field strength and the current density, taking the place of the velocity and the vorticity in equation (6.5.7b), respectively, from which this name is obtained.

A common solution procedure for incompressible flows is to evolve the vorticity $\boldsymbol{\omega}$ by a transport equation (Chapter 7) and then to compute the velocity field induced by the vorticity using the Biot-Savart integral (6.5.7). Once $\mathbf{u}_V$ has been determined, a harmonic field $\mathbf{v}$ is introduced to satisfy any boundary conditions that may be imposed, such as uniform flow at infinity or no penetration at the surface of a solid body. Of course, only a single boundary condition (e.g., the no-penetration condition) can be used on a boundary to determine $\mathbf{v}$. Other boundary conditions, such as the no-slip condition at the interface of a rigid solid for a viscous flow, must be satisfied implicitly via the evolution of the vorticity field.

Example 6.5.1. Write an integral for the stream function ψ in a two-dimen-sional, incompressible flow with no boundaries but with nonzero vorticity.

SOLUTION. For two-dimensional, unbounded, incompressible flow in the x-y plane, the stream function is related to the vector potential by $\boldsymbol{\beta} = \psi \mathbf{e}_z$ and the vorticity is given by $\boldsymbol{\omega} = \omega \mathbf{e}_z$, where $\mathbf{e}_z$ is a unit vector in the z-direction. Substituting these results into (6.5.4) and using the appropriate Green's function from (6.5.5) for a two-dimensional flow give an integral for the stream function as

$$\psi(x, y, t) = -\frac{1}{2\pi} \int_A \ln(r)\omega(x', y', t)\, da'. \tag{6.5.8}$$

6.6 FAR-FIELD ASYMPTOTIC FORM FOR VELOCITY IN A FLOW EXTENDING TO INFINITY

In solving flows in a space with no boundaries or flows external to some rigid body, we frequently encounter the problem of determining the limiting value of certain integrals of the velocity field over a surface S in the flow, which may for the present be taken as a circle in two dimensions or a sphere in three dimensions, as the radius of S approaches infinity. In order to determine the behavior of such integrals, it is necessary to know the asymptotic behavior of the velocity field far from the region of nonzero vorticity. The asymptotic far-field form for velocity can be obtained from the Biot-Savart equation (6.5.7); however, the arguments and results differ somewhat for two- and three-dimensional flows.

For three-dimensional flow, the part of the velocity field that is generated by the vorticity can be written as

$$\mathbf{u}(\mathbf{x}, t) = \frac{1}{4\pi} \int_V \nabla\left(\frac{1}{r}\right) \times \boldsymbol{\omega}(\mathbf{x}', t)\, dv'. \tag{6.6.1}$$

Denoting the distance from a fixed origin O by $s \equiv |\mathbf{x}|$ and defining $\mathbf{r} \equiv \mathbf{x} - \mathbf{x}'$, we can expand $\nabla(1/r)$ in a Taylor series about $\mathbf{r} = \mathbf{x}$ to obtain

$$\begin{aligned}\nabla\left(\frac{1}{r}\right) &= \nabla\left(\frac{1}{s}\right) + (\mathbf{r} - \mathbf{x}) \cdot \nabla\left[\nabla\left(\frac{1}{r}\right)\right]_{r=s} + O\left(\frac{1}{s^4}\right)\\ &= \nabla\left(\frac{1}{s}\right) - \mathbf{x}' \cdot \nabla\left[\nabla\left(\frac{1}{r}\right)\right]_{r=s} + O\left(\frac{1}{s^4}\right).\end{aligned} \tag{6.6.2}$$

Substituting this Taylor series expansion into the Biot-Savart equation (6.6.1) gives

$$\mathbf{u}(\mathbf{x},t) = \frac{1}{4\pi}\nabla\left(\frac{1}{s}\right) \times \int_V \boldsymbol{\omega}(\mathbf{x}',t)\,dv' - \frac{1}{4\pi}\frac{\partial}{\partial x_i}\left[\nabla\left(\frac{1}{r}\right)\right]_{r=s} \times \int_V x_i'\boldsymbol{\omega}(\mathbf{x}',t)\,dv' + O\left(\frac{1}{s^4}\right), \tag{6.6.3}$$

where summation over the repeated index i is performed in the last term.

Since the vorticity is a solenoidal vector field, it follows that

$$\nabla \cdot (x_i\boldsymbol{\omega}) = \nabla x_i \cdot \boldsymbol{\omega} = \delta_{ij}\omega_j = \omega_i. \tag{6.6.4}$$

The divergence theorem can be used, for any three-dimensional region V, to write

$$\int_V \omega_i\,dv = \int_V \nabla \cdot (x_i\boldsymbol{\omega})\,dv = \int_S x_i\boldsymbol{\omega} \cdot \mathbf{n}\,da, \tag{6.6.5}$$

where S is the bounding surface of V with outward unit normal $\mathbf{n}$. This integral approaches zero as S approaches infinity provided that the vorticity magnitude ω decreases at least as fast as $\omega \approx s^{-n}$, $n \geq 4$. Assuming that the vorticity is sufficiently compact to satisfy this restriction, the first term on the right-hand side of (6.6.3) must vanish in a three-dimensional flow. The second term on the right-hand side of (6.6.3) does not in general vanish, so we conclude that

$$|\mathbf{u}| \sim O\left(\frac{1}{s^3}\right) \quad \text{as } s \to \infty. \tag{6.6.6}$$

For two-dimensional flows, the Biot-Savart equation can be written as

$$\mathbf{u}(\mathbf{x},t) = -\frac{1}{2\pi}\int_A \nabla(\ln r) \times \boldsymbol{\omega}(\mathbf{x}',t)\,da'. \tag{6.6.7}$$

Expanding $\nabla(\ln r)$ in a Taylor series about $\mathbf{r} = \mathbf{x}$ gives

$$\begin{aligned}\nabla(\ln r) &= \nabla(\ln s) + (\mathbf{r} - \mathbf{x}) \cdot \nabla[\nabla(\ln r)]_{r=s} + O\left(\frac{1}{s^3}\right)\\ &= \nabla(\ln s) - \mathbf{x}' \cdot \nabla[\nabla(\ln r)]_{r=s} + O\left(\frac{1}{s^3}\right),\end{aligned} \tag{6.6.8}$$

which when substituted into (6.6.7) yields

$$\mathbf{u}(\mathbf{x},t) = -\frac{1}{2\pi}\nabla(\ln s) \times \int_A \boldsymbol{\omega}(\mathbf{x}',t)\,da' + \frac{1}{2\pi}\frac{\partial}{\partial x_i}[\nabla(\ln r)]_{r=s} \times \int_A x_i'\boldsymbol{\omega}(\mathbf{x}',t)\,da' + O\left(\frac{1}{s^3}\right). \tag{6.6.9}$$

The integral in the first term on the right-hand side of (6.6.9) can be expressed in terms of the circulation Γ_∞ about a circuit encompassing the entire flow field as

$$\int_A \boldsymbol{\omega}(\mathbf{x}, t)\, da' = \Gamma_\infty \mathbf{e}_z. \tag{6.6.10}$$

In two dimensions, the total circulation Γ_∞ either may or may not vanish for different flows. For cases where Γ_∞ does not vanish, such as for a single patch of uniform vorticity, the expansion (6.6.9) is dominated by the first term on the right-hand side, such that

$$|\mathbf{u}| \sim O\left(\frac{1}{s}\right) \quad \text{as } s \to \infty. \tag{6.6.11}$$

For flows in which the total circulation Γ_∞ does vanish, such as would be the case for uniform flow past a fixed body in a viscous fluid, the asymptotic form for the velocity magnitude is controlled by the second term in (6.6.9), giving

$$|\mathbf{u}| \sim O\left(\frac{1}{s^2}\right) \quad \text{as } s \to \infty. \tag{6.6.12}$$

A similar argument can be used for the part of the velocity that is generated by the dilatation rate using the Green's function solution (6.5.6). In three dimensions, the magnitude of $\mathbf{u}_D$ is found to approach zero as $1/s^3$ if the net volumetric production rate (the integral of dilatation rate over all space) vanishes and as $1/s^2$ if the volumetric production rate does not vanish. In two dimensions, the magnitude of $\mathbf{u}_D$ varies as $1/s^2$ if the net volumetric production rate vanishes and as $1/s$ if the volumetric production rate does not vanish.

BIBLIOGRAPHY

Aris, R. (1962). *Vectors, Tensors and the Basic Equations of Fluid Mechanics*, Prentice-Hall, Englewood Cliffs, NJ (reprinted Dover Publications, New York, 1989).

Batchelor, G.K. (1967). *An Introduction to Fluid Dynamics*, Cambridge University Press, Cambridge.

Kellogg, O.D. (1929). *Foundations of Potential Theory*, Springer Publications, New York (reprinted Dover Publications, New York, 1954).

PROBLEMS

1. Consider the flow of an irrotational, incompressible fluid in two dimensions along a plane rigid surface, where the flow is in the x-direction and the surface normal is oriented in the y-direction. Prove that the shear stress T_{12} vanishes at the surface, even when the fluid viscosity does not vanish. Comment on the applicability of the assumption of irrotationality in viscous fluids.
2. For the two-dimensional velocity fields listed below, determine whether each of the representations $\mathbf{u} = \nabla\phi$ and $\mathbf{u} = \nabla\psi \times \mathbf{e}_z$ exist, and if so, find expressions for the scalars ϕ and ψ.

(a) $\mathbf{u} = cx\mathbf{e}_x - cy\mathbf{e}_y$

(b) $\mathbf{u} = cy\mathbf{e}_x - cx\mathbf{e}_y$

(c) $\mathbf{u} = cy\mathbf{e}_x + cx\mathbf{e}_y$

3. Using the Helmholtz representation, the vorticity field can be decomposed as

$$\boldsymbol{\omega} = \nabla\xi + \nabla \times \mathbf{q},$$

where $\nabla \cdot \mathbf{q} = 0$.

(a) Show that the gradient term $\nabla\xi$ makes no contribution to the velocity field, as given by the Biot-Savart integral (6.5.7).

(b) Show that for any space with no boundaries and for which $\xi \to 0$ as $r \to \infty$, the scalar ξ must vanish everywhere.

4. Show that the representation $\mathbf{u} = \nabla\phi + \lambda\nabla\eta$, where ϕ, γ, and η are Monge potentials, can be used to write $\mathbf{u} \cdot \boldsymbol{\omega} = \nabla\phi \cdot (\nabla\lambda \times \nabla\eta)$.

5. Consider a flow of an incompressible fluid in a fixed region V on whose boundary S the no-penetration condition applies. Prove that $\int_V \mathbf{u}\, dv = 0$.

6. Show that the velocity magnitude approaches zero as $|\mathbf{u}| \sim 1/s^n$, where $n = 2$ in two-dimensional flows and $n = 3$ three-dimensional flows, when the flow is generated by rate of dilatation, with no vorticity and zero total mass supply.

7. Find the limiting values for the following integrals as the radius r_0 of the circle C with outward unit normal $\mathbf{n}$ approaches infinity in an unbounded, two-dimensional, incompressible flow field with zero velocity at infinity and zero vorticity outside of some finite region $\mathfrak{R}$.

(a) $\displaystyle\int_C \mathbf{u} \cdot \mathbf{u}\, d\ell$ with $\Gamma_\infty \neq 0$

(b) $\displaystyle\int_C \mathbf{u} \cdot \mathbf{n}\, d\ell$ with $\Gamma_\infty \neq 0$

(c) $\displaystyle\int_C \mathbf{u} \cdot \mathbf{n}\, d\ell$ with $\Gamma_\infty = 0$

(d) $\displaystyle\int_C \omega^2\, d\ell$ with $\Gamma_\infty = 0$

8. Find the limiting values for the following integrals as the radius r_0 of the sphere S with outward unit normal $\mathbf{n}$ approaches infinity in an unbounded, three-dimensional, incompressible flow field with zero velocity at infinity and zero vorticity outside of some finite region $\mathfrak{R}$.

(a) $\displaystyle\int_S \mathbf{u} \cdot \mathbf{u}\, da$

(b) $\displaystyle\int_S r^2\mathbf{u} \cdot \mathbf{n}\, da$

(c) $\displaystyle\int_S \mathbf{u} \cdot \mathbf{n}\, da$

CHAPTER 7

VORTICITY TRANSPORT THEOREMS

For a large variety of fluid flows, the vorticity magnitude is zero or nearly zero outside of some compact region, such as a boundary layer about a body surface, a wake behind a blunt body, or a shear layer between two fluid layers moving at different speeds. The vorticity evolution in these high-vorticity regions determines the dynamics of the fluid flow as a whole, since for an incompressible flow, the velocity at any point can be determined kinematically from the vorticity field via the Biot-Savart equation. Furthermore, as we shall see, the pressure plays no role in the vorticity transport equation for uniform-density fluid. By posing the equation of motion in terms of vorticity, the pressure is immediately eliminated from the transport equations and the incompressibility condition is automatically satisfied. In this chapter, the vorticity transport equation is derived and mechanisms that give rise to vorticity variations are discussed. Several general theorems governing vorticity evolution are derived, including a number of invariant quantities associated with the vorticity field.

7.1 VORTICITY TRANSPORT EQUATION

The momentum conservation equation (4.2.9), together with the expression (4.5.4) for the stress tensor in an incompressible, inviscid fluid, can be expressed as

$$\frac{\partial \mathbf{u}}{\partial t} + (\mathbf{u} \cdot \nabla)\mathbf{u} = \mathbf{b} - \frac{1}{\rho}\nabla p. \tag{7.1.1}$$

From the vector identity (2.6.6), the convective acceleration term in (7.1.1) can be rewritten as $(\mathbf{u} \cdot \nabla)\mathbf{u} = \nabla \kappa + \boldsymbol{\omega} \times \mathbf{u}$, where κ is the kinetic energy per unit mass, such that (7.1.1) becomes

$$\frac{\partial \mathbf{u}}{\partial t} + \boldsymbol{\omega} \times \mathbf{u} = \mathbf{b} - \nabla \kappa - \frac{\nabla p}{\rho}. \tag{7.1.2}$$

The *vorticity transport equation* is obtained by taking the curl of (7.1.2), while making use of vector identities (2.6.5) and (2.6.8), yielding

$$\frac{D\boldsymbol{\omega}}{Dt} = (\boldsymbol{\omega} \cdot \nabla)\mathbf{u} + \frac{\nabla \rho \times \nabla p}{\rho^2} + \nabla \times \mathbf{b}. \tag{7.1.3}$$

In writing the vorticity transport equation (7.1.3), we use the incompressibility condition (3.4.8), the kinematic condition that vorticity must be divergence free, and the definition of the material derivative.

The three terms on the right-hand side in (7.1.3) can be associated with three different mechanisms for change of vorticity in an inviscid fluid. The first term, $(\boldsymbol{\omega} \cdot \nabla)\mathbf{u}$, is known as the *vortex stretching* term, and it is associated with change in vorticity due to a velocity gradient in a direction parallel to the vorticity vector. The significance of this term can be illustrated by the simple flow considered in Example 7.1.1. Vortex stretching is of fundamental importance in a wide variety of three-dimensional fluid flows. In particular, stretching of tubelike vortex structures is the principal mechanism for vorticity enhancement in turbulent flows. The vortex stretching term vanishes for all two-dimensional flows, since both the velocity and the velocity gradient vanish in the direction of the vorticity vector (i.e., normal to the plane of motion). There exist other flows in which the vorticity and velocity vectors are everywhere orthogonal, such as axisymmetric flows without swirl, but in which the vortex stretching term does not vanish since the velocity gradient does not vanish in the direction of the vorticity vector.

The second term on the right-hand side of (7.1.3) is the *baroclinic vorticity generation* term. This term describes the generation of vorticity that arises when surfaces of constant pressure (isobaric surfaces) become oriented differently than surfaces of constant density (isopycnal surfaces). For the case of a *barotropic* fluid, $p = p(\rho)$, the isobaric and isopycnal surfaces are always parallel and the baroclinic generation term vanishes. However, in many applications the pressure is also affected by temperature and species concentration of the fluid constituents, in which case the formation of a pressure gradient along the isopycnals is not unusual. The baroclinic generation term is responsible for formation of buoyancy-driven upwelling and currents in the oceans and atmosphere, and it plays an important role in many combustion and heat transfer problems.

The last term in (7.1.3) is vorticity generation by a nonconservative body force. We recall that all conservative forces, such as gravity, can be written as the gradient of some potential function, or $\mathbf{b} = \nabla\lambda$. From the vector identity (2.6.1), the $\nabla \times \mathbf{b}$ term vanishes for all conservative body forces. For this reason, gravity cannot produce motion of a fluid directly through vorticity generation but must produce motion indirectly via the pressure gradient in the baroclinic generation term at fluid interfaces or in regions of density gradient. Any flow without density variation (including that at a fluid-fluid interface) is unaffected by gravity. Other types of body forces, such as electromagnetic forces, are nonconservative and do directly alter the fluid vorticity.

Example 7.1.1 Vortex Stretching. We consider a columnar vortex confined between two flat plates, as illustrated in Figure 7.1. It is supposed that the vorticity is uniform within the vortex core and can be written in cylindrical polar coordinates (R, α, z) as

$$\boldsymbol{\omega}(\mathbf{x}, t) = \begin{cases} \omega(t)\mathbf{e}_z & \text{for } R \leq \sigma(t), \\ 0 & \text{for } R > \sigma(t), \end{cases} \tag{7.1.4}$$

where $\sigma(t)$ is the vortex core radius. Letting $L(t)$ denote the distance between the two parallel plates, we superpose an axisymmetric straining flow with velocity components

$$v_R = -\frac{cR}{2}, \qquad v_\alpha = 0, \qquad v_z = cz \tag{7.1.5}$$

such that $dL/dt = cL$, where c is called the *straining rate*. The flow (7.1.5) is irrotational, so the vorticity is unaffected by the superposition of the straining flow at time $t = 0$. The vorticity transport equation (7.1.3) for this flow, assuming uniform density and no body force, reduces to

$$\frac{d\omega}{dt} = c\omega, \tag{7.1.6}$$

where we use the total derivative notation since ω is a function only of time in this example. Stretching of the vortex ($c > 0$) results in an increase of the vorticity magnitude within the vortex core, and conversely compression of the vortex ($c < 0$) (assuming that the vortex remains columnar) results in a decrease in vorticity mag-

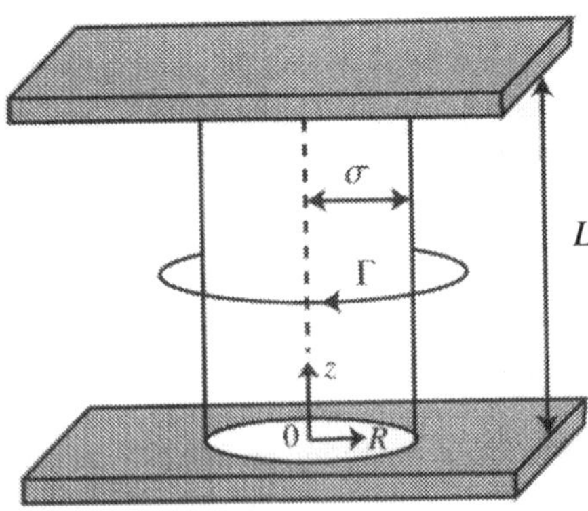

Figure 7.1 A columnar vortex with core radius σ extending between two infinite parallel plates separated by a distance $L(t)$.

nitude. It is obvious by comparing the evolution equations for L and ω for this flow that the two are related according to

$$\frac{L(t)}{L(0)} = \frac{\omega(t)}{\omega(0)}. \tag{7.1.7}$$

It is shown in Section 7.2 that this result is an example of a more general relationship between the change in vorticity magnitude and the stretch of a material line segment lying parallel to the vortex line.

From the vorticity transport equation (7.1.3), with the last two terms omitted, it is clear that fluid particles that are initially irrotational remain irrotational for all time. This observation implies that the vortex core must be a material region such that the volume $V = \pi\sigma^2 L$ of the vortex core is constant. Given the relationship (7.1.7) between the vorticity magnitude and the length L, it must also be true that the vortex circulation $\Gamma = \pi\sigma^2\omega$ is independent of time (which is also a consequence of a more general result presented in Section 7.3).

7.2 MOTION OF VORTEX LINES

In the case where the density is everywhere uniform and the body force is conservative, the inviscid vorticity transport equation takes the form

$$\frac{D\boldsymbol{\omega}}{Dt} = (\boldsymbol{\omega} \cdot \nabla)\mathbf{u}. \tag{7.2.1}$$

A general solution of (7.2.1) was obtained by Cauchy in terms of the deformation gradient tensor $\mathbf{F}$ in the form

$$\boldsymbol{\omega} = \mathbf{F} \cdot \boldsymbol{\omega}_0, \tag{7.2.2}$$

where $\boldsymbol{\omega}_0(\boldsymbol{\xi})$ is the initial value of vorticity on a material point P at position $\boldsymbol{\xi}$ in the reference configuration. Using the result (3.2.7) for the material derivative of $\mathbf{F}$ and substituting (7.2.2) into the left-hand side of (7.2.1) yield

$$\frac{D\omega_i}{Dt} = \frac{\partial u_i}{\partial x_j} F_{jA}\omega_{0,A}, \tag{7.2.3}$$

which is equal to the vortex stretching term on the right-hand side of (7.2.1).

One consequence of the solution (7.2.2) is that if a fluid is initially irrotational at a material point P such that $\boldsymbol{\omega}_0(P) = 0$, it will remain irrotational at P for all time. By extension, if a flow is initially everywhere irrotational, it will always be everywhere irrotational. The vortex stretching term is not capable of generating new vorticity in a flow in which the vorticity initially vanishes.

Another consequence of (7.2.2) is expressed in the following theorem.

Theorem 7.2.1. Vortex lines are material lines.

In some reference configuration, the vorticity at a fluid particle P is given by $\boldsymbol{\omega}_0$. We consider a material line segment $d\boldsymbol{\xi}$ centered at P and oriented parallel to $\boldsymbol{\omega}_0$, so that

$$d\boldsymbol{\xi} = C\boldsymbol{\omega}_0, \tag{7.2.4}$$

where C is a constant. At some later time t, the material line segment at P is given by $d\mathbf{x} = \mathbf{F} \cdot d\boldsymbol{\xi}$. Substituting (7.2.4) into this expression and using the Cauchy solution (7.2.2) yield

$$d\mathbf{x} = C\boldsymbol{\omega}. \tag{7.2.5}$$

This result implies that the line segment $d\mathbf{x}$ remains oriented along the vortex line at the material point P at any time t. The material line formed of all infinitesimal material line segments along a vortex line at some initial time hence remains coincident with the vortex line for all subsequent times.

Taking the magnitude of (7.2.4) and (7.2.5) gives

$$d\xi = C\omega_0, \qquad dx = C\omega. \tag{7.2.6}$$

Eliminating the constant C gives

$$\frac{\omega}{\omega_0} = \Lambda, \tag{7.2.7}$$

where Λ is the stretch of a material line segment oriented along the vortex line, defined in (3.3.2). When (7.2.7) is applied to the flow considered in Example 7.1.1, we immediately obtain the result (7.1.7).

The fact that vortex lines are material lines places a number of restrictions on possible processes in inviscid fluids. For instance, it is not possible for a vortex line to "break" and reconnect with another vortex line unless the velocity field becomes singular at the point at which this break occurs. The question of whether an initially smooth velocity field can develop singularities within the fluid in some finite time is still an unresolved research topic (Beale et al., 1984). In the absence of such singularities, however, a number of commonly observed flow processes are impossible in an inviscid fluid.

For instance, Figure 1.3 shows a computational result for the impact of two vortex tubes of opposite sign vorticity (at a vortex Reynolds number $\mathrm{Re} \equiv \Gamma/\nu = 1500$) which are driven together by an instability of the vortex pair (Section 16.8). When the vortex tubes collide, the vortex lines in each tube appear to break and reconnect to those within the opposite tube, forming a series of looplike structures. The process by which vortex lines are severed and reconnect to other vortex lines is not possible in an inviscid fluid. Instead, vortex tubes in an inviscid fluid continue to deform in such situations with no cutting or reconnection of the loops taking place (Kerr, 1993).

Another example of the severing of vortex lines occurs when a thin body (such as a blade) passes through the core of a vortex structure (Krishnamoorthy, 1997;

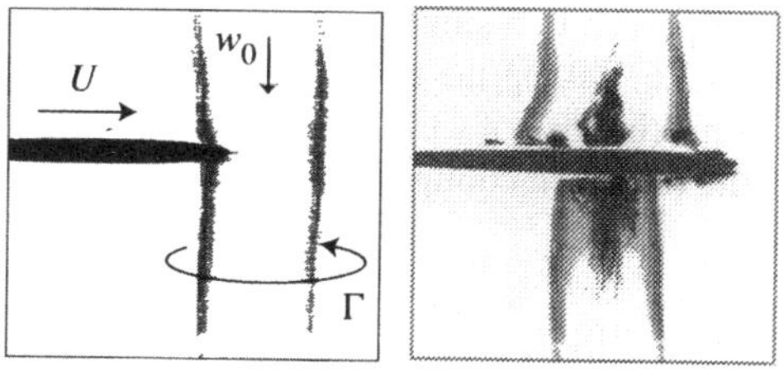

(*a*) (*b*)

Figure 7.2 Experimental flow visualization showing cutting of a columnar vortex by a thin blade. The image is obtained using laser-induced fluorescence in a vertical imaging plane. The gray streaks indicate the edges of the columnar vortex, and the dark patches in (*b*) indicate secondary vorticity that has become entrained into the columnar vortex core. (Krishnamoorthy, 1997)

Marshall and Krishnamoorthy, 1997). A set of experimental pictures illustrating the vortex "cutting" process is shown in Figure 7.2. The blade is traversed toward the vortex structure and is sufficiently thin that it penetrates into the vortex core. During the initial stages of this penetration, the vorticity within the core deforms about the blade leading edge. However, after some time viscous interaction causes the vortex lines within the primary vortex to break and reconnect with those within the blade boundary layer, giving the appearance that the vortex has been "cut" by the blade. In an inviscid fluid, the vortex lines within the primary vortex can never be cut and they will continue indefinitely wrapping about the blade leading edge (Marshall and Grant, 1996).

7.3 PRESERVATION OF CIRCULATION

The three vortex laws of Helmholtz are summarized as follows.

Theorem 7.3.1 (Helmholtz Vortex Laws)

(a) The strength of a vortex tube is uniform along the tube.
(b) Vortex lines are material lines.
(c) The strenth of a vortex tube is invariant in time.

The first of these laws applies to all fluid flows, as proven in Section 3.5. The second law, which applies only to inviscid fluids, is proven in Section 7.2. The third law, which applies to inviscid fluids with uniform density, is a consequence of the following theorem due to Kelvin.

Theorem 7.3.2 (Kelvin Circulation Theorem). The circulation about a material closed circuit is independent of time in a uniform-density, inviscid flow.

This theorem can be proved in a similar manner to that used to derive the transport theorem in Section 3.6. Taking the time derivative of the circulation Γ about a material circuit C and making use of (3.2.1) to write the integral about the position C_0 occupied by C in the reference configuration yield

$$\begin{aligned}\frac{d\Gamma}{dt} &= \frac{d}{dt}\int_C u_i dx_i \\ &= \frac{d}{dt}\int_{C_0} u_i F_{iA} d\xi_A \\ &= \int_{C_0} \frac{D}{Dt}(u_i F_{iA}) d\xi_A \\ &= \int_{C_0} \left(\frac{Du_i}{Dt} F_{iA} + u_i \frac{\partial u_i}{\partial x_j} F_{jA}\right) d\xi_A,\end{aligned}$$

where (3.2.7) is used to expand the material derivative of $\mathbf{F}$ in writing the last of the above expressions. If the dummy indices i and j are exchanged in the last expression, F_{iA} can be factored out, such that again using $dx_i = F_{iA} d\xi_A$ we obtain

$$\frac{d\Gamma}{dt} = \int_C \left(\frac{D\mathbf{u}}{Dt} + \nabla\kappa\right) \cdot d\mathbf{x}. \tag{7.3.1}$$

Substituting the momentum equation (7.1.1) into (7.3.1) yields

$$\frac{d\Gamma}{dt} = \int_C \left(\mathbf{b} - \frac{1}{\rho}\nabla p + \nabla\kappa\right) \cdot d\mathbf{x}. \tag{7.3.2}$$

In the case that the body force is conservative, so that $\mathbf{b} = \nabla\lambda$, and the fluid has uniform density, this result reduces to

$$\frac{d\Gamma}{dt} = \int_C \nabla\left(\lambda + \kappa - \frac{p}{\rho}\right) \cdot d\mathbf{x}. \tag{7.3.3}$$

If A denotes an open surface with boundary C at time t and with unit normal $\mathbf{n}$, Stokes's theorem can be used to write (7.3.3) as

$$\frac{d\Gamma}{dt} = \int_A \left[\nabla \times \nabla\left(\lambda + \kappa - \frac{p}{\rho}\right)\right] \cdot \mathbf{n}\, da, \tag{7.3.4}$$

from which, after using the vector identity (2.6.1), it follows that $d\Gamma/dt$ must vanish.

7.4 VORTICITY MEASURES AND INVARIANTS

A variety of integral quantities involving the vorticity field and other vector fields have special significance in characterizing and constraining the motion of inviscid fluids. In this section we discuss a number of these quantities for three-dimensional flows.

7.4.1 Linear and Angular Impulse

The *linear impulse* **P** and *angular impulse* **L** of an inviscid fluid flow are defined as the resultant force and momentum impulse per unit mass, respectively, necessary to generate the motion of the fluid from rest. The mathematical expressions for the linear and angular impulse are given by

$$\mathbf{P} = \frac{1}{2}\int_V \mathbf{x} \times \boldsymbol{\omega}\, dv, \qquad \mathbf{L} = -\frac{1}{2}\int_V r^2 \boldsymbol{\omega}\, dv, \tag{7.4.1}$$

where V denotes the entire space occupied by the fluid. It is assumed in writing (7.4.1) that the vorticity decays sufficiently fast at infinity that the above integrals exist.

An alternative form for the angular impulse in a space with $\boldsymbol{\omega} \cdot \mathbf{n} = 0$ on the boundary S of V is given by

$$\mathbf{L} = \frac{1}{3}\int_V \mathbf{x} \times (\mathbf{x} \times \boldsymbol{\omega})\, dv. \tag{7.4.2}$$

The form (7.4.2) can be derived by using the vector identity (2.6.9) to write

$$\mathbf{x} \times (\mathbf{x} \times \boldsymbol{\omega}) = -r^2 \boldsymbol{\omega} + (\mathbf{x} \cdot \boldsymbol{\omega})\mathbf{x}, \tag{7.4.3}$$

where $r^2 = \mathbf{x} \cdot \mathbf{x}$. It follows from the chain rule that

$$\frac{\partial}{\partial x_k}(r^2 x_i \omega_k) = 2x_k x_i \omega_k + r^2 \omega_i, \tag{7.4.4}$$

so that combining (7.4.3) and (7.4.4) gives

$$-\tfrac{1}{2} r^2 \boldsymbol{\omega} = \tfrac{1}{3}\mathbf{x} \times (\mathbf{x} \times \boldsymbol{\omega}) - \tfrac{1}{6}\nabla \cdot (\boldsymbol{\omega} r^2 \mathbf{x}). \tag{7.4.5}$$

Substituting (7.4.5) into $(7.4.1)_2$ and using the divergence theorem give

$$\mathbf{L} = \frac{1}{3}\int_V \mathbf{x} \times (\mathbf{x} \times \boldsymbol{\omega})\, dv - \frac{1}{6}\int_S r^2 \mathbf{x}(\boldsymbol{\omega} \cdot \mathbf{n})\, da. \tag{7.4.6}$$

If $\boldsymbol{\omega} \cdot \mathbf{n} = 0$ on S, the last integral in (7.4.6) vanishes and the equivalence of the form (7.4.2) to $(7.4.1)_2$ is proved.

In an unbounded fluid having uniform density and subject to conservative body forces and in which the velocity vanishes at infinity and there exist no immersed solid boundaries, both **P** and **L** are independent of time and are thus known as *invariant* quantities. Invariant quantities provide a way to characterize a fluid flow by a measure that does not change in time. Invariant quantities are furthermore useful in providing a measure of error in numerical computations and in construction of analytical models of vorticity dynamics.

To show that **P** is invariant under the conditions stated above, we take the time derivative of **P** to obtain

$$\frac{d\mathbf{P}}{dt} = \frac{1}{2}\int_V \left(\mathbf{u} \times \boldsymbol{\omega} + \mathbf{x} \times \frac{D\boldsymbol{\omega}}{Dt}\right) dv. \tag{7.4.7}$$

Substituting the vorticity transport equation (7.2.1) into (7.4.7) and transforming to indicial notation, we obtain

$$\begin{aligned}
\frac{dP_i}{dt} &= \frac{1}{2}\int_V \left(\varepsilon_{ijk}u_j\omega_k + \varepsilon_{ijk}x_j\omega_\ell \frac{\partial u_k}{\partial x_\ell}\right) dv \\
&= \frac{1}{2}\int_V \left[2\varepsilon_{ijk}u_j\omega_k + \frac{\partial}{\partial x_\ell}(\varepsilon_{ijk}x_j\omega_\ell u_k)\right] dv \\
&= \int_V \varepsilon_{ijk}u_j\omega_k \, dv + \frac{1}{2}\int_S n_\ell \varepsilon_{ijk}x_j\omega_\ell u_k \, da,
\end{aligned} \tag{7.4.8}$$

where the surface S in the last integral is the boundary of V such that as S approaches infinity (for an unbounded region), this integral vanishes. Substituting the definition of vorticity, $\boldsymbol{\omega} = \nabla \times \mathbf{u}$, into the volume integral in the last expression above yields

$$\begin{aligned}
\frac{dP_i}{dt} &= \int_V \varepsilon_{ijk}\varepsilon_{k\ell m}u_j \frac{\partial u_m}{\partial x_\ell} dv \\
&= \int_V (\delta_{i\ell}\delta_{jm} - \delta_{im}\delta_{j\ell})u_j \frac{\partial u_m}{\partial x_\ell} dv \\
&= \int_V \left[\frac{\partial}{\partial x_i}\left(\frac{1}{2}u_j u_j\right) - \frac{\partial}{\partial x_j}(u_j u_i)\right] dv,
\end{aligned} \tag{7.4.9}$$

where in the second term of the last expression we use the fact that velocity is divergence free for an incompressible fluid. The generalized Green's theorem (2.7.1) can be used to transform the terms in the last expression in (7.4.9) to integrals over the surface S, both of which vanish as S approaches infinity. The proof that the angular impulse **L** is invariant is similar to that given above for the linear impulse and is left as an exercise.

If the derivation given above were repeated for a fluid with a nonconservative body force **b** per unit volume included in the vorticity transport equation, the resulting expressions for variation of **P** and **L** with time would be

$$\frac{d\mathbf{P}}{dt} = \int_V \mathbf{b}\, dv, \qquad \frac{d\mathbf{L}}{dt} = \int_V \mathbf{x} \times \mathbf{b}\, dv. \tag{7.4.10}$$

Hence, the nonconservative body force and its curl are the source of variation of **P** and **L**, respectively. Integrating (7.4.10) over time and assuming that $t = 0$ corresponds to a state of rest, so that $\mathbf{P}(0) = \mathbf{L}(0) = 0$, yield

$$\mathbf{P}(t) = \int_V \left(\int_0^t \mathbf{b}\, dt \right) dv, \qquad \mathbf{L}(t) = \int_V \left(\int_0^t \mathbf{x} \times \mathbf{b}\, dt \right) dv. \tag{7.4.11}$$

This result leads to the physical interpretation of **P** and **L** given in the beginning of this section.

7.4.2 Kinetic Energy

The kinetic energy is defined by

$$T \equiv \frac{1}{2} \int_V \mathbf{u} \cdot \mathbf{u}\, dv, \tag{7.4.12}$$

where for later convenience density is omitted. In order that the above integral exist in an unbounded flow, it is necessary that the velocity approach zero sufficiently fast at infinity. It is sometimes useful to write kinetic energy in terms of an integral over the vorticity field. This can be performed in two different ways for an unbounded flow. One such form is obtained using the chain rule and the identity $\partial x_i / \partial x_j = \delta_{ij}$ to write

$$\frac{\partial}{\partial x_j}(x_i u_i u_j) = u_i u_i + x_i u_j \frac{\partial u_i}{\partial x_j}, \tag{7.4.13a}$$

$$\frac{1}{2} \frac{\partial}{\partial x_i}(x_i u_j u_j) = \frac{3}{2} u_j u_j + x_i u_j \frac{\partial u_j}{\partial x_i}. \tag{7.4.13b}$$

Subtracting (7.4.13a) from (7.4.13b) yields

$$\frac{1}{2} u_j u_j = -x_i u_j \left(\frac{\partial u_i}{\partial x_j} - \frac{\partial u_j}{\partial x_i} \right) + \frac{1}{2} \frac{\partial}{\partial x_i}(x_i u_j u_j) - \frac{\partial}{\partial x_j}(x_i u_j u_i). \tag{7.4.14}$$

Integrating (7.4.14) over the space V, the last two terms can be converted into integrals over the boundary S of V using the generalized Green's theorem (2.7.1), which vanish as S approaches infinity for an unbounded space. The first term on the right-hand side of (7.4.14) can be expressed in terms of the vorticity tensor W_{ij} as

$$x_i u_j \left(\frac{\partial u_i}{\partial x_j} - \frac{\partial u_j}{\partial x_i} \right) = 2 x_i u_j W_{ij}. \tag{7.4.15}$$

Making use of the relationship (3.3.18) between the vorticity tensor and the vorticity vector, we can alternatively write (7.4.15) as

$$x_i u_j \left(\frac{\partial u_i}{\partial x_j} - \frac{\partial u_j}{\partial x_i} \right) = \varepsilon_{ijk} x_i \omega_k u_j. \tag{7.4.16}$$

Substituting (7.4.14)–(7.4.16) into (7.4.12) yields the kinetic energy as an integral over the scalar triple product of the velocity, vorticity, and position vectors as

$$T = \int_V \mathbf{u} \cdot (\mathbf{x} \times \boldsymbol{\omega})\, dv. \tag{7.4.17}$$

A second form for the kinetic energy is obtained with use of the vector identity (2.6.7), which states that for any two vectors $\mathbf{u}$ and $\boldsymbol{\beta}$, we can write

$$\nabla \cdot (\mathbf{u} \times \boldsymbol{\beta}) = \boldsymbol{\beta} \cdot (\nabla \times \mathbf{u}) - \mathbf{u} \cdot (\nabla \times \boldsymbol{\beta}). \tag{7.4.18}$$

If $\mathbf{u}$ is the velocity vector (such that $\boldsymbol{\omega} = \nabla \times \mathbf{u}$) and $\boldsymbol{\beta}$ is the vector potential (such that $\mathbf{u} = \nabla \times \boldsymbol{\beta}$ for an incompressible fluid), then (7.4.18) becomes

$$\nabla \cdot (\mathbf{u} \times \boldsymbol{\beta}) = \boldsymbol{\beta} \cdot \boldsymbol{\omega} - \mathbf{u} \cdot \mathbf{u}. \tag{7.4.19}$$

Integrating (7.4.19) over V, the divergence theorem can be used to convert the left-hand side to an integral over the surface S of V, which vanishes as S approaches infinity, so that the kinetic energy reduces to

$$T = \frac{1}{2} \int_V \boldsymbol{\beta} \cdot \boldsymbol{\omega}\, dv. \tag{7.4.20}$$

The issue of invariance of T can be addressed by taking the time derivative of (7.4.1) and using the momentum equation (7.1.1), yielding

$$\frac{dT}{dt} = \int_V \mathbf{u} \cdot \frac{D\mathbf{u}}{Dt}\, dv = \int_V \mathbf{u} \cdot \nabla \left(\lambda - \frac{p}{\rho}\right) dv, \tag{7.4.21}$$

where we assume that the body force is conservative, such that $\mathbf{b} = \nabla\lambda$. Since $\mathbf{u}$ is divergence free, we can write

$$\frac{dT}{dt} = \int_V \nabla \cdot \left[\left(\lambda - \frac{p}{\rho}\right)\mathbf{u}\right] dv = \int_S \left(\lambda - \frac{p}{\rho}\right)(\mathbf{n} \cdot \mathbf{u})\, da, \tag{7.4.22}$$

where $\mathbf{n}$ is the outward unit normal of S. The result (7.4.22) implies that the kinetic energy T is invariant for any flow in which $\mathbf{u} \cdot \mathbf{n} = 0$ on all boundaries S of V and, for external flows, in which $\mathbf{u}$ approaches zero at infinity faster than r^{-2} in three dimensions and r^{-1} in two dimensions.

7.4.3 Helicity

The helicity J of a fluid flow is defined as the integral over all space of the scalar product of the velocity and vorticity vectors, or

$$J \equiv \int_V \mathbf{u} \cdot \boldsymbol{\omega}\, dv. \tag{7.4.23}$$

Helicity is an interesting topological measure of the vorticity field and in particular can be related to the degree of "knottedness" of vortex lines. Further information on interpretation of helicity is given by Moffat (1969).

Invariance of helicity can be examined by taking the time derivative of (7.4.23) and using both the momentum equation (7.1.1) and the vorticity equation (7.2.1), to obtain

$$\begin{aligned}\frac{dJ}{dt} &= \int_V \left(\frac{Du_i}{Dt}\omega_i + u_i\frac{D\omega_i}{Dt}\right) dv\\ &= \int_V \left[\omega_i \frac{\partial}{\partial x_i}\left(\lambda - \frac{p}{\rho}\right) + u_i\omega_j\frac{\partial u_i}{\partial x_j}\right] dv.\end{aligned} \tag{7.4.24}$$

Since the vorticity is divergence free, it can be taken inside the derivatives in (7.4.24) to write

$$\frac{dJ}{dt} = \int_V \left\{\frac{\partial}{\partial x_i}\left[\left(\lambda - \frac{p}{\rho}\right)\omega_i\right] + \frac{1}{2}\frac{\partial}{\partial x_j}(\omega_j u_i u_i)\right\} dv. \tag{7.4.25}$$

The generalized Green's theorem (2.7.1) can be used to write both of the terms inside the integral in (7.4.25) as surface integrals, giving

$$\frac{dJ}{dt} = \int_S \left(\lambda + \kappa - \frac{p}{\rho}\right)(\mathbf{n}\cdot\boldsymbol{\omega})\, da, \tag{7.4.26}$$

where κ is the kinetic energy per unit mass.

The result (7.4.26) indicates that helicity is invariant in any inviscid flow in which $\mathbf{n}\cdot\boldsymbol{\omega} = 0$ on all boundaries S of V. One consequence of the observation that vortex lines are material lines is that vorticity must always vanish on any material surface on which it initially vanishes. Thus, if the boundary S of V is material and if $\mathbf{n}\cdot\boldsymbol{\omega}$ initially vanishes on S, then $\mathbf{n}\cdot\boldsymbol{\omega}$ always vanishes on S and invariance of helicity is preserved.

7.4.4 Enstrophy

It is shown in Section 6.6 that the integral of vorticity over all space can be written as

$$\int_V \boldsymbol{\omega}\, dv = \int_S \mathbf{x}(\boldsymbol{\omega}\cdot\mathbf{n})\, da. \tag{7.4.27}$$

This integral vanishes for any space in which $\mathbf{n}\cdot\boldsymbol{\omega} = 0$ on the boundary S. An alternative measure of the vorticity field is the integral of the square of the vorticity, or $\omega^2 = \boldsymbol{\omega}\cdot\boldsymbol{\omega}$. The *enstrophy* is defined by the integral

$$E \equiv \frac{1}{2}\int_V \omega^2\, dv. \tag{7.4.28}$$

Taking the scalar product of the vorticity transport equation (7.2.1) with $\boldsymbol{\omega}$ and integrating over V give an expression for the rate of change of enstrophy as

$$\frac{dE}{dt} = \int_V \omega_i \omega_j D_{ij} \, dv, \tag{7.4.29}$$

where D_{ij} are the components of the rate of deformation tensor, defined in (3.2.5). Recalling the result (3.3.6), we can rewrite (7.4.29) as

$$\frac{dE}{dt} = \int_V \omega^2 \left(\frac{\dot{\Lambda}}{\Lambda} \right) dv, \tag{7.4.30}$$

where Λ is the stretch of material line segments oriented along the vortex lines. The result (7.4.30) can alternatively be derived with use of the relation (7.2.7) between the local stretch along a vortex line and the vorticity magnitude.

While we have shown that vortex stretching cannot generate vorticity at a fluid particle at which there is initially no vorticity, it is clear from (7.4.30) that stretching can change the enstrophy of the flow. A schematic indicating typical enstrophy variation with time in isotropic turbulence is given in Figure 7.3, showing an initial increase in enstrophy due to vortex stretching and an eventual decay as the stretching is counteracted by viscous diffusion (see also Lesieur, 1997).

Another interesting property of enstrophy arises from the identity

$$\frac{\partial}{\partial x_i} \left(u_j \frac{\partial u_i}{\partial x_j} \right) = \frac{\partial u_j}{\partial x_i} \frac{\partial u_i}{\partial x_j} = D_{ij} D_{ij} - W_{ij} W_{ij}. \tag{7.4.31}$$

By straightforward calculation using the relationship (3.3.18) between the vorticity tensor and the vorticity vector, it follows that $W_{ij} W_{ij} = \omega^2$. Integrating (7.4.31) over V and using Green's theorem (2.7.1) yields

$$E = \frac{1}{2} \int_V D_{ij} D_{ij} \, dv - \frac{1}{2} \int_S n_i u_j \frac{\partial u_i}{\partial x_j} \, da. \tag{7.4.32}$$

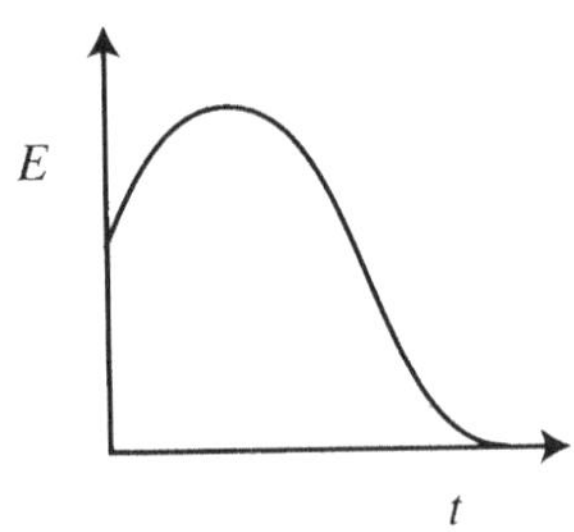

Figure 7.3 Schematic of the enstrophy variation with time in isotropic turbulence.

The surface integral in (7.4.32) vanishes for any unbounded flow in which the velocity vanishes at infinity or for any bounded flow (including viscous flows) in which the velocity vanishes at the boundary S of the flow domain. In a viscous flow, the first term on the right-hand side of (7.4.32) is equal to the net energy dissipation rate divided by 4μ.

7.5 VORTICITY INVARIANTS IN TWO-DIMENSIONAL FLOWS

In this section, the limiting forms for two-dimensional flows of the integral quantities introduced in Section 7.4 are examined, and some new integral invariants are introduced that apply specifically to two-dimensional flows. All of the invariants discussed in the previous section for three-dimensional flows have analogies in two dimensions except for the helicity, which is identically zero in two dimensions. However, some of the integrals introduced in the previous section become unbounded in certain two-dimensional flows.

7.5.1 Total Vorticity

For three-dimensional flows, the integral of the vorticity over all space vanishes, provided that $\boldsymbol{\omega} \cdot \mathbf{n} = 0$ on the domain boundary. For two-dimensional flows, the vorticity vector is orthogonal to the plane of motion, and this result no longer holds. If we consider motion in the x-y plane, denoted hereinafter by A, the vorticity can be written as $\boldsymbol{\omega} = \omega \mathbf{e}_z$. The integral of ω over A, given by

$$\Gamma_\infty = \int_A \omega \, da, \tag{7.5.1}$$

is the total circulation of the flow. That the total circulation is an invariant of motion is obvious from the fact that the vorticity equation in a two-dimensional flow, with uniform density and conservative body force, reduces simply to $D\omega/Dt = 0$. The integral over ω raised to any power is also invariant, provided that the integral exists. A special case of particular interest is the enstrophy E, given for two-dimensional flows by

$$E = \frac{1}{2} \int_A \omega^2 \, da. \tag{7.5.2}$$

For the remainder of this section, it is important to recall the results (6.6.11) and (6.6.12), which state that the magnitude of the velocity induced by vorticity decays as $1/r^2$ when $\Gamma_\infty = 0$ and as $1/r$ when $\Gamma_\infty \neq 0$ in an unbounded space as the radius r of the point under consideration approaches infinity.

7.5.2 Linear and Angular Impulse and Related Invariants

The x- and y-components of the linear impulse in a two-dimensional flow with $\Gamma_\infty = 0$ are given by

$$P_1 = \int_A y\omega\, da, \qquad P_2 = -\int_A x\omega\, da. \tag{7.5.3}$$

The expressions (7.5.3) differ by a factor of $\frac{1}{2}$ from the expression $(7.4.1)_1$ for impulse in a three-dimensional flow. This difference arises as a consequence of the fact that the vortex lines are not closed in two-dimensional flows. The angular impulse is given by the same expression $(7.4.1)_2$ as in three-dimensional flows; however, the alternative expression (7.4.2) for angular momentum is not valid for two-dimensional flows.

The proof of the invariance of $\mathbf{P}$ and $\mathbf{L}$ in Section 7.4.1 still applies for two-dimensional flows. However, the interpretation of these quantities as representing the linear and angular impulse required to generate the flow from rest is dependent upon the fact that the values of $\mathbf{P}$ and $\mathbf{L}$ are independent of the choice of origin of the coordinate system. In other words, if we displace the position vector $\mathbf{x}$ by an amount $\mathbf{x}_0$, the value of the linear and angular impulse should not be affected. In three dimensions, this property of $\mathbf{P}$ is ensured by the fact that the integral of vorticity over all space vanishes. Performing this coordinate displacement in two dimensions and denoting the quantity $\mathbf{P}$ in the displaced coordinate system as $\mathbf{P}'$ yield

$$\mathbf{P}' = \int_A (\mathbf{x} - \mathbf{x}_0) \times \boldsymbol{\omega}\, da = \mathbf{P} - \mathbf{x}_0 \times \Gamma_\infty \mathbf{e}_z. \tag{7.5.4}$$

Thus for the case $\Gamma_\infty = 0$, $\mathbf{P}$ is independent of the choice of the origin of the coordinates, and our previous interpretation for this quantity still applies. On the other hand, for the case $\Gamma_\infty \neq 0$, $\mathbf{P}$ is dependent on the choice of coordinates through the last term in (7.5.4). This term is independent of time due to the invariance of Γ_∞ and so does not affect the invariance property of P, but our former interpretation of P no longer applies. Similar results apply with respect to $\mathbf{L}$.

For the case $\Gamma_\infty \neq 0$, we define a *center of vorticity* $\mathbf{x}_c$ by

$$\Gamma_\infty \mathbf{x}_c \equiv \int_A \mathbf{x}\omega\, da \tag{7.5.5}$$

and a *radius of gyration* D by

$$\Gamma_\infty D^2 \equiv \int_A \left[(x - x_c)^2 + (y - y_c)^2\right]\omega\, da. \tag{7.5.6}$$

The radius of gyration is a measure of the dispersion of vorticity about the centroid $\mathbf{x}_c$. Invariance of both $\mathbf{x}_c$ and D follows directly from the invariance properties of Γ_∞, $\mathbf{P}$, and $\mathbf{L}$.

7.5.3 Interaction Energy

When $\Gamma_\infty = 0$, the result of Section 7.4.2 that the kinetic energy is an invariant of motion and that it can be expressed in the alternative form (7.4.20) still holds. Writing the vector potential in terms of the stream function ψ as $\boldsymbol{\beta} = \psi \mathbf{e}_z$, (7.4.20) reduces in two dimensions to

$$T = \frac{1}{2}\int_A \psi\omega\, da. \tag{7.5.7}$$

The other alternative form (7.4.17) for kinetic energy applies only to three-dimensional flows.

On the other hand, when $\Gamma_\infty \neq 0$, the kinetic energy is infinite for flow in an unbounded domain. We suppose for the moment that the domain A is a finite simply connected region with a bounding curve C. Using the result (7.4.19) in (7.4.12) and again equating the z-component of the vector potential to the stream function yield

$$T = \frac{1}{2}\int_A \psi\omega\, da - \frac{1}{2}\int_C \psi\mathbf{u}\cdot d\mathbf{x}. \tag{7.5.8}$$

Since at large distances $|\mathbf{u}| \sim \Gamma_\infty/2\pi r$ and $\psi \sim -\Gamma_\infty(\ln r)/2\pi$, the integral over C in (7.5.8) becomes unbounded as C is moved out to infinity. Choosing C to be a circle of radius R, we can write

$$T \sim \frac{1}{2}\int_A \psi\omega\, da + \frac{1}{4\pi}(\ln R)\Gamma_\infty^2 \text{ as } R \to \infty. \tag{7.5.9}$$

The last term in (7.5.9) is the part of T that becomes singular in an unbounded domain for $\Gamma_\infty \neq 0$. This singular term depends not on the manner in which the vorticity is distributed, but on the total amount of vorticity in A. The first term on the right-hand side of (7.5.9) is thus the only part of the kinetic energy that depends on the distribution of vorticity. This term is called the *interaction energy* H, defined by

$$H \equiv \frac{1}{2}\int_A \psi\omega\, da. \tag{7.5.10}$$

That H is an invariant of motion follows from the fact that both T and the last (singular) term in (7.5.9) are invariants.

7.6 VORTICITY TRANSPORT AND INVARIANTS IN VISCOUS FLOWS

While the topic under consideration in this text is the flow of inviscid fluids, it is important to note how some of the results in this chapter are affected by the presence of viscosity, if for no other reason than to emphasize the differences between the flow of inviscid fluids and that of real fluids. In a viscous fluid, an additional term occurs on the right-hand side of the vorticity transport equation of the form $\nu\nabla^2\boldsymbol{\omega}$, where ν is the kinematic viscosity. For the special case of a two-dimensional flow with uniform density and a conservative body force, the vorticity transport equation in a viscous flow becomes

$$\frac{D\omega}{Dt} = \nu \nabla^2 \omega. \tag{7.6.1}$$

Equation (7.6.1) has the same form as that for heat transfer, so that we say that viscosity acts to *diffuse* the vorticity. Similar to Fourier's expression for the heat flux, the *vorticity flux* in two-dimensional flows is given by $-\nu\nabla\omega$.

Viscous diffusion does not generate new vorticity within the flow, but rather it acts to spread out the existing vorticity. New vorticity is generated at the boundary between a fluid and a solid or two immiscible fluids of different density and is subsequently diffused into the flow by viscous diffusion. For instance, in the problem of uniform flow past an immersed solid body, a vortex sheet is generated at the surface of the body in an inviscid flow, the strength of which is equal in magnitude to the slip velocity at the body surface. Without viscosity, there is no mechanism for this vortex sheet to spread into the flow, so it is trapped on the body surface and the exterior flow remains irrotational. In a viscous flow, the surface vorticity diffuses normal to the body surface out into the flow, where it is subsequently convected away.

In this chapter we have shown the three vortex laws of Helmholtz to play a fundamental role in constraining the motion of inviscid fluids. It is of interest to examine the implication on these vortex laws of the addition of viscous forces. The first vortex law, Theorem 3.5.2, states that the strength of a vortex tube is uniform along the vortex tube. This law is purely kinematic, resulting from the fact that vorticity is divergent free, and it therefore applies equally to inviscid and viscous flows. The second vortex law, Theorem 7.2.1, states that vortex lines are material lines. This statement is derived in Section 7.2 based on the Cauchy solution to the inviscid vorticity transport theorem. In the presence of the viscous diffusion term, the Cauchy solution no longer applies, and Theorem 7.2.1 is no longer valid. This difference between inviscid flows and that of real (viscous) fluids makes possible the various vortex reconnection and vortex line cutting processes illustrated in Section 7.2. The third vortex law, Theorem 7.3.1, states that the strength of a vortex tube remains constant in time. This theorem is also derived based on the assumption of inviscid flow and is no longer generally valid for a viscous fluid. However, adding the viscous term in the momentum equation, the result (7.3.3) becomes

$$\frac{d\Gamma}{dt} = -\int_C [\nabla \times (\nu \boldsymbol{\omega})] \cdot d\mathbf{x}. \tag{7.6.2}$$

The last term in (7.6.2) leads to a diffusive transport of vorticity across the bounding curve C of the material surface A spanning the vortex tube, thereby causing a change in the circulation about C. However, in the special case that $\nabla \times \boldsymbol{\omega} = 0$ everywhere on the curve C, the circulation about C will be unchanged. This might occur, for instance, in the case of a tubular shaped region of vorticity that is surrounded by an effectively irrotational fluid, which is a fairly common vorticity structure in many

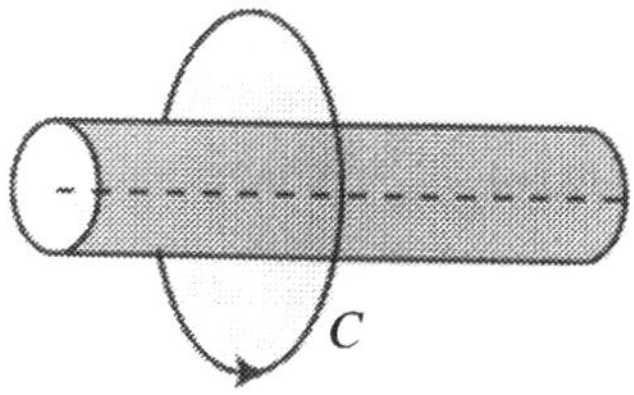

Figure 7.4 A columnar vortex surrounded by a circuit C lying in the irrotational flow outside of the vortex core.

fluid flows. When the circuit C is chosen to surround this vorticity tube but to lie outside of the tube in the irrotational part of the flow (as shown in Figure 7.4), the result (7.6.2) implies that the strength of the vortex will not change in time, even though viscous diffusion may cause a progressive outward dispersion of the vorticity field from the central axis of the tube.

Aside from the three vortex laws discussed above, we have also derived a number of invariant quantities for inviscid flows involving integrals over the vorticity field, including the linear and angular impulse, the kinetic energy, the helicity, and, for two-dimensional flows, the total vorticity. In a three-dimensional viscous flow, the linear and angular impulses remain invariant in an unbounded flow, but the kinetic energy is no longer invariant due to viscous energy dissipation and work performed by bodies immersed in the flow. Helicity is also no longer invariant in viscous flows. In a two-dimensional flow, both the total vorticity and the linear impulse are invariant for viscous flows in an unbounded domain (with no material interfaces), but the angular impulse is invariant only for the case $\Gamma_\infty = 0$. For nonzero values of Γ_∞, the angular impulse varies linearly with time in accordance to

$$\frac{dL}{dt} = -2\nu\Gamma_\infty, \tag{7.6.3}$$

corresponding to increased dispersion of the vorticity field. Consequently, while the vorticity centroid remains invariant in a two-dimensional, unbounded viscous flow, the radius of gyration increases monotonically as

$$D^2(t) = 4\nu t + D^2(0). \tag{7.6.4}$$

BIBLIOGRAPHY

Beale, J.T., T. Kato, and A. Majda (1984). "Remarks on the breakdown of smooth solutions for the 3-D Euler equations," *Communications in Mathematical Physics* **94**, 61–66.

Casey, J., and P.M. Naghdi (1991). "On the Lagrangian description of vorticity," *Archive for Rational Mechanics and Analysis* **115**, 1–14.

Kerr, R.M. (1993). "Evidence for a singularity of the three-dimensional, incompressible Euler equations," *Physics of Fluids A* **5**(7), 1725–1746.

Krishnamoorthy, S. (1997). "Normal vortex-body interaction in three-dimensional viscous flow," Ph.D. dissertation, University of Iowa, Iowa City.

Lesieur, M. (1997). *Turbulence in Fluids*, 3rd ed., Kluwer Publishers, Dordrecht, The Netherlands.

Marshall, J.S., and J.R. Grant (1996). "Penetration of a blade into a vortex core: vorticity response and unsteady blade forces," *Journal of Fluid Mechanics* **306**, 83–109.

Marshall, J.S., and S. Krishnamoorthy (1997). "On the instantaneous cutting of a columnar vortex with non-zero axial flow," *Journal of Fluid Mechanics* **351**, 41–74.

Moffat, K. (1969). "The degree of knottedness of tangled vortex lines," *Journal of Fluid Mechanics* **35**, 117–129.

Truesdell, C.A. (1954). *The Kinematics of Vorticity*, Indiana University Press, Bloomington, IN.

PROBLEMS

1. Show that the vorticity transport equation for an inviscid, uniform-density fluid can be written in the following two alternative forms.

   (a) $\dfrac{D\boldsymbol{\omega}}{Dt} = \boldsymbol{\omega} \cdot \mathbf{D}$, where $\mathbf{D}$ is the rate of deformation tensor

   (b) $\dfrac{D\boldsymbol{\omega}}{Dt} = \boldsymbol{\omega} \cdot (\nabla \mathbf{u})^{\mathrm{T}}$

2. Consider the flow of an inviscid incompressible fluid through an axisymmetric contraction in a tube with upstream radius b_1 and downstream radius b_2, as shown in Figure 7.5. The upstream velocity field is written in cylindrical polar coordinates (R, α, z) as

$$u_z = U_1, \qquad u_R = 0, \qquad u_\alpha = \omega_1 R$$

and the downstream velocity field as

$$u_z = U_2, \qquad u_R = 0, \qquad u_\alpha = \omega_2 R.$$

   (a) Determine ω_2 and U_2 as functions of ω_1, U_1, b_1, and b_2.

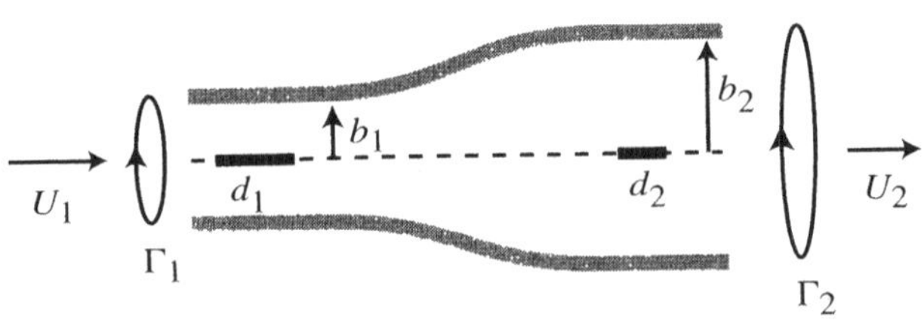

Figure 7.5 Swirling flow in a diverging tube, showing a line segment aligned along the symmetry axis.

(b) A small material line segment lying on the tube axis has length d_1 in the upstream part of the tube. Determine the downstream length d_2 of the line segment.

3. A vortex ring with circular core cross section has initial core radius $a(0) = a_0$, initial ring radius $b(0) = b_0$, and vorticity that is initially distributed uniformly over a cross section of the core with magnitude $\omega(0) = \omega_0$. The ring is assumed to be very thin, such that $a_0 \ll b_0$. A line source with strength m and radial velocity $u_R = m/2\pi R$ is placed along the axis of the vortex ring, as shown in Figure 7.6. Assuming that the core cross section remains nearly circular for all time $t > 0$, determine how the following quantities vary in time and explain the reasons for each of your answers.

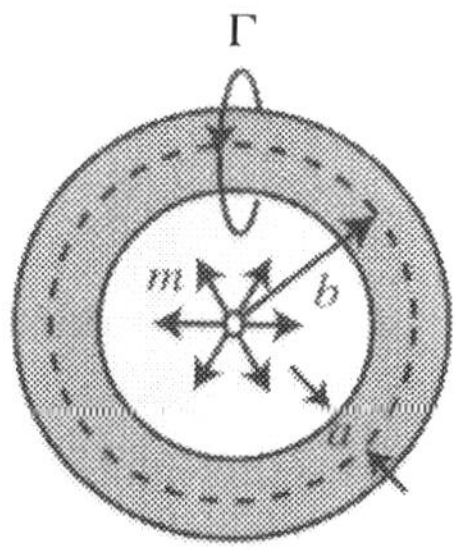

Figure 7.6 A vortex ring of ring radius b with a line source placed along the symmetry axis.

(a) The circumference $C = 2\pi b$ of the core axis
(b) The vorticity magnitude $\omega(t)$ at the center of rotation of the ring cross section
(c) The integral $\int_A \boldsymbol{\omega} \cdot \mathbf{n}\, da$, where A is a cross section of the vortex core and $\mathbf{n}$ is a unit normal to A

4. A small hairpin vortex is aligned at 45° to the direction of the mean shear flow $\mathbf{u} = S(y)y\mathbf{e}_x$ in a turbulent boundary layer, as shown in Figure 7.7. The dynamics of the vortex are such that it remains aligned in this direction for some interval of time T.

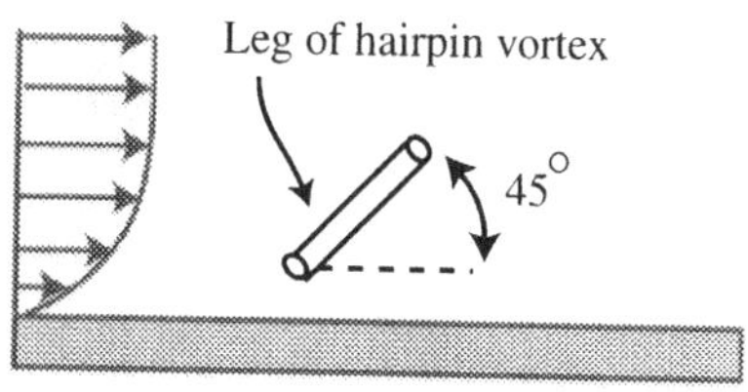

Figure 7.7 A hairpin vortex aligned at 45° to the direction of mean shear in a turbulent boundary layer.

(a) Determine the rate of stretching of the vortex by the mean shear flow as a function of distance y above the wall. Would the vortex be stretched more rapidly if it were aligned in a different direction? Why or why not?

(b) Assuming that the flow is approximately inviscid (i.e., that the vortex lies within the turbulence inertial range), estimate the rate at which the maximum vorticity increases within the vortex.

5. Consider a two-dimensional fluid flow in the x-y plane with no internal or external boundaries and in which the vorticity $\boldsymbol{\omega} = \omega \mathbf{e}_z$ vanishes outside of a compact region. Let B denote the integral

$$B \equiv \int_A \omega \, da,$$

where the integration is over the whole x-y plane. Prove that B is invariant for an incompressible, uniform-density, *viscous* fluid.

6. Consider the integral

$$\mathbf{G} \equiv \int_V \mathbf{u} \times \boldsymbol{\omega} \, dv,$$

where the integration is over the entire volume occupied by the fluid.

(a) Show that for a *three-dimensional* space in which the vorticity is compact and which has no internal or external boundaries, the integral $\mathbf{G}$ vanishes identically in an incompressible fluid.

(b) Does $\mathbf{G}$ also vanish for a *two-dimensional* incompressible fluid in which the vorticity is compact and which has no internal or external boundaries?

7. Prove that the angular impulse $\mathbf{L}$ is invariant in a three-dimensional, inviscid, incompressible flow with no rigid boundaries.

8. Calculate the linear impulse $\mathbf{P}$ and enstrophy E of a vortex ring, with vorticity field $\boldsymbol{\omega} = \Gamma \delta(R - b) \delta(z) \mathbf{e}_\alpha$ in cylindrical polar coordinates (R, α, z).

9. Calculate the angular impulse $\mathbf{L}$ and interaction energy H of a two-dimensional Rankine vortex with vorticity field

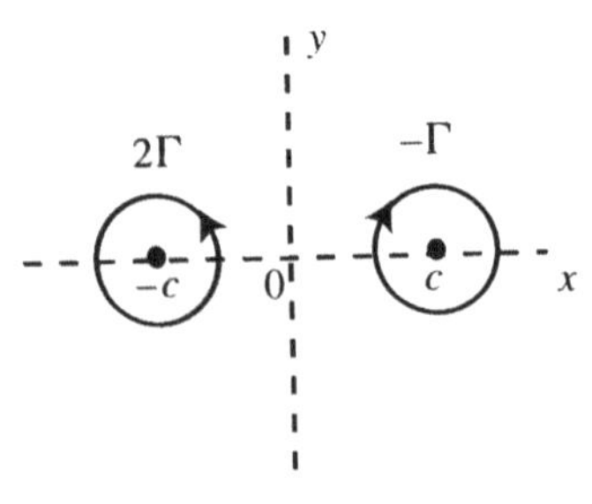

Figure 7.8 A vortex pair with unequal strength.

$$\omega = \begin{cases} \omega_0, & r \leq a, \\ 0, & r > a. \end{cases}$$

10. Consider a pair of point vortices located at $(x, y) = (\pm c, 0)$ with strengths 2Γ and $-\Gamma$, as shown in Figure 7.8. Calculate the center of vorticity x_C, radius of gyration D, linear impulse $\mathbf{P}$, angular impulse $\mathbf{L}$, and enstrophy E.

CHAPTER 8

PRESSURE THEOREMS

It is demonstrated in Chapter 7 that when the equations of motion for an incompressible fluid with uniform density are written in the velocity-vorticity formulation, it is not necessary to know the pressure field in order to solve for the fluid flow, subject to appropriate boundary conditions. However, knowledge of pressure is necessary in order to evaluate many of the consequences of fluid flows, for instance for determining fluid forces and moments on immersed bodies or for determining sound generation by the flow. Furthermore, in many important classes of problems involving flows with moving material interfaces, either the boundary conditions themselves or the location of the boundary depends on the pressure field. Water wave propagation is a common example of the former type of problems, whereas fluid interaction with flexible or compliant structures is an important example of the latter type of problem. Of course, when solving for fluid flow in the primitive-variable (velocity-pressure) formulation, the pressure plays an essential role in ensuring satisfaction of the incompressibility constraint. This chapter examines several theorems involving the pressure field which illustrate the role of pressure for incompressible flows and provide different means for pressure solution.

8.1 PRESSURE POISSON EQUATION

The governing equation for pressure is obtained by taking the divergence of the momentum equation (7.1.2), which for an incompressible fluid with uniform density and conservative body force ($\mathbf{b} = \nabla\lambda$) gives

$$\nabla^2\left(\frac{p}{\rho} + \kappa - \lambda\right) = \nabla \cdot (\mathbf{u} \times \boldsymbol{\omega}). \tag{8.1.1}$$

Equation (8.1.1) holds for both inviscid and viscous flows, since the divergence of the viscous term $\mu\nabla^2\mathbf{u}$ vanishes for an incompressible flow. If the velocity and vorticity fields are known, (8.1.1) provides a Poisson equation for the pressure field. The combination within parentheses on the left-hand side of (8.1.1) is sometimes called the Bernoulli coefficient and is denoted by

$$B(\mathbf{x}, t) \equiv \frac{p}{\rho} + \kappa - \lambda. \tag{8.1.2}$$

Equation (8.1.1) indicates that the $\mathbf{u} \times \boldsymbol{\omega}$ field is a source of the function $B(\mathbf{x}, t)$ and that $B(\mathbf{x}, t)$ is a harmonic function (i.e., a solution of the Laplace equation) for any flow in which $\mathbf{u} \times \boldsymbol{\omega}$ is divergence free. One example of such a case is an irrotational flow, where $\boldsymbol{\omega}$ vanishes everywhere, which is examined in Section 8.2.

The pressure equation (8.1.1) is frequently used in numerical computation methods, together with the momentum equation (7.1.1), in place of the incompressibility condition $\nabla \cdot \mathbf{u} = 0$, which is employed in the derivation of (8.1.1) (Ferziger and Perić, 1996). For any pressure and velocity field that satisfy (8.1.1) and the momentum equation (7.1.1), the value of $\nabla \cdot \mathbf{u}$ must be independent of time at any fixed spatial point. If the velocity field is initially divergence free, then it must remain divergence free for any $(p, \mathbf{u})$ satisfying (8.1.2) and (7.1.1). For this reason, the pressure Poisson equation (8.1.1) is often thought of as being equivalent to the continuity equation (3.4.8) in numerical computations.

8.2 BERNOULLI THEOREM

For any irrotational flow of an inviscid fluid with conservative body force and uniform density, the form (7.1.2) of the momentum equation reduces to

$$\nabla\left(p + \rho\kappa - \rho\lambda + \rho\frac{\partial\phi}{\partial t}\right) = 0, \tag{8.2.1}$$

where ϕ is the velocity potential. The terms within parentheses in (8.2.1) are uniform in space, so that

$$p + \rho\kappa - \rho\lambda + \rho\frac{\partial\phi}{\partial t} = F(t), \tag{8.2.2}$$

where $F(t)$ is an unknown function of time that is chosen to be compatible with the boundary and initial conditions. Equation (8.2.2) is typically called the *unsteady Bernoulli equation*, and it provides an explicit expression for the pressure in an incompressible flow whenever the velocity is known. In the special case of a *steady* irrotational flow, (8.2.2) reduces to a requirement that the Bernoulli coefficient is constant in both time and space, or

$$B \equiv \frac{p}{\rho} + \kappa - \lambda = \text{const.} \tag{8.2.3}$$

8.3 BOUNDARY INTEGRAL EQUATION FOR PRESSURE

We consider in this section a body that is completely immersed in a fluid flow, where the body has bounding surface S with outward unit normal $\mathbf{n}$ which is pointing out of the body into the fluid (Figure 8.1). The fluid flow external to the body has a uniform velocity $\mathbf{U}_\infty(t)$ at infinity, and for convenience we assume that there is no body force acting on the fluid. The momentum equation at infinity is

$$\rho \frac{d\mathbf{U}_\infty}{dt} = -\nabla p_\infty \tag{8.3.1}$$

such that p_∞ varies linearly with position along the direction of the uniform flow. A function P is defined by

$$P(\mathbf{x}, t) \equiv \frac{1}{\rho}(p - p_\infty) + \tfrac{1}{2}(\mathbf{u} \cdot \mathbf{u} - \mathbf{U}_\infty \cdot \mathbf{U}_\infty). \tag{8.3.2}$$

Using (8.3.1) and (8.3.2), the momentum equation (7.1.2) at any point $\mathbf{x}$ in the flow can be written as

$$\frac{\partial(\mathbf{u} - \mathbf{U}_\infty)}{\partial t} + \nabla P + \boldsymbol{\omega} \times \mathbf{u} = 0. \tag{8.3.3}$$

Taking the divergence of (8.3.3) yields

$$\nabla^2 P = \nabla \cdot (\mathbf{u} \times \boldsymbol{\omega}), \tag{8.3.4}$$

whereas evaluating (8.3.3) on the surface S and taking the scalar product with the unit normal $\mathbf{n}$ gives

$$\left.\frac{\partial P}{\partial n}\right|_S = \mathbf{n} \cdot \left[-\frac{\partial(\mathbf{u} - \mathbf{U}_\infty)}{\partial t} + \mathbf{u} \times \boldsymbol{\omega} \right]. \tag{8.3.5}$$

We now recall Green's second identity (2.7.8), which written for the scalar function P defined by (8.3.2) and a Green's function $G(\mathbf{x} - \mathbf{x}')$ has the form

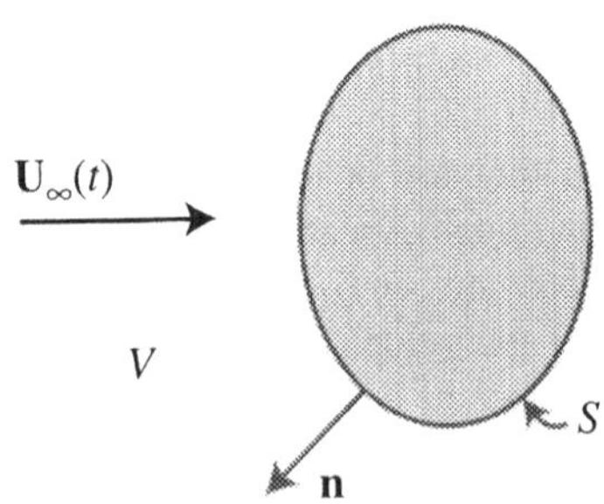

Figure 8.1 Uniform flow $\mathbf{U}_\infty(t)$ past an immersed body with bounding surface S and outward unit normal $\mathbf{n}$. The fluid occupies a region V outside of the body.

$$\int_V (G\nabla'^2 P' - P'\nabla'^2 G)\, dv' = \int_S \left(P'\frac{\partial G}{\partial n'} - G\frac{\partial P'}{\partial n'} \right) da'. \tag{8.3.6}$$

The integration in (8.3.6) is performed over variables evaluated at $\mathbf{x}'$ and the pressure is evaluated at some other location $\mathbf{x}$. The sign of the last term in (8.3.6) is opposite that in (2.7.8) since the unit normal $\mathbf{n}$ points into the region V occupied by the fluid, whereas the normal $\mathbf{n}$ points out of the region V in (2.7.8). For a three-dimensional space, the Green's function of the Poisson equation is $G = -1/(4\pi r)$, where $r = |\mathbf{x} - \mathbf{x}'|$. From the definition (6.3.5) of the Green's function, we can write

$$\int_V P'\nabla^{2'} G\, dv' = \begin{cases} P(\mathbf{x}, t) & \text{for } \mathbf{x} \in V_{\text{int}}, \\ \frac{1}{2}P(\mathbf{x}, t) & \text{for } \mathbf{x} \in S, \\ 0 & \text{for } \mathbf{x} \notin V. \end{cases} \tag{8.3.7}$$

Differentiation of the Green's function yields

$$\frac{\partial G}{\partial n'} = \mathbf{n}' \cdot \nabla' G = -\frac{1}{4\pi}\frac{\mathbf{n}' \cdot \mathbf{r}}{r^3}, \tag{8.3.8}$$

where $\mathbf{r} \equiv \mathbf{x} - \mathbf{x}'$. Substituting (8.3.4)–(8.3.5) and (8.3.7)–(8.3.8) into (8.3.6) and evaluating the Green's identity for a point $\mathbf{x}$ lying on the surface S give

$$2\pi P(\mathbf{x}, t) - \int_S P'\frac{\mathbf{n}' \cdot \mathbf{r}}{r^3}\, da' = \int_S \frac{1}{r}\mathbf{n}' \cdot \left[\frac{\partial(\mathbf{U}_B - \mathbf{U}_\infty)}{\partial t} - \mathbf{u}' \times \boldsymbol{\omega}' \right] da' - \int_V \frac{1}{r}\nabla' \cdot (\mathbf{u}' \times \boldsymbol{\omega}')\, dv', \tag{8.3.9}$$

where $\mathbf{n} \cdot \mathbf{U}_B$ is the prescribed normal velocity at the surface of the body. The surface integrals in (8.3.9) and in the following equations in this section are to be interpreted in the sense of Cauchy principal-value integrals, and as such they do not include the finite contribution of the singularity at $\mathbf{x} = \mathbf{x}'$. Using the chain rule and the divergence theorem, we can write

$$\int_V \nabla'\left(\frac{1}{r}\right) \cdot (\mathbf{u}' \times \boldsymbol{\omega}')\, dv' = -\int_S \frac{1}{r}\mathbf{n}' \cdot (\mathbf{u}' \times \boldsymbol{\omega}')\, da' - \int_V \frac{1}{r}\nabla' \cdot (\mathbf{u}' \times \boldsymbol{\omega}')\, dv', \tag{8.3.10}$$

where the negative sign on the surface integral in (8.3.10) arises from the fact that $\mathbf{n}$ points into the region V. Using the fact that $\nabla'(1/r) = \mathbf{r}/r^3$, the result (8.3.10) can be used to write (8.3.9) as

$$2\pi P(\mathbf{x}, t) - \int_S P'\frac{\mathbf{n}' \cdot \mathbf{r}}{r^3}\, da' = \int_S \frac{1}{r}\mathbf{n}' \cdot \frac{\partial(\mathbf{U}_B - \mathbf{U}_\infty)}{\partial t}\, da' + \int_V \frac{\mathbf{r}}{r^3} \cdot (\mathbf{u}' \times \boldsymbol{\omega}')\, dv'. \tag{8.3.11}$$

The result (8.3.11) has the form of a Fredholm integral equation of the second kind for the quantity P (Tricomi, 1957), from which the pressure can be readily attained. Because this integral equation involves integration of the unknown quantity P only over the boundary S of the domain V, it is one of a class of equations generally known as *boundary integral equations*. Once the value of P is obtained on the boundary S by solution of (8.3.11), its value can readily be found at any point interior to V simply by replacing the 2 in the first term of (8.3.11) by 4 and performing the surface and volume integrals over known quantities. This result can also be generalized to viscous flows, for which the boundary integral equation for P has a form similar to (8.3.11) with the addition of one surface integral on the right-hand side that depends on the viscosity.

The last integral in (8.3.11) indicates that the vector $\mathbf{u} \times \boldsymbol{\omega}$ provides a volume source for variation of the quantity P. For an inviscid fluid, the vorticity in this term includes both the free vorticity interior to the domain V and the bounded vorticity on S due to slip at the body surface. It is helpful to separate these two effects, particularly since within the bounded vortex sheet on S the vorticity has infinite magnitude. The vortex sheet strength $\boldsymbol{\gamma}$, with magnitude $\gamma = |\boldsymbol{\gamma}|$, is defined as the integral of vorticity over the thickness of the bounded vortex sheet on S. As shown in Chapter 11, the magnitude γ of the vortex sheet strength is equal to the magnitude of the slip velocity on the body surface. The volume integral in (8.3.11) can be decomposed in terms of integrals over the free and bounded vorticity, which for the special case of a fixed body has the form

$$\int_V \frac{\mathbf{r}}{r^3} \cdot (\mathbf{u}' \times \boldsymbol{\omega}')\, dv' = \int_{V_{\text{int}}} \frac{\mathbf{r}}{r^3} \cdot (\mathbf{u}' \times \boldsymbol{\omega}')\, dv' - 2\pi \left(\frac{\gamma^2}{2}\right) + \int_S \frac{\gamma'^2}{2} \frac{\mathbf{n}' \cdot \mathbf{r}}{r^3}\, da'. \tag{8.3.12}$$

Here V_{int} denotes the interior of the region V, where the integral over V_{int} in (8.3.12) includes only the free vorticity. Substituting (8.3.12) into (8.3.11) gives a boundary integral equation for quantity $\overline{P} \equiv P + \frac{1}{2}\gamma^2$ at the surface S of a fixed body as

$$2\pi \overline{P}(\mathbf{x}, t) - \int_S \overline{P}' \frac{\mathbf{n}' \cdot \mathbf{r}}{r^3}\, da' = -\int_S \frac{1}{r} \mathbf{n}' \cdot \frac{\partial \mathbf{U}_\infty}{\partial t}\, da' + \int_{V_{\text{int}}} \frac{\mathbf{r}}{r^3} \cdot (\mathbf{u}' \times \boldsymbol{\omega}')\, dv'. \tag{8.3.13}$$

Techniques for numerical solution of (8.3.13) are discussed by Marshall and Grant (1996).

Since the velocity $\mathbf{u}$ in the definition (8.3.2) of P is evaluated underneath the vortex sheet in (8.3.11), for a fixed body we have $\overline{P} = (p - p_\infty)/\rho + (\gamma^2 - \mathbf{U}_\infty \cdot \mathbf{U}_\infty)/2$ for $\mathbf{x} \in S$. By way of comparison, the Bernoulli coefficient B, defined in (8.1.3), for a point just above the vortex sheet at the body surface is simply $B = p/\rho + \gamma^2/2$, so that $\overline{P}$ represents the difference between the Bernoulli coefficient of a point just above the surface S and a point at infinity. For the special case of steady irrotational flow, both of the integrals on the right-hand side of (8.3.13) vanish, so the solution of (8.3.13) is simply $\overline{P} = 0$. This solution corresponds to the result, obtained in Section 8.2, that the Bernoulli coefficient is uniform in space for a steady irrotational flow. For unsteady irrotational flows, the surface integral on the right-hand side of (8.3.13) alters $\overline{P}$ by a harmonic function, since the Laplacian of this term vanishes for any point $\mathbf{x}$ in the interior of V.

8.4 SPECIAL RESULTS FOR STEADY, TWO-DIMENSIONAL FLOWS

For any two-dimensional flow, the relationship (6.5.2) between the vector potential and the vorticity field can be rewritten in terms of the stream function as

$$\nabla^2\psi = -\omega. \tag{8.4.1}$$

It is recalled that for an inviscid two-dimensional flow, the vorticity transport equation reduces simply to $D\omega/Dt = 0$, which requires the value of vorticity to be independent of time at a given fluid particle. In a steady flow, the boundary conditions for the pathlines do not change in time, so the vorticity transport equation implies that vorticity must also be uniform along a pathline. Since streamlines and pathlines are coincident for a steady flow, we conclude that the vorticity is a function only of the stream function, or $\omega = \omega(\psi)$, in any steady, two-dimensional, inviscid flow. Onc can therefore rewrite (8.4.1) as

$$\nabla^2\psi = -\omega(\psi), \tag{8.4.2}$$

which provides a closed, nonlinear equation for the stream function for a specified vorticity field $\omega(\psi)$.

For any steady flow with uniform density and conservative body force, the form (7.1.2) of the momentum equation can be written as

$$\nabla B = \mathbf{u} \times \boldsymbol{\omega}, \tag{8.4.3}$$

where B is the Bernoulli coefficient defined in (8.1.2). From (8.4.3), the vector ∇B is normal to both $\mathbf{u}$ and $\boldsymbol{\omega}$, so that for a steady flow

$$\frac{DB}{Dt} = \mathbf{u} \cdot \nabla B = 0. \tag{8.4.4}$$

Equation (8.4.4) implies that B does not change with time on a fluid particle, and for a steady flow it follows that B is uniform on a pathline. Since pathlines and streamlines are coincident for steady flows, the Bernoulli coefficient must be uniform along a streamline, or $B = B(\psi)$. The gradient of B can be written as

$$\nabla B = \frac{dB}{d\psi}\nabla\psi. \tag{8.4.5}$$

Using the representation $\mathbf{u} = \nabla\psi \times \mathbf{e}_z$ for velocity in a two-dimensional incompressible flow, together with the vector identity (2.6.9), we can write

$$\mathbf{u} \times \boldsymbol{\omega} = -\omega \nabla \psi. \tag{8.4.6}$$

Substituting (8.4.5) and (8.4.6) into (8.4.3) and equating the coefficients of $\nabla\psi$ gives an equation for variation of the Bernoulli coefficient with the stream function in any steady, two-dimensional, inviscid flow as

$$\frac{dB}{d\psi} = -\omega(\psi). \tag{8.4.7}$$

In the special case that ω has a uniform value ω_0, (8.4.7) can be integrated to yield

$$B + \psi\omega_0 = \frac{p}{\rho} + \kappa - \lambda + \psi\omega_0 = \text{const.} \tag{8.4.8}$$

This form of the Bernoulli equation is applicable to problems involving a steady irrotational motion superimposed on some flow having constant vorticity, such as uniform shear flow or rigid rotation.

BIBLIOGRAPHY

Marshall, J. S. and J. R. Grant (1996). "Penetration of a blade into a vortex core: vorticity response and unsteady blade forces," *Journal of Fluid Mechanics* **306**, 83–109.

Ferziger, J.H. and M. Peric (1996). *Computational Methods for Fluid Dynamics*, Springer-Verlag: Berlin.

Tricomi, F.G. (1957). *Integral Equations*, Interscience Publishers: New York (reprinted Dover Publications, New York, 1985).

PROBLEMS

1. Prove that the minimum value of pressure in an irrotational, incompressible flow must occur on the boundary S of the flow domain V.
2. A crude description of flow in a blood vessel (Figure 8.2) is given by a plug-flow model with pulsating velocity field $\mathbf{u} = g(kz - \sigma t)R\mathbf{e}_R + [W + A\sin(kz - \sigma t)]\mathbf{e}_z$.
   (a) Find an expression for the function $g(kz - \sigma t)$ such that the flow is incompressible.

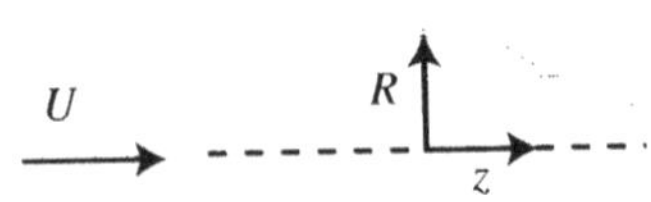

Figure 8.2 Pulsating flow in a blood vessel.

(b) Show that the vorticity vanishes on the symmetry axis $R = 0$. Determine the pressure variation $p(kz - \sigma t)$ on $R = 0$.

3. Consider a three-dimensional flow with no rigid boundaries in which the vorticity field is prescribed and the velocity vanishes at infinity. The velocity field can be determined, in principle, from the Biot-Savart integral (6.5.7). Starting with the Poisson equation $\nabla^2 B = \nabla \cdot (\mathbf{u} \times \boldsymbol{\omega})$, derive a Green's function solution for the Bernoulli coefficient B.

4. A rotating circular tank of radius b is filled with a volume V of fluid. The tank is allowed to rotate for a long time, until the flow is a two-dimensional rigid-body motion in any horizontal cross section of the tank. Derive an equation for the height $h(r)$ of the fluid interfacc.

5. For a two-dimensional, steady flow with uniform vorticity ω_0, show that the difference in the value of the Bernoulli coefficient B between two streamlines ψ_1 and ψ_2 is proportional to the volumetric flow rate Q between the streamlines.

CHAPTER 9

TWO-DIMENSIONAL POTENTIAL FLOWS

All real fluid flows are three dimensional. In many problems, however, it is reasonable to assume that over all or part of the problem domain the flow is approximately two dimensional, such that in a planar slice of the flow (hereinafter designated as the x-y plane) both the velocity component normal to this plane and the normal gradient of velocity everywhere vanish. The assumption of two-dimensional flow not only results in one less spatial dimension, which simplifies tasks such as computational solution of the flow field, but also makes possible a number of analytical solution techniques that do not apply to three-dimensional flows in general. These special solution techniques for two-dimensional flows are presented in the current chapter for irrotational flows and in Chapter 11 for flows with nonzero vorticity. In addition to these special solution techniques for two-dimensional flows, a number of more general three-dimensional solution techniques, such as boundary-integral methods and the slender-body approximation, can be specialized to two-dimensional flows.

9.1 ANALOGY BETWEEN POTENTIAL FLOWS AND ANALYTIC FUNCTIONS OF A COMPLEX VARIABLE

Letting x and y denote the usual position vectors, the complex position z is given by $z \equiv x + iy$, where $i \equiv \sqrt{-1}$. A general complex-valued function $f(x, y)$ can be decomposed into its real and imaginary parts as

$$f(x, y) = g(x, y) + ih(x, y), \tag{9.1.1}$$

where g and h are real valued functions. In general, the function f can be written as a function of the complex position z and its complex conjugate $\bar{z} \equiv x - iy$. The complex function f is said to be *analytic* at a point z_0 if the derivative df/dz exists at z_0 and if the value of this derivative is independent of the direction in which it is calculated. A necessary and sufficient condition for the function f in (9.1.1) to be analytic is that the real and imaginary parts of f satisfy the *Cauchy-Riemann equations*

$$\frac{\partial g}{\partial x} = \frac{\partial h}{\partial y}, \qquad \frac{\partial g}{\partial y} = -\frac{\partial h}{\partial x}. \tag{9.1.2}$$

The proof of (9.1.2) is straightforward and can be found in any standard text on complex variable theory. In general, any differentiable complex function f that can be expressed as a function only of z (and is hence independent of $\bar{z}$) satisfies the above conditions and is analytic.

Turning now to fluid dynamics, we recall that for any two-dimensional, incompressible, potential flow, the irrotationality and incompressibility conditions are

$$\frac{\partial u}{\partial y} - \frac{\partial v}{\partial x} = 0, \qquad \frac{\partial u}{\partial x} + \frac{\partial v}{\partial y} = 0. \tag{9.1.3}$$

The velocity can be written in terms of a scalar potential to satisfy the irrotationality condition identically, and alternatively the velocity can be written in terms of a stream function to satisfy the incompressibility condition identically. For a flow that satisfies both equations of (9.1.3), it follows that

$$u = \frac{\partial \phi}{\partial x} = \frac{\partial \psi}{\partial y}, \qquad v = \frac{\partial \phi}{\partial y} = -\frac{\partial \psi}{\partial x}. \tag{9.1.4}$$

We now consider the two complex functions defined by

$$F \equiv \phi(x, y) + i\psi(x, y), \qquad W \equiv u(x, y) - iv(x, y), \tag{9.1.5}$$

which are called the *complex potential* and the *complex velocity*, respectively. From the results (9.1.3) and (9.1.4), it is clear that both F and W satisfy the Cauchy-Riemann equations (9.1.2). It follows that F and W are analytic and can therefore be written as functions of the complex position z only, or $F = F(z)$ and $W = W(z)$. Furthermore, from the definitions (9.1.5) it immediately follows that F and W are related by

$$W = \frac{dF}{dz}. \tag{9.1.6}$$

The square of the magnitude of a complex variable f is the product of f and its complex conjugate $\bar{f}$. From its definition, we see that the square of the magnitude of

W is simply twice the kinetic energy per unit mass, or

$$\kappa = \tfrac{1}{2} W \bar{W}. \tag{9.1.7}$$

In polar coordinates, the complex position is given by

$$z = re^{i\theta} = r\cos\theta + ir\sin\theta. \tag{9.1.8}$$

The complex velocity can be written in terms of the radial and azimuthal velocity components, u_r and u_θ, as

$$W = (u_r - iu_\theta)e^{-i\theta}. \tag{9.1.9}$$

The equation (9.1.4) for stream function and potential function becomes

$$u_r = \frac{\partial\phi}{\partial r} = \frac{1}{r}\frac{\partial\psi}{\partial\theta}, \qquad u_\theta = \frac{1}{r}\frac{\partial\phi}{\partial\theta} = -\frac{\partial\psi}{\partial r}. \tag{9.1.10}$$

It is established in this section that any analytic function of a complex variable can be associated with a two-dimensional potential flow, and vice versa, by the definitions of F and W given in (9.1.5). One approach to solving fluid flow problems is to select some analytic function for F and then to compute the corresponding velocity components using (9.1.4). Of course, this inverse approach requires a bit of luck (as well as some experience) to guess the correct analytic function corresponding to any given physical flow problem, and the approach is therefore usually limited to fairly simple problems. A much more versatile use of the analogy between analytic functions and two-dimensional potential flow solutions is to employ the powerful theorems regarding transformation of analytic functions in order to transform a difficult fluid flow problem into an easier problem. These "conformal transformation" techniques are discussed later in this chapter, after first introducing some simple fluid flow solutions.

9.2 SOME BASIC POTENTIAL FLOWS

The Helmholtz representation (6.4.1) for the velocity field u allows us to write $\mathbf{u} = \mathbf{v} + \nabla\alpha + \nabla\times\boldsymbol{\beta}$, where $\mathbf{v}$ is a harmonic function and α and $\boldsymbol{\beta}$ are induced from the rate of dilatation and vorticity fields, respectively. The Green's function solution for α and $\boldsymbol{\beta} = \beta\mathbf{e}_z$ in a two-dimensional flow are obtained from (6.5.3) and (6.5.4) as

$$\alpha(\mathbf{x}, t) = \frac{1}{2\pi}\int_A \ln\left|\mathbf{x} - \mathbf{x}'\right| \Delta(\mathbf{x}', t)\, da', \tag{9.2.1}$$

$$\beta(\mathbf{x}, t) = -\frac{1}{2\pi}\int_A \ln\left|\mathbf{x} - \mathbf{x}'\right| \omega(\mathbf{x}', t)\, da'. \tag{9.2.2}$$

In the following, the above results are used to derive a number of elementary solutions for fluid flows in an unbounded domain.

9.2.1 Uniform Flow

Some of the simplest fluid flows correspond to velocity fields in which both the vorticity and dilatation everywhere vanish, such that $\alpha = \beta = 0$ in (6.4.1). We refer to this case as a harmonic flow, since the velocity vector everywhere satisfies Laplace's equation. Harmonic flows are completely determined from the boundary conditions, such that if the velocity vanishes at the boundaries of the fluid domain it must vanish everywhere. One simple harmonic flow is *uniform flow*, for which the velocity vector is equal to a constant **U**. The velocity magnitude is denoted by U and the angle that the velocity vector makes with the positive x-axis is denoted by a, such that the velocity components can be written as

$$u = U\cos(a), \qquad v = U\sin(a). \tag{9.2.3}$$

Substituting (9.2.3) into the definition (9.1.5) of the complex velocity and integrating (9.1.6) over z give the complex potential for a uniform flow as

$$F(z) = Ue^{-ia}z. \tag{9.2.4}$$

The stream function ψ and the scalar potential for ϕ uniform flow are obtained from (9.2.4) and the definition (9.1.5) of the complex potential as

$$\phi = U(x\cos a + y\sin a), \qquad \psi = U(y\cos a - x\sin a). \tag{9.2.5}$$

The streamlines and equipotential lines, which are shown in Section 6.1 to be orthogonal to each other, are sketched in Figure 9.1*a* for a uniform flow.

9.2.2 Plane Straining Flow

Another common example of harmonic flow in an unbounded space is plane straining flow, for which the velocity components vary linearly with distance as

$$u = cx, \qquad v = -cy, \tag{9.2.6}$$

where the parameter c is called the *straining rate*. The complex potential for this flow is obtained from (9.1.5) and (9.1.6) as

$$F(z) = \tfrac{1}{2}cz^2, \tag{9.2.7}$$

and the potential function and stream function are

$$\phi = \tfrac{1}{2}c(x^2 - y^2), \qquad \psi = cxy. \tag{9.2.8}$$

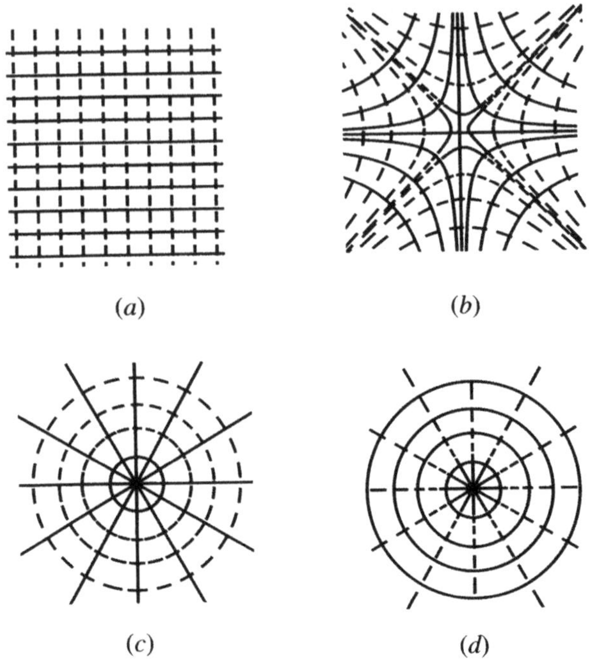

Figure 9.1 Streamlines (solid) and equipotential lines (dashed) for (*a*) uniform flow, (*b*) plane straining flow, (*c*) a point source, and (*d*) a point vortex.

The streamlines and equipotential lines for plane straining flow are shown in Figure 9.1*b*.

9.2.3 Source/Sink

In other two-dimensional flows, the vorticity or dilatation rate vanishes everywhere except on certain points or lines within the flow field, at which the dilatation rate or vorticity is singular. We also refer to such flows as incompressible potential flows, since the support of Δ and ω (i.e., the region on which these fields are nonzero) has zero area. The flow field is obtained by substituting the (singular) distribution of Δ or ω into (9.2.1) and (9.2.2) and integrating. One example of such a flow is a source or sink, which is a two-dimensional point singularity in dilatation rate. For a source at (x_0, y_0), the dilatation rate is given by

$$\Delta(x - x_0, y - y_0) = m\delta(x - x_0, y - y_0), \tag{9.2.9}$$

where m is called the *source strength* and is equal to the volumetric discharge from the source per unit length into the plane of the flow. Equation (9.2.1) yields

$$\alpha = \frac{m}{2\pi} \ln\left[(x - x_0)^2 + (y - y_0)^2\right]^{1/2}. \tag{9.2.10}$$

Since $\beta = 0$ for a flow with no vorticity, and setting the harmonic function $\mathbf{v}$ equal to zero, the velocity components are

$$u = \frac{m(x - x_0)}{2\pi[(x - x_0)^2 + (y - y_0)^2]}, \qquad v = \frac{m(y - y_0)}{2\pi[(x - x_0)^2 + (y - y_0)^2]}. \tag{9.2.11}$$

Substituting (9.2.11) into (9.1.5) and integrating over z gives the complex potential for a source as

$$F(z) = \frac{m}{2\pi} \ln(z - z_0). \tag{9.2.12}$$

The potential function and stream function for a source are obtained from (9.2.11) and (9.1.4) as

$$\phi = \frac{m}{2\pi} \ln\left[(x - x_0)^2 + (y - y_0)^2\right]^{1/2}, \qquad \psi = \frac{m}{2\pi} \tan^{-1}\left(\frac{y - y_0}{x - x_0}\right). \tag{9.2.13}$$

The above results take on a simpler form when expressed in polar coordinates, with the origin of the coordinate system coinciding with the source location. The velocity components become purely radial,

$$u_r = \frac{m}{2\pi r}, \qquad u_\theta = 0, \tag{9.2.14}$$

and the potential function and stream function reduce to

$$\phi = \frac{m}{2\pi} \ln r, \qquad \psi = \frac{m}{2\pi}\theta. \tag{9.2.15}$$

The streamlines and equipotential lines for a source, shown in Figure 9.1c, are radial lines emanating from the center and concentric circles about the center, respectively.

9.2.4 Vortex

A two-dimensional point vortex is a point singularity of the vorticity field of the form

$$\omega(x - x_0, y - y_0) = \Gamma\delta(x - x_0, y - y_0), \tag{9.2.16}$$

where Γ is the vortex strength. It follows from (3.5.10) that Γ also corresponds to the circulation about any circuit surrounding the point (x_0, y_0). From (9.2.2), the solution for β becomes

$$\beta = -\frac{\Gamma}{2\pi} \ln\left[(x - x_0)^2 + (y - y_0)^2\right]^{1/2}. \tag{9.2.17}$$

The corresponding velocity components are

$$u = -\frac{\Gamma(y - y_0)}{2\pi[(x - x_0)^2 + (y - y_0)^2]}, \qquad v = \frac{\Gamma(x - x_0)}{2\pi[(x - x_0)^2 + (y - y_0)^2]}, \tag{9.2.18}$$

and the complex potential is

$$F(z) = -\frac{i\Gamma}{2\pi} \ln(z - z_0). \tag{9.2.19}$$

The potential function and stream function for a vortex are obtained from (9.2.18) and (9.1.4) as

$$\phi = \frac{\Gamma}{2\pi} \tan^{-1}\left(\frac{y - y_0}{x - x_0}\right), \qquad \psi = -\frac{\Gamma}{2\pi} \ln\left[(x - x_0)^2 + (y - y_0)^2\right]^{1/2}. \tag{9.2.20}$$

In polar coordinates, the velocity components become

$$u_r = 0, \qquad u_\theta = \frac{\Gamma}{2\pi r}, \tag{9.2.21}$$

and the potential function and stream function are

$$\phi = \frac{\Gamma\theta}{2\pi}, \qquad \psi = -\frac{\Gamma}{2\pi} \ln r. \tag{9.2.22}$$

The streamlines and equipotential lines for a vortex, shown in Figure 9.1*d*, are the reverse of those for a source.

9.3 SUPERPOSITION OF BASIC FLOWS

The sum of any two analytic functions is another analytic function. Correspondingly, the sum of any two potential flow solutions is another potential flow solution. This principle of *superposition* can be applied in general to combine solutions of any linear differential equation and follows directly from the basic property $L[a + b] = L[a] + L[b]$ for a linear differential or integral operator $L[\cdot]$ acting on two functions a and b. The superposition principle is frequently used to combine basic potential flow solutions to obtain slightly more complex solutions, as demonstrated in the examples given in this section.

Example 9.3.1 Doublet. A doublet is created at a point z_0 by adding a source of strength m at $z = z_0 - \varepsilon$ to a sink of strength $-m$ at $z = z_0 + \varepsilon$ and then taking the limit $\varepsilon \to 0$ and $m \to \infty$ as $m\varepsilon/\pi \to \mu$, where μ is called the doublet strength. This process is most easily performed using the complex potential (9.2.12) for the source/sink to write

$$F(z) = \frac{m}{2\pi} \ln(z - z_0 + \varepsilon) - \frac{m}{2\pi} \ln(z - z_0 - \varepsilon). \tag{9.3.1}$$

For small ε

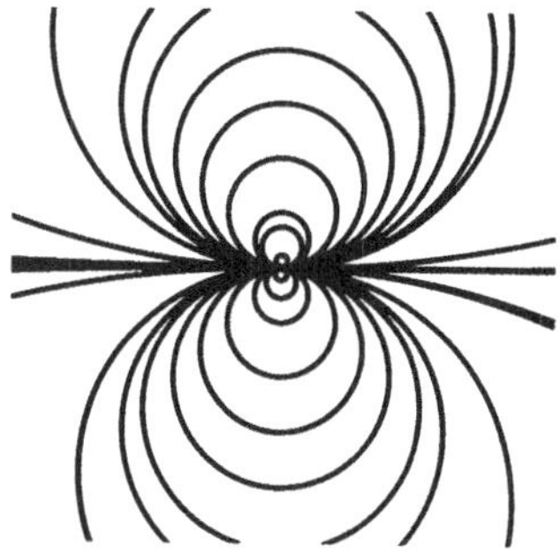

Figure 9.2 Streamlines for a doublet.

$$\ln(z - z_0 + \varepsilon) - \ln(z - z_0 - \varepsilon) = \frac{2\varepsilon}{z - z_0} + O(\varepsilon^2), \tag{9.3.2}$$

so that taking the limit described above yields the complex potential for a doublet as

$$F(z) = \frac{\mu}{z - z_0}. \tag{9.3.3}$$

The potential function and stream function for a doublet are

$$\phi = \mu \frac{x - x_0}{(x - x_0)^2 + (y - y_0)^2}, \qquad \psi = -\mu \frac{y - y_0}{(x - x_0)^2 + (y - y_0)^2}. \tag{9.3.4}$$

A plot of the streamlines for a doublet is given in Figure 9.2. Because the sink and the source composing the doublet are of equal and opposite strength, there is no net mass flow from the singularity at the doublet center. All of the fluid that is emitted from the source is eventually entrained back into the sink.

The complex velocity is obtained by differentiating (9.3.3) as

$$W(z) = -\frac{\mu}{(z - z_0)^2}. \tag{9.3.5}$$

The velocity components for the doublet are expressed in polar coordinates, using (9.1.9) and (9.3.5), as

$$u_r = -\frac{\mu}{r^2} \cos\theta, \qquad u_\theta = -\frac{\mu}{r^2} \sin\theta. \tag{9.3.6}$$

Whereas the velocity field generated by a source or a vortex decays in proportion to $1/r$ as $r \to \infty$, that due to a doublet decays in proportion to $1/r^2$.

Example 9.3.2 Vortex Pair. A vortex pair is formed by superposing two vortices of equal and opposite strength. For instance, consider a vortex with strength Γ at $z = -\varepsilon_i$ and a vortex with the opposite strength $-\Gamma$ at $z = \varepsilon_i$. The complex potential for this flow is

$$F(z) = -\frac{i\Gamma}{2\pi}\ln(z + i\varepsilon) + \frac{i\Gamma}{2\pi}\ln(z - i\varepsilon). \tag{9.3.7}$$

The stream function for the vortex pair flow is obtained from (9.2.20) as

$$\psi = -\frac{\Gamma}{4\pi}\ln\left|\frac{x^2 + (y+\varepsilon)^2}{x^2 + (y-\varepsilon)^2}\right|. \tag{9.3.8}$$

The streamlines of the vortex pair in a frame traveling with the vortices have the form of coaxial closed circuits with the centers of the two vortices as limiting points (Figure 9.3).

Each vortex induces a motion of the other vortex with speed

$$V = \frac{\Gamma}{2\pi b}, \tag{9.3.9}$$

where $b = 2\varepsilon$ is the separation distance between the vortex centers. This velocity corresponds simply to the velocity induced by one vortex at the center of the other vortex if the other vortex were not present. It is clear from symmetry that a single two-dimensional point vortex can induce no velocity on itself.

If the two vortices are now allowed to approach each other ($\varepsilon \to 0$), (9.3.7) becomes

$$F(z) = -\frac{i\Gamma}{2\pi}[\ln(z + i\varepsilon) - \ln(z - i\varepsilon)] \sim \frac{\Gamma\varepsilon}{\pi z}. \tag{9.3.10}$$

Letting the vortex strengths increase as the separation distance decreases, such that $\Gamma\varepsilon/\pi \to \mu$, (9.3.10) becomes

$$F(z) = \frac{\mu}{z}, \tag{9.3.11}$$

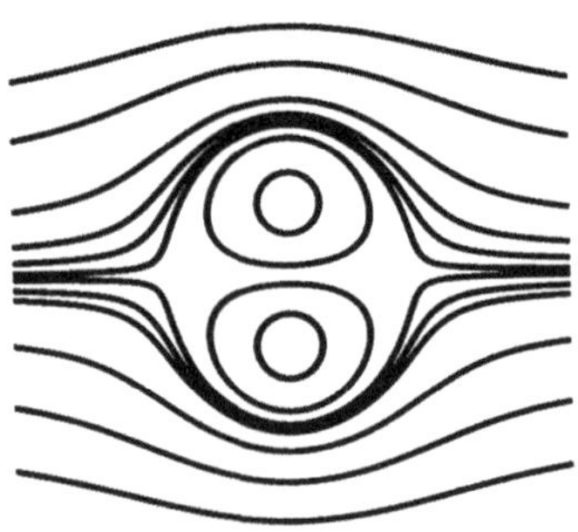

Figure 9.3 Streamlines for a vortex pair in a frame traveling with the vortices.

which is simply the complex potential for a doublet centered at $z = 0$. We thus find that a doublet may be formed either as a limit of an approaching source and sink or as a limit of two approaching vortices of opposite sign.

Example 9.3.3 Uniform Flow Past a Circle. Uniform flow past a circle may be formed by superposing a doublet with a uniform stream, such that the complex potential becomes

$$F(z) = Uz + \frac{\mu}{z}. \tag{9.3.12}$$

The complex velocity is

$$W(z) = U - \frac{\mu}{z^2}, \tag{9.3.13}$$

and the velocity components in polar coordinates are

$$u_r = \left(U - \frac{\mu}{r^2}\right)\cos\theta, \qquad u_\theta = -\left(U + \frac{\mu}{r^2}\right)\sin\theta. \tag{9.3.14}$$

Since the radial velocity component vanishes at the surface $r = a$ of the circle, the doublet strength is set as

$$\mu = a^2 U, \tag{9.3.15}$$

such that (9.3.12) becomes

$$F(z) = U\left(z + \frac{a^2}{z}\right). \tag{9.3.16}$$

Streamlines for both uniform flow past a circle and for uniform flow past a doublet of strength set according to (9.3.15) are shown in Figure 9.4. The flows outside of the boundary $r = a$ of the circle are identical in the two cases. Of course, in the case of flow past a doublet there is also a flow inside the curve $r = a$; however, this part of

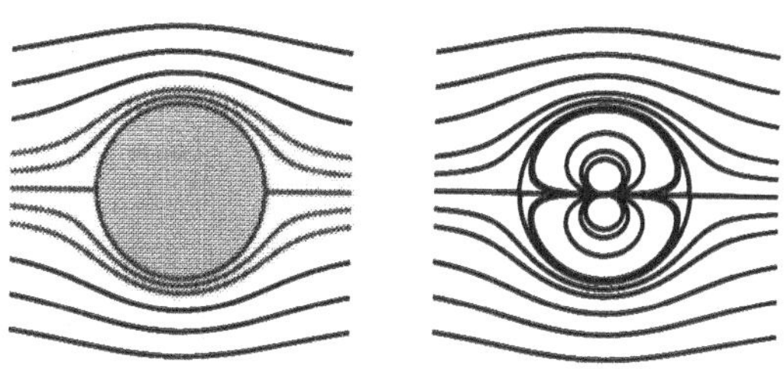

(*a*) (*b*)

Figure 9.4 Streamlines for uniform flow past (*a*) a circle and (*b*) a doublet.

the flow is of no consequence. The basic idea of superposition methods for solution of flow past a fixed body is to replace the actual body with some distribution of Δ or ω either inside or on the surface of the body, such that the surface of the body coincides with a streamline of the superposed flow. For a moving body, we must instead appeal to the more general condition that the normal velocity to the body surface in the superposed flow be the same as the prescribed normal velocity of the body.

A similar method can be used to generate flow past a variety of oval shapes using a superposition of uniform flow and either a source/sink pair or a vortex pair with some finite separation distance. For instance, for the case of a source/sink pair at $z = \pm\varepsilon$ the stream function becomes

$$\psi = Uy + \frac{m}{2\pi}\tan^{-1}\left(\frac{y}{x+\varepsilon}\right) - \frac{m}{2\pi}\tan^{-1}\left(\frac{y}{x-\varepsilon}\right), \qquad (9.3.17)$$

and for a pair of vortices propagating to the left with centers at $z = \pm\varepsilon i$ the stream function becomes

$$\psi = Uy + \frac{\Gamma}{4\pi}\ln\left|\frac{x^2+(y+\varepsilon)^2}{x^2+(y-\varepsilon)^2}\right|. \qquad (9.3.18)$$

Streamlines for these two cases are shown in Figures 9.5 and 9.6 for different values of the parameters $m/\pi U\varepsilon$ and $\Gamma/\pi U\varepsilon$. The source pair is observed to generate a closed body for all values of $m/\pi U\varepsilon$, with the body becoming thinner and shorter as $m/\pi U\varepsilon$ decreases. The vortex pair generates a closed body only for values of $\Gamma/\pi U\varepsilon$ greater than unity, where this critical value corresponds to a condition at which the separating streamline passes through the stagnation point at the midpoint between the two vortex centers (Figure 9.6*b*). For $\Gamma/\pi U\varepsilon$ less than unity, a separate region of closed streamlines forms surrounding each vortex (Figure 9.6*c*).

Example 9.3.4 Swirling Sink Flow. Another example of the superposition method is given for the case of a sink that is swirling about its center. Superposing a vortex of strength Γ with a sink of strength m, both located at the origin, yields the complex

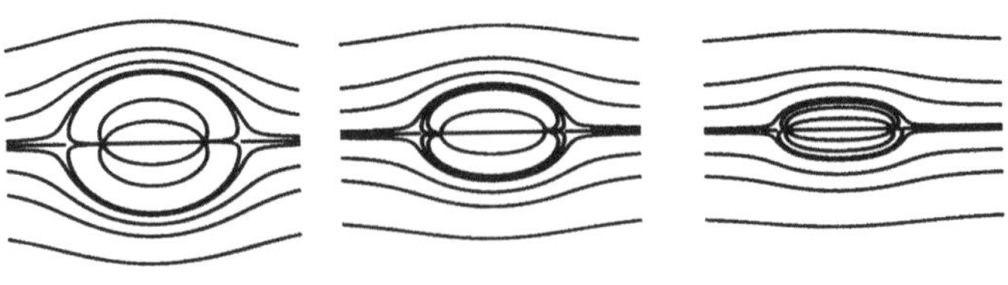

(*a*) (*b*) (*c*)

Figure 9.5 Streamlines for a source pair for values of the parameter $m/\pi U\varepsilon$ of (*a*) 0.5, (*b*) 1.0, and (*c*) 2.0.

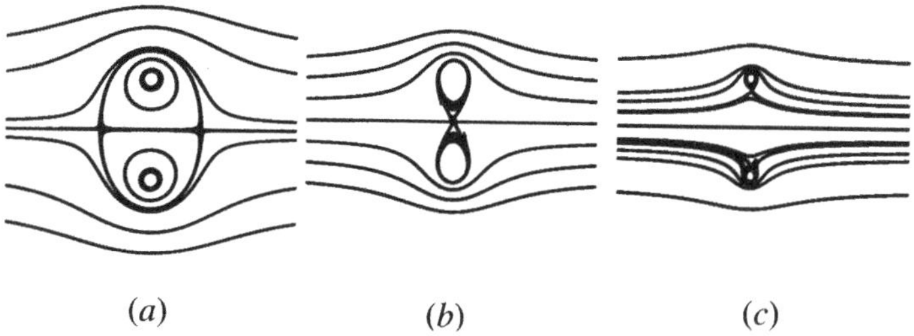

Figure 9.6 Streamlines for a vortex pair for values of the parameter $\Gamma/\pi U\varepsilon$ of (a) 2.0, (b) 1.0 and (c) 0.5.

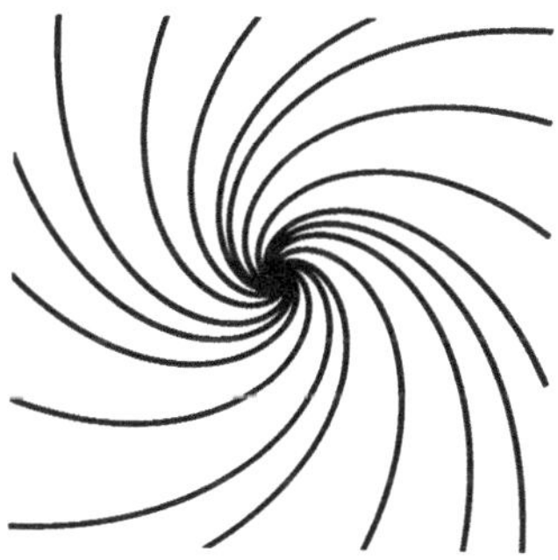

Figure 9.7 Streamlines for swirling sink flow with $\Gamma/m = -1$.

potential

$$F(z) = \frac{m - i\Gamma}{2\pi} \ln z. \tag{9.3.19}$$

The stream function for this flow is given in polar coordinates by

$$\psi = \frac{1}{2\pi}(m\theta - \Gamma \ln r), \tag{9.3.20}$$

and the streamlines are plotted in Figure 9.7. The inward spiraling shape of the streamlines is qualitatively similar to particle paths that are observed in a bathtub drain.

9.4 CIRCLE THEOREM

It is shown in Section 9.3 that the superposition method can be used to generate the velocity field for uniform flow past a circle as well as a number of other closed body shapes. However, for the case of nonuniform background flow past a circle, it is not at all clear how to proceed using the superposition method. A theorem is introduced in the present section that allows a circle to be inserted into an arbitrary irrotational background flow.

Theorem 9.4.1 (Circle Theorem). Let $f(z)$ be the complex potential of a flow with no rigid boundaries and with no singularities within a distance a of the origin. A circular cylinder can be introduced into the flow field, centered at the origin, by forming the complex potential

$$F(z) = f(z) + \bar{f}\left(\frac{a^2}{z}\right). \tag{9.4.1}$$

In the above theorem, the overbar denotes the complex conjugate. The proof of the theorem is straightforward. Since $|z|^2 = z\bar{z} = a^2$ on the perimeter of the circle, the complex potential (9.4.1) can be rewritten as

$$F(z) = f(z) + \bar{f}(\bar{z}). \tag{9.4.2}$$

It follows from (9.4.2) that $F(z)$ is real valued on $|z| = a$ and hence has zero stream function (i.e., $|z| = a$ is a streamline).

Example 9.4.1 Uniform Flow Past a Circle. The complex potential for uniform flow is $f(z) = Uz$. Since $f(z)$ does not explicitly depend upon the imaginary number i, the complex conjugate of $f(z)$ is the same as the original function. From (9.4.1), the complex potential for uniform flow past a circle is

$$F(z) = U\left(z + \frac{a^2}{z}\right), \tag{9.4.3}$$

which is the same as the result (9.3.16) obtained using the superposition method.

Example 9.4.2 Circle Immersed in a Straining Flow. The complex potential for straining flow is given in (9.2.7) as $f(z) = cz^2/2$, where c is the straining rate. Substituting this expression for $f(z)$ into (9.4.1) gives the complex potential for a circle immersed in a straining flow as

$$F(z) = \frac{c}{2}\left(z^2 + \frac{a^4}{z^2}\right). \tag{9.4.4}$$

Unlike the uniform-flow case, this result might not have been particularly easy to guess using the superposition method.

In general, the circle theorem is restricted to background flows that contain no rigid boundaries, since the location of the boundary would be changed by the introduction of a circle. However, it was noted by Levin (1954) that the circle theorem may be used together with rigid boundaries for a certain special case, as expressed in the following theorem.

Theorem 9.4.2. Let a curve C_1 be mapped onto another curve C_2 by the mapping $\zeta = a^2/z$ such that if z lies on C_1, then ζ lies on C_2 and vice versa. If both C_1 and C_2 are streamlines of the background flow $f(z)$, then they are also streamlines of the flow $F(z)$ defined by (9.4.1).

Example 9.4.3 Uniform Flow Past a Compound Cylinder. The circle C_1 defined by $|z - b|^2 = b^2 - a^2$ is invariant under the transformation $\zeta = a^2/z$. From Theorem 9.4.2, in any background flow in which this curve is a streamline, the curve remains a streamline after introducing a circle of radius a by the transformation (9.4.1). For instance, the complex potential for uniform flow past the circle C_1 at an orientation angle α with respect to the x-axis is

$$f(z) = U\left[(z - b)e^{-i\alpha} + \frac{b^2 - a^2}{z - b}e^{i\alpha}\right]. \tag{9.4.5}$$

If a circle of radius a is added at the origin using (9.4.1), the complex potential becomes

$$F(z) = U\left[(z - b)e^{-i\alpha} + \frac{b^2 - a^2}{z - b}e^{i\alpha} + \left(\frac{a^2}{z} - b\right)e^{i\alpha} + \frac{b^2 - a^2}{(a^2/z) - b}e^{-i\alpha}\right]. \tag{9.4.6}$$

When $b > a$, the result (9.4.6) gives the complex potential for uniform flow past a compound cylinder bounded by the curve $|z|^2 = a^2$ for $x < 0$ and the curve $|z - b|^2 = b^2 - a^2$ for $x > 0$. Streamlines for this flow with $b/a = 2$ are plotted in Figure 9.8.

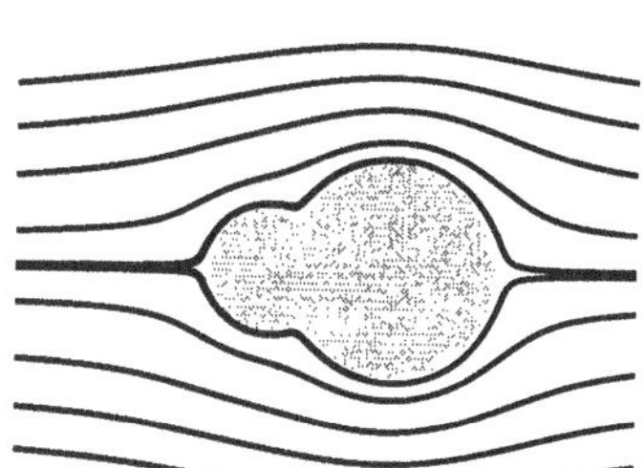

Figure 9.8 Streamlines for uniform flow past a compound cylinder with $b/a = 2$.

9.5 METHOD OF IMAGES

In solving fluid flow problems involving some distribution of dilatation rate or vorticity in the vicinity of an immersed body, it is always possible to replace the body by some distribution of dilatation rate or vorticity inside or on the surface of the body, which is called the *image set* (or simply the *image*) of the external dilatation rate or vorticity distribution. The image set is defined such as to satisfy the no-penetration condition on the body surface. As it has been defined here, the image set is not unique, and in fact for any problem there exists an infinite variety of admissible image sets. A solution of a flow using the image set in place of the actual body is said to employ the *method of images*. However, except for very simple problems, the selection of the image set is not at all obvious and is usually just as difficult as solution of the original problem. The method of images in the general form is therefore more of a conceptual aid than a solution method. However, if additional restrictions are placed on the location of the images in order to make the image set unique, then the method of images can be a very powerful basis for numerical solution methods. For instance, in the various boundary integral methods presented in Chapter 12, the restriction that the images lie on the surface of the body allows us to form an integral equation for the image strength. In the current section, a few examples of the method of images are given for problems involving vorticity and dilatation rate near surfaces of simple shapes, including a plane and a circular cylinder.

Example 9.5.1 Vortex and Source Near a Flat Plate. The problem of a source of strength m located a distance b above a plane wall is illustrated in Figure 9.9a. By symmetry, it is clear that an admissible image set for this problem would consist of a source of the same strength m located at a distance b below the plane $y = 0$ such that the actual problem of flow with a single source in the upper half x-y plane (Figure 9.9a) is replaced by a different problem involving flow in the entire x-y plane with two sources (Figure 9.9b). The complex potential for the problem in Figure 9.9b

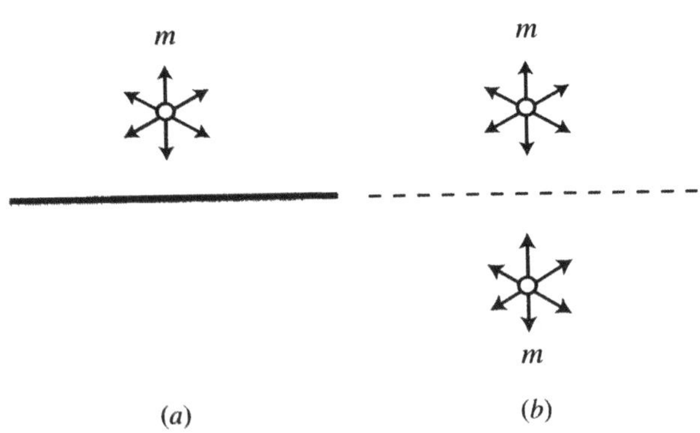

Figure 9.9 Source above a flat wall, showing (a) the actual problem and (b) the image problem.

is obtained using superposition as

$$F(z) = \frac{m}{2\pi}\ln(z - ib) + \frac{m}{2\pi}\ln(z + ib). \tag{9.5.1}$$

The complex velocity is obtained by differentiating (9.5.1) as

$$W(z) = \frac{m}{2\pi}\left(\frac{1}{z - ib} + \frac{1}{z + ib}\right) \tag{9.5.2}$$

such that along the center line $y = 0$, $W(z)$ becomes

$$W(x, 0) = \frac{m}{\pi}\frac{x}{x^2 + b^2}. \tag{9.5.3}$$

The complex velocity W is real valued on $y = 0$, which is consistent with the no-penetration condition at the wall.

For the problem of a vortex of strength Γ located a distance b above a plane wall (Figure 9.10*a*), symmetry again indicates that an appropriate image set is another vortex with *opposite* strength $-\Gamma$ located at a distance b below the plane $y = 0$ (Figure 9.10*b*). The resulting flow is simply that of a vortex pair, as discussed in Section 9.3.2, having the complex potential

$$F(z) = -\frac{i\Gamma}{2\pi}\ln(z - ib) + \frac{i\Gamma}{2\pi}\ln(z + ib). \tag{9.5.4}$$

The vortex propagates along the wall, due to the velocity induced by its image, at a rate

$$V = \frac{\Gamma}{4\pi b}. \tag{9.5.5}$$

Example 9.5.2 Vortex and Source Near a Circular Cylinder. The problem of a source of strength m positioned at a distance b from the surface of a circle of radius

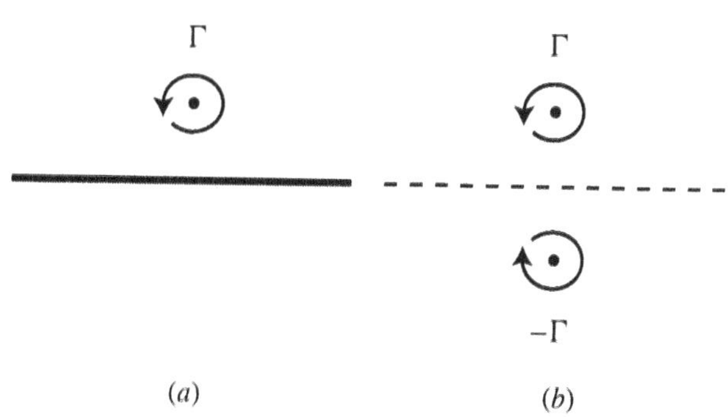

Figure 9.10 Vortex above a flat wall, showing (*a*) the actual problem and (*b*) the image problem.

a can be readily solved using the circle theorem. The complex potential for a single source at $z = -(a + b)$ is

$$f(z) = \frac{m}{2\pi} \ln[z + (a + b)], \tag{9.5.6}$$

so the complex potential for the source and the circle together is obtained using (9.4.1) as

$$F(z) = \frac{m}{2\pi} \ln[z + (a + b)] + \frac{m}{2\pi} \ln\left[\frac{a^2}{z} + (a + b)\right]. \tag{9.5.7}$$

Equation (9.5.7) can be rewritten as

$$F(z) = \frac{m}{2\pi} \ln[z + (a + b)] + \frac{m}{2\pi} \ln\left[z + \frac{a^2}{a + b}\right] - \frac{m}{2\pi} \ln(z) + \frac{m}{2\pi} \ln(a + b). \tag{9.5.8}$$

The first term in (9.5.8) is the actual source at $z = -(a + b)$. The second and third terms, which constitute an image set of the source with respect to the circle, consist of a source located inside the circle at $z = -a^2/(a + b)$ and a sink at the center of the circle. The image source and sink are of equal and opposite strength, respectively, to the original source, such that the net mass outflow from the image set vanishes. The equivalent problem with the circle replaced by the image set is illustrated in Figure 9.11*a*, and the streamlines for this flow are plotted in Figure 9.11*b*. The last term in (9.5.8) is independent of z and is therefore of no dynamical significance.

Repeating this problem with a vortex of strength Γ in place of the source, the complex potential can be written with use of the circle theorem as

$$F(z) = -\frac{i\Gamma}{2\pi} \ln[z + (a + b)] + \frac{i\Gamma}{2\pi} \ln\left[\frac{a^2}{z} + (a + b)\right]. \tag{9.5.9}$$

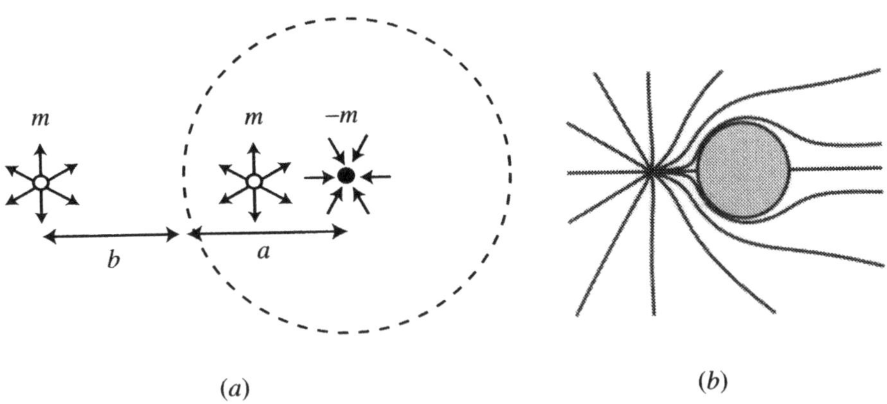

Figure 9.11 Schematic showing (*a*) the image problem and (*b*) streamlines for a source of strength m located at a distance b from the surface of a circle of radius a.

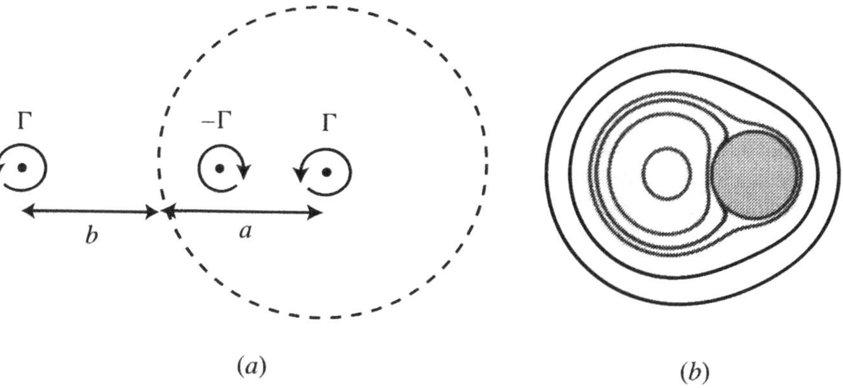

Figure 9.12 Schematic showing (a) the image problem and (b) streamlines for a vortex of strength Γ located at a distance b from the surface of a circle of radius a.

After rearranging, (9.5.9) becomes

$$F(z) = -\frac{i\Gamma}{2\pi}\ln[z + (a+b)] + \frac{i\Gamma}{2\pi}\ln\left[z + \frac{a^2}{a+b}\right] - \frac{i\Gamma}{2\pi}\ln(z) + \frac{i\Gamma}{2\pi}\ln(a+b). \tag{9.5.10}$$

The first term of (9.5.10) is the original vortex of strength Γ at $z = -(a+b)$. The second and third terms form the image set of the vortex in the circle, which consists of a counterrotating vortex of strength $-\Gamma$ at $z = -a^2/(a+b)$ and a corotating vortex of strength Γ at the center of the circle such that the net circulation about the circle vanishes. The equivalent problem with the circle replaced by the image set is illustrated in Figure 9.12a, and the streamlines are plotted in Figure 9.12b. The last term in (9.5.10) does not contribute to the velocity field since it is independent of z.

One interesting aspect of this problem is that the image set of the vortex induces a propagation velocity on the vortex in the azimuthal direction with magnitude $V = \Gamma a^2/2\pi b(2a+b)(a+b)$ in the clockwise sense. The vortex therefore travels in a circular path in a clockwise direction around the circle with period $P = 4\pi^2 b(2a+b)(a+b)^2/\Gamma a^2$.

9.6 CONFORMAL TRANSFORMATION OF FLUID FLOWS

A transformation of a complex variable entails a mapping from one complex plane to another of the form $\zeta = f(z, \bar{z})$. In a *conformal transformation*, the mapping function f is analytic such that the transformation can be expressed as $\zeta = f(z)$, where the function $f(z)$ is differentiable. Conformal transformations have the property that the angle between any two small adjoining line segments is preserved under the transformation, except at certain points called *critical points*. A critical point is a point at which the derivative of the mapping function vanishes, or $df/dz = 0$. Any smooth curve can be constructed from an infinite number of small connected line segments, where the angle between any two adjoining line segments approaches π as the number of line segments approaches infinity. The angle conservation property of conformal transformations thus implies that any smooth curve remains smooth under a conformal transformation unless the curve passes through a critical point of the transformation. The angle conservation property of conformal transformations is illustrated in the following example.

Example 9.6.1 Elliptic Coordinates. A transformation from one orthogonal coordinate system to another orthogonal system in two dimensions can be represented by a conformal transformation. For instance, the transformation

$$z = f(\zeta) = c\cosh(\zeta) \tag{9.6.1}$$

can be associated with a transformation from Cartesian coordinates to elliptic coordinates. In particular, if $\zeta = \xi + i\eta$ and $z = x + iy$, then separating (9.6.1) into its real and imaginary parts gives

$$\begin{aligned} x + iy &= c\cosh(\xi + i\eta) \\ &= c\cosh\xi\cos\eta + ic\sinh\xi\sin\eta \end{aligned} \tag{9.6.2}$$

so that

$$x = c\cosh\xi\cos\eta, \qquad y = c\sinh\xi\sin\eta. \tag{9.6.3}$$

The transformation (9.6.1) maps the strip $\{\xi, \eta : 0 \le \xi < \infty,\ 0 \le \eta \le 2\pi\}$ in the ζ-plane into the entire z-plane. Eliminating η from (9.6.3) using the identity $\cos^2\eta + \sin^2\eta = 1$ gives

$$\frac{x^2}{c^2\cosh^2\xi} + \frac{y^2}{c^2\sinh^2\xi} = 1. \tag{9.6.4}$$

Similarly, eliminating ξ from (9.6.3) using the identity $\cosh^2\xi - \sinh^2\xi = 1$ gives

$$\frac{x^2}{c^2\cos^2\eta} - \frac{y^2}{c^2\sin^2\eta} = 1. \tag{9.6.5}$$

Lines of constant ζ and constant η in the ζ-plane thus map into ellipses and hyperbolas, respectively, in the z-plane, as shown in Figure 9.13. The semimajor axis a and semiminor axis b of an ellipse with $\xi = \xi_0 =$ const are given by

$$a = c\cosh\xi_0, \qquad b = c\sinh\xi_0, \tag{9.6.6}$$

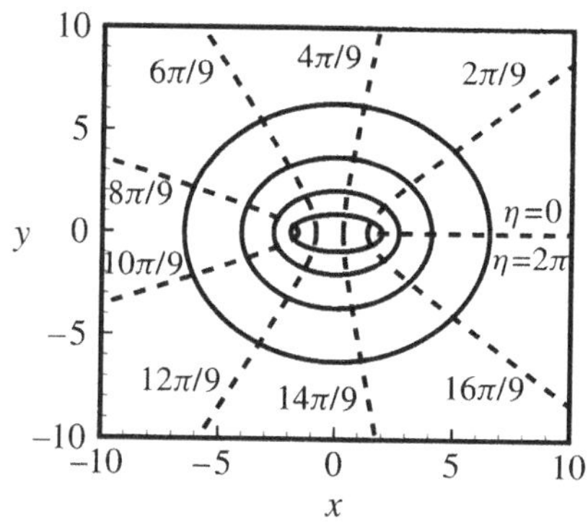

Figure 9.13 Constant lines of the elliptical coordinates ξ and η. Lines of constant ξ (solid) are ellipses, given for $\xi = 0.5, 1.0, 1.5, 2.0$. Lines of constant η (dashed) are hyperbolas, given for nine even increments over the interval $0 \leq \eta < 2\pi$ with η increasing in the counterclockwise direction.

so that $a^2 - b^2 = c^2$. Both the lines of constant ζ and constant η in the ζ-plane and the confocal ellipses and hyperbolas to which these lines map onto in the z-plane are orthogonal, which follows from the angle conservation property of conformal transformations.

The first derivative of the mapping function in (9.6.1) is given by

$$\begin{aligned}\frac{df}{d\zeta} &= c\sinh(\zeta) \\ &= c(\sinh\xi\cos\eta + i\cosh\xi\sin\eta)\end{aligned} \tag{9.6.7}$$

so that the transformation exhibits critical points (at which $df/d\zeta$ vanishes) within the strip $0 \leq \eta \leq 2\pi$ at (ξ, η) equal to $(0, 0)$, $(0, \pi)$, and $(0, 2\pi)$. These points are labeled A, B, and C, respectively, in Figure 9.14a. The first and the last of these critical points map onto the focus point $(x, y) = (c, 0)$ of the family of ellipses in

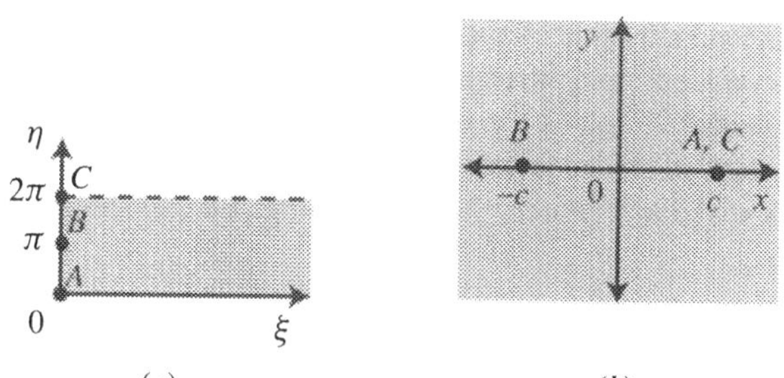

Figure 9.14 Sketch showing the mapping of critical points A, B and C by the transformation $z = c\cosh(\zeta)$ in (a) the ζ-plane and (b) the z-plane.

the z-plane and the middle critical point maps onto the other focus $(x, y) = (-c, 0)$, as shown in Figure 9.14*b*. In the limit of an ellipse passing through the two foci, such that the semimajor axis $a = c$, (9.6.6) indicates that $\xi_0 = 0$ and the semiminor axis vanishes. This limiting ellipse of infinite aspect ratio a/b reduces to a line segment with end points at the two critical points.

9.6.1 Transformation of the Complex Potential and Complex Velocity

The usefulness of conformal transformations for solution of fluid flows lies in their ability to transform a difficult problem into a problem for which the solution is known. Hence, it is important to understand how the variables that characterize a fluid flow, such as the complex potential and the strengths of sources and vortices, change under a conformal transformation. The mapping of these quantities can be determined with aid of the following theorem.

Theorem 9.6.1. A conformal transformation maps a harmonic function into another harmonic function.

Any harmonic function $\phi = \hat{\phi}(x, y)$ satisfies the Laplace equation in the z-plane, or

$$\frac{\partial^2 \hat{\phi}}{\partial x^2} + \frac{\partial^2 \hat{\phi}}{\partial y^2} = 0. \tag{9.6.8}$$

If this function is now mapped into another function $\phi = \tilde{\phi}(\xi, \eta)$ upon applying the conformal transformation $\zeta = f(z)$, then Theorem 9.6.1 requires that $\tilde{\phi}$ is harmonic in the ζ-plane, or that

$$\frac{\partial^2 \tilde{\phi}}{\partial \xi^2} + \frac{\partial^2 \tilde{\phi}}{\partial \eta^2} = 0. \tag{9.6.9}$$

The theorem can be proved simply by substituting the transformations $\xi = \xi(x, y)$ and $\eta = \eta(x, y)$ and noting that $\xi(x, y)$ and $\eta(x, y)$ satisfy the Cauchy-Riemann equations (9.1.2) since $f(z)$ is analytic.

Since both the velocity potential ϕ and the stream function ψ are harmonic functions in any two-dimensional, irrotational, incompressible flow, it is necessary that the complex potential $F \equiv \phi + i\psi$ be harmonic both before and after the transformation. Satisfaction of this requirement is guaranteed by the property of conformal transformations stated in Theorem 9.6.1 if we write the transformation for the complex potential as

$$F = \hat{F}(z) = \tilde{F}(\zeta). \tag{9.6.10}$$

Hence, the complex potential of flow in the z-plane can be obtained by substituting $\zeta = f(z)$ into the expression $\tilde{F}(\zeta)$ for the complex potential in the ζ-plane, and vice versa.

This method of changing the complex potential from one flow into that for another by a coordinate change does not work for the complex velocity. In particular, in the z-plane the complex velocity is defined by $\hat{W}(z) = d\hat{F}(z)/dz$ and in the ζ-plane it is similarly defined by $\tilde{W}(\zeta) = d\tilde{F}(\zeta)/d\zeta$. Using (9.6.10) and the chain rule, the transformation rule for the complex velocity becomes

$$\hat{W}(z) = \tilde{W}(\zeta)\frac{d\zeta}{dz}. \tag{9.6.11}$$

9.6.2 Transformation of Source and Vortex Strengths

The change in source and vortex strengths under a conformal transformation can be related to the transformation properties of the integral of the complex velocity about a closed circuit. We consider a closed circuit C on the x-y plane that bounds a region A, where the outward unit normal of C on the x-y plane is denoted by $\mathbf{n}$. The magnitude of an increment of length along C is denoted by $d\ell$, which is assumed to have a positive value when C is traversed in a counterclockwise direction. The integral of the complex velocity about C can be decomposed as

$$\begin{aligned}\int_C W\,dz &= \int_C (u - iv)(dx + i\,dy)\\ &= \int_C (u\,dx + v\,dy) + i\int_C (u\,dy - v\,dx)\\ &= \int_C \mathbf{u}\cdot d\mathbf{x} + i\int_C \mathbf{u}\cdot\mathbf{n}\,d\ell.\end{aligned} \tag{9.6.12}$$

Using the Stokes theorem for the first integral in the last equation in (9.6.12) and the two-dimensional divergence theorem for the second integral, together with the definitions of vorticity and dilatation rate, yields the result

$$\int_C W dz = \int_A \omega\,da + i\int_A \Delta\,da. \tag{9.6.13}$$

The real part of the integral of W about a closed circuit C is the circulation about C and the imaginary part is the net volumetric flow rate generated within C.

The change in the integral in (9.6.13) during a conformal transformation can be determined from the known transformation property (9.6.11) for W. In particular, if C_z denotes the position of the curve C in the z-plane and C_ζ denotes the mapping of this same curve in the ζ-plane, then (9.6.11) can be used to write

$$\int_{C_z} \hat{W}(z)\,dz = \int_{C_z} \tilde{W}(\zeta)\frac{d\zeta}{dz}\,dz = \int_{C_\zeta} \tilde{W}(\zeta)\,d\zeta. \tag{9.6.14}$$

Hence, the integral in (9.6.13) is invariant to a conformal transformation. It follows that both the net vorticity and the net dilatation rate within any region A in the z-plane are the same as the respective quantities for the mapping of region A in the ζ-plane. In particular, letting A be chosen as some infinitesimal neighborhood surrounding a two-dimensional point source or vortex, a consequence of this result is that both the source and vortex strengths must be invariant under conformal transformation.

9.7 TRANSFORMATION $\zeta = z^n$

The conformal transformation

$$\zeta = f(z) = z^n \tag{9.7.1}$$

can be used to map a wedge of angle $\theta = \pi/n$ with vertex at the origin in the z-plane onto the real axis in the ζ-plane. The transformation has a critical point at the origin for $n > 1$, since $df/dz = nz^{n-1}$ vanishes at $z = 0$ whenever $n > 1$. For cases with $n < 1$, the inverse transformation $z = g(\zeta) = \zeta^{1/n}$ has a critical point at the origin, since $dg/d\zeta = (1/n)\zeta^{(1-n)/n}$ vanishes at $\zeta = 0$ whenever $0 < n < 1$.

The case of uniform flow in the ζ-plane, with complex potential $\tilde{F}(\zeta) = U\zeta$ and complex velocity $\tilde{W}(\zeta) = U$, is mapped in the z-plane into a flow with complex potential

$$\hat{F}(z) = Cz^n, \tag{9.7.2}$$

where $C = U^n$ is a constant. The corresponding complex velocity in the z-plane can be obtained either by differentiating $\hat{F}(z)$ in (9.7.2) or by the transformation relation (9.6.11) for W, yielding

$$\hat{W}(z) = nCz^{n-1}. \tag{9.7.3}$$

The streamlines of the flow associated with (9.7.2) are plotted in Figure 9.15 for several values of n. For the case $n = 1$, ζ is the same as z and the uniform flow is unchanged. For $n > 1$, the real axis is mapped onto a wedge with angle less than π. As indicated by (9.7.3), the velocity field exhibits a stagnation point at the corner of

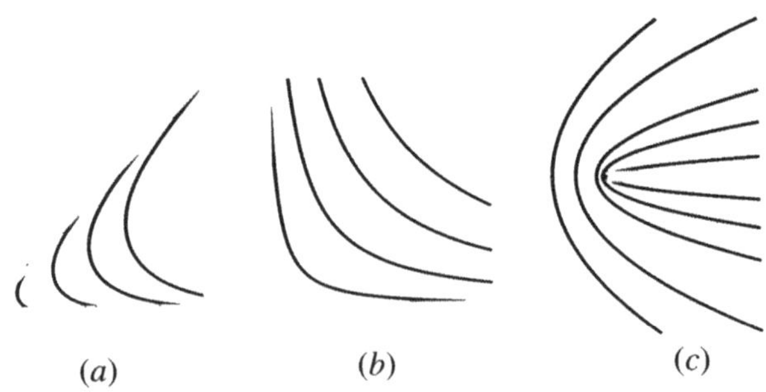

Figure 9.15 Streamlines for flow generated using the transformation $\zeta = z^n$ for cases with (*a*) $n = 4$, (*b*) $n = 2$, and (*c*) $n = \frac{1}{2}$.

the wedge for such cases. The particular case of $n = 2$ reduces to the plane straining flow (or *stagnation-point flow*) described in Section 9.2.2. For $\frac{1}{2} < n < 1$, the real axis is mapped onto a wedge with angle greater than π and the velocity exhibits a singularity at the corner of the wedge. In the limiting case $n = \frac{1}{2}$, the real axis is mapped onto a semi-infinite line and the resulting flow field corresponds to flow around the edge of a thin flat plate, in the vicinity of the edge. Of course, in any real fluid flow, cases with $\frac{1}{2} < n < 1$ would be accompanied by separation of the boundary layer at the corner of the wedge. It should also be noted that none of the flows described by (9.7.2), except the identity transformation $n = 1$, correspond to a full solution of the problem of uniform flow past a wedge, since the velocity field does not approach the uniform flow condition for large z. Nevertheless, it can be shown (Batchelor, 1967, p. 413) that the flow field given by (9.7.2) corresponds to the leading term in the asympotic expansion of the flow in the vicinity of the corner of the wedge.

Example 9.7.1 Source or Vortex in a 90° Corner. The problem of a source located at a position $z_0 = a + ia$ in a 90° corner, centered at the origin, is illustrated in Figure 9.16a. The flow field for this problem is solved by using the transformation $\zeta = z^2$ to map the two perpendicular boundaries of the 90° corner onto the real axis.

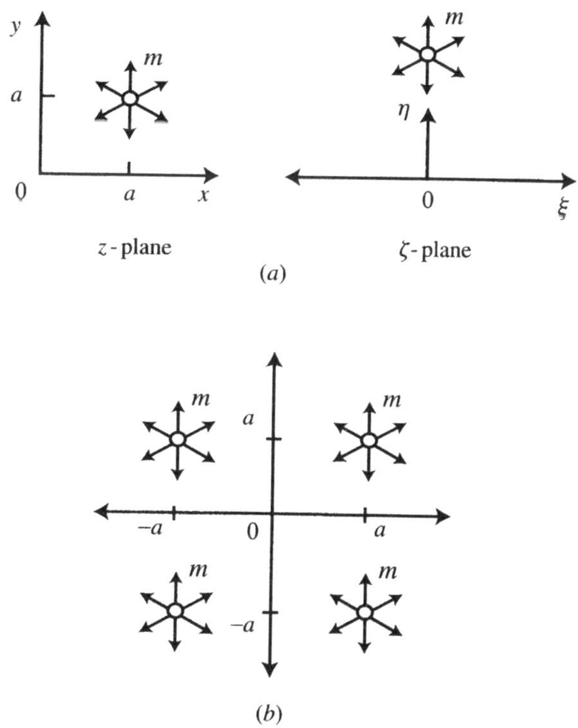

Figure 9.16 Source placed in a 90° corner: (a) z-plane and ζ-plane; (b) image problem.

In the ζ-plane, the flow reduces to that of a source over an infinite flat plate, which is solved using the method of images in Example 9.5.1. The source is translated during this mapping to the point $\zeta_0 = 2ia^2$, and we recall that the strength of a source is invariant under any conformal transformation. The complex potential in the ζ-plane is given by (9.5.1) as

$$\tilde{F}(\zeta) = \frac{m}{2\pi}\ln(\zeta - 2ia^2) + \frac{m}{2\pi}\ln(\zeta + 2ia^2). \tag{9.7.4}$$

Transforming back onto the z-plane gives the complex potential for a source near a 90° corner as

$$\hat{F}(z) = \frac{m}{2\pi}\ln(z^2 - 2ia^2) + \frac{m}{2\pi}\ln(z^2 + 2ia^2). \tag{9.7.5}$$

The result (9.7.5) can be rewritten, after noting that

$$z^2 - 2ia^2 = [z - a(1+i)][z + a(1+i)],$$
$$z^2 + 2ia^2 = [z - a(1-i)][z + a(1-i)],$$

as

$$\hat{F}(z) = \frac{m}{2\pi}\ln[z - (a+ia)] + \frac{m}{2\pi}\ln[z + (a+ia)] + \frac{m}{2\pi}\ln[z - (a-ia)] + \frac{m}{2\pi}\ln[z + (a-ia)]. \tag{9.7.6}$$

The four terms in (9.7.6) can be associated with the original source and three image sources, as shown in Figure 9.16*b*.

The problem of a vortex in a 90° corner can be solved in a similar manner, and the resulting complex potential is

$$\hat{F}(z) = -\frac{i\Gamma}{2\pi}\ln[z - (a+ia)] - \frac{i\Gamma}{2\pi}\ln[z + (a+ia)] + \frac{i\Gamma}{2\pi}\ln[z - (a-ia)] + \frac{i\Gamma}{2\pi}\ln[z + (a-ia)]. \tag{9.7.7}$$

The location and sense of the three image vortices corresponding to the last three terms in (9.7.7) are shown in Figure 9.17. The three image vortices induce a velocity on the original vortex that causes the vortex to advect downward and horizontally outward with time. For instance, when the vortex is located at $z_0 = a + ia$, the components of the induced velocity $\mathbf{u}_I$ on the vortex are $u_I = \Gamma/8\pi a$ and $v_I = -\Gamma/8\pi a$. If a similar approach to that used to obtain (9.7.7) is used for a vortex at any location $z_0 = a + ib$, the induced velocity of the vortex can be obtained as a function of time. The path of the vortex can be obtained by integration of the advection equation

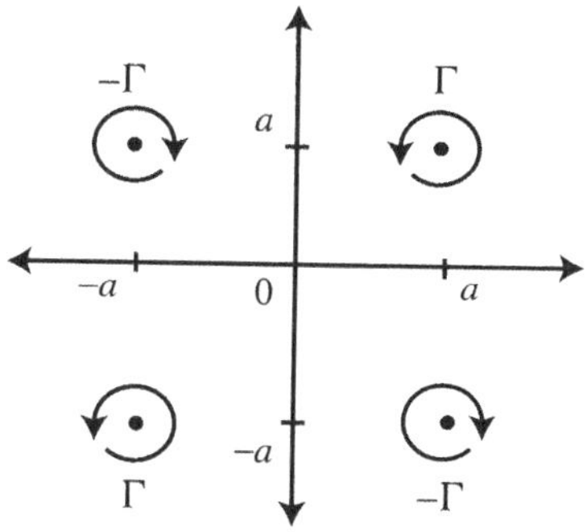

Figure 9.17 Image problem for a vortex placed in a 90° corner.

$$\frac{d\mathbf{x}}{dt} = \mathbf{u}_I(\mathbf{x}), \tag{9.7.8}$$

where the induced velocity $\mathbf{u}_I$ of the vortex is a function only of the current vortex position $\mathbf{x}$. The solution for a vortex in a 90° corner is identical to the solution for a pair of opposite-strength vortices propagating toward an infinite plane wall, and a plot showing the path of the vortex pair is given in Figure 9.18. The problem of a vortex pair impinging head-on into an infinite plane wall is sometimes used as a simple model for the inviscid interaction of aircraft wake vortices with the ground during landing and take-off.

Example 9.7.2 Flow Around a Sharp Edge. Uniform flow in the x-direction can be mapped into flow around a sharp edge at $z = 0$ using the transformation $\zeta = z^{1/2}$. The complex potential for uniform flow in the ζ-plane is $\tilde{F}(\zeta) = U\zeta$, so that application of the above mapping yields

$$\hat{F}(z) = Cz^{1/2}, \tag{9.7.9}$$

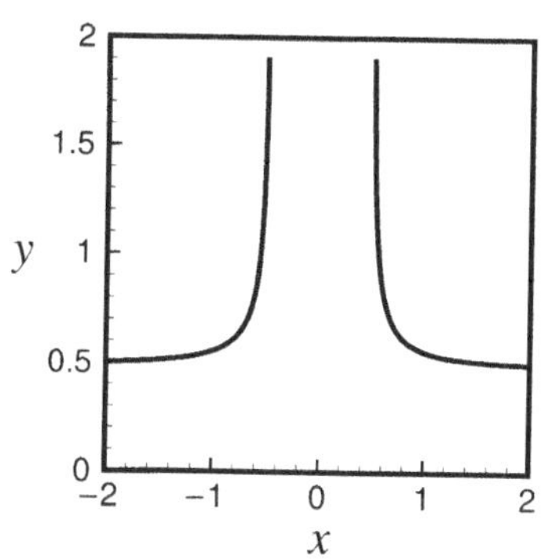

Figure 9.18 Path of a vortex pair approaching a flat wall.

where C is a constant. The stream function and potential function for this flow are given by $\phi = cr^{1/2}\cos(\theta/2)$ and $\psi = cr^{1/2}\sin(\theta/2)$. Streamlines are shown in Figure 9.15c. The complex velocity in the z-plane is

$$W = \frac{C}{2z^{1/2}}. \tag{9.7.10}$$

The velocity components in the radial and azimuthal directions are $u_r = (C/2r^{1/2})\cos(\theta/2)$ and $u_\theta = -(C/2r^{1/2})\sin(\theta/2)$. An important feature of this result is that the velocity magnitude exhibits a $1/r^{1/2}$ singularity as $r \to 0$ at the sharp edge, which results from the abrupt turning of the flow around the edge.

9.8 JOUKOWSKI TRANSFORMATION

The Joukowski transformation is defined by

$$z = f(\zeta) = \zeta + \frac{c^2}{\zeta}, \tag{9.8.1}$$

where c is a real valued parameter. Of course, (9.8.1) can also be written in inverse form, with z and ζ switched, but the form given above is most useful for applications where the physical problem is usually described in the z-plane. The gradient of the Joukowski mapping function is

$$\frac{df}{d\zeta} = 1 - \frac{c^2}{\zeta^2}, \tag{9.8.2}$$

which vanishes at the two critical points $\zeta = \pm c$, corresponding to $z = \pm 2c$. The mapping of a circle $\zeta = be^{i\varphi}$, where $b > c$, surrounding the two critical points as φ varies from 0 to 2π is shown in Figure 9.19. Under the transformation (9.8.1), the

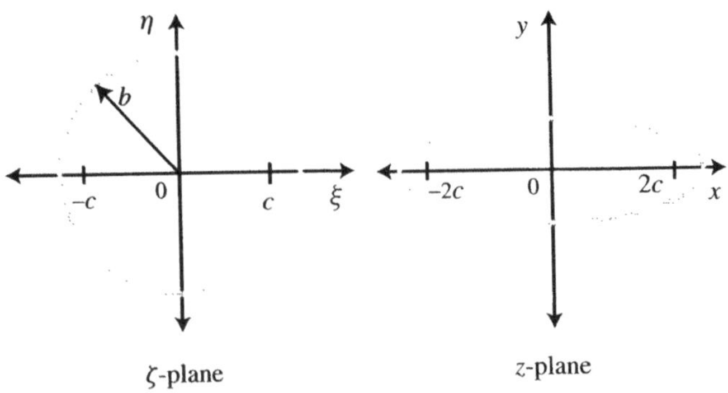

Figure 9.19 Mapping of a circle with radius $b > c$ into an ellipse by the Joukowski transformation.

circle is mapped onto the curve given by

$$z = be^{i\varphi} + \frac{c^2}{b}e^{-i\varphi}. \tag{9.8.3}$$

Breaking (9.8.3) into its real and imaginary parts gives

$$x = \left(b + \frac{c^2}{b}\right)\cos\varphi, \qquad y = \left(b - \frac{c^2}{b}\right)\sin\varphi. \tag{9.8.4}$$

Eliminating the parameter φ in (9.8.4) with use of the identity $\cos^2\varphi + \sin^2\varphi = 1$ yields

$$\frac{x^2}{(b + c^2/b)^2} + \frac{y^2}{(b - c^2/b)^2} = 1, \tag{9.8.5}$$

which is the equation of an ellipse. Since a circle and an ellipse are both smooth curves, the angle between any two adjoining line segments is equal to π in both cases and is hence preserved under the transformation.

If instead we examine the mapping of a smooth curve that passes through the critical points $\zeta = \pm c$ of the transformation, the angle between line segments ending at the critical points is not preserved. For instance, the circle $\zeta = ce^{i\varphi}$, shown in Figure 9.20*a*, is mapped into

$$\zeta = ce^{i\varphi} + ce^{-i\varphi} = 2c\cos\varphi. \tag{9.8.6}$$

Since ζ in (9.8.6) is everywhere real valued, the mapping of the circle in Figure 9.20*a* lies on the line segment spanning from $-2c$ to $2c$ in the z-plane, as shown in Figure 9.20*b*. The angle between adjoining line segments is everywhere preserved except at the two critical points, at which the angles are increased from π to 2π during the transformation.

One important feature of the Joukowski transformation is that for large values of ζ, (9.8.1) reduces to the identity transformation $z = \zeta$. A consequence of this prop-

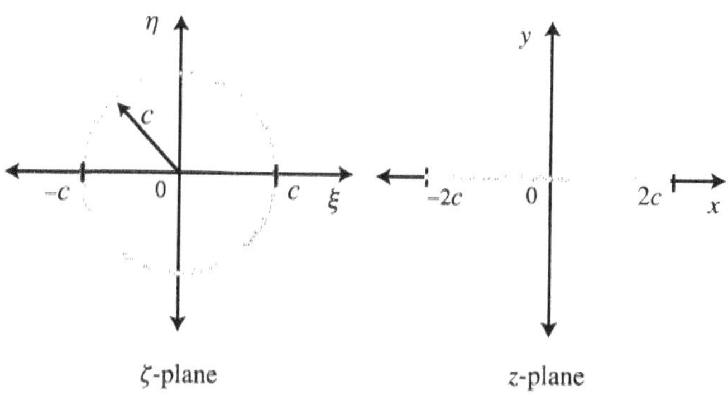

Figure 9.20 Mapping of a circle passing through the critical points at $\xi = \pm c$ into a line segment by the Joukowski transformation.

erty is that the flow field at infinity is unchanged by the Joukowski transformation, so if the flow field satisfies the boundary condition at infinity in the ζ-plane, it will also do so in the z-plane.

Example 9.8.1 Uniform Flow Past an Ellipse. We have noted that a circle in the ζ-plane with radius $R > c$, centered at the origin, is transformed into an ellipse in the z-plane with semimajor axis $a = R + c^2/R$ and semiminor axis $b = R - c^2/R$. The complex potential for uniform flow at an angle α to the horizontal past the circle in the ζ-plane is

$$\tilde{F}(\zeta) = U\left(\zeta e^{-i\alpha} + \frac{R^2}{\zeta} e^{i\alpha}\right). \tag{9.8.7}$$

The complex potential $\hat{F}(z)$ for flow past an ellipse is obtained by substituting into (9.8.7) the inverse mapping

$$\zeta = g(z) = \tfrac{1}{2}[z \pm (z^2 - 4c^2)^{1/2}], \tag{9.8.8}$$

where the sign is chosen such that $\zeta = z$ when $c = 0$. The parameters c and R are obtained as functions of the semimajor and semiminor axes a and b as

$$R = \tfrac{1}{2}(a + b), \qquad c = \tfrac{1}{2}(a^2 - b^2)^{1/2}. \tag{9.8.9}$$

The streamlines for flow past an ellipse with aspect ratio $a/b = 2$ are shown in Figure 9.21 for cases with $\alpha = 0, \pi/2$.

Example 9.8.2 Uniform Flow Past a Flat Plate. The solution for uniform flow at an angle α to the horizontal past a flat plate of length ℓ lying along the horizontal is obtained as a special case of the solution of Example 9.8.1, for which the major and minor semiaxes are chosen as $a = \ell/2$ and $b = 0$. Equation (9.8.9) gives the parameters c and R in this case as

$$R = c = \ell/4. \tag{9.8.10}$$

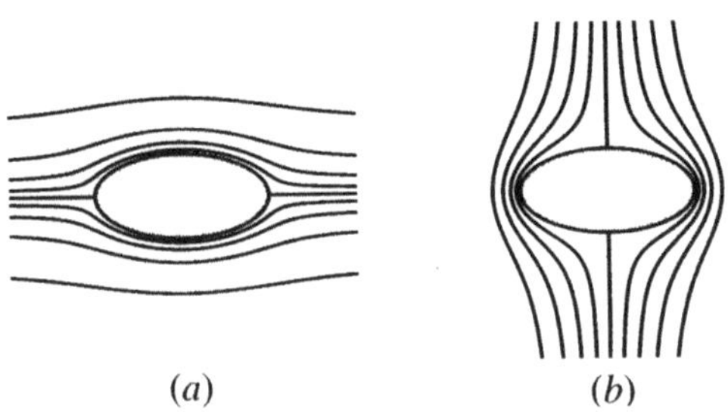

Figure 9.21 Streamlines for flow past an ellipse with aspect ratio $a/b = 2$ for cases with orientation angle (a) $\alpha = 0$ and (b) $\alpha = \pi/2$.

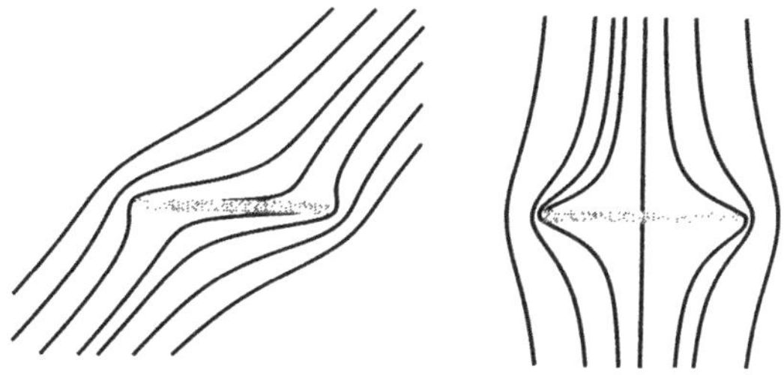

Figure 9.22 Streamlines for uniform flow past a flat plate with orientation angle (a) $\alpha = \pi/4$ and (b) $\alpha = \pi/2$.

The streamlines of the flow are given for cases with $\alpha = \pi/4, \pi/2$ in Figure 9.22. The flow fields shown in Figure 9.22 are rather unphysical, since in any real flow the boundary layer separates at the sharp edges of the plate. A number of approaches can be used to account for this effect, the choice of which depends on whether angle α is small or large. This issue is discussed further in Sections 9.10 and 10.5.

9.9 SCHWARZ-CHRISTOFFEL EQUATION

The Schwarz-Christoffel equation provides a general procedure for constructing a transformation that maps the boundary of any simple closed polygon onto the real axis. The interior points of the polygon are mapped by the transformation onto the upper half of the complex plane.

Before stating the precise form of this procedure, it is useful to first review some of the properties of polygons. We call a set of N straight lines, denoted by C, a *simple closed polygon* if C satisfies the following two properties:

(a) It is possible to travel from any given point on C to any other point on C without leaving C (i.e., the lines are connected).

(b) The set of lines C divides the points of the plane into two disjoint sets, called the *interior* and the *exterior* points of the polygon. Any two interior points can be joined by a path that does not intersect C, as can any two exterior points; however, it is impossible to travel from an interior to an exterior point (or vice versa) without crossing C.

Two common polygons are the rectangle ($N = 4$) and the triangle ($N = 3$), for which the above properties are obviously satisfied. Often in fluid dynamics problems, we wish to consider cases of flow involving a boundary that extends to infinity. Such boundaries can be constructed as limiting cases of simple closed polygons by allowing one or more of the vertices of the polygon to extend to infinity, and in this limiting sense these shapes continue to satisfy the second property above. A number of examples of simple closed polygons that extend to infinity and are of interest in

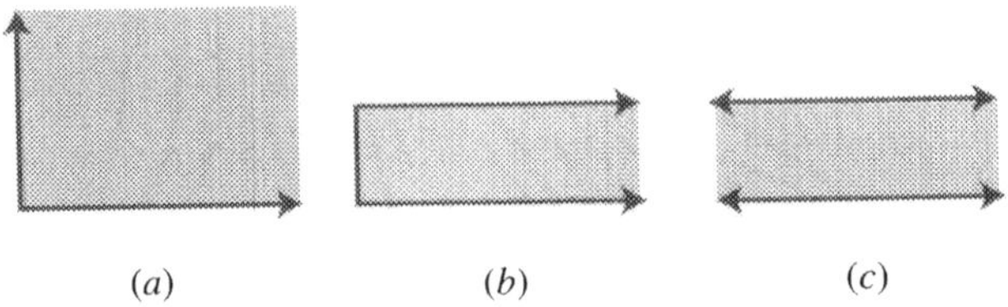

Figure 9.23 Some shapes that can be formed from the interior of a simple closed polygon with one or more vertices taken to infinity: (*a*) 90° corner; (*b*) semi-infinite strip; (*c*) infinite strip.

fluid dynamics problems are shown in Figure 9.23. For instance, the wedge shape in Figure 9.23*a* can be constructed by allowing two vertices of a triangle to extend to infinity. The semi-infinite strip in Figure 9.23*b* can be constructed by allowing the right two vertices of a rectangle to extend to infinity, or alternatively allowing one vertex of a triangle to extend to infinity. The infinite strip in Figure 9.23*c* can similarly be constructed by allowing all of the vertices of a rectangle to extend to infinity.

Denoting the interior angles of a simple closed polygon at each vertex n by θ_n, the closure property requires that the θ_n satisfy

$$\sum_{n=1}^{N} \theta_n = (N-2)\pi. \tag{9.9.1}$$

The sum (9.9.1) includes any possible vertices at infinity for which the interior angle may approach either zero or some finite value. For instance, if the semi-infinite strip in Figure 9.23*b* is constructed from a rectangle ($N = 4$), then all interior angles are $\pi/2$. However, if it is constructed from a triangle ($N = 3$), then the interior angle of the vertex at infinity is zero and that of the other two vertices is $\pi/2$. Either construction satisfies (9.9.1).

A simple precursor to the Schwarz-Christoffel procedure is the transformation $\zeta = z^{\pi/\theta_1}$ discussed in Section 9.7. We recall that this transformation maps a wedge with interior angle θ_1 and vertex located at the origin $z = \zeta = 0$ onto the real axis. In the case that the vertex is located at some other point $z = A_1$ which maps onto the point $\zeta = a_1$ in the ζ-plane, the transformation is given by

$$\zeta - a_1 = (z - A_1)^{\pi/\theta_1}. \tag{9.9.2}$$

Equation (9.9.2) can be rearranged as

$$z - A_1 = (\zeta - a_1)^{\theta_1/\pi}. \tag{9.9.3}$$

Taking the derivative of (9.9.3) with respect to ζ gives the equivalent form

$$\frac{dz}{d\zeta} = \frac{\theta_1}{\pi}(\zeta - a_1)^{\theta_1/\pi - 1}. \tag{9.9.4}$$

The basic idea of the Schwarz-Christoffel procedure is to apply an extension of the transformation (9.9.4) such that a critical point of the transformation coincides with every vertex of the polygon. We know that for any analytic mapping function, the angles between adjoining line segments of the polygon boundary are invariant except at the critical points. The change in angle between adjoining line segments ending at a critical point depends only on the power to which the polynomial in ζ having a zero at the critical point is raised in the expression for the derivative $dz/d\zeta$ [e.g., the exponent $\theta_1/\pi - 1$ in (9.9.4)]. Thus, if (9.9.4) were multiplied by any analytic function $g(\zeta)$ that is nowhere equal to zero on the wedge with interior angle θ_1, the wedge would still be mapped onto the real axis by the resulting transformation.

With these considerations, it is evident that a transformation that maps any simple closed polygon with N vertices in the z-plane onto the real axis in the ζ-plane can be constructed by extending (9.9.4) as

$$\frac{dz}{d\zeta} = K(\zeta - a_1)^{\theta_1/\pi - 1}(\zeta - a_2)^{\theta_2/\pi - 1} \cdots (\zeta - a_N)^{\theta_N/\pi - 1}, \tag{9.9.5}$$

where a_n are the locations on the real axis in the ζ-plane to which the vertices of the polygon are mapped by the transformation and K is some constant which may be complex. In addition to mapping the polygon boundary onto the real axis, the transformation obtained by integration of (9.9.5) maps the interior of the polygon onto the upper half of the complex ζ-plane. Equation (9.9.5) is called the *Schwarz-Christoffel equation*, and a proof of the properties of transformations obtained by integration of this equation can be found in standard texts on complex variable theory (e.g., Ablowitz and Fokas, 1997).

Two remarks are helpful prior to giving examples for the use of (9.9.5) in fluid flow problems. The first remark is that only the constant K and exactly three of the coefficients a_n, $n = 1, \ldots, N$, can be chosen arbitrarily for any given problem. The remaining a_n coefficients must be selected so as to make the polygon of the desired shape. Depending on the problem, choice of appropriate values for the constrained a_n coefficients can sometimes be difficult.

The second remark is that when one of the vertices of the polygon extends to infinity, the factor corresponding to that vertex can be omitted from (9.9.5). For instance, if $a_1 \to \infty$ in (9.9.5), the constant K can be selected as $K = K^*/(-a_1)^{\theta_1/\pi - 1}$. Since $(\zeta - a_1)^{\theta_1/\pi - 1}/(-a_1)^{\theta_1/\pi - 1} \to 1$ as $a_1 \to \infty$ for any finite ζ, (9.9.5) reduces to

$$\frac{dz}{d\zeta} = K^*(\zeta - a_2)^{\theta_2/\pi - 1} \cdots (\zeta - a_N)^{\theta_N/\pi - 1}, \tag{9.9.6}$$

which is the same as (9.9.5) without the factor involving a_1.

Example 9.9.1 Flow into a Channel through a Narrow Slit in the Wall. The problem of flow into an infinite channel of width h through a narrow slit in one wall, which may be modeled by a source of strength m, is shown in Figure 9.24a. This problem is solved by the following procedure: (1) use the Schwarz-Christoffel equa-

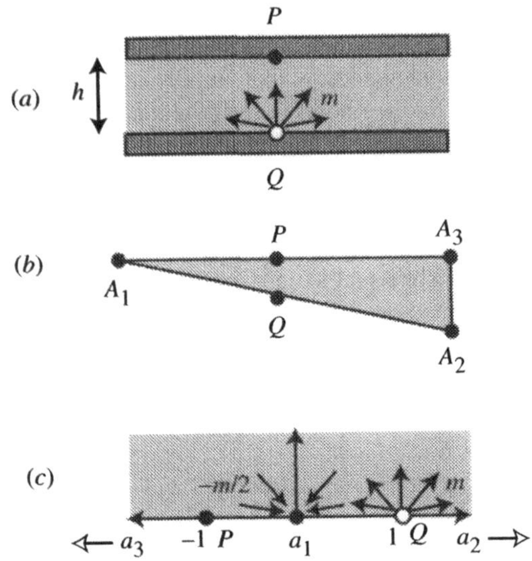

Figure 9.24 Flow into an infinite channel through a narrow slit in the wall: (a) schematic of problem; (b) schematic showing vertices of triangle used to construct the infinite channel in the z-plane; (c) mapping of vertices onto real axis of the ζ-plane.

tion to construct a transformation that maps the infinite channel onto the real axis of the ζ-plane, (2) solve the corresponding problem in the ζ-plane, and (3) transform back to the physical z-plane to obtain a solution that satisfies all boundary conditions of the problem. There are several ways in which the infinite channel can be mapped onto the real axis, so we need to make some careful choices. One way of constructing the infinite channel is to start with a triangle, with one vertex A_1 on the left and two vertices A_2 and A_3 on the right. The interior angle at A_1 approaches zero as the vertex approaches negative infinity, and the interior angles at A_2 and A_3 approach $\pi/2$. The source of strength m is located at the origin Q, and the point directly across the channel from Q is labeled P. See Figure 9.24b.

We are now free to choose the constant K and the three coefficients a_1, a_2, and a_3 in (9.9.5) corresponding to the desired mappings of the three vertices A_1, A_2, and A_3. The lower half of the strip is mapped onto the positive real axis in the ζ-plane and the upper half of the strip is mapped onto the negative real axis, with A_1 mapped to $\zeta = 0$, as shown in Figure 9.24c. This can be accomplished by letting $a_1 = 0$, $a_2 \to \infty$, and $a_3 \to -\infty$. Since a_2 and a_3 are infinite, we need only account for the factor involving a_1 in (9.9.5), which with $\theta_1 = 0$ becomes

$$\frac{dz}{d\zeta} = K\zeta^{-1}. \tag{9.9.7}$$

Integrating (9.9.7) over ζ gives the transformation as

$$z = K \ln \zeta + B, \tag{9.9.8}$$

where K and B are undetermined constants. For convenience, K and B are set such that the points Q and P in the z-plane map onto the points $\zeta = 1$ and $\zeta = -1$, respectively, although other choices are also acceptable. With these choices, (9.9.8) gives

$$0 = K \ln 1 + B, \qquad ih = K \ln(-1) + B,$$

where we recall that the principal value of $\ln(-1)$ is $\ln(e^{i\pi}) = i\pi$. Solving for K and B gives $B = 0$, $K = h/\pi$. Solving (9.9.8) for ζ, the transformation becomes

$$\zeta = e^{\pi z/h}. \tag{9.9.9}$$

The flow field in the z-plane can be taken as being generated by a source of strength m at the origin and a sink of strength $-m/2$ at both $x \to \infty$ and $x \to -\infty$. In the ζ-plane, the source of strength m is mapped to $\zeta = 1$ and the sink of strength $-m/2$ at $x \to -\infty$ is mapped to $\zeta = 0$, as indicated in Figure 9.24c. The other sink continues to lie at infinity and hence does not contribute to the flow field. The complex potential due to the sink of strength $-m/2$ and the source of strength m is

$$\tilde{F}(\zeta) = \frac{m}{2\pi} \ln(\zeta - 1) - \frac{m}{4\pi} \ln \zeta. \tag{9.9.10}$$

The problem is now essentially solved, since it remains only to substitute the expression (9.9.9) for ζ into (9.9.10) to obtain the complex potential in the z-plane. However, the result can be put into a nice form by first rewriting (9.9.10) as

$$\tilde{F}(\zeta) = \frac{m}{2\pi} \ln(\zeta^{1/2} - \zeta^{-1/2})$$

and then substituting (9.9.9) to obtain

$$\begin{aligned} F(z) &= \frac{m}{2\pi} \ln(e^{\pi z/2h} - e^{-\pi z/2h}) \\ &= \frac{m}{2\pi} \ln\left[\sinh\left(\frac{\pi z}{2h}\right)\right] + C, \end{aligned} \tag{9.9.11}$$

where the constant $C = (m/2\pi)\ln 2$ in (9.9.11) has no dynamical significance.

Example 9.9.2 Uniform Flow Past a Semi-Infinite Rectangular Body. The problem of uniform flow past a semi-infinite rectangular body of width $2h$ is sketched in Figure 9.25a. We consider a polygon consisting of the line segment $\overline{A_1A_2}$ on the symmetry surface of the flow and line segments $\overline{A_2A_3}$ and $\overline{A_3A_4}$ on one-half of the rectangular body, where A_1 and A_4 are taken to $-\infty$ and ∞, respectively. The interior of the polygon is taken to be the region above the body surface in the positive half of the z-plane, so the interior angles at A_2 and A_3 are $\theta_2 = \pi/2$ and $\theta_3 = 3\pi/2$, respectively. In the ζ-plane, A_1 and A_4 extend to infinity, so these vertices do not need to be included in the transformation (9.9.5). The vertex A_2 is mapped to $a_2 = 1$ and the vertex A_3 is mapped to $a_3 = -1$ in the ζ-plane, as shown in Figure 9.25b.

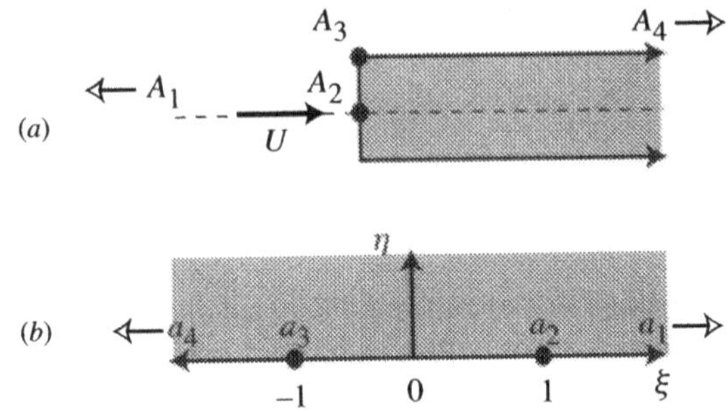

Figure 9.25 Vertices of the mapping used to solve the problem of uniform flow past a semi-infinite rectangular body: (a) z-plane; (b) ζ-plane.

With these choices, the Schwarz-Christoffel equation (9.9.5) reduces to

$$\frac{dz}{d\zeta} = K(\zeta - 1)^{-1/2}(\zeta + 1)^{1/2}. \tag{9.9.12}$$

Integrating (9.9.12) over ζ gives

$$z = K[(\zeta^2 - 1)^{1/2} + \cosh^{-1}\zeta] + B, \tag{9.9.13}$$

where K and B are undetermined constants. The vertices A_2 and A_3 in the z-plane, located at $z = 0$ and $z = ih$, respectively, map onto the points $a_2 = 1$ and $a_3 = -1$ in the ζ-plane. Since $\cosh^{-1}\zeta = \ln[\zeta + (\zeta^2 - 1)^{1/2}]$, K and B in (9.9.13) must satisfy

$$0 = K \ln 1 + B, \qquad ih = K\ln(-1) + B. \tag{9.9.14}$$

Setting $\ln(-1)$ equal to its principal value $i\pi$ gives $B = 0$ and $K = h/\pi$.

The flow in the ζ-plane must satisfy the no-penetration condition on the real axis, and it must reduce to the condition of uniform flow in the z-plane as z approaches infinity. A uniform flow $\tilde{F}(\zeta) = -V\zeta$ in the ζ-plane obviously satisfies the first of these conditions. To see that it also satisfies the second, we note that ζ can be considered a function of z to write

$$\hat{F}(z) = -V\zeta(z), \tag{9.9.15}$$

which upon taking the derivative gives the complex velocity as

$$\hat{W}(z) = -V\frac{d\zeta}{dz} = -\frac{V\pi}{h}\left(\frac{\zeta - 1}{\zeta + 1}\right)^{1/2}. \tag{9.9.16}$$

For $\zeta \to \pm\infty$, the complex velocity $\hat{W}(z) \to -\pi V/h$. The condition that the velocity approach a uniform flow of speed U as $z \to \pm\infty$ (which for this transformation corresponds to $\zeta \to \pm\infty$) is thus satisfied by setting the constant $V = -hU/\pi$. The resulting velocity potential in the z-plane is therefore given in implicit form as

$$\hat{F}(z) = \frac{hU}{\pi}\zeta(z), \tag{9.9.17}$$

where the function $\zeta(z)$ is obtained by solution of (9.9.13). An alternative form of this solution is obtained by writing $\zeta = \cosh p$, where p is a parametric variable, in which case the transformation (9.9.13) becomes

$$z = \frac{h}{\pi}(p + \sinh p) \tag{9.9.18}$$

and the complex potential becomes

$$\hat{F}(z) = \frac{hU}{\pi}\cosh p. \tag{9.9.19}$$

9.10 FREE-STREAMLINE THEORY

Several classes of fluid flow problems involve steady flow with an interface between a liquid and a gas phase. Unless the gas has a large velocity relative to that of the liquid or the radius of curvature of the interface is so small that surface tension plays an important role, it is a common approximation to neglect the gas inertia and to assume that the pressure is uniform along the interface between the two phases. Some classes of flows for which such an approximation is useful include the motion of a liquid jet surrounded by gas (such as liquid emanating from an orifice into the atmosphere), cavitating liquid flow past a solid body with a vapor bubble behind the body, and steady channel flow problems involving a liquid streaming past an obstacle (such as channel flow under a sluice gate). Examples of specific flow problems in each of these three classes are illustrated in Figure 9.26.

The liquid-gas interface is a streamline of the flow, since it is always composed of the same fluid particles and the flow is steady, so that in two dimensions the stream function ψ is uniform on the interface. If the pressure is assumed to be uniform on the interface, the streamline is called a *free streamline*. In the remainder of this section, the effect of gravity is neglected, which is a reasonable assumption when the height change h over the flow is much less than the dynamic head κ/g, where κ is the kinetic energy per unit mass and g is the gravitational acceleration. Under these assumptions, the Bernoulli theorem (8.2.3) requires that κ also be uniform along a free streamline.

Problems involving free streamlines are difficult to solve analytically, since the position of the free streamlines is not a priori known, and these problems often require the use of numerical computation. A common numerical method for problems of this type, which can also be used for unsteady flows, utilizes a doublet distribution along the free streamline and solves the flow field using a boundary-integral approach. Numerical methods of this type are discussed in Chapter 15. However, there does exist an important class of two-dimensional problems, involving free stream-

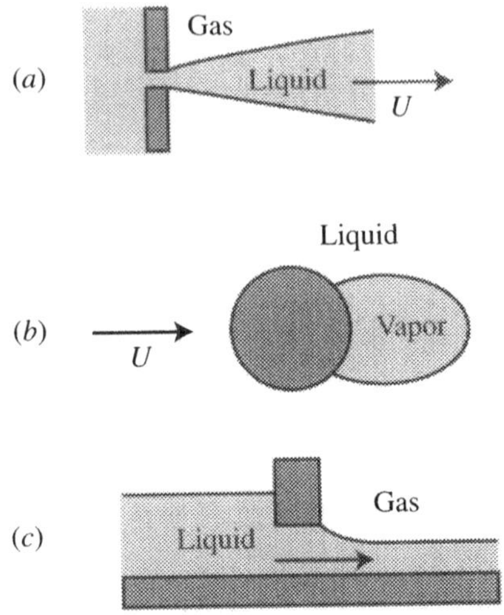

Figure 9.26 Examples of problems where the assumption of constant pressure on a liquid-gas interface is useful: (*a*) liquid jet emitted in the atmosphere; (*b*) cavitation bubble behind a body; (*c*) flow past a sluice gate.

lines attached to bodies constructed from straight line segments, which are analytically tractable, and methods for solution of such problems are the topic of the current section.

The method for analytic solution of problems with free streamlines is based on the introduction of a new complex variable Ω, defined by

$$\Omega \equiv -\ln W, \tag{9.10.1}$$

where $W = dF/dz$ is the complex velocity. Writing W in polar form as $W = qe^{-i\lambda}$, where $q = \sqrt{2\kappa}$ is the velocity magnitude, this new variable can be rewritten as

$$\Omega = \ln\left(\frac{1}{q}\right) + i\lambda. \tag{9.10.2}$$

The real part of Ω is uniform on a free streamline. When the problem is transformed to the Ω-plane (called the *hodograph plane*), the free streamlines appear as straight vertical lines over which only the imaginary part of Ω varies. Over the rectilinear surfaces of the body on which the no-penetration condition is applied, the velocity must be oriented tangent to the body surface, so that along these surfaces the polar angle λ of the complex velocity (and hence the imaginary part of Ω) is uniform. The body surface is thus transformed into horizontal lines in the hodograph plane. The boundary of the flow domain in the Ω-plane, composed of both the surface of the solid body and free streamlines, consists of a polygon formed of a collection of horizontal and vertical lines.

It is also the case that the boundary of the flow domain in the plane of the complex potential F consists of straight boundaries which coincide with either free streamlines and no-penetration surfaces (on each of which the stream function ψ is uniform) or inlets and outlets over which the velocity is assumed to be oriented in the normal direction (such that the velocity potential ϕ is uniform). Once the boundary of the problem is determined in the Ω-plane, the Schwartz-Christoffel procedure can be used to map the flow boundary onto the real axis in the plane of some transformed variable ζ such that the interior of the flow domain is mapped onto the upper half of the complex ζ-plane.

The solution procedure is best understood by considering specific examples; however, we offer the following summary. The procedure typically starts by making a sketch of the original problem in the physical z-plane, consisting of free streamlines, straight no-penetration boundaries, and inlets and exits. Two additional sketches are made of the problem in the Ω-plane and in the F-plane, in both of which the flow boundary has the form of a polygon. The Schwartz-Christoffel equation (9.9.5) is then used to construct a transformation of the flow boundary in the Ω-plane onto the real axis in the plane of a transformation variable ζ. A second use of the Schwartz-Christoffel equation (9.9.5) enables one to determine a conformal transformation between F and ζ such that the flow boundary in the F-plane is mapped onto the real axis in the ζ-plane. From these two conformal transformations, both F and Ω can be written as functions of the transformation variable ζ such that by eliminating ζ a relationship between F and Ω is obtained. The expression (9.10.1), with $W = dF/dz$, is substituted into this relationship to obtain a first-order ordinary differential equation for $F(z)$ which is integrated to obtain the final solution.

Example 9.10.1 Liquid Flow through an Orifice into the Atmosphere. When liquid flows through an orifice of thickness $2d$ into a gas atmosphere (in the absence of gravity), it is observed that far downstream of the orifice the liquid jet approaches an asymptotic thickness $2b$, where $b < d$. In this example, the free-streamline theory is used to solve for the contraction ratio b/d.

A sketch showing the liquid jet profile and a set of six points $\{A', B', C', A, B, C\}$ in the physical (z) plane is given in Figure 9.27a. The points B and B' are placed at the corners $(x, y) = (\pm d, 0)$ of the orifice. The points A and A' are taken to $\pm\infty$ in the x-direction along the wall separating the liquid and the gas, and the points C and C' are taken to $-\infty$ on the jet surface far downstream of the orifice. The boundaries of the liquid jet correspond to the streamlines $\pm\psi_0$. By the relationship (6.2.8) between change in value of the stream function and volumetric flow rate in two-dimensional flows, the stream function value $\pm\psi_0$ at the liquid jet boundary is related to the asymptotic fluid speed U far downstream of the orifice by

$$\psi_0 = -Ub. \tag{9.10.3}$$

The flow in the physical z-plane is now transformed into the hodograph Ω-plane using the transformation

$$\Omega \equiv -\ln\left(\frac{W}{U}\right). \tag{9.10.4}$$

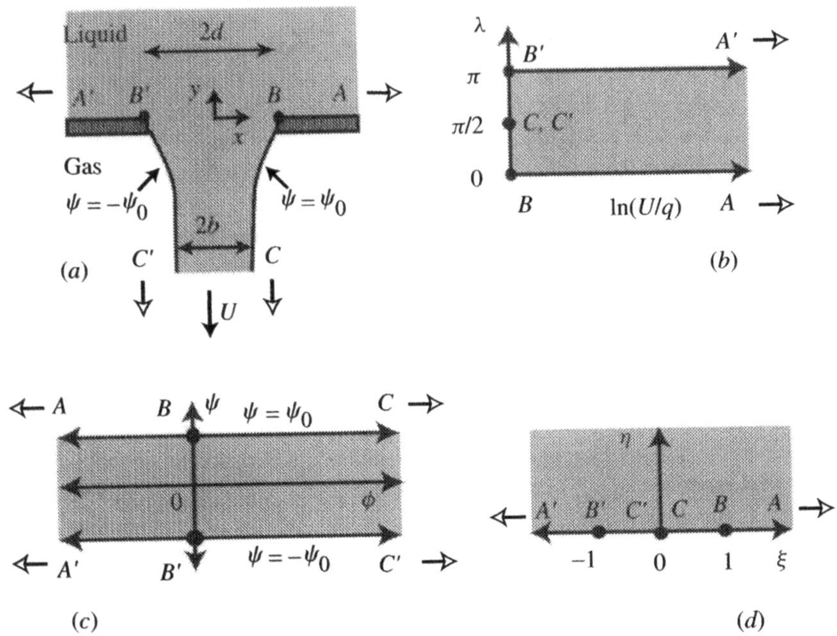

Figure 9.27 Transformation planes used in solving the problem of two-dimensional flow through an orifice: (*a*) physical z-plane; (*b*) Ω-plane; (*c*) F-plane; (*d*) ζ-plane.

The fact that the fluid speed q is constant along the free streamlines implies that $q = U$ [or $\ln(q/U) = 0$] along the curves BC and $B'C'$. These curves thus map into vertical line segments in the Ω-plane with zero real part. Similarly, along the wall separating the liquid and the gas the orientation of the velocity vector is constant, so that the lines AB and $A'B'$ map into horizontal line segments in the hodograph plane with $\lambda = 0$ and $\lambda = \pi$, respectively. The liquid flow maps into a semi-infinite strip in the hodograph plane, as shown in Figure 9.27*b*. Since the liquid flow is bounded by curves with constant stream function, the flow field can be mapped in the plane of the complex potential F into an infinite strip bounded by the horizontal lines $\psi = \psi_0$ and $\psi = -\psi_0$ (Figure 9.27*c*).

The flow field in both the Ω- and F-planes are mapped into the upper half plane of some other complex variable ζ using the Schwarz-Christoffel equation (9.9.5). The points B and B' in the Ω-plane are mapped onto the points $\zeta = 1$ and $\zeta = -1$, respectively. Since the interior angle of the semi-infinite strip at both B and B' is $\pi/2$, (9.9.5) gives

$$\frac{d\Omega}{d\zeta} = K(\zeta - 1)^{-1/2}(\zeta + 1)^{-1/2} = K(\zeta^2 - 1)^{-1/2}, \tag{9.10.5}$$

where K is an undetermined constant. Integrating over ζ yields

$$\Omega = K \cosh^{-1}(\zeta) + L, \tag{9.10.6}$$

where L is a constant of integration. The choice that the points B and B' map onto $\zeta = \pm 1$ requires that $K = 1$ and $L = 0$. Solving for ζ from (9.10.6) gives

$$\zeta = \cosh(\Omega). \tag{9.10.7}$$

In the F-plane, the interior angles of points A and A' on the left-hand side of the strip and of points C and C' on the right-hand side of the strip are $\pi/2$. If C and C' are mapped onto $\zeta = 0$, the Schwarz-Christoffel equation (9.9.5) becomes

$$\frac{dF}{d\zeta} = K'\zeta^{-1}. \tag{9.10.8}$$

Integrating (9.10.8) yields

$$F = K' \ln(\zeta) + L'. \tag{9.10.9}$$

The condition that points B and B' map onto $\zeta = \pm 1$ requires that the constants K' and L' are set as $K' = -2\psi_0/\pi$ and $L' = i\psi_0$. Solving for ζ from (9.10.9) gives

$$\zeta = \exp\left[-\frac{\pi}{2}\left(\frac{F}{\psi_0} - i\right)\right]. \tag{9.10.10}$$

Eliminating ζ between (9.10.7) and (9.10.10) and using (9.10.4) for Ω give an implicit expression for the complex potential $F(z)$ as

$$F(z) = i\psi_0 - \frac{2\psi_0}{\pi} \ln\left(\cosh\left\{\ln\left[U\left(\frac{dF}{dz}\right)^{-1}\right]\right\}\right). \tag{9.10.11}$$

The contraction ratio b/d can be obtained by integration of dz along one of the free streamlines bordering the liquid jet. Using chain rule, dz can be expanded as

$$dz = \frac{dz}{dF}\frac{dF}{d\zeta}d\zeta. \tag{9.10.12}$$

The transformation (9.10.8) gives

$$\frac{dF}{d\zeta} = -\frac{2\psi_0}{\pi\zeta}. \tag{9.10.13}$$

Also from the fact that $q = U$ on the free streamlines (so that $W/U = e^{-i\lambda} = e^{-\Omega}$), we can write

$$\frac{dz}{dF} = \frac{1}{W} = \frac{1}{U}e^{\Omega}. \tag{9.10.14}$$

From the transformation (9.10.7), (9.10.14) becomes

$$\frac{dz}{dF} = \frac{1}{U}\exp[\cosh^{-1}(\zeta)] = \frac{1}{U}[\zeta + (\zeta^2 - 1)^{1/2}]. \tag{9.10.15}$$

Substituting (9.10.13) and (9.10.15) into (9.10.12) gives

$$dz = -\frac{2\psi_0}{\pi\zeta U}[\zeta + (\zeta^2 - 1)^{1/2}]d\zeta. \tag{9.10.16}$$

Integration of (9.10.16) over the free streamline between points C and B gives

$$z_B - z_C = -\frac{2\psi_0}{\pi U}\int_0^1 \left[1 + \frac{(\zeta^2 - 1)^{1/2}}{\zeta}\right] d\zeta. \tag{9.10.17}$$

Taking the real part of (9.10.17), where $(\zeta^2 - 1)^{1/2}$ is purely imaginary in the range $\zeta = (0, 1)$ and $\zeta = \cosh(i\lambda) = \cos(\lambda)$ is real on the free streamline, we obtain

$$x_B - x_C = -\frac{2\psi_0}{\pi U}. \tag{9.10.18}$$

Setting $x_B = d$ and $x_C = b$ and substituting the expression (9.10.3) for ψ_0 yield a solution for the contraction ratio as

$$\frac{b}{d} = \frac{\pi}{\pi + 2} \cong 0.611. \tag{9.10.19}$$

This value is in good agreement with experimental data for sufficiently deep liquid layers.

BIBLIOGRAPHY

Ablowitz, M.J. and A.S. Fokas (1997). *Complex Variables*, Cambridge University Press: New York.

Batchelor, G.K. (1967). *An Introduction to Fluid Dynamics*, Cambridge University Press: Cambridge.

Lamb, H. (1932). *Hydrodynamics*, Cambridge University Press: Cambridge (reprinted Dover Publications, New York, 1945).

Levin, E. (1954). "Note on the circle theorem of hydrodynamics," *Quarterly of Applied Mathematics* **12** (3), 315–316.

Milne-Thompson, L.M. (1940). "Hydrodynamical images," *Proceedings of the Cambridge Philosophical Society* **36**, 246–247.

Milne-Thompson, L.M. (1968). *Theoretical Hydrodynamics*, 5th ed., Macmillan & Company, London.

PROBLEMS

1. Derive the Cauchy-Riemann equations (9.1.2) by using the property that the derivative of a complex function $f = g(x, y) + ih(x, y)$ is independent of direction in the complex plane, so that $\partial f/\partial x = \partial f/\partial(iy)$.
2. Show that any analytic function $f = g(x, y) + ih(x, y)$ can be written as a function of the complex position z alone, independent of the complex conjugate $\bar{z}$. Take the derivative of f with respect to $\bar{z}$ and show that this derivative vanishes when f is analytic.
3. Prove that if the no-penetration condition is satisfied on the surface of a body in the complex ζ-plane, then it is also satisfied in the complex z-plane provided that $\zeta = f(z)$ is a conformal transformation.
4. The velocity field about the nose region of an elongated body can be approximated by superposing uniform flow and a source.
   (a) Find the velocity field for this flow. Determine the distance of the stagnation point upstream of the source and the body thickness as functions of the uniform-flow speed U and source strength m.
   (b) Does the velocity field approach the stagnation point flow of the form (9.7.2), with $n = 2$, close to the stagnation point?
5. Consider the flow induced by a vortex of strength $-\Gamma$ located at a distance a from a plane wall in the presence of a uniform flow U, as shown in Figure 9.28. Determine the location of the stagnation points on the wall as a function of aU/Γ. What is the maximum value of aU/Γ for which a stagnation point occurs on the wall?

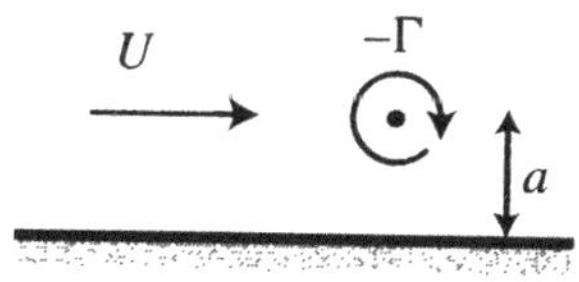

Figure 9.28 Uniform flow along an infinite wall with a vortex of strength $-\Gamma$.

6. A pair of counterrotating vortices of strength $\pm\Gamma$ propagates toward a circular cylinder, as shown in Figure 9.29. The cylinder has radius a and the initial vortex

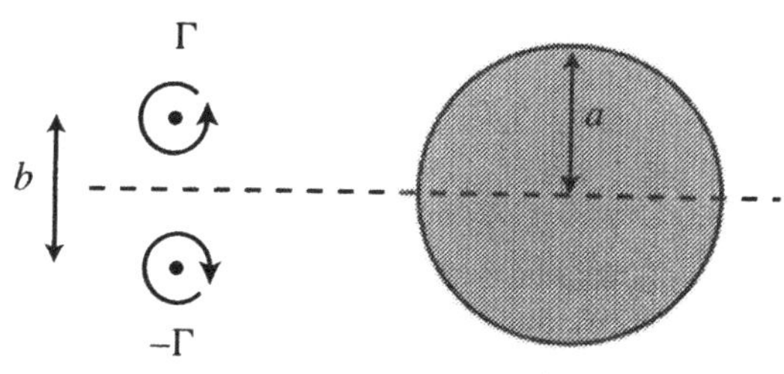

Figure 9.29 A vortex pair propagating toward a circle of radius a.

separation distance is b (far upstream from the cylinder). Compute and plot the path of the vortices as the pair approaches and passes by the cylinder.

7. Consider the problem of uniform flow with speed U past a doublet of strength μ located at a distance a from a flat plate, as shown in Figure 9.30. Plot the shape of the zero-streamline surface that encloses the doublet for different values of μ/Ua.

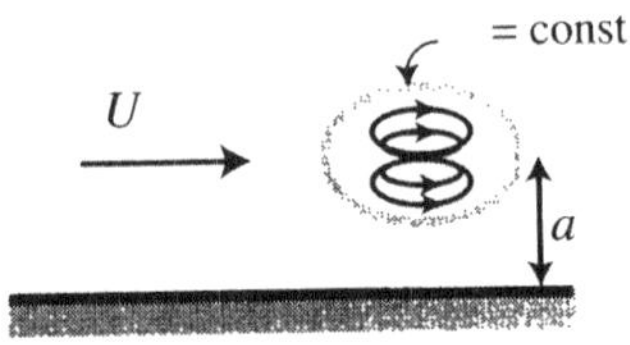

Figure 9.30 Uniform flow past a doublet located at a distance a from a flat wall.

8. Use the Joukowski transformation to obtain a solution for the complex potential for the problem of uniform, irrotational flow past a vertical flat plate of length a which is oriented as shown in Figure 9.31. Sketch the streamlines in the irrotational flow solution.

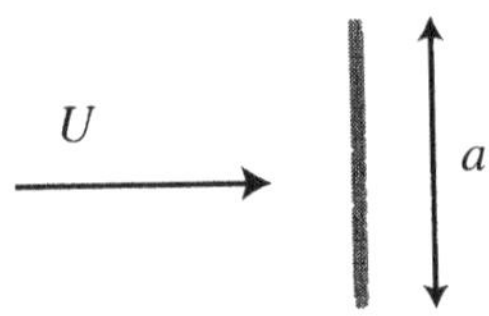

Figure 9.31 Uniform flow past a vertical flat plate of length a.

9. An air bubble rising in a liquid adapts a flattened shape due to pressure differences on the bubble surface. Consider a two-dimensional model of this process in which the bubble is restricted to maintain an elliptical shape and the gas pressure inside the bubble is assumed to be uniform. The terminal rise velocity of the bubble in the liquid bath is U and the uniform gas-liquid surface tension is γ. Sketch the shape of the bubble relative to the rise direction (draw the rise direction with an arrow) and estimate the aspect ratio of the bubble in an equilibrium state. Provide a complete explanation of your reasoning in obtaining this estimate.

10. Consider the transformation $\zeta = a^2/z$, where $z = x + iy$, $\zeta = \xi + i\eta$, and a is a positive, real constant. Is this transformation conformal? Where are the critical points? Determine what shape a circle of radius a in the z-plane maps to in the ζ-plane. Where are the points interior and exterior to this circle mapped?

11. Consider the two-dimensional problem of a vortex of strength Γ located at a point $z = a/2$ inside a circular cylinder of radius a (Figure 9.32). The vortex is known to rotate in a circular path about the center of the circular cylinder. Use the transformation given in Problem 10 to solve for the period of rotation of the

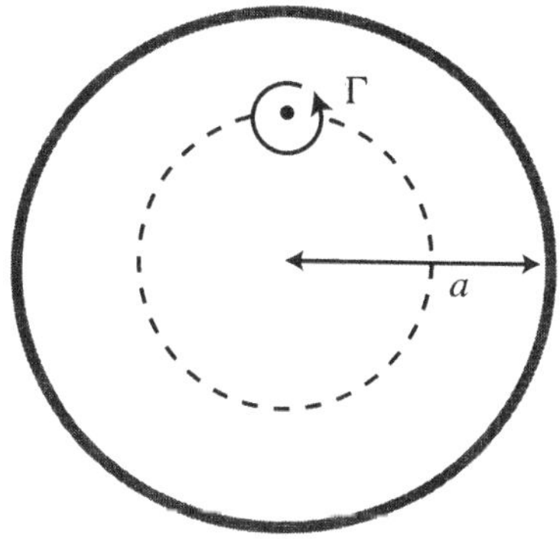

Figure 9.32 Vortex located inside a circular cylinder.

vortex. Your solution should have no singularities interior to the circle $|z| = a$ except at the vortex location. Also note that any vortex at the circle center has no effect on satisfaction of the no-penetration condition on the surface $|z| = a$ of the circular cylinder.

12. Solve the problem of a source of strength m located midway in a channel of thickness h (Figure 9.33) using the Schwarz-Christoffel transformation.

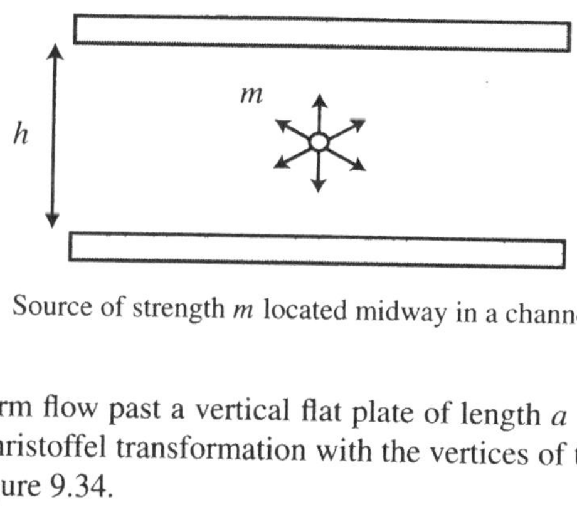

Figure 9.33 Source of strength m located midway in a channel of width h.

13. Solve for uniform flow past a vertical flat plate of length a (Figure 9.31) using the Schwarz-Christoffel transformation with the vertices of the polygon located as shown in Figure 9.34.

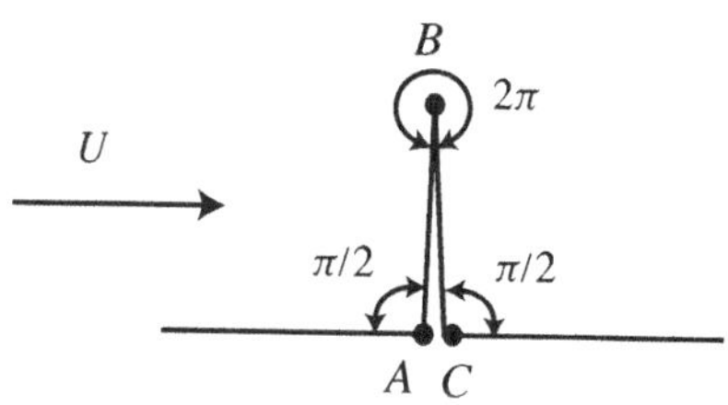

Figure 9.34 Location of vertices A, B, and C used for mapping flow past a flat plate into uniform flow along the real axis using the Schwarz-Christoffel transformation.

CHAPTER 10

FORCES ON BODIES IN TWO-DIMENSIONAL FLOWS

The assumption of two-dimensional flow not only simplifies solution of the velocity field, but in many cases it can be used to derive simple methods for determination of forces acting on an immersed body. In this chapter, we consider several methods for determining the force acting on a body in an irrotational flow. One of these force theorems makes use of the analogy between potential flows and analytic functions of a complex variable to express the force and moment on a body as integrals in the complex plane. For this reason, we start the chapter with a brief summary of theorems dealing with integration in the complex plane.

10.1 INTEGRATION IN THE COMPLEX PLANE

Integrals of functions in the complex plane are often much easier to evaluate than integrals of real-valued functions. This observation is based on two theorems, stated below, in which integration is performed over a closed circuit C in the complex plane and the integrand $f(z)$ is a function that is analytic everywhere but at a finite number of points.

Theorem 10.1.1 (Cauchy-Goursat Theorem). If $f(z)$ is analytic at all points inside and on a closed contour C, then

$$\int_C f(z)\,dz = 0. \qquad (10.1.1)$$

Theorem 10.1.2 (Residue Theorem). If $f(z)$ is analytic within and on a closed curve C, except at a finite number of singular points $z_1, \ldots, z_N$, then

$$\int_C f(z)\,dz = 2\pi i \sum_{n=1}^{N} R_n, \tag{10.1.2}$$

where R_n is the residue of $f(z)$ at z_n.

The residue R_n is most commonly calculated using the formula

$$R_n = \lim_{z \to z_n} \frac{1}{(m-1)!} \frac{d^{m-1}}{dz^{m-1}}[(z - z_n)^m f(z)], \tag{10.1.3}$$

which applies in cases where the singularity in the integrand $f(z)$ has the form of the mth order pole $g(z)/(z - z_n)^m$, where the function $g(z)$ is analytic at $z = z_n$. In the special case of a simple pole ($m = 1$), (10.1.3) reduces to

$$R_n = \lim_{z \to z_n} (z - z_n) f(z). \tag{10.1.4}$$

In general, $f(z)$ can be expanded in an annular region $r_0 < r < r_1$ bounded by the curves C_0 and C_1 around a singularity at z_0 (Figure 10.1) in a Laurent series

$$f(z) = \cdots + \frac{b_2}{(z - z_0)^2} + \frac{b_1}{z - z_0} + a_0 + a_1(z - z_0) + a_2(z - z_0)^2 + \cdots, \tag{10.1.5}$$

where the coefficients are given by

$$a_n = \frac{1}{2\pi i} \int_{C_0} \frac{f(\xi)}{(\xi - z_0)^{n+1}}\,d\xi, \qquad n = 0, 1, 2, \ldots,$$

$$b_n = \frac{1}{2\pi i} \int_{C_1} \frac{f(\xi)}{(\xi - z_0)^{-n+1}}\,d\xi, \qquad n = 1, 2, \ldots,$$

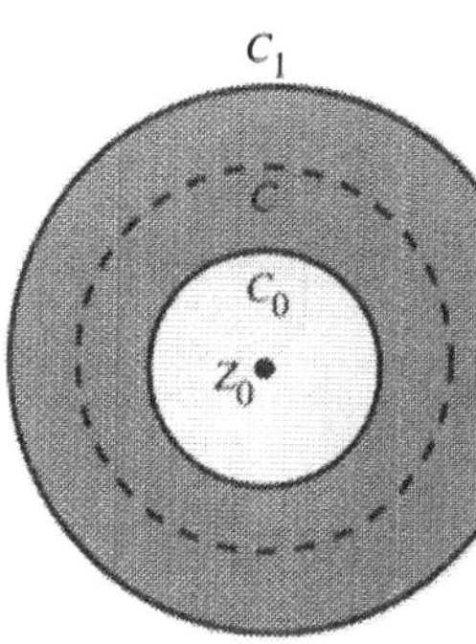

Figure 10.1 Annular region surrounding a singular point z_0 within which the Laurent series converges.

and where C denotes any closed curve lying in the annulus between C_0 and C_1. When all of the coefficients b_n vanish, the Laurent series reduces to the usual Taylor series. The residue of $f(z)$ at z_n is equal to the b_1 coefficient of the Laurent series expansion of $f(z)$ around z_n.

10.2 BLASIUS FORCE AND MOMENT LAWS

For the case of steady, irrotational flow about an immersed body, the force and moment on the body can be expressed in terms of integrals in the complex plane, which can be conveniently evaluated using the residue theorem. We consider a two-dimensional, fixed body with bounding contour C_B and outward unit normal $\mathbf{n}_B$, about which is drawn a closed contour C with outward normal $\mathbf{n}$, such that the complex velocity is everywhere analytic in the region A lying between C_B and C (Figure 10.2). Application of the momentum equations for a steady, inviscid fluid to the region A yields

$$\int_C \rho\mathbf{u}(\mathbf{u}\cdot\mathbf{n})\,d\ell = -\int_C p\mathbf{n}\,d\ell + \int_{C_B} p\mathbf{n}_B\,d\ell, \tag{10.2.1a}$$

$$\int_C \rho(\mathbf{x}\times\mathbf{u})(\mathbf{u}\cdot\mathbf{n})\,d\ell = -\int_C p(\mathbf{x}\times\mathbf{n})\,d\ell + \int_{C_B} p(\mathbf{x}\times\mathbf{n}_B)\,d\ell. \tag{10.2.1b}$$

The force $\mathbf{F}$ and the moment $\mathbf{M}$ acting on the body per unit length normal to the plane of motion are given for an inviscid fluid by

$$\mathbf{F} = -\int_{C_B} p\mathbf{n}_B\,d\ell, \qquad \mathbf{M} = -\int_{C_B} p\mathbf{x}\times\mathbf{n}_B\,d\ell. \tag{10.2.2}$$

Solving for $\mathbf{F}$ and $\mathbf{M}$ from (10.2.1) yields expressions for the force and moment as integrals over the contour C as

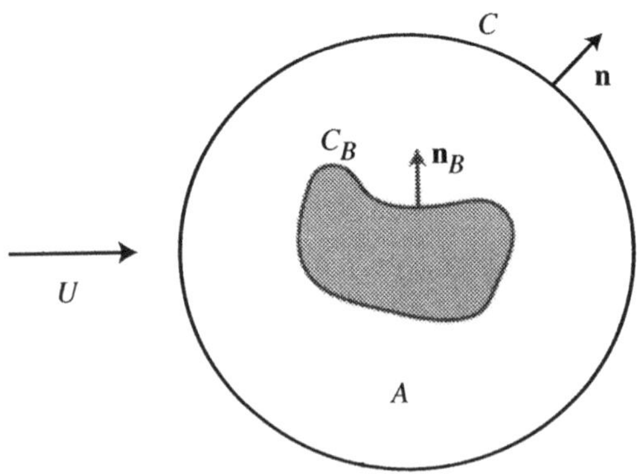

Figure 10.2 Uniform flow past an immersed body with bounding contour C_B and another contour C surrounding the body.

$$\mathbf{F} = -\int_C [p\mathbf{n} + \rho\mathbf{u}(\mathbf{u}\cdot\mathbf{n})]\, d\ell, \tag{10.2.3a}$$

$$\mathbf{M} = -\int_C [p(\mathbf{x}\times\mathbf{n}) + \rho(\mathbf{x}\times\mathbf{u})(\mathbf{u}\cdot\mathbf{n})]\, d\ell. \tag{10.2.3b}$$

Substituting the Bernoulli equation (8.2.3) for irrotational, steady flow with no body force into (10.2.3) gives

$$\mathbf{F} = \rho \int_C \left[\frac{1}{2}(\mathbf{u}\cdot\mathbf{u})\mathbf{n} - \mathbf{u}(\mathbf{u}\cdot\mathbf{n})\right] d\ell, \tag{10.2.4a}$$

$$\mathbf{M} = \rho \int_C \left[\frac{1}{2}(\mathbf{u}\cdot\mathbf{u})(\mathbf{x}\times\mathbf{n}) - (\mathbf{x}\times\mathbf{u})(\mathbf{u}\cdot\mathbf{n})\right] d\ell. \tag{10.2.4b}$$

In writing (10.2.4), the two-dimensional form of the integral identities (2.7.1) and (2.7.5) is used to write

$$\int_C B\mathbf{n}\, d\ell = \int_A \nabla B\, da = 0, \qquad \int_C B\mathbf{x}\times\mathbf{n}\, d\ell = B\int_A \nabla\times\mathbf{x}\, da = 0, \tag{10.2.5}$$

where B is the Bernoulli constant. The drag and lift are equated with the x- and y-components of the force, so that $\mathbf{F} = D\mathbf{e}_x + L\mathbf{e}_y$. The components of the displacement vector $d\mathbf{x}$ are written for counterclockwise integration about C in terms of its length $d\ell \equiv |d\mathbf{x}|$ and the components n_1 and n_2 of the unit normal vector as

$$dx = -n_2\, d\ell, \qquad dy = n_1\, d\ell. \tag{10.2.6}$$

The integral (10.2.4) yields expressions for the drag and lift as

$$D = \rho \int_C \left[\frac{1}{2}(v^2 - u^2)n_1 - uvn_2\right] d\ell, \tag{10.2.7a}$$

$$L = \rho \int_C \left[\frac{1}{2}(u^2 - v^2)n_2 - uvn_1\right] d\ell. \tag{10.2.7b}$$

In terms of the square of the complex velocity, $W^2 = (u^2 - v^2) - 2uvi$, (10.2.7) becomes

$$D - iL = -\frac{\rho}{2}\int_C W^2(n_1 + in_2)\, d\ell. \tag{10.2.8}$$

The result (10.2.6) can be used to write $(n_1 + in_2)\, d\ell = -i dz$. Substituting this result into (10.2.8) yields the *Blasius force law*

$$D - iL = \frac{\rho i}{2}\int_C W^2\, dz. \tag{10.2.9}$$

Both the drag and lift can thus be obtained by evaluating the residues of W^2 within a closed contour C surrounding the body.

Similarly, writing the expression for the moment in (10.2.4) in component form with $M = \mathbf{M} \cdot \mathbf{e}_z$ gives

$$M = \rho \int_C \left\{ \left[\frac{1}{2}(u^2 - v^2)y - xuv \right] n_1 + \left[\frac{1}{2}(u^2 - v^2)x - yuv \right] n_2 \right\} d\ell. \tag{10.2.10}$$

Noting that the real part of $zW^2\,dz$ is given by

$$\mathrm{Re}\,\{zW^2\,dz\} = [2uvx - (u^2 - v^2)y]n_1 - [x(u^2 - v^2) + 2uvy]n_2, \tag{10.2.11}$$

the expression (10.2.10) can be written as

$$M = -\frac{\rho}{2}\mathrm{Re}\left\{ \int_C zW^2\,dz \right\}. \tag{10.2.12}$$

The result (10.2.12) is known as the *Blasius moment law.*

Example 10.2.1 Force on Circular Cylinder with Circulation Γ in Uniform Stream. As an example of the use of the Blasius force theorem, we consider the problem of the force acting on a circular cylinder of radius a immersed in a uniform stream of speed U. We admit the possibility of nonzero swirl around the cylinder such that the circulation about any circuit C enclosing the cylinder is Γ. This swirl can be generated by placing a line vortex along the cylinder axis. The complex potential for the flow field is obtained by adding the expressions in (9.2.19) and (9.3.16) as

$$F = U\left(z + \frac{a^2}{z}\right) - \frac{i\Gamma}{2\pi}\ln(z), \tag{10.2.13}$$

and the associated complex velocity is

$$W = \frac{dF}{dz} = U\left(1 - \frac{a^2}{z^2}\right) - \frac{i\Gamma}{2\pi z}. \tag{10.2.14}$$

The azimuthal velocity at the body surface is obtained from (10.2.14) by setting $z = ae^{i\theta}$ as

$$u_\theta\,|_{r=a} = \frac{\Gamma}{2\pi a} - 2U\sin\theta. \tag{10.2.15}$$

The stagnation points for this flow can be obtained by setting the azimuthal velocity on the surface equal to zero, giving

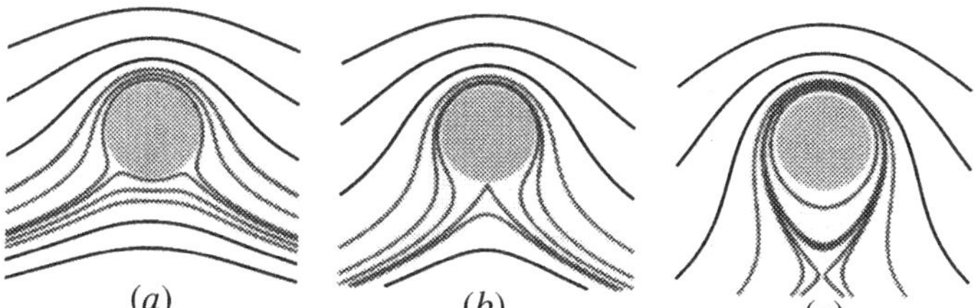

Figure 10.3 Streamlines for uniform flow past a circular cylinder with a central vortex, for cases with Γ/aU of (a) -3π, (b) -4π, and (c) -5π.

$$\sin\theta_{\text{stag}} = \frac{\Gamma}{4\pi a U}. \tag{10.2.16}$$

Streamlines for this flow are plotted in Figure 10.3 for three different values of Γ/aU. With $\Gamma = 0$, the stagnation points are located at the leading and trailing edges of the body. Negative circulation causes the streamlines to move downward, as shown in Figure 10.3a. When $|\Gamma/aU| \geq 4\pi$, the stagnation points move off the body surface, as shown in Figure 10.3c.

Substituting (10.2.14) into the Blasius force law (10.2.9) gives

$$D - iL = \frac{\rho i}{2}\int_C \left[U^2\left(1 - \frac{2a^2}{z^2} + \frac{a^4}{z^4}\right) - \frac{i\Gamma U}{\pi z}\left(1 - \frac{a^2}{z^2}\right) - \frac{\Gamma^2}{4\pi^2 z^2}\right] dz. \tag{10.2.17}$$

The integrand of (10.2.17) can be rearranged as

$$f(z) = \frac{U^2 a^4}{z^4} + \frac{i a^2 \Gamma U}{\pi z^3} - \left(2U^2 a^2 + \frac{\Gamma^2}{4\pi^2}\right)\frac{1}{z^2} - \frac{i\Gamma U}{\pi z} + U^2, \tag{10.2.18}$$

which has the form of a Laurent series. The residue of the integral in (10.2.17) is equal to the b_1 coefficient of the Laurent series in (10.2.18), or $R(z = 0) = -i\Gamma U/\pi$. The same result can be obtained using formula (10.1.3) with $m = 4$. From the residue theorem, we can write (10.2.17) as

$$D - iL = \rho U \Gamma i. \tag{10.2.19}$$

The result (10.2.19) indicates that in the absence of either the uniform stream or swirl there is no force acting on the body. This observation might also have been deduced from the symmetry of the flow field. In the combined presence of swirl and the uniform stream, the front-back symmetry of the velocity field is maintained, so the only force on the body is a lift force given by $L = -\rho U \Gamma$. This result is a special case of the Kutta-Joukowski theorem, which is derived for bodies of arbitrary shape in Section 10.4.

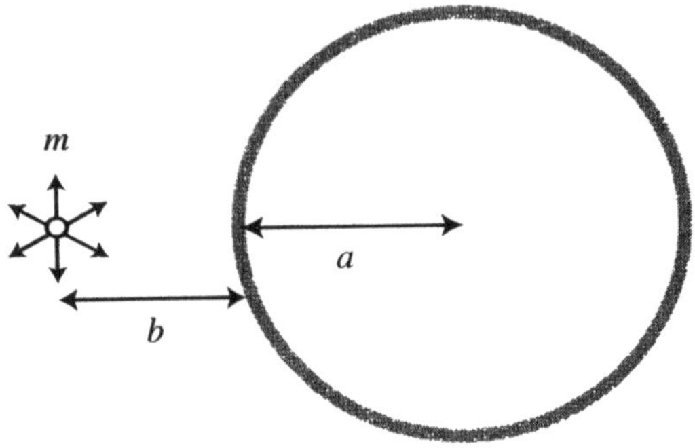

Figure 10.4 A source of strength m located at a distance b outside of a circular cylinder of radius a.

10.3 LAGALLY'S THEOREM

We now apply the Blasius integral law to compute the force on a circular cylinder of radius a due to a source located at a distance b from the cylinder surface, as shown in Figure 10.4. This problem is solved in Example 9.5.2 using the circle theorem, and the complex potential is given in (9.5.8) as

$$F(z) = \frac{m}{2\pi}\ln[z+(a+b)] + \frac{m}{2\pi}\ln\left[z+\frac{a^2}{a+b}\right] - \frac{m}{2\pi}\ln(z). \tag{10.3.1}$$

The complex velocity is

$$W(z) = \frac{m}{2\pi[z+(a+b)]} + \frac{m}{2\pi\left[z+a^2/(a+b)\right]} - \frac{m}{2\pi z}, \tag{10.3.2}$$

so the Blasius integral law (10.2.5) becomes

$$\begin{aligned} D - iL = \frac{\rho m^2 i}{2}\int_C \Bigg(&\frac{1}{4\pi^2[z+(a+b)]^2} + \frac{1}{4\pi^2\left[z+a^2/(a+b)^2\right]^2} + \frac{1}{4\pi^2 z^2} \\ &+ \frac{1}{2\pi^2[z+(a+b)]\left[z+a^2/(a+b)^2\right]} - \frac{1}{2\pi^2 z\left[z+a^2/(a+b)^2\right]} \\ &- \frac{1}{2\pi^2 z[z+(a+b)]}\Bigg)dz \end{aligned} \tag{10.3.3}$$

for any closed circuit C surrounding the circle $|z| = a$ but not enclosing the source at $z = -(a + b)$. The integrand of (10.3.3) is singular within the region enclosed by the circuit C at points $z_1 = 0$ and $z_2 = -a^2/(a + b)$. The first three terms in the integrand of (10.3.3) are each already in the form of a Laurent series, in which the b_1 term vanishes, so these terms do not contribute to the residues at the singular points. The second set of three terms in the integrand in (10.3.3) are of the form of first-order poles. Application of (10.1.4) at the two singular points gives

$$R_1 = -\frac{a+b}{2\pi^2 a^2} - \frac{1}{2\pi^2(a+b)}, \qquad R_2 = \frac{a+b}{2\pi^2(2ab+b^2)} + \frac{a+b}{2\pi^2 a^2}. \tag{10.3.4}$$

Using the residue theorem, the result (10.3.3) becomes

$$D - iL = \frac{\rho m^2}{2\pi}\left(\frac{1}{a+b} - \frac{a+b}{2ab+b^2}\right) = -\frac{\rho m^2 a^2}{2\pi b(a+b)(2a+b)}, \tag{10.3.5}$$

which indicates that there is a suction force on the cylinder attracting it toward the source.

The result (10.3.5) can be written in a different form by noting that the velocity u_I induced at the location of the source by the image source and sink within the cylinder is obtained from (10.3.2) as

$$u_I = \frac{m}{2\pi}\left(\frac{1}{a+b} - \frac{a+b}{2ab+b^2}\right). \tag{10.3.6}$$

Comparing (10.3.5) and (10.3.6), we can write the force on the cylinder as

$$D = \rho m u_I. \tag{10.3.7}$$

This result is a special case of the following theorem.

Theorem 10.3.1 (Lagally's Theorem). The force $\mathbf{F}$ per unit axial length exerted on a cylinder of arbitrary cross-sectional shape due to a source of strength m located outside the cylinder is related to the velocity $\mathbf{u}_I$ induced by the cylinder image set at the location of the source by

$$\mathbf{F} = \rho m \mathbf{u}_I. \tag{10.3.8}$$

A general proof of Lagally's theorem, which applies to both two- and three-dimensional flows, is given in Section 12.8.

10.4 D'ALEMBERT'S PARADOX AND THE KUTTA-JOUKOWSKI LIFT LAW

Consider a steady uniform flow with speed U past a fixed cylinder of arbitrary cross-sectional shape with a circulation Γ about any closed circuit enclosing the cylinder.

Let C_B denote the contour of the body cross section and C denote another closed contour enclosing the cylinder, with unit normal vectors $\mathbf{n}_B$ and $\mathbf{n}$, respectively, as shown in Figure 10.2. Application of the steady-state control-volume momentum equation to the region bounded by circuits C_B and C gives the expression (10.2.1). Substitution of the Bernoulli equation for pressure yields the integral (10.2.4) for force. The velocity field is decomposed as the sum of the mean flow and a perturbation field as

$$\mathbf{u} = \mathbf{U} + \mathbf{u}'. \tag{10.4.1}$$

Substituting (10.4.1) into (10.2.4) gives

$$\mathbf{F} = \rho \int_C \left\{ \left[\mathbf{U} \cdot \mathbf{u}' + \frac{1}{2} \mathbf{u}' \cdot \mathbf{u}' \right] \mathbf{n} - [\mathbf{u}'(\mathbf{U} \cdot \mathbf{n}) + \mathbf{u}'(\mathbf{u}' \cdot \mathbf{n})] \right\} d\ell. \tag{10.4.2}$$

Derivation of (10.4.2) uses the identities given in (10.2.5) and the fact that $\int_C \mathbf{u}' \cdot \mathbf{n}\, d\ell = 0$, which follows from the continuity restriction $\nabla \cdot \mathbf{u}' = 0$. Alternatively, using the definition (2.6.9) of the vector triple product, (10.4.2) can be rewritten as

$$\mathbf{F} = -\rho \int_C \left[\mathbf{U} \times (\mathbf{u}' \times \mathbf{n}) - \frac{1}{2} (\mathbf{u}' \cdot \mathbf{u}')\mathbf{n} + \mathbf{u}'(\mathbf{u}' \cdot \mathbf{n}) \right] d\ell. \tag{10.4.3}$$

The velocity perturbation $\mathbf{u}'$ induced by the body can be represented as the velocity field induced by some dilatation rate and vorticity distribution within or on the surface of the body. The integral of the rate of dilatation over the entire flow field vanishes since the body is fixed and the no-penetration condition is satisfied on the body surface. The integral of vorticity over the flow field is equal to the circulation Γ. From the results (6.6.9) and (6.6.11), the perturbation velocity $\mathbf{u}'$ asymptotically approaches the flow induced by a point vortex of strength Γ as $r \to \infty$, or

$$\mathbf{u}' \sim \frac{\Gamma}{2\pi r} \mathbf{e}_\theta. \tag{10.4.4}$$

Letting the contour C approach infinity, the integral over the second and third terms on the right-hand side of (10.4.3) goes to zero. The integral over the first term in (10.4.3) is most easily evaluated using the special case that the contour C is circular such that substitution of the asymptotic expression (10.4.4) into (10.4.3) gives

$$\mathbf{F} = -\frac{\Gamma\rho}{2\pi} \int_C \mathbf{U} \times (\mathbf{e}_\theta \times \mathbf{e}_r)\, d\theta = -\rho\Gamma\mathbf{U} \times \mathbf{e}_z, \tag{10.4.5}$$

where $\mathbf{e}_z$ is the unit vector normal to the plane of motion.

For the case that the circulation about the body vanishes, the result (10.4.5) provides a proof of the *D'Alembert theorem*, stated below.

Theorem 10.4.1 (D'Alembert Theorem). In a steady, uniform flow of an inviscid incompressible fluid past an immersed stationary body, about which there is no circulation, the net force exerted on the body by the fluid is zero.

In the more general case of nonzero circulation about the body, the result (10.4.5) yields the *Kutta-Joukowski theorem* stated below. Both the D'Alembert and Kutta-Joukowski theorems apply to bodies of arbitrary cross-sectional shape.

Theorem 10.4.2 (Kutta-Joukowski Theorem). In a steady, uniform flow with speed U of an inviscid incompressible fluid past a stationary body, with circulation Γ about the body, the lift force exerted on the body is given by $L = -\rho \Gamma U$.

10.5 APPLICATION TO TWO-DIMENSIONAL AIRFOILS

In this section, we examine the problem of uniform flow with speed U and angle of attack α past a symmetric airfoil with a sharp trailing edge (Figure 10.5). This problem models the nearly two-dimensional flow that occurs near the midsection of an airplane wing. Our principal interest is to determine an expression for the lift force exerted by the flow on the airfoil. The Kutta-Joukowski theorem indicates that the lift force is related to the circulation about a circuit surrounding the airfoil, which is shown in Example 10.2.1 to also be related to the location of the stagnation points on the body surface.

Two different possibilities for the streamlines in the uniform flow past an airfoil are shown in Figure 10.6, in each of which there are two stagnation points on the body surface. In Figure 10.6a, the rear stagnation point is located on the upper sur-

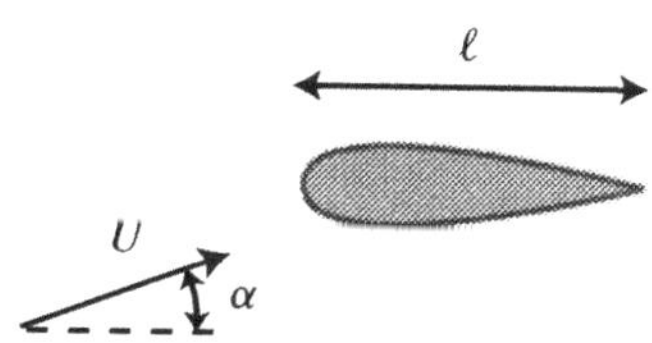

Figure 10.5 Uniform flow with angle of attack α past a symmetric Joukowski airfoil.

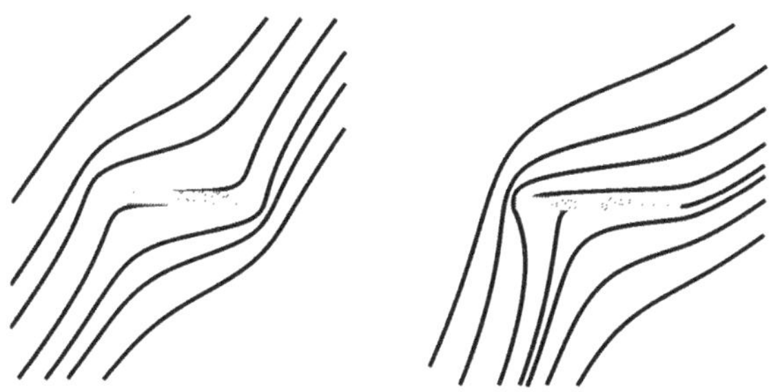

(*a*) (*b*)

Figure 10.6 Streamlines for two cases in the uniform flow past an airfoil: (*a*) rear stagnation point on the upper airfoil surface; (*b*) rear stagnation point at the trailing edge.

face of the airfoil and the velocity magnitude is nonzero at the sharp trailing edge. The no-penetration boundary condition requires that the surface velocity be everywhere tangent to the body surface. If the airfoil trailing edge is assumed to have infinite curvature, then fluid particles traveling along the body surface must turn infinitely fast as they advect around the sharp trailing edge, implying a singularity in the fluid velocity at this point. This type of sharp-edge singularity is demonstrated in Example 9.7.2, for which the velocity magnitude approaches infinity in proportion to $1/r^{1/2}$ as the distance r to the sharp edge approaches zero. In the case shown in Figure 10.6*b*, the rear stagnation point is coincident with the airfoil trailing edge, so there is no singularity in the velocity field. In flow experiments with sufficiently small angle of attack (less than about 8°–12°, depending on the airfoil shape), the boundary layer is observed to remain attached to the body surface until the vorticity separates at the trailing edge. The streamlines shown in Figure 10.6*b* are therefore more consistent with flow around actual airfoils, both because the nonphysical velocity singularity at the trailing edge is avoided and because the streamline pattern in this figure mimics the effects of boundary layer separation at the trailing edge. These observations are summarized in the *Kutta condition*, which may be stated as follows: *an irrotational flow past a body with a sharp trailing edge at small angle of attack adjusts itself so that the rear stagnation point coincides with the trailing edge*. The Kutta condition can be used to determine the circulation about the body, from which the lift force on the body can be obtained from the Kutta-Joukowski theorem.

Example 10.5.1 Flat-Plate Airfoil. We consider a flat-plate airfoil of length ℓ immersed in a uniform flow of speed U at angle of attack α to the plane of the plate, as shown in Figure 10.7*a*. This flat-plate airfoil can be mapped into a circle using the

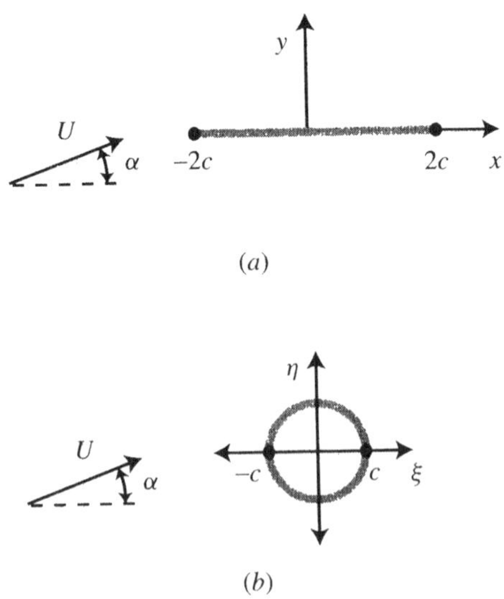

Figure 10.7 Uniform flow past a flat-plate airfoil: (*a*) physical z-plane; (*b*) transformed ζ-plane.

Joukowski transformation $z = \zeta + c^2/\zeta$, where $c = \ell/4$. The resulting flow in the ζ-plane, shown in Figure 10.7b, consists of uniform flow at angle of attack α past a circular cylinder of radius $a = c$. The complex potential for the flow in the ζ-plane is obtained using the circle theorem and the complex potential (9.2.4) for uniform flow as

$$\tilde{F}(\zeta) = U\left(\zeta e^{-i\alpha} + \frac{a^2}{\zeta}e^{i\alpha}\right) - \frac{i\Gamma}{2\pi}\ln(\zeta). \tag{10.5.1}$$

The strength of the vortex at the cylinder center is adjusted such that the rear stagnation point is located at $\zeta = c$, which is mapped into the trailing edge of the airfoil by the Joukowski transformation. The complex velocity for this flow in the ζ-plane is

$$\tilde{W}(\zeta) = U\left(e^{-i\alpha} - \frac{a^2}{\zeta^2}e^{i\alpha}\right) - \frac{i\Gamma}{2\pi\zeta}. \tag{10.5.2}$$

Setting $\tilde{W}(c) = 0$ yields an expression for the circulation about the airfoil as

$$\Gamma = -4\pi c U \sin\alpha. \tag{10.5.3}$$

Substitution of (10.5.3) into the Kutta-Joukowski theorem yields the lift coefficient $C_L \equiv L/\frac{1}{2}\rho\ell U^2$ for the airfoil as

$$C_L = 2\pi \sin\alpha. \tag{10.5.4}$$

In accordance with the Kutta-Joukowski theorem, the lift force on the airfoil is oriented normal to the direction of the free-stream velocity.

Example 10.5.2 Symmetric Joukowski Airfoil. The symmetric Joukowski airfoil, shown in Figure 10.8a, has a blunt leading edge and a sharp trailing edge. This geometry is typical of the wing section used for most commercial aircraft and helicopter rotors. The maximum thickness of the airfoil is denoted by t and the chord length is denoted by ℓ. The Joukowski transformation $z = \zeta + c^2/\zeta$ maps this airfoil into a circle in the ζ-plane centered at $\zeta = -m$ and of radius $a = c + m$, shown in Figure 10.8b. The part of the circle passing through the critical point $\zeta = c$ is mapped into the sharp trailing edge. The other critical point $\zeta = -c$ is contained within the circle, giving the airfoil a blunt leading edge. The displacement m of the circle center along the negative real axis is related to the airfoil thickness by $m/c = (4/3\sqrt{3})(t/\ell)$, and the transformation constant c is related to the chord length by $c/\ell = 1/4 + O(t/\ell)^2$, where t/ℓ is assumed to be small. The complex potential for uniform flow at angle of attack α past the circle in the ζ-plane is

$$\tilde{F}(\zeta) = U\left[(\zeta + m)e^{-i\alpha} + \left(\frac{a^2}{\zeta + m}\right)e^{i\alpha}\right] - \frac{i\Gamma}{2\pi}\ln(\zeta + m), \tag{10.5.5}$$

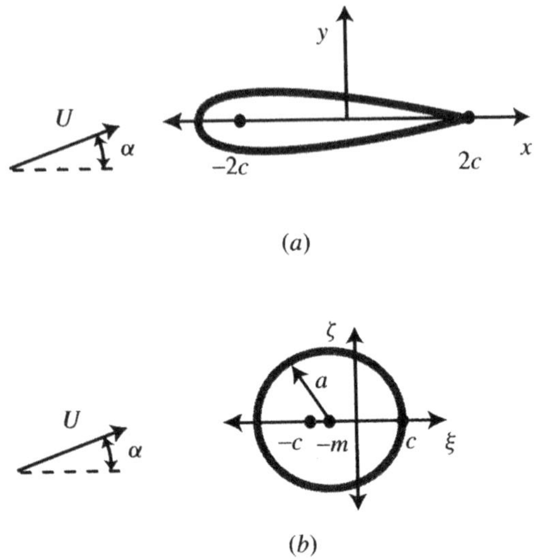

Figure 10.8 Uniform flow past a symmetric Joukowski airfoil: (*a*) physical *z*-plane; (*b*) transformed ζ-plane.

and the complex velocity in the ζ-plane is

$$\tilde{W}(\zeta) = U\left[e^{-i\alpha} - \frac{a^2}{(\zeta+m)^2}e^{i\alpha}\right] - \frac{i\Gamma}{2\pi(\zeta+m)}. \qquad (10.5.6)$$

The Kutta condition requires that $\tilde{W}(c) = 0$, from which (10.5.6) yields a solution for the circulation about the airfoil as

$$\Gamma = -4\pi a U \sin\alpha. \qquad (10.5.7)$$

Using the Kutta-Joukowski theorem and the substitution

$$a = \left(\frac{\ell}{4}\right)\left[1 + \left(\frac{4}{3\sqrt{3}}\right)\left(\frac{t}{\ell}\right)\right],$$

valid through $O(t/\ell)$, gives an expression for the lift coefficient as

$$C_L = 2\pi\left(1 + \frac{4}{3\sqrt{3}}\frac{t}{\ell}\right)\sin\alpha. \qquad (10.5.8)$$

This result reduces to that for the flat-plate airfoil as $t \to 0$. Although finite thickness increases the airfoil lift, in practice it is necessary to keep t/ℓ small in order to avoid boundary layer separation from the low-pressure (suction) side of the airfoil.

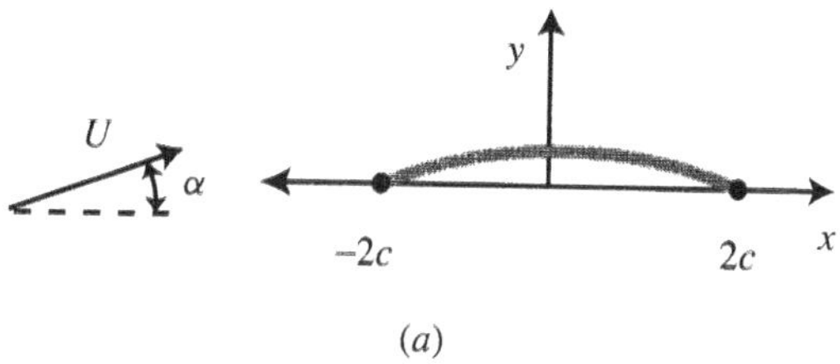

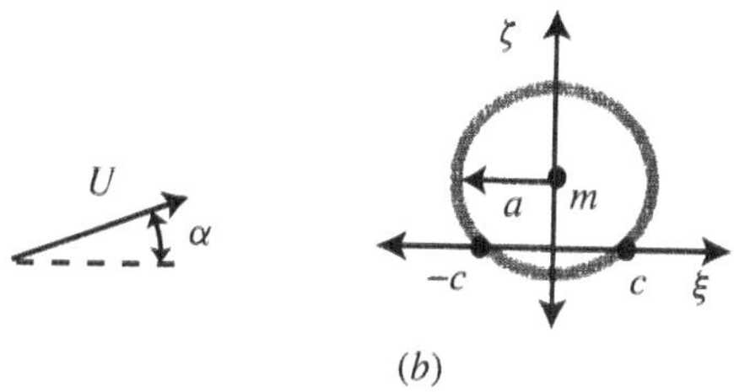

Figure 10.9 Uniform flow past a cambered airfoil: (*a*) physical z-plane; (*b*) transformed ζ-plane.

Example 10.5.3 Cambered-Plate Airfoil. A cambered-plate airfoil, shown in Figure 10.9*a*, consists of a plate formed in the shape of a circular arc, with sharp leading and trailing edges. This geometry is typical of many fan and propeller blades. The maximum displacement of the airfoil is denoted by h and the chord length is denoted by ℓ. The Joukowski transformation $z = \zeta + c^2/\zeta$ maps this airfoil into a circle in the ζ-plane centered at $\zeta = im$ and with radius $a = [c^2 + m^2]^{1/2}$, as shown in Figure 10.9*b*. The parts of the circle passing through the critical points at $\zeta = c$ and $\zeta = -c$ are mapped onto the sharp trailing and leading edges of the airfoil, respectively. The displacement m of the circle is related to the airfoil camber length h by $h = 2m$, and the Joukowski transformation constant c is related to the chord length by $c = \ell/4$. The complex potential for uniform flow at angle of attack α past the circle in the ζ-plane is

$$\tilde{F}(\zeta) = U\left[(\zeta - im)e^{-i\alpha} + \left(\frac{a^2}{\zeta - im}\right)e^{i\alpha}\right] - \frac{i\Gamma}{2\pi}\ln(\zeta - im), \qquad (10.5.9)$$

and the complex velocity in the ζ-plane is

$$\tilde{W}(\zeta) = U\left[e^{-i\alpha} - \frac{a^2}{(\zeta - im)^2}e^{i\alpha}\right] - \frac{i\Gamma}{2\pi(\zeta - im)}. \qquad (10.5.10)$$

The Kutta condition requires that $\tilde{W}(c) = 0$, from which (10.5.10) yields a solution for the circulation about the airfoil as

$$\Gamma = -4\pi c U \left(\sin\alpha + \frac{m}{c}\cos\alpha \right) \tag{10.5.11}$$

For $m/c \ll 1$, the Taylor series expansions for $\sin(m/c)$ and $\cos(m/c)$ give

$$\begin{aligned} \sin\left(\alpha + \frac{m}{c}\right) &= \cos\left(\frac{m}{c}\right)\sin\alpha + \sin\left(\frac{m}{c}\right)\cos\alpha \\ &= \sin\alpha + \left(\frac{m}{c}\right)\cos\alpha + O\left(\frac{m}{c}\right)^2, \end{aligned}$$

so (10.5.11) can be approximated with error $O(m/c)^2$ as

$$\Gamma \cong -4\pi c U \sin\left(\alpha + \frac{m}{c}\right). \tag{10.5.12}$$

Using the Kutta-Joukowski theorem and the substitutions $c = \ell/4$ and $m = h/2$ gives an expression for the lift coefficient as

$$C_L \cong 2\pi \sin\left(\alpha + \frac{2h}{\ell}\right). \tag{10.5.13}$$

This result reduces to that for the flat-plate airfoil as $h \to 0$. While nonzero camber increases the airfoil lift, the ratio h/ℓ must be kept small in order to avoid boundary layer separation from the airfoil.

BIBLIOGRAPHY

Ashley, H. and Landahl, M. (1965). *Aerodynamics of Wings and Bodies*, Addison-Wesley Publishing Company, Reading, Massachusetts (reprinted Dover Publications, New York, 1985).

Churchill, R.V., J.W. Brown, and R.F. Verhey (1974). *Complex Variables and Applications*, 3rd ed. McGraw-Hill, New York.

Currie, I.G. (1992). *Fundamental Mechanics of Fluids*, 2nd ed. McGraw-Hill, New York.

Milne-Thompson, L.M. (1958). *Theoretical Aerodynamics*, 4th ed. Macmillan and Company, New York (reprinted Dover Publications, New York, 1966).

PROBLEMS

1. Compute the surface pressure on a circular cylinder in a uniform flow U about which there is a circulation Γ. Compute the lift force by integrating $-p\sin\theta$ over the surface of the cylinder and compare your result to the lift given by the Kutta-Joukowski theorem.
2. The near wake behind a circular cylinder in a uniform flow at Reynolds number approximately in the range $10 < \mathrm{Re} < 40$ exhibits a stationary vortex pair

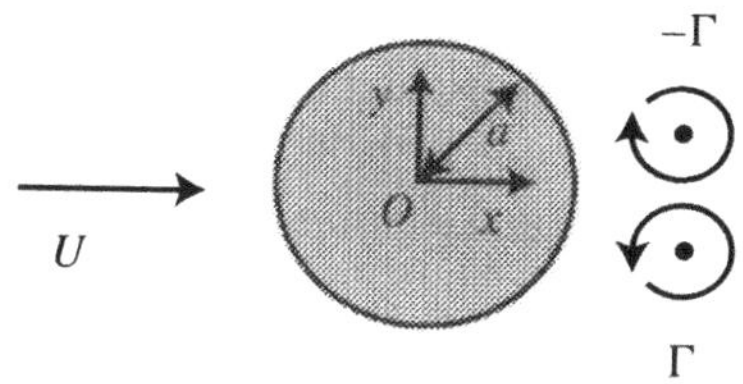

Figure 10.10 Stationary Föppl vortices in the wake of a circular cylinder placed in a uniform stream.

(known as Föppl vortices), as shown in Figure 10.10. The cylinder radius is a, the uniform flow speed is U, and the vortex location is $(c, \pm d)$.

(a) Derive the values of the vortex positions for which the flow field is in equilibrium.

(b) Use the Blasius force theorem to derive an expression for the drag force induced on the cylinder by these vortices in an inviscid flow.

3. Uniform flow past a general, two-dimensional body can be approximated, to arbitrary accuracy, by replacing the body with a set of N vortices of strength Γ_k located at z_k, $k = 1, \ldots, N$, and M point sources of strength m_k located at $\hat{z}_k$, $k = 1, \ldots, M$, where all vortices and sources are positioned either within or on the surface of the body. For a fixed body about which there is a circulation Γ, $\sum_{k=1}^{M} m_k = 0$ and $\sum_{k=1}^{N} \Gamma_k = \Gamma$. Use the Blasius force law for this system to develop an alternative derivation of the Kutta-Joukowski theorem.

4. Consider the problem of a rotating cylinder with radius a and angular velocity Ω such that the velocity field outside of the cylinder has the same form as that induced by a vortex with azimuthal velocity $v = a\Omega$ at location $r = a$. Introduce a vortex of strength Γ located at a position $z = 2a$ outside of the cylinder, as shown in Figure 10.11. For what value of Ω is the vortex held in a constant position? Use the Blasius theorem to determine the force on the cylinder in this equilibrium state.

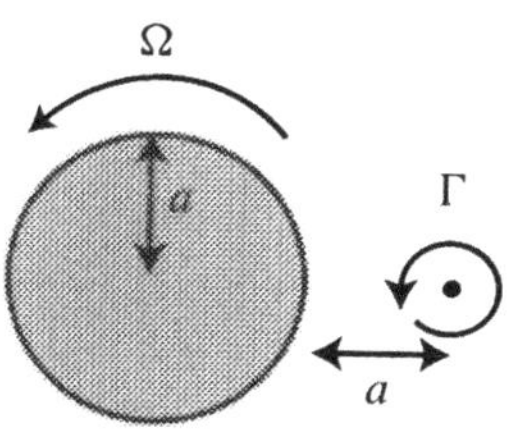

Figure 10.11 Vortex of strength Γ placed outside of a cylinder of diameter a that is spinning with rotation rate Ω.

5. Determine the shape of a Joukowski airfoil using the Joukowski transformation of the circle indicated in Example 10.5.2.

6. Consider the problem of uniform flow U oriented at an angle of 10° to a flat-plate airfoil. Determine the location on the airfoil of all stagnation points both (a) with enforcement of the Kutta condition at the trailing edge and (b) with the condition of no circulation around the plate. Determine the net force and moment acting on the airfoil in the former case.
7. A point vortex is located at a distance b upstream of the leading edge of a flat-plate airfoil, as shown in Figure 10.12. The vortex circulation is Γ_0, and a uniform flow is present with speed U in the x-direction. The chord length of the foil is ℓ and the foil is at zero angle of attack to the free stream.

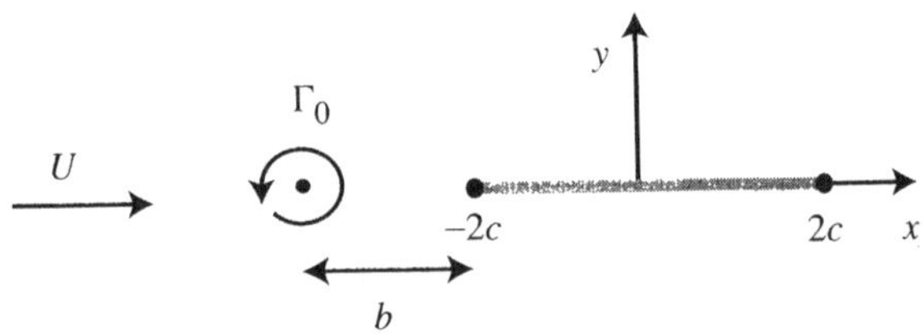

Figure 10.12 Point vortex located at a distance b upstream of the leading edge of a flat-plate airfoil.

(a) Use the Joukowski transformation $z = \zeta + c^2/\zeta$ to map the airfoil into a circle in the complex ζ-plane. Derive the relationship between the constant c in the Joukowski transformation and the airfoil chord length ℓ. Determine where the vortex maps to under this transformation in terms of the constants b and ℓ.

(b) For the case where $b = \ell/2$, determine the necessary circulation about the airfoil such that the Kutta condition is satisfied at the trailing edge.

CHAPTER 11

TWO-DIMENSIONAL FLOWS WITH VORTICITY

Inviscid flows with nonzero vorticity are commonly used as models for a wide range of high Reynolds number fluid flow problems. Two-dimensional models of inviscid vorticity transport are important for the study of applications such as the vortex street wake downstream of a cylinder, Kelvin-Helmholtz roller vortices that form in unstable shear layers, dynamic stall of pitching airfoils, and mesoscale oceanic vortices. The restriction to two dimensions decreases the number of independent variables, with associated decrease in number of computational points for numerical solutions. This restriction also enables the use of certain solution methods that cannot be used in the more general case of three-dimensional flows. The current chapter reviews the major solution methods and derives a few exact solutions for rotational two-dimensional flows.

11.1 SYSTEMS OF POINT VORTICES

Systems of point vortices are used as a crude model for the bulk motion of coherent vortices in two-dimensional turbulence, and they are the basis for more general numerical methods in which point vortex clouds are used to discretize a continuous vorticity distribution. The complex potential for a single vortex is given in (9.2.19). Taking the derivative with respect to the complex variable z yields the complex velocity for a single point vortex centered at z_0 as

$$W = -\frac{i\Gamma}{2\pi(z - z_0)}. \tag{11.1.1}$$

A point vortex induces no velocity on itself. In a system of many point vortices, each vortex will be advected by the induced velocity from all of the other vortices. The equation of motion for a vortex under the influence of $N - 1$ surrounding vortices is

$$\frac{d\bar{z}_k}{dt} = -\sum_{\substack{j=1 \\ j\neq k}}^{N} \frac{i\Gamma_j}{2\pi(z_k - z_j)}, \tag{11.1.2}$$

where an overbar denotes the complex conjugate.

Vorticity invariants for two-dimensional inviscid flows in a domain with no solid boundaries are derived in Section 7.5. The invariants include the total vorticity, the interaction energy (which is the nonsingular part of the kinetic energy), and the linear and angular impulse. For systems of point vortices, the vorticity field is given by a set of delta functions as

$$\omega(\mathbf{x}, t) = \sum_{j=1}^{N} \Gamma_j \delta(\mathbf{x} - \mathbf{x}_j). \tag{11.1.3}$$

The corresponding stream function is obtained by substituting (11.1.3) into the Green's function solution (6.5.8), giving

$$\psi(\mathbf{x}, t) = -\sum_{j=1}^{N} \frac{\Gamma_j}{2\pi} \ln|\mathbf{x} - \mathbf{x}_j|. \tag{11.1.4}$$

The integral of vorticity over the flow domain is equal to the sum of the vortex strengths, or

$$\Gamma_\infty = \sum_{j=1}^{N} \Gamma_j. \tag{11.1.5}$$

Invariance of Γ_∞ can be satisfied identically by holding the strength of all point vortices constant.

The linear impulse, $\mathbf{P} = P_1\mathbf{e}_x + P_2\mathbf{e}_y$, is given by (7.5.3) as

$$P_1 = \int_A y\omega da, \qquad P_2 = -\int_A x\omega da.$$

Substituting the vorticity distribution (11.1.3) and performing the integration over the delta functions give the linear impulse for a system of point vortices as

$$P_1 = \sum_{j=1}^{N} \Gamma_j y_j, \qquad P_2 = -\sum_{j=1}^{N} \Gamma_j x_j. \tag{11.1.6}$$

The angular impulse given by $(7.4.1)_2$ can similarly be reduced with use of the vorticity distribution (11.1.3) as

$$L = -\frac{1}{2}\sum_{j=1}^{N}(x_j^2 + y_j^2)\Gamma_j. \tag{11.1.7}$$

The kinetic energy in a two-dimensional flow is given by (7.5.7). It is shown in Chapter 7 that the kinetic energy is infinite in any two-dimensional flow for which $\Gamma_\infty \neq 0$. For a system of point vortices, the infinite value of kinetic energy in an unbounded system occurs both due to the flow far away from the vortices and due to the velocity singularity at the vortex centers. To examine the divergence of kinetic energy, we temporarily replace the unbounded domain A by a domain A' that is bounded by a large circle of radius R enclosing all point vortices and small circles of radius ε about each point vortex center. Substituting (11.1.3) and (11.1.4) for vorticity and stream function into (7.5.7) and letting $R \to \infty$ and $\varepsilon \to 0$ gives an asymptotic form for kinetic energy of a system of point vortices as

$$T \sim -\frac{1}{4\pi}\sum_{k=1}^{N}\sum_{\substack{j=1\\ j\neq k}}^{N}\Gamma_j\Gamma_k \ln|\mathbf{x}_k - \mathbf{x}_j| + \frac{1}{4\pi}(\ln R)\Gamma_\infty^2 - \frac{1}{4\pi}\left(\sum_{j=1}^{N}\Gamma_j^2\right)\ln\varepsilon. \tag{11.1.8}$$

The asymptotic form above is derived using the expression (7.5.8) for kinetic energy in a bounded domain and using leading-order approximations for ψ and $\mathbf{u}$ in parts of the flow boundary far from the point vortices (as $R \to \infty$) and close to the point vortex centers (as $\varepsilon \to 0$). The last two terms in (11.1.8) are invariant since the vortex strengths are constant. The interaction energy H is set equal to the first term on the right-hand side of (11.1.8), or

$$H = -\frac{1}{4\pi}\sum_{k=1}^{N}\sum_{\substack{j=1\\ j\neq k}}^{N}\Gamma_j\Gamma_k \ln|\mathbf{x}_k - \mathbf{x}_j|, \tag{11.1.9}$$

which must be invariant due to the invariance of T.

It is sometimes advantageous to express these invariants using the complex variable $z = x + iy$. Equivalent expressions for the invariants (11.1.6), (11.1.7), and (11.1.9) in complex variable form are given by

$$P = -i\sum_{j=1}^{N}\Gamma_j z_j, \qquad L = -\frac{1}{2}\sum_{j=1}^{N}\Gamma_j z_j \bar{z}_j,$$

$$H = -\frac{1}{4\pi}\sum_{k=1}^{N}\sum_{\substack{j=1\\ j\neq k}}^{N}\Gamma_j\Gamma_k \ln|z_k - z_j|. \tag{11.1.10}$$

Here P is a complex function whose real and imaginary parts are P_1 and P_2, respectively.

The equation (11.1.2) governing the motion of a system of point vortices can be shown to have the same form as Hamilton's equations:

$$\frac{dq_j}{dt} = \frac{\partial H}{\partial p_j}, \qquad \frac{dp_j}{dt} = -\frac{\partial H}{\partial q_j}, \qquad j = 1, \ldots, N, \tag{11.1.11}$$

where $q_j \equiv \Gamma_j^{1/2} x_j$ and $p_j \equiv \Gamma_j^{1/2} y_j$. The vortex interaction energy H is called the *Hamiltonian* of the system (11.1.11), and in analogy to a similar set of equations that governs the dynamics of systems rigid particles, q_j and p_j are called the *generalized coordinates* and *generalized momenta* of the system, respectively.

Hamilton's equations are known to govern the motion of a large number of nondissipative dynamical systems. Hamiltonian systems have been extensively studied and are known to exhibit a number of unique properties (Ozorio de Almeida, 1988; Whittaker, 1937). In some cases, the properties of Hamiltonian systems can be used to restrict the dynamics of point vortex systems without actually solving for the motion of each vortex. As one example, Aref and Pomphrey (1980) use a theorem from dynamics to prove that any system with three or fewer equal-strength vortices is integrable, so that chaotic motion of the group of vortices can occur only for systems of at least four vortices. More information on the dynamics of point vortex systems is given in the review article by Aref (1983).

11.2 CONFORMAL TRANSFORMATION OF UNIFORM-VORTICITY PATCHES

Many important vortex dynamics problems involve deformation of a patch of vorticity by an external flow, for which case representation of the vortex patch by a single point vortex would be inappropriate. Such problems are particularly important for determination of stability of a vortex that is deformed by an external straining or shearing flow. In large-scale oceanic flows, for which a quasi-two-dimensional assumption is appropriate, straining of coherent vortex patches by the velocity induced by other vortex patches is a principal mechanism leading to eddy merger. This process leads to the inverse energy cascade characteristic of two-dimensional turbulent flows in which the energy associated with small-scale fluctuations is passed to progressively larger scales of the turbulence. Problems involving deformation of vortex patches can often be effectively modeled using the approximation of constant vorticity within the patch. This section and the next will introduce different approaches for solution of vortex patches with uniform vorticity distribution, which may be used as a basis for either analytical or numerical solution methods.

One approach for solution of flow induced by a uniform vorticity patch of arbitrary shape is to use the conformal transformation approach described in Chapter 9 to map the distorted patch contour onto the unit circle. We consider a vortex patch with vorticity $\omega = \omega_0$ inside the patch and boundary C in the z-plane. The curve C is mapped onto the unit circle $|\zeta| = 1$ by the conformal transformation $z = f(\zeta)$, where the function $f(\zeta)$ is yet to be determined. We construct a function

$\Phi(z) = K(z) + G(z)$, which is defined such that $\Phi(z) = \bar{z}$ on C, where $K(z)$ and $G(z)$ are called *Schwarz functions* and are defined such that $K(z)$ is analytic inside C and $G(z)$ is analytic outside C. In principle, $\Phi(z)$ can be related to the conformal mapping $f(z)$. Consider now a complex velocity field given by

$$W \equiv u - iv = \begin{cases} -\frac{1}{2}i\omega_0 G & \text{outside } C, \\ \frac{1}{2}i\omega_0(K - \bar{z}) & \text{inside } C. \end{cases} \tag{11.2.1}$$

The function W is analytic outside C (where the flow is irrotational), whereas inside C the vorticity is ω_0. Continuity of velocity on the vortex boundary is guaranteed by the condition $\Phi(z) = \bar{z}$ on C, so that $K - \bar{z} = \Phi - \bar{z} - G = -G$ on C. Once the velocity components (U, V) on the boundary C are known, the evolution of a point $z = Z(t)$ on C is given by

$$\frac{dZ}{dt} = U + iV, \tag{11.2.2}$$

which follows from the fact that the boundary of the patch is a material surface.

In general, it is difficult to find functions G and K that satisfy the above conditions. Two examples of the use of this method are given below. Further discussion of this method is given in the text by Saffman (1992).

Example 11.2.1 Rankine's Circular Vortex. A rather trivial solution of the method described above can be obtained for a vortex patch that has the form of a circle of radius a centered at the origin of the coordinate system. The patch boundary C can be mapped onto the unit circle by the function $\zeta = z/a$. Since $z = a^2/\bar{z}$ on C, a suitable choice for the functions G and K that satisfies the restriction $\Phi \equiv G + K = \bar{z}$ on C is

$$G = \frac{a^2}{z}, \qquad K = 0. \tag{11.2.3}$$

Substituting these functions into (11.2.1) gives the complex velocity as

$$W \equiv u - iv = \begin{cases} -i\omega_0 a^2/(2z) & \text{outside } C, \\ -i\omega_0\bar{z}/2 & \text{inside } C. \end{cases} \tag{11.2.4}$$

Writing this result in terms of the polar velocity components (u_r, u_θ) and using $\Gamma = \omega_0\pi a^2$ gives $u_r = 0$ and

$$u_\theta = \begin{cases} \Gamma/2\pi r & \text{outside } C, \\ \Gamma r/2\pi a^2 & \text{inside } C. \end{cases} \tag{11.2.5}$$

The velocity has the same form as a point vortex outside C and is a solid-body rotation inside C, as illustrated in Figure 11.1.

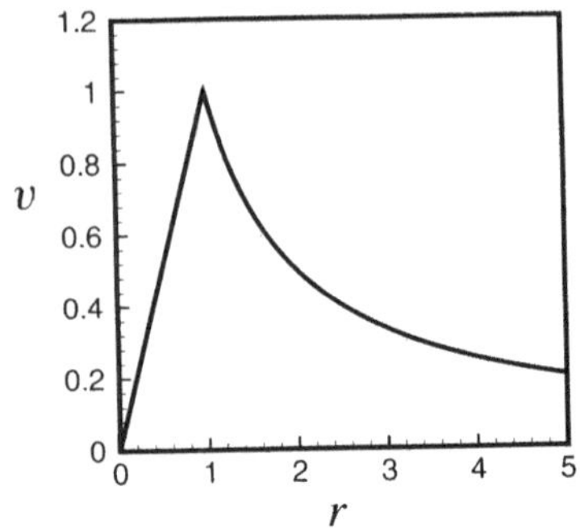

Figure 11.1 Azimuthal velocity profile for a Rankine vortex.

Example 11.2.2 Kirchhoff's Elliptic Vortex. A vortex patch of elliptical form, with major semiaxis a and minor semiaxis b, was shown by Kirchhoff (see Lamb, 1932, p. 232) to rotate without change of form about its centroid with angular velocity

$$\Omega = \frac{ab\omega_0}{(a+b)^2}. \tag{11.2.6}$$

This solution can be obtained (following Saffman, 1992, p. 167) by mapping the boundary C of the patch onto the unit circle using a transformation of Joukowski form,

$$z = \alpha\zeta + \frac{\beta}{\zeta}, \qquad \zeta = \frac{z + (z^2 - 4\alpha\beta)^{1/2}}{2\alpha}. \tag{11.2.7}$$

The point $\zeta = 1$ is mapped onto the end of the major axis of the elliptical patch by setting

$$\alpha = \tfrac{1}{2}(a+b)e^{i\phi}, \qquad \beta = \tfrac{1}{2}(a-b)e^{i\phi}, \tag{11.2.8}$$

where $\phi(t)$ is the angle between the major axis of the ellipse and the x coordinate axis. The area of the elliptical patch is given by

$$A = \pi ab = \pi(\alpha\bar{\alpha} - \beta\bar{\beta}), \tag{11.2.9}$$

where an overbar denotes the complex conjugate. Taking the complex conjugate of (11.2.7) gives

$$\bar{z} = \bar{\alpha}\bar{\zeta} + \frac{\bar{\beta}}{\bar{\zeta}}. \tag{11.2.10}$$

Since $\zeta\bar{\zeta} = 1$ on the boundary C, which corresponds to the unit circle in the ζ-plane, evaluating (11.2.10) on C gives

$$\bar{z}|_C = \frac{\bar{\alpha}}{\zeta} + \bar{\beta}\zeta. \tag{11.2.11}$$

Solving for $\bar{\alpha}$ from (11.2.9) and substituting into (11.2.11) give

$$\bar{z}|_C = \frac{\bar{\beta}}{\alpha}\left(\alpha\zeta + \frac{\beta}{\zeta}\right) + \frac{ab}{\alpha\zeta}. \tag{11.2.12}$$

The term in parentheses in (11.2.12) is equal to z from the transformation (11.2.7), so that

$$\bar{z}|_C = \frac{\bar{\beta}}{\alpha}z + \frac{ab}{\alpha\zeta}. \tag{11.2.13}$$

Since $ab/\alpha\zeta$ is analytic outside the vortex patch and $\bar{\beta}z/\alpha$ is everywhere analytic, the Schwarz functions $K(z)$ and $G(z)$ can be chosen as

$$K(z) = \frac{\bar{\beta}z}{\alpha}, \qquad G(z) = \frac{ab}{\alpha\zeta}. \tag{11.2.14}$$

It follows from (11.2.13) that the function $\Phi \equiv K + G$ satisfies the restriction $\Phi = \bar{z}$ on C. The velocity inside the vortex patch is obtained using (11.2.1) as

$$u - iv = \frac{i\omega_0}{2}(K - \bar{z}) = -\frac{\omega_0}{2}(Cy + iDx), \tag{11.2.15}$$

where

$$C = 1 + \left(\frac{a-b}{a+b}\right)e^{-2i\phi}, \qquad D = 1 - \left(\frac{a-b}{a+b}\right)e^{-2i\phi}. \tag{11.2.16}$$

The velocity outside the vortex patch is given by

$$u - iv = -\frac{i\omega_0}{2}G = -\frac{i\omega_0 ab}{2\alpha\zeta}. \tag{11.2.17}$$

We let the boundary of the vortex patch be represented parametrically by $z = Z(t, s)$, where s is a parameter measuring location along the curve C defined such that the unit circle in the ζ-plane can be expressed as $\zeta = e^{is}$. Since the boundary of the vortex patch is a material surface, we can write $d(Z - z)/dt = 0$ on C. Using the facts that $dz/dt = u + iv$ and $dZ/dt = \partial Z/\partial t + (\partial Z/\partial s)(ds/dt)$, the boundary condition on C becomes

$$\frac{\partial Z}{\partial t} - (u + iv)\big|_C = \frac{\partial Z}{\partial s}\frac{ds}{dt}. \tag{11.2.18}$$

Taking the complex conjugate of (11.2.17) yields

$$(u + iv)\big|_C = \frac{i\omega_0 ab}{2\bar{\alpha}} e^{is}. \tag{11.2.19}$$

From (11.2.7) and the parameterization $\zeta = e^{is}$, it follows that

$$\frac{\partial Z}{\partial t} = \dot{\alpha} e^{is} + \dot{\beta} e^{-is}, \qquad \frac{\partial Z}{\partial s} = i(\alpha e^{is} - \beta e^{-is}). \tag{11.2.20}$$

Substituting (11.2.19) and (11.2.20) into the boundary condition (11.2.18) yields

$$\left(\dot{\alpha} - \frac{i\omega_0 ab}{2\bar{\alpha}}\right) e^{is} + \dot{\beta} e^{-is} = [i\alpha e^{is} - i\beta e^{-is}]\frac{ds}{dt}. \tag{11.2.21}$$

The boundary condition (11.2.21) must hold for all values of s. Since ds/dt is real-valued, we can independently set to zero the sum of the coefficients of terms multiplied by e^{is} and that of terms multiplied by e^{-is}. Using the resulting expressions to eliminate ds/dt yields an equation that restricts the change of α and β with time as

$$\frac{\dot{\alpha}}{\alpha} - \frac{i\omega_0 ab}{2\alpha\bar{\alpha}} = -\frac{\dot{\beta}}{\beta}. \tag{11.2.22}$$

Using the definitions (11.2.8) for α and β, (11.2.22) is found to have the unique solution

$$\dot{a} = 0, \qquad \dot{b} = 0, \qquad \dot{\phi} = \frac{\omega_0 ab}{4\alpha\bar{\alpha}} \equiv \Omega, \tag{11.2.23}$$

which is identical to the result (11.2.6) given previously. An alternative derivation of the elliptic vortex patch solution using elliptic coordinates is given by Lamb (1932, p. 232).

The velocity $\mathbf{u}_r$ of fluid particles relative to the rotating axes of the ellipse is related to the fluid velocity $\mathbf{u}$ relative to fixed axes by

$$u_r = u + \Omega y, \qquad v_r = v - \Omega x. \tag{11.2.24}$$

Substituting the velocity field (11.2.15) inside the vortex patch into (11.2.24) and using the result (11.2.23) yield an expression for the relative velocity components as

$$u_r = \dot{x}_r = -\frac{a\Omega}{b} y_r, \qquad v_r = \dot{y}_r = \frac{b\Omega}{a} x_r, \tag{11.2.25}$$

where (x_r, y_r) are coordinates measured relative to the rotating ellipse axes. Integrating (11.2.25) yields

$$x_r = c_1 a \cos(\Omega t + c_2), \qquad y_r = c_1 b \sin(\Omega t + c_2), \tag{11.2.26}$$

where c_1 and c_2 are constants of integration. This result indicates that the particle paths are ellipses *relative to the rotating axes of the elliptical patch.* The particle paths in terms of absolute coordinates (x, y) can be obtained by the transformation

$$\begin{aligned} x &= x_r \cos(\Omega t) - y_r \sin(\Omega t) \\ &= \tfrac{1}{2}c_1(a+b)\cos(2\Omega t + c_2) + \tfrac{1}{2}c_1(a-b)\cos(c_2), \end{aligned} \tag{11.2.27a}$$

$$\begin{aligned} y &= x_r \sin(\Omega t) + y_r \cos(\Omega t) \\ &= \tfrac{1}{2}c_1(a+b)\sin(2\Omega t + c_2) + \tfrac{1}{2}c_1(a-b)\sin(c_2). \end{aligned} \tag{11.2.27b}$$

The absolute paths of particles inside the vortex patch are circles, about which the particles advect with angular velocity 2Ω, which is twice the angular velocity of the patch contour.

11.3 CONTOUR INTEGRATION METHOD FOR UNIFORM-VORTICITY PATCHES

An expression for stream function in a two-dimensional incompressible flow is given by (6.5.8) as

$$\psi(x, y, t) = -\frac{1}{2\pi}\int_{A^*} \ln(r)\omega(x', y', t)\, da', \tag{11.3.1}$$

where A^* denotes all of space. In the current section, we apply this result to a patch of vorticity occupying a finite region A with boundary C. The vorticity is assumed to be equal to a uniform constant ω_0 within the vortex patch and is zero outside of the patch. For this case, (11.3.1) reduces to an integral over the interior of the patch as

$$\psi(x, y, t) = -\frac{\omega_0}{2\pi}\int_A \ln(r)\, da'. \tag{11.3.2}$$

Taking the gradient of (11.3.2) and using the fact that $\nabla f(\mathbf{x}-\mathbf{x}') = -\nabla' f(\mathbf{x}-\mathbf{x}')$ for any function f, where ∇ and ∇' are the delta operators with respect to the unprimed and primed coordinate systems, respectively, yield

$$\nabla\psi = -\frac{\omega_0}{2\pi}\int_A \nabla(\ln|\mathbf{x}-\mathbf{x}'|)\, da' = \frac{\omega_0}{2\pi}\int_A \nabla'(\ln|\mathbf{x}-\mathbf{x}'|)\, da'. \tag{11.3.3}$$

The last integral can be transformed into an integral over the boundary C of the patch using Green's theorem (2.7.1), giving

$$\nabla\psi = \frac{\omega_0}{2\pi}\int_C \ln|\mathbf{x}-\mathbf{x}'|\mathbf{n}'\, d\ell', \tag{11.3.4}$$

where $\mathbf{n}'$ is the outward unit normal and $d\ell'$ is an element of arc length of C evaluated at point $\mathbf{x}'$. For integration in the counterclockwise direction

$$n_1' d\ell' = dy', \qquad n_2' d\ell' = -dx', \tag{11.3.5}$$

so the velocity components are given by

$$\begin{aligned} u &= \frac{\partial \psi}{\partial y} = -\frac{\omega_0}{2\pi} \int_C \ln |\mathbf{x} - \mathbf{x}'| \, dx', \\ v &= -\frac{\partial \psi}{\partial x} = -\frac{\omega_0}{2\pi} \int_C \ln |\mathbf{x} - \mathbf{x}'| \, dy'. \end{aligned} \tag{11.3.6}$$

The result (11.3.6) gives rise to a numerical method for evaluating the evolution of a vortex patch, called the *contour dynamics method*, in which the boundary C is discretized by some number N computational points. Since C is a material curve, the equation of motion of a computational point n with position $\mathbf{x}_n(t)$ is given by

$$\frac{d\mathbf{x}_n}{dt} = \mathbf{u}(\mathbf{x}_n, t), \tag{11.3.7}$$

where the velocity $\mathbf{u}(\mathbf{x}_n, t)$ at point n is obtained by evaluating the integrals in (11.3.6). A minor complication that arises when evaluating the integrals numerically is that the integrands are singular at points where $\mathbf{x} = \mathbf{x}'$; however, the singularity is integrable and can be handled by analytically evaluating the integral over segments where the singularity occurs.

A more serious type of singularity occurs when the contours of two different vortex patches, or two different parts of the contour of a single vortex patch, come close to each other. For such situations, a "contour surgery" approximation developed by Dritchel (1988) is often used to smooth the singularity and connect nearby sections of the contours. The contour dynamics method can also be used for a vortex patch with variation of vorticity within the patch by approximating the patch using a series of constant-vorticity levels, with a contour bordering each level. In such cases, computational points are placed on each contour and evolved using integrals similar to (11.3.6), with ω_0 replaced by the vorticity jump across the contour. A review of contour dynamics methods is given by Pullin (1992).

An example computation obtained using the contour dynamics method to compute the self-induced rotation of an initially elliptical vortex patch about its centroid is shown in Figure 11.2. The figure also shows the pathlines of two fluid particles located on the patch boundary. During the time that the patch boundary completes one-half rotation about its centroid, the fluid particles on the patch boundary traverse a complete circle. Like most Lagrangian methods, in which the computational points move as material points, there is no numerical dissipation of vorticity in the contour dynamics method. There is therefore no observed spreading of the vortex core area and the vortex strength remains constant.

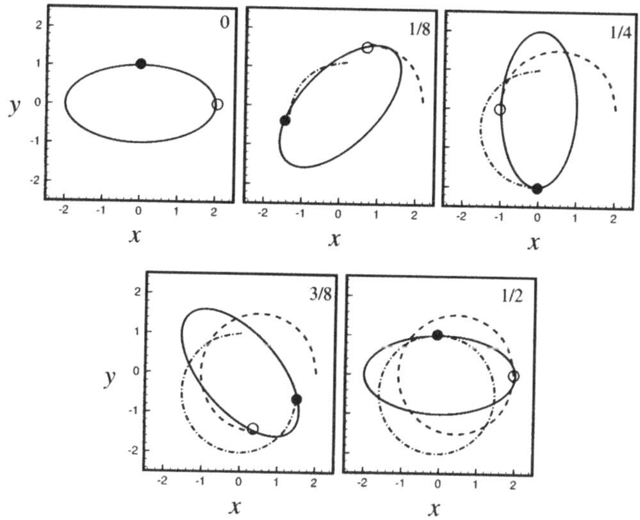

Figure 11.2 Self-induced rotation of an elliptical vortex patch about its centroid with aspect ratio 2, computed using a contour dynamics method. The solid line indicates the patch contour, and the dashed and dashed-dotted lines are pathlines of fluid particles whose current position is marked by filled and open circles, respectively.

11.4 DISCRETE-VORTEX NUMERICAL METHOD: BASIC METHOD

Discrete-vortex methods provide a numerical computation approach for solution of general incompressible flows. The basic idea of these methods is to evolve vorticity on a set of Lagrangian (or advected) computational points via solution of the vorticity transport equation and to solve for the velocity field by solution of the Biot-Savart integral. In order to perform the Biot-Savart integration, it is necessary to introduce some representation of the vorticity field based on the vorticity value at the computational points. Discrete-vortex methods can be categorized according to the type of vorticity representation as either singular or continuous. A singular vortex method uses a vorticity representation consisting of singular points, such as point vortices, or perhaps regularized singularities, such as Rankine vortices with small cores. Continuous vortex methods use a continuous vorticity support based on overlapping elements (e.g., "blobs"), each of which has Gaussian or some other smooth variation of vorticity within its core. The vorticity elements are highly overlapping and smooth in continuous vortex methods, whereas they infrequently overlap and are either singular or piecewise smooth in singular vortex methods. While vortex methods have been used for solution of two- and three-dimensional motions of both inviscid and viscous fluids (Cottet and Koumoutsakos, 2000; Leonard, 1985), the discussion in the current section is restricted to two-dimensional inviscid flows.

Lagrangian vortex methods offer a number of advantages compared to other methods for computation of inviscid flows, such as finite-difference, finite-element, or finite-volume methods. In particular, Lagrangian vortex methods possess the following attributes: (1) computational points need only be introduced in regions where there is significant vorticity (which is a small part of the flow field in many problems), (2) external flows can be solved without truncation to a finite domain, (3) Lagrangian methods can be formulated that exhibit little or no numerical dissipation, and (4) the calculations are self-adaptive, since the computational points are carried to regions of intense vorticity by the flow. The above observations also apply to contour dynamics methods, discussed in the previous section. Contour dynamics methods have the additional advantage that only the boundary of a uniform-vorticity patch needs to be discretized, whereas solution of vortex patch motion with a discrete-vortex method requires points to be placed in the entire interior of the patch. On the other hand, discrete-vortex methods are much simpler to implement than contour dynamics methods for problems with spatial variation of vorticity, and discrete vortex methods do not exhibit the long-time numerical instability that occurs in contour dynamics computations (Baker, 1990). (This instability can be controlled with the contour surgery method.) Furthermore, vortex methods are extendable to three-dimensional and viscous flows, whereas contour dynamics methods have a much more restricted range of applications.

Lagrangian vortex methods follow the vorticity on a set of N computational points, whose position $\mathbf{x}_n(t)$ moves with the local fluid velocity according to

$$\frac{d\mathbf{x}_n}{dt} = \mathbf{u}(\mathbf{x}_n, t). \tag{11.4.1}$$

The vorticity and velocity fields are evolved by solution of the Biot-Savart integral and the vorticity transport equation, which for two-dimensional inviscid flow have the form

$$\frac{D\omega}{Dt} = 0, \tag{11.4.2}$$

$$\mathbf{u}(\mathbf{x}, t) = -\frac{1}{2\pi}\int_A \frac{\mathbf{r} \times \omega(\mathbf{x}', t)\mathbf{e}_z}{r^2}\, da', \tag{11.4.3}$$

where $\mathbf{r} = \mathbf{x} - \mathbf{x}'$, $r = |\mathbf{r}|$, and $\mathbf{e}_z$ is a unit vector normal to the plane of the flow.

In order to integrate the Biot-Savart equation, it is necessary to interpolate the vorticity in the space in-between the computational points. This interpolation is represented symbolically as

$$\omega(\mathbf{x}, t) = \sum_{n=1}^{N} \Gamma_n(t) f_n(\mathbf{x} - \mathbf{x}_n). \tag{11.4.4}$$

Substituting (11.4.4) into (11.4.3) and integrating, the Biot-Savart integral becomes

$$\mathbf{u}(\mathbf{x}, t) \cong -\frac{1}{2\pi}\sum_{n=1}^{N} \Gamma_n(t) \int_A \frac{\mathbf{r} \times f_n(\mathbf{x}' - \mathbf{x}_n)\mathbf{e}_z}{r^2}\, da'. \tag{11.4.5}$$

The function $f_n(\mathbf{x} - \mathbf{x}_n)$, called the core or basis function, specifies how vorticity is distributed in space for each element. The basis function is normalized such that

$$\int_A f_n(\mathbf{x} - \mathbf{x}_n)\, da = 1. \tag{11.4.6}$$

Singular vortex methods employ basis functions that are singular, or highly localized, such as a point vortex or a Rankine vortex with small core. Continuous vortex methods employ smooth basis functions, such as a Gaussian, with a large amount of element overlap. It is also desirable that the integral over the basis function in (11.4.5) possess an analytic solution.

The element strength, or amplitude, $\Gamma_n(t)$ represents the integral over all space of the vorticity associated with a given vorticity element. Integrating (11.4.4) over A and using the normalization (11.4.6) yield the identity

$$\sum_{n=1}^{N} \Gamma_n = \int_A \omega\, da. \tag{11.4.7}$$

As noted in Section 7.5, the right-hand side of (11.4.7) is invariant in time. Furthermore, for singular vorticity support (as used in singular vortex methods) this integral is invariant for all material regions A.

The standard approach for setting the element amplitude is to assign some constant area h_n^2 to each vortex element, such that the element amplitude is related to the vorticity ω_n at computational point n by

$$\Gamma_n = \omega_n h_n^2. \tag{11.4.8}$$

The vorticity transport equation (11.4.2) requires that vorticity be constant at a material point, so (11.4.8) implies that the element amplitudes are constant in time. This approach has the advantage that invariance of the integral on the right-hand side of (11.4.7) is guaranteed.

On the other hand, as pointed out by Beale (1986), the region of area h_n^2 associated with each element becomes highly deformed as the elements are advected in the flow, with the result that over time the vorticity representation (11.4.4) for continuous vortex methods can become increasingly noisy, leading to substantial differences between the vorticity representation and the actual vorticity distribution along with subsequent increase in velocity calculation error. An alternative approach designed to handle this problem is to refit the element amplitudes at each time step such that the vorticity given by the representation (11.4.4) is constant with time on the computational points. This procedure results in a $N \times N$ matrix equation for Γ_n of the form

$$\omega_m = \sum_{n=1}^{N} f_{mn} \Gamma_n, \qquad m = 1, \ldots, N, \tag{11.4.9}$$

where $f_{mn} \equiv f_n(\mathbf{x}_m - \mathbf{x}_n)$. For highly overlapping elements, the matrix equation (11.4.9) is ill-conditioned and the exact solution for Γ_n yields a noisy vorticity representation when substituted into (11.4.4). An iterative method that yields an approximate solution of (11.4.9) in which the noise in the exact solution is filtered out was developed by Marshall and Grant (1996), based on analogous procedures that have been used for solution of other ill-conditioned problems (such as solution of the heat equation backwards in time). In this procedure, the element amplitude is *temporarily* assumed to be uniform over a set of elements $Q(m)$ located in the neighborhood of element m. Letting $P(m)$ denote the remaining elements, (11.4.9) can be approximated as

$$\omega_m = \Gamma_m^{(q+1)} \sum_{n \in Q(m)} f_{mn} + \sum_{n \in P(m)} \Gamma_n^{(q)} f_{mn}, \tag{11.4.10}$$

where q is an iteration index. The set $Q(m)$ is typically chosen to include the nearest 5–15 neighbors of each element m, although the solution is found not to be particularly sensitive to the choice of the set $Q(m)$. The iteration (11.4.10) converges quickly, with relative error in Γ_n of less than 10^{-6} in 6–12 iterations. The vorticity error from the iteration procedure (11.4.10) is of order $\ell^2 \nabla^2 \omega$, where ℓ is the typical minimum distance between adjacent computational points.

An example computation using the basic vortex method, with adaptive amplitude fitting, is performed to examine the mutually induced distortion and eventual merger of a pair of corotating vortex patches. The vortex initial condition is shown schematically in Figure 11.3. It is found empirically that two initially circular patches of radius a with uniform vorticity and equal strengths will eventually merge when the initial separation distance d, measured from the vortex core centers, is less than the critical value $d_{\text{crit}}/a = 3.3$ (Waugh, 1992). The vortex evolution is shown over a time series in Figure 11.4 for a case with $d/a = 3$, where the patch interior is indicated by black or gray shading for vorticity originating from the left-hand or right-hand patch, respectively. The merger of the two vortex patches is shown in this figure over five even time increments. At the end of the merger process, most of the vorticity is located in a central patch, which is surrounded by spiraling vorticity arms.

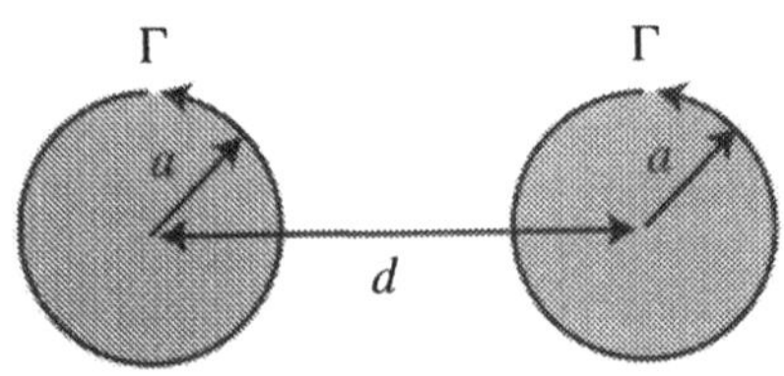

Figure 11.3 Separation distance d for a pair of initially circular corotating vortices with core radius a.

Figure 11.4 Merger of a pair of corotating vortices of equal strength, computed using the discrete-vortex method with adaptive amplitude fitting. The vortices initially have uniform vorticity with $d/a = 3$. The vorticity field is plotted as a black or gray region, indicating vorticity originating from the left- or right-hand vortex, respectively, over five equal time increments.

11.5 DISCRETE-VORTEX NUMERICAL METHOD: ACCELERATION TECHNIQUES

One difficulty with vortex methods is that the number of computations required to compute the velocity on each of N computational points varies as $O(N^2)$, which can become prohibitive for large N. Two approaches have been developed in the literature to accelerate the velocity computation, both of which reduce the number of computations per time step to $O(N \ln N)$ or less for sufficiently large values of N.

11.5.1 Vortex-in-Cell Method

The basic idea of the vortex-in-cell method is to replace the $O(N^2)$ solution of the Biot-Savart integral (11.4.3) by a fast Fourier transform solution of the Poisson equation

$$\nabla^2 \psi = -\omega. \tag{11.5.1}$$

The vortex-in-cell procedure requires a grid to be placed over the flow field, through which the Lagrangian computation points move. As in any grid-based method, suitable boundary conditions must be selected at the sides of this grid. Because it is desired to employ fast Fourier transform methods, a uniform grid is usually employed in the vortex-in-cell method, which makes the method difficult to use for geometrically complicated domains.

Once the grid is selected, the vortex-in-cell method employs four steps for computation of the velocity field. In the first step, the vorticity is interpolated from the Lagrangian computational points to the surrounding grid points. This interpolation is performed using an area-weighted interpolation procedure illustrated in Figure 11.5. We consider a Lagrangian computational point n located in a grid cell with four grid nodes, labeled $i = \{1, 2, 3, 4\}$. Horizontal and vertical lines are drawn passing through the Lagrangian point n, forming four subregions of the grid cell. The areas of these subregions are labeled $A_1, \ldots, A_4$, such that, for example, the region with area A_1 is opposite grid node 1. If A denotes the area of the entire grid cell

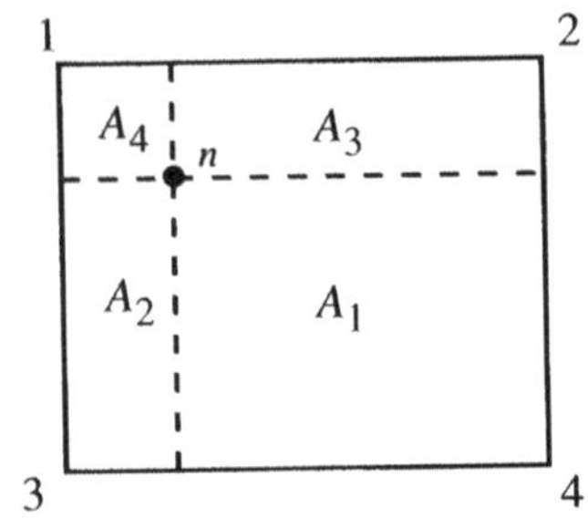

Figure 11.5 Interpolation procedure used in the vortex-in-cell method.

($A = \sum_{i=1}^{4} A_i$), the vorticity contributed to the ith surrounding grid node by the nth Lagrangian element is $A_i \Gamma_n / A^2$. The total vorticity ω_i on grid node i is given by the formula

$$\omega_i = \sum_{n=1}^{N} \frac{A_i^n \Gamma_n}{A^2}, \tag{11.5.2}$$

where A_i^n is the area associated with grid node i corresponding to a Lagrangian element n lying in a neighboring grid cell.

The second step is to solve the Poisson equation (11.5.1) for the stream function using a fast Fourier transform approach. The third step is to solve for the velocity components at the grid nodes from the definition of stream function,

$$u = \frac{\partial \psi}{\partial y}, \qquad v = -\frac{\partial \psi}{\partial x}, \tag{11.5.3}$$

which is usually done using a centered finite-difference procedure. The final step is to interpolate the velocity back onto the Lagrangian computational points. This interpolation is performed by again using an area weighting procedure, such that if u_i^n are the velocity components at the four nodes of the grid cell containing the Lagrangian point n, then the velocity on the Lagrangian point is given by

$$u_n = \sum_{i=1}^{4} \frac{u_i^n A_i^n}{A}. \tag{11.5.4}$$

The Lagrangian computational points are then advected according to (11.4.1) and the procedure is repeated.

The vortex-in-cell method is fast and relatively simple to implement. When the grid is sufficiently fine to resolve the vortex structures, the method is capable of efficiently and accurately computing the motion of a large number of Lagrangian vortex elements. However, this method introduces some of the negative aspects associated with use of a fixed grid, including necessity of assigning boundary conditions at the grid edges, which is inconvenient for unbounded flows, and the need to evaluate stream function on all grid nodes rather than just the nodes located near regions of significant vorticity. The requirement that the grid be uniform and the boundary conditions periodic makes it difficult to use this method for flows past solid bodies. It is also important to note that the interpolation procedure used in the vortex-in-cell method introduces a numerical dissipation that acts both to artificially spread and to dissipate the vorticity field. Like similar errors introduced by truncation of the nonlinear convective acceleration term of the Euler equations with fixed-grid methods, the numerical dissipation can be controlled with adequate resolution of the vorticity field, but it does place restrictions on the ability of the method for simulation of truly inviscid flows.

The so-called *method of local corrections* (Almgren et al., 1994; Anderson, 1986) can be used to correct for most of the interpolation errors in the vortex-in-cell method. The main idea behind this approach is to compute the velocity field induced only by those vortex elements sufficiently distant from the point at which the velocity is desired using the vortex-in-cell method and to compute the effect of the nearby elements directly. The velocity induced by the distant elements is obtained by first computing the entire flow field using the standard vortex-in-cell method and then subtracting off the flow induced by the nearby elements.

11.5.2 Multipole Expansion Method

The multipole expansion acceleration method employs an ordering of the Lagrangian points via a treelike box structure. The box trees used in vortex methods typically have a single box covering all the Lagrangian points as the first generation, which is divided into some number m boxes to yield the next generation, which are in turn divided to generate a subsequent generation, and so forth. The division process is continued until a specified number of Lagrangian points are contained in the highest generation (i.e., smallest) boxes. Each box (except the first-generation box) has a parent box and each box (except the highest generation boxes) has some number m offspring boxes.

The basic idea of multipole expansion acceleration methods is to replace the point-point interaction approach used in the sum (11.4.5) by a box-point interaction approach. In this procedure, the contribution of all "source" points contained in a sufficiently distant box to the velocity at a given "target" point is computed based only on the distance between the box centroid and the target point and on the "moments" of the source points within the box (which are independent of the location of the target point). The box-point interaction can be computed to arbitrary accuracy using the multipole expansion (which is a variant of the Taylor expansion). For sufficiently large number of Lagrangian points N, the box-point multipole expansion scheme can be shown to require $O(N \ln N)$ computations per time step. Some investigators go even further to use a box-box interaction scheme, in which the contribution of a box of source points is computed at the center of a box of target points, and then the velocity at the individual Lagrangian target points is obtained by a Taylor series. It is possible with this local expansion approach to reduce the number of computations

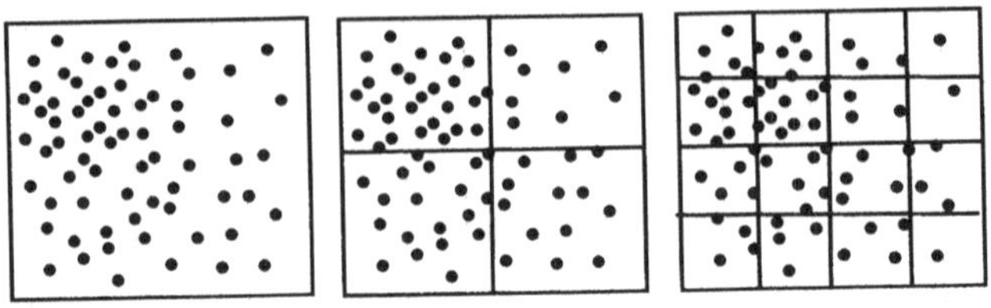

Level 1 Level 2 Level 3

Figure 11.6 Formation of a regular box structure for a multipole expansion acceleration method.

per time step to $O(N)$; however, it is questionable whether the time savings obtained by the local expansion approach justifies the additional velocity error and coding complications introduced.

The main distinction between different multipole expansion methods lies in the boxing procedures. Two main types of boxing procedures are frequently used, which we shall refer to as *regular* and *adaptive*. In a regular boxing procedure, the large box containing all of the source points is divided into a prescribed number m offspring boxes of equal size. These boxes are in turn each divided into m equal-size offspring boxes, and so forth, as illustrated in Figure 11.6. Since the spatial distribution of the Lagrangian points is not uniform, some of the boxes may reach the specified number of Lagrangian points sooner than others, so that the number of generations varies from one part of the tree to another. This approach has the advantage that it is a simple matter to determine the neighboring boxes of a given box, which reduces the computer memory requirement. In an adaptive boxing procedure, the first-generation box is divided into two offspring boxes that each have equal number of Lagrangian points (within one). The offspring boxes may be of different sizes, and the division is typically performed by bisecting the box along the median Lagrangian point in the coordinate direction in which the distance separating the points is largest. Each of these offspring boxes are divided into two subsequent offspring boxes of equal number of points, and so forth. This procedure is illustrated in Figure 11.7. The adaptive boxing procedure has the advantage that each box in a given generation

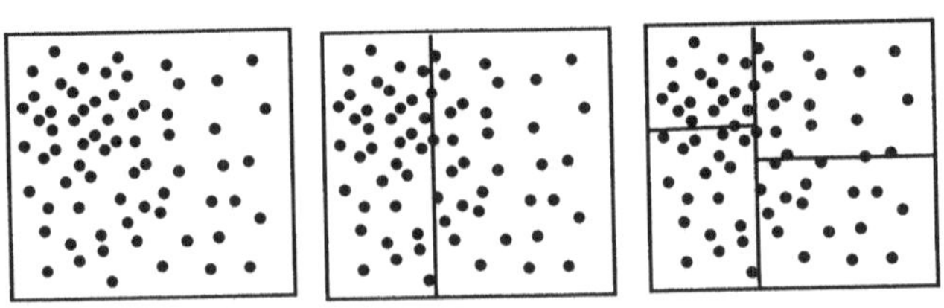

Level 1 Level 2 Level 3

Figure 11.7 Formation of an adaptive box structure for a multipole expansion acceleration method.

has approximately the same number of Lagrangian points, which minimizes the total number of boxes and aids in load balancing for implementation of this method on parallel computers (Gharakhani and Ghoniem, 1996).

The contribution $\Delta\mathbf{u}_\ell$ from source points in a box ℓ to the velocity at a target point $\mathbf{x}_p$ is given by

$$\Delta\mathbf{u}_\ell(\mathbf{x}_p, t) = \frac{1}{2\pi}\sum_{m=0}^{\infty}\sum_{n=0}^{\infty}\frac{(-1)^{m+n}}{m!\,n!}I_{\ell,mn}\frac{\partial^{m+n}}{\partial x^m \partial y^n}\left(\frac{\mathbf{e}_z \times \mathbf{r}}{r^2}\right), \tag{11.5.5}$$

where $r \equiv |\mathbf{r}| \equiv |\mathbf{c}_\ell - \mathbf{x}_p|$ is the distance between the centroid $\mathbf{c}_\ell = \xi_\ell \mathbf{e}_x + \eta_\ell \mathbf{e}_y$ of box ℓ and the target point. The symbol $I_{\ell,mn}$ denotes the moments of box ℓ, defined as a sum over the N_ℓ Lagrangian points in box ℓ by

$$I_{\ell,mn} = \sum_{q=1}^{N_\ell}\Gamma_q(x_q - \xi_\ell)^m(y_q - \eta_\ell)^n. \tag{11.5.6}$$

The box moments $I_{\ell,mn}$ are independent of the location of the target point and can thus be computed once and stored at each time step.

While the moments of the highest generation (smallest) boxes are computed by the sum (11.5.6), moments of boxes in previous generations of the tree structure can be more efficiently computed by summing the contributions of the offspring boxes, using the binomial formula to expand the moments of the difference terms in (11.5.6). For instance, the difference term in the x-direction can be expanded from the center of box ℓ to that of a parent box i as

$$(x - \xi_1)^m = [(x - \xi_\ell) + (\xi_\ell - \xi_i)]^m = \sum_{r=0}^{m}\binom{m}{r}(\xi_\ell - \xi_i)^r(x - \xi_\ell)^{m-r}, \tag{11.5.7}$$

where we use the standard notation

$$\binom{m}{r} \equiv \frac{m!}{(m-r)!\,r!}.$$

Using this approach, the contribution of offspring box ℓ to the moment of parent box i is given by

$$\sum_{r=0}^{m}\sum_{s=0}^{n}\binom{m}{r}\binom{n}{s}(\xi_\ell - \xi_i)^r(\eta_\ell - \eta_i)^s I_{\ell,(m-r)(n-s)}. \tag{11.5.8}$$

Similarly, the derivatives in (11.5.5) can be computed directly (which is sufficient for low-order terms) or obtained by recurrence relationships (Chen and Marshall, 1999).

The multipole expansion procedure is used only for boxes that lie at greater than a critical distance d_{crit} from the target point. To optimize the acceleration method, it is desirable to perform the box-point interaction with as large a box as possible and using as few terms in the multipole expansion as possible while maintaining the error below a specified value. We suppose that the multipole expansion (11.5.5) is truncated to include only those indices satisfying $m + n \leq P$, where P is called the "order" of the expansion. A theoretical upper bound for the absolute error of the multipole expansion of order P is given by Salmon and Warren (1994) as

$$E_P \leq \frac{1}{(d-b)^2}\left[(P+2)\frac{B_{P+1}}{d^{P+1}} - (P+1)\frac{B_{P+2}}{d^{P+2}}\right], \tag{11.5.9}$$

where b is a length scale associated with the box, d is the distance between the box center and the target point, and the moment magnitudes B_k are defined for a box ℓ by

$$B_k \equiv \sum_{q=1}^{N_\ell} |\Gamma_q||\mathbf{x}_q - \mathbf{c}_\ell|^k. \tag{11.5.10}$$

A slightly weaker (but simpler to implement) error bound is derived from (11.5.9) by Winckelmans et al. (1995) as

$$E_P \leq \frac{B_0}{(d-b)^2}\left[(P+2)\left(\frac{B_2}{B_0 b^2}\right)^{(P+1)/2}\left(\frac{b}{d}\right)^{P+1} - (P+1)\left(\frac{B_2}{B_0 b^2}\right)^{P}\left(\frac{b}{d}\right)^{P+2}\right]. \tag{11.5.11}$$

The error bound (11.5.11) has the advantage that it depends only on the zero- and second-order moment magnitudes, B_0 and B_2, which are easy to compute. The Newton-Raphson iteration method can be used to solve for the critical distance from (11.5.11) with a prescribed value of absolute error. Empirical tests yield typical error values about an order of magnitude lower than given by these theoretical upper bounds. An optimized multipole expansion acceleration procedure can be constructed by using these theoretical error bounds, or empirically adjusted modifications, to adaptively determine the critical distance d_{crit} for each source box and to select the combination of the largest source box and the lowest order of the multipole expansion required to compute the velocity at a given target point to within a prescribed accuracy.

Results of an example computation for the instability and roll-up of a shear layer is plotted at two different times in Figure 11.8. These results are obtained using a vortex method with the adaptive amplitude fitting procedure (11.4.9) and an adaptive multipole expansion acceleration method. The vorticity field is interpolated for plotting purposes by fitting a triangular mesh to the Lagrangian computational points. Figure 11.8 shows the roll-up of the shear layer into vortical structures that are connected by thin, spiraling vorticity sheets. Both vortex methods and contour dynamics methods are well suited for resolving these thin vorticity sheets, since the Lagrangian

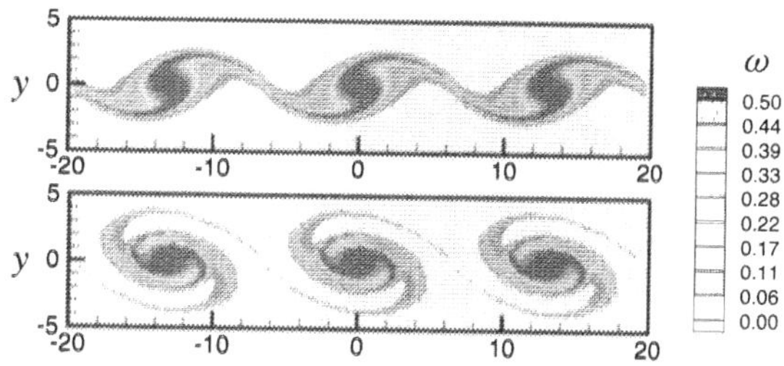

Figure 11.8 Vorticity contours for a computation of the roll up of a shear layer with Gaussian initial vorticity distribution of the form $\omega = \exp(-y^2)$ using a discrete vortex method. Periodic boundary conditions are enforced by adding five periods of the vorticity field in the Biot-Savart integration on each side of the computed region.

points are advected into these sheets by the flow and the flow calculation has no numerical dissipation.

11.6 VORTEX SHEETS

Thin sheets of vorticity are often observed in flow visualization experiments and numerical flow computations in both two- and three-dimensional flows. For instance, boundary layer separation usually results in ejection of a thin vorticity sheet away from the body. Similarly, roll-up of the vorticity sheet behind an airplane wing leads to formation of the tip vortex structures. Vortex sheets also provide a useful tool in several computational methods to enforce boundary conditions at a flow discontinuity. In this section, we examine some aspects of the dynamics of vortex sheets in two dimensions.

11.6.1 Induced Velocity

A vortex sheet can be constructed by a limiting process in which a vorticity layer with thickness 2ε is compressed to zero thickness. The vorticity is zero outside the layer and within the layer has magnitude $\omega = \gamma/2\varepsilon$. As the layer thickness shrinks to zero, the vorticity increases in proportion to $1/\varepsilon$. The vortex sheet is characterized by the parameter γ, called the *sheet strength*, defined by

$$\gamma = \lim_{\varepsilon \to 0} \int_{-\varepsilon}^{\varepsilon} \omega \, dy. \tag{11.6.1}$$

In the limit as the sheet thickness shrinks to zero, the vorticity within the flat vortex sheet can be expressed as $\omega = \gamma\delta(y)$. Substituting this vorticity distribution into the Biot-Savart equation (6.5.7a) and integrating over the sheet thickness yields the velocity induced by a flat vortex sheet as

$$\mathbf{u}(\mathbf{x}, t) = \frac{1}{2\pi}\boldsymbol{\gamma} \times \int_C \frac{\mathbf{r}}{r^2}\, dx', \qquad (11.6.2)$$

where $\boldsymbol{\gamma} \equiv \gamma \mathbf{e}_z$ and $\mathbf{r} = \mathbf{x} - \mathbf{x}'$. Writing $\mathbf{r} = (x - x')\mathbf{e}_x + y\mathbf{e}_y$, the integral in (11.6.2) becomes

$$\int_C \frac{\mathbf{r}}{r^2}\, d\xi' = \mathbf{e}_x \int_{-\infty}^{\infty} \frac{(x - x')dx'}{(x - x')^2 + y^2} + \mathbf{e}_y \int_{-\infty}^{\infty} \frac{y\, dx'}{(x - x')^2 + y^2}. \qquad (11.6.3)$$

The first integral on the right-hand side of (11.6.3) vanishes due to symmetry. The second integral can be rewritten with the change of variable $\eta = (x - x')/y$, giving

$$\int_{-\infty}^{\infty} \frac{y\, dx'}{(x - x')^2 + y^2} = \pm \int_{-\infty}^{\infty} \frac{d\eta}{1 + \eta^2} = \pm\pi, \qquad (11.6.4)$$

where the plus sign corresponds to $y > 0$ and the minus sign to $y < 0$. Writing the unit normal to the vortex sheet as $\mathbf{n} = \mathbf{e}_y$, the induced velocity field is obtained as $\mathbf{u} = \pm\frac{1}{2}\boldsymbol{\gamma} \times \mathbf{n}$, which yields the net velocity jump across the vortex sheet as

$$\langle \mathbf{u} \rangle = \boldsymbol{\gamma} \times \mathbf{n}. \qquad (11.6.5)$$

This result for the velocity jump across a vortex sheet holds also for the general case of a curved vortex sheet with variable sheet strength, provided that the sheet curvature and strength vary smoothly over the sheet surface. This fact follows from the observation that if n is distance between a point P where the velocity is computed and the closest point S on the vortex sheet, then most of the velocity at P is induced from a section of length $c = O(n)$ on either side of the vortex sheet from S. As the point P approaches the sheet at S, the length c becomes increasingly small and the vortex sheet appears increasingly linear over this length. This approximate argument becomes exact in the limit $n \to 0$, leading to the result (11.6.5).

11.6.2 Equivalence of Vortex and Doublet Sheets

The velocity induced by a general vortex sheet occupying a curve C is given by

$$\mathbf{u}(\mathbf{x}, t) = \int_C \boldsymbol{\gamma}' \times \nabla G d\xi' = -\nabla \times \int_C G\boldsymbol{\gamma}'\, d\xi', \qquad (11.6.6)$$

where $G(\mathbf{x} - \mathbf{x}')$ is the Green's function of the Poisson equation in the space on which the vortex sheet is contained (e.g., the x-y plane). Green's functions for two- and three-dimensional spaces are given in (6.5.5). Equation (11.6.6) also holds in three dimensions if the curve C with element of length $d\xi$ is replaced by a surface S with element of area da. The velocity potential of a doublet sheet with strength distribution $\mu(\xi)$ along a curve C, with doublet axis aligned along the unit normal $\mathbf{n}$

of C, can be obtained by superposing point doublets as

$$\phi(\mathbf{x}, t) = \int_C \mu'\mathbf{n}' \cdot \nabla' G d\xi'. \tag{11.6.7}$$

In this subsection, we show that the velocity induced by a doublet sheet can be expressed in a form identical to that induced by a vortex sheet, and we derive an expression relating the vortex and doublet sheet strengths.

Taking the gradient of (11.6.7) and noting that the derivative with respect to the unprimed variable passes through the integral, the induced velocity from the doublet sheet is written as

$$\mathbf{u}(\mathbf{x}, t) = -\nabla\left[\nabla \cdot \int_C G\mu'\mathbf{n}' \, d\xi'\right]. \tag{11.6.8}$$

Recalling the vector identity

$$\nabla^2\mathbf{a} = \nabla(\nabla \cdot \mathbf{a}) - \nabla \times (\nabla \times \mathbf{a}), \tag{11.6.9}$$

and letting $\mathbf{a} \equiv \int_C \mu'\mathbf{n}' G(r) d\xi'$, the Laplacian of $\mathbf{a}$ is given by

$$\nabla^2\mathbf{a} \equiv \int_C \mu'\mathbf{n}'\nabla^2 G d\xi' = \int_C \mu'\mathbf{n}'\delta(r)\, d\xi', \tag{11.6.10}$$

where $\delta(r)$ is the Dirac delta and we use the property (6.3.5) of the Green's function. The result (11.6.10) implies that $\nabla^2\mathbf{a}$ vanishes at points not on the doublet sheet. Using (11.6.9), the expression (11.6.8) for the induced velocity from the doublet sheet can be written as

$$\mathbf{u}(\mathbf{x}, t) = -\nabla \times \left[\nabla \times \int_C G\mu'\mathbf{n}' \, d\xi'\right] = -\nabla \times \int_C \mu'\nabla G \times \mathbf{n}' \, d\xi'. \tag{11.6.11}$$

Use of the identity $\nabla G = -\nabla' G$ and integration by parts yield

$$\mathbf{u}(\mathbf{x}, t) = -\nabla \times \int_C G\nabla' \times (\mu'\mathbf{n}')\, d\xi'. \tag{11.6.12}$$

The expression (11.6.12) for the velocity induced by a doublet sheet is the same as the expression (11.6.6) for velocity induced by a vortex sheet if the vortex sheet strength is set as

$$\boldsymbol{\gamma} = \nabla \times (\mu\mathbf{n}). \tag{11.6.13}$$

An alternative form of (11.6.13) is obtained by taking the vector product with the unit normal $\mathbf{n}$, giving

$$\mathbf{n} \times \boldsymbol{\gamma} = \mathbf{n} \times (\nabla\mu \times \mathbf{n}) + \mu\mathbf{n} \times (\nabla \times \mathbf{n}). \tag{11.6.14}$$

The first term on the right-hand side of (11.6.14) is the projection of $\nabla\mu$ tangent to the vortex sheet, which is equal simply to $\nabla\mu$ since the doublet strength is defined only along C. The second term on the right-hand side of (11.6.14) can be expanded using the vector identity (2.6.6) as

$$\mathbf{n} \times (\nabla \times \mathbf{n}) = \tfrac{1}{2}\nabla(\mathbf{n} \cdot \mathbf{n}) - (\mathbf{n} \cdot \nabla)\mathbf{n}. \tag{11.6.15}$$

The first term on the right-hand side of (11.6.15) vanishes because $\mathbf{n}$ is a unit vector and the second term vanishes because the gradient of $\mathbf{n}$ is tangent to the surface S_B. Hence, (11.6.14) reduces to

$$\mathbf{n} \times \boldsymbol{\gamma} = \nabla\mu. \tag{11.6.16}$$

The relationships (11.6.13) and (11.6.16) hold in both two- and three-dimensional flows.

11.6.3 Equation of Motion of Vortex Sheet

The equation of motion of a vortex sheet is obtained by evaluating (11.6.6) at a point $\mathbf{X}(\xi) = X(\xi)\mathbf{e}_x + Y(\xi)\mathbf{e}_y$ on the sheet using the two-dimensional Green's function, where ξ is a parameter measuring distance along the sheet and time dependence of $\mathbf{X}$ is implicit. Since the vortex sheet moves as a material curve, we have

$$\frac{d\mathbf{X}}{dt} = -\frac{1}{2\pi}\int_C \frac{[\mathbf{X}(\xi) - \mathbf{X}(\xi')] \times \boldsymbol{\gamma}(\xi')}{|\mathbf{X}(\xi) - \mathbf{X}(\xi')|^2}\, d\xi'. \tag{11.6.17}$$

This integral, and those that follow in this section, is interpreted in the sense of a Cauchy principal value, so that the singularity at $\xi = \xi'$ makes no contribution.

It is sometimes convenient to write this result in complex variable form in terms of the complex position $Z = X + iY$ and its complex conjugate $\bar{Z} = X - iY$. Writing (11.6.17) in component form gives

$$\frac{dX}{dt} = -\frac{1}{2\pi}\int_C \frac{[Y(\xi) - Y(\xi')]\gamma(\xi')}{[(X(\xi) - X(\xi')]^2 + [Y(\xi) - Y(\xi')]^2}\, d\xi', \tag{11.6.18a}$$

$$\frac{dY}{dt} = \frac{1}{2\pi}\int_C \frac{[X(\xi) - X(\xi')]\gamma(\xi')}{[(X(\xi) - X(\xi')]^2 + [Y(\xi) - Y(\xi')]^2}\, d\xi'. \tag{11.6.18b}$$

Combining these terms gives the equation of motion of the vortex sheet as

$$\frac{d\bar{Z}}{dt}(\xi) = -\frac{i}{2\pi}\int_C \frac{\gamma(\xi')d\xi'}{Z(\xi) - Z(\xi')}. \tag{11.6.19}$$

The equation (11.6.19) can be alternatively expressed in terms of the integrated strength $\Gamma(P_1, P_2)$ between two points P_1 and P_2, defined by

$$\Gamma(P_1, P_2) = \int_{P_1}^{P_2} \gamma(\xi)\, d\xi, \tag{11.6.20}$$

or in differential form

$$d\Gamma = \gamma(\xi)\, d\xi. \tag{11.6.21}$$

The position Z can be expressed as a function of either the parameters ξ or Γ as $Z = \hat{Z}(\xi) = \tilde{Z}(\Gamma)$. Using (11.6.21), and dropping the tilde on Z, the equation (11.6.19) for motion of the vortex sheet can be written as

$$\frac{d\bar{Z}}{dt}(\Gamma) = -\frac{i}{2\pi} \int_C \frac{d\Gamma'}{Z(\Gamma) - Z(\Gamma')}, \tag{11.6.22}$$

which is known as the *Birkhoff-Rott equation* (Birkhoff, 1962; Rott, 1956).

Example 11.6.1 Vortex Sheet Roll-up. The strength of a vortex sheet shed by an aircraft wing of span length $2a$ (with elliptic loading) is given approximately by

$$\gamma(\xi) = \frac{2\xi U}{\sqrt{a^2 - \xi^2}}, \qquad -a \le \xi \le a. \tag{11.6.23}$$

Near the ends of the sheet γ approaches infinity asymptotically as

$$\gamma \sim Ax^{-1/2}, \tag{11.6.24}$$

where $x \equiv a - \xi$ and A is a constant. A simplified model that examines the roll-up of the vortex sheet near the end point $\xi = \pm a$ was proposed by Kaden (1931), consisting of a semi-infinite vortex sheet (Figure 11.9*a*) spanning the line $x \ge 0$ with strength variation (11.6.24). The integrated sheet strength in the initial solution is $\Gamma(x) = 2Ax^{1/2}$, so the sheet position at $t = 0$ can be expressed in terms of Γ as

$$Z(\Gamma, 0) = \frac{\Gamma^2}{4A^2}, \qquad 0 \le \Gamma < \infty. \tag{11.6.25}$$

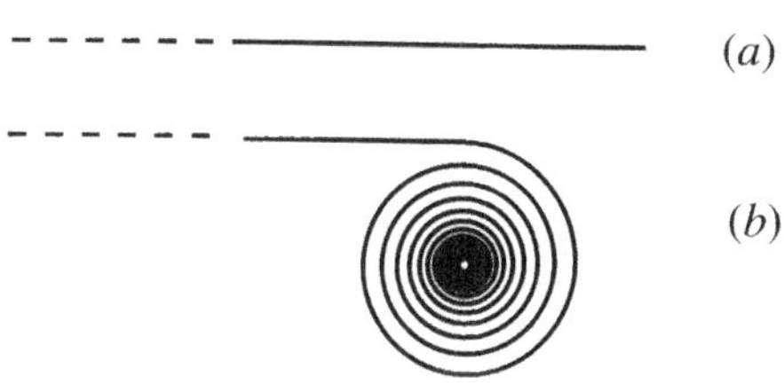

Figure 11.9 Kaden solution for roll-up of a vortex sheet: (*a*) initial condition and (*b*) long-time similarity solution.

A similarity solution of the Birkhoff-Rott equation (11.6.22) for $t > 0$ can be obtained in the form

$$Z(\Gamma, t) = (At)^{2/3} f(\eta), \tag{11.6.26}$$

where $\eta(\Gamma, t) \equiv \Gamma/(A^4 t)^{1/3}$ is the similarity variable and $f(\eta)$ is a complex-valued function. Substituting (11.6.26) into (11.6.22) gives the ordinary integro-differential equation

$$2\bar{f} - \eta \frac{d\bar{f}}{d\eta} = -\frac{3i}{2\pi} \int_0^\infty \frac{d\eta'}{f(\eta) - f(\eta')}, \qquad 0 \leq \eta < \infty. \tag{11.6.27}$$

The initial condition at $t = 0$ reduces to the asymptotic expression $f(\eta) \sim \eta^2/4$ as $\eta \to \infty$.

At long time (corresponding to $\eta \to 0$), the vortex rolls up into a spiral centered at some point $z = f_0(At)^{2/3}$, where f_0 is a complex constant, as illustrated schematically in Figure 11.9b. A long-time asymptotic solution for $f(\eta)$ in the region near the center of this spiral is given by $f(\eta) = f_0 + R(\eta)e^{i\phi(\eta)}$, where

$$R(\eta) \sim \frac{\eta^2}{4\lambda}, \qquad \phi(\eta) - \phi_0 \sim \frac{8\lambda^2}{\pi \eta^3}, \tag{11.6.28}$$

and λ and ϕ_0 are constants (see Saffman, 1992, for additional details). Eliminating η between the two equations in (11.6.28) gives

$$R \sim \frac{\lambda^{1/3}}{\pi^{2/3}(\phi - \phi_0)^{2/3}}. \tag{11.6.29}$$

Substituting this long-time solution into (11.6.26), the radius r of the spiral is found to vary with the angle parameter ϕ (where $0 \leq \phi < \infty$) as

$$r \sim C \left(\frac{t}{\phi - \phi_0} \right)^{2/3}, \tag{11.6.30}$$

where $C \equiv (A^2\lambda/\pi^2)^{1/3}$ is a constant. This long-time solution indicates that as the vortex sheet spirals around its tip, the radius decreases ever more gradually, such that infinitely many spirals occur as $r \to 0$. The sheet strength γ correspondingly decreases as ϕ increases, due to stretching of the sheet, such that $\gamma \to 0$ at the spiral center.

Before closing this section, we note that numerical solution of vortex sheet motions poses some unique challenges. In particular, it was shown analytically by Moore (1979) that an initially smooth vortex sheet develops a curvature singularity in finite time. This singularity has since been examined and verified by numerical computations (e.g., Krasny, 1986; Vanden-Broeck, 1993).

11.7 VORTEX SHEET REPRESENTATION OF FLAT-PLATE AIRFOILS

A method for solution of two-dimensional flow past an airfoil is developed in Section 10.5 using conformal mapping. Flow past thin bodies, such as an airfoil, can alternatively be solved by placing singular distributions of vorticity or dilatation rate along the centerline of the body, in the form of a vortex or source sheet. In the current section, we examine a solution for flow past thin lifting bodies obtained by placing a vortex sheet along the length of the body. This method is more flexible than the conformal mapping approach, since it can be used for bodies of arbitrary shape and it can be extended to apply to three-dimensional flows. The method can be extended to account for nonzero body thickness (so long as the thickness is much less than the body length) by also placing a source sheet along the body centerline. Flow past bodies of finite thickness is examined in the discussion of "slender body theory" in Section 12.5.

We consider a thin airfoil lying on a curve $y = y_B(x)$ along the interval $(0, c)$, as shown in Figure 11.10. A uniform flow incident on the body has speed U and angle of attack α, where α is assumed to be small. The no-penetration condition on the airfoil imposes the kinematic boundary condition

$$u\frac{dy_B}{dx} = v \qquad \text{on } y = y_B(x). \tag{11.7.1}$$

This boundary condition is enforced by placing a vortex sheet on the airfoil surface C, given by $y = y_B(x)$. The velocity field, including the uniform flow and the velocity induced by the vortex sheet, is given by

$$\mathbf{u}(\mathbf{x}) = \mathbf{U} - \frac{1}{2\pi}\int_C \frac{[\mathbf{x} - \mathbf{x}'(\xi)] \times \boldsymbol{\gamma}(\xi)}{|\mathbf{x} - \mathbf{x}'(\xi)|^2}\, d\xi, \tag{11.7.2}$$

where $\boldsymbol{\gamma} = \gamma\mathbf{e}_z$ is the vortex sheet strength and $\mathbf{x}'(\xi)$ is a point on C identified by the parameter ξ. Letting $\mathbf{U} = U\cos\alpha\mathbf{e}_x + U\sin\alpha\mathbf{e}_y$, the components of (11.7.2) are

$$u(\mathbf{x}) = U\cos\alpha - \frac{1}{2\pi}\int_C \frac{[y - y'(\xi)]\gamma(\xi)}{[x - x'(\xi)]^2 + [y - y'(\xi)]^2}\, d\xi, \tag{11.7.3a}$$

$$v(\mathbf{x}) = U\sin\alpha + \frac{1}{2\pi}\int_C \frac{[x - x'(\xi)]\gamma(\xi)}{[x - x'(\xi)]^2 + [y - y'(\xi)]^2}\, d\xi. \tag{11.7.3b}$$

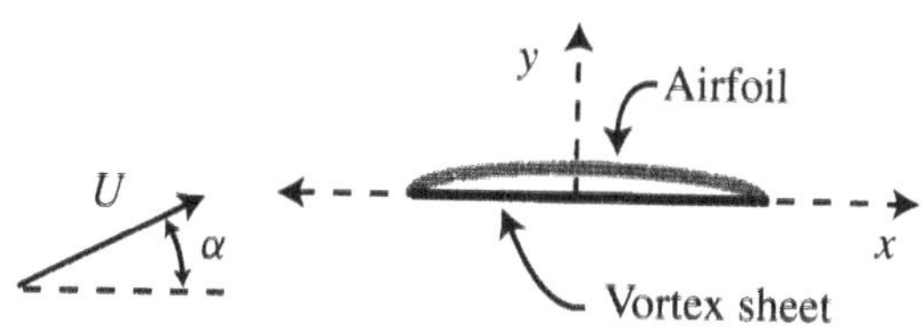

Figure 11.10 Flow at angle of attack α past a thin airfoil lying on the curve $y = y_B(x)$. The flow field is produced by placing a vortex sheet on the interval $(0, c)$ on the x-axis.

The Kutta condition requires that there be no velocity discontinuity at the trailing edge of the airfoil. Since the jump in tangential velocity across the vortex sheet is proportional to the vortex sheet strength, the Kutta condition can be enforced by requiring that γ vanishes at the trailing edge.

Substituting (11.7.3) into the boundary condition (11.7.1) yields an integral equation for the vortex sheet strength γ. However, this integral equation is complicated and admits no known exact analytical solution for arbitrary curves C. The problem can be simplified by assuming that the airfoil deviates only slightly from a flat plate. Letting $y_{\max}$ denote the maximum deviation of the airfoil from the x-axis, we define a small parameter ε by $\varepsilon \equiv y_{\max}/c$. The airfoil is assumed to have small deflection from a flat plate and small slope, such that $\varepsilon \ll 1$ and $dy_B/dx = O(\varepsilon)$. Under these conditions, the boundary condition (11.7.1) can be displaced to apply on the x-axis using the Taylor series approximation $f(y_B) = f(0) + y_B(\partial f/\partial y)|_{y=0} + \cdots$, where each subsequent term in the Taylor series is smaller by $O(\varepsilon)$ than the previous term. To leading order in ε, the boundary condition (11.7.1) can be applied at $y = 0$, rather than at $y = y_B(x)$. For a point lying at position x on the x-axis and for small values of α, the velocity components (11.7.3) become

$$u \cong U, \qquad v \cong U\alpha + \frac{1}{2\pi}\int_0^c \frac{\gamma(\xi)}{x - x'(\xi)}\, d\xi. \tag{11.7.4}$$

Substituting (11.7.4) into the kinematic boundary condition (11.7.1), evaluated at $y = 0$, yields an integral equation for $\gamma(x)$ as

$$\frac{1}{2\pi}\int_0^c \frac{\gamma(x')}{x - x'}\, dx' = U\frac{dy_B}{dx} - \alpha U. \tag{11.7.5}$$

The Kutta condition requires that at the trailing edge $\gamma(c) = 0$. The vortex sheet strength is singular at the leading edge, with the limit

$$\gamma(x) \sim A_0 U \left(\frac{c}{x}\right)^{1/2} \quad \text{as } x \to 0, \tag{11.7.6}$$

where A_0 is a constant to be determined. In searching for a solution to the integral equation (11.7.5), it is convenient to define a modified vortex sheet strength $\eta(x)$ by

$$\eta(x) \equiv \gamma(x) - A_0 U \left(\frac{c - x}{x}\right)^{1/2}, \tag{11.7.7}$$

such that $\eta(0) = \eta(c) = 0$. A solution for the integral equation is obtained using the change of independent variable

$$x = \frac{c}{2}(1 + \cos\theta), \qquad 0 \le \theta \le \pi, \tag{11.7.8}$$

such that (11.7.7) can be written using the half-angle theorem for tangent as

$$\eta(\theta) = \gamma(\theta) - A_0 U \tan(\theta/2). \tag{11.7.9}$$

We seek a solution for $\eta(\theta)$ in terms of the Fourier sine series

$$\eta(\theta) = U \sum_{n=1}^{\infty} A_n \sin(n\theta), \tag{11.7.10}$$

where the factor U is used to make the coefficients A_n dimensionless. Substituting (11.7.9) and (11.7.10) into the integral equation (11.7.5) gives

$$\frac{1}{2\pi} \int_0^c \left[A_0 \tan\left(\frac{\theta'}{2}\right) + \sum_{n=1}^{\infty} A_n \sin(n\theta') \right] \frac{(-\sin\theta')\,d\theta'}{\cos\theta - \cos\theta'} = \frac{dy_B}{dx} - \alpha. \tag{11.7.11}$$

The integrals over θ' in (11.7.11) can be expressed in terms of the integral I_n, defined by

$$I_n = \int_0^{\pi} \frac{\cos(n\theta')d\theta'}{\cos(\theta') - \cos(\theta)} = \pi \frac{\sin(n\theta)}{\sin\theta}, \qquad 0 \le \theta \le \pi, \tag{11.7.12}$$

giving

$$\frac{1}{2\pi} A_0(-I_0 + I_1) + \frac{1}{4\pi} \sum_{n=1}^{\infty} A_n(-I_{n-1} + I_{n+1}) = \alpha - \frac{dy_B}{dx} \tag{11.7.13}$$

Using (11.7.12), this reduces to an expression for the coefficients A_n as

$$A_0 + \sum_{n=1}^{\infty} A_n \cos(n\theta) = 2\left(\alpha - \frac{dy_B}{dx}\right). \tag{11.7.14}$$

This result has the form of a Fourier cosine series and can be inverted using the orthogonality properties of $\cos(n\theta)$ to yield expressions for the coefficients A_n as

$$A_0 = 2\alpha - \frac{2}{\pi} \int_0^{\pi} \frac{dy_B}{dx} d\theta,$$

$$A_n = -\frac{4}{\pi} \int_0^{\pi} \frac{dy_B}{dx} \cos(n\theta)\,d\theta, \qquad n = 1, 2, \ldots. \tag{11.7.15}$$

The airfoil lift, obtained from the Kutta-Joukowski law, is proportional to the total circulation of the vortex sheet, such that

$$L = \rho U \int_0^c \gamma(x)\,dx, \tag{11.7.16}$$

and the lift coefficient is

$$C_L \equiv \frac{L}{\frac{1}{2}\rho U^2 c} = \frac{2}{Uc}\int_0^c \gamma(x)\,dx. \tag{11.7.17}$$

Substituting into (11.7.17) the solution in (11.7.9) and (11.7.10) for $\gamma(x)$ and using the expression (11.7.15) for the coefficients give

$$C_L = \pi\left(A_0 + \frac{1}{2}A_1\right) = 2\pi\alpha - 2\int_0^\pi (1+\cos\theta)\frac{dy_B}{dx}\,d\theta. \tag{11.7.18}$$

The first term on the right-hand side of (11.7.18) agrees with the exact solution (10.5.4) obtained from conformal mapping for a flat-plate airfoil (letting $\sin\alpha \cong \alpha$ for small angle of attack), and the second term accounts for the influence of airfoil curvature on the lift force.

11.8 FLOW WITH UNIFORM BACKGROUND VORTICITY

The potential flow idealization discussed in Chapters 9 and 10 is extended in the current section for cases with uniform, but nonzero, background vorticity. This idealization might be used, for instance, to account for the effect of ambient shear on the force on a cylindrical body.

Two-dimensional flows with uniform background vorticity $\omega = \omega_0 = \text{const}$ are governed by the Poisson equation

$$\nabla^2\psi = -\omega_0. \tag{11.8.1}$$

Since ω_0 is constant, (11.8.1) is linear, and its solution can be decomposed as the sum of a particular solution $\psi_1(\mathbf{x}, t)$ and a homogeneous solution $\psi_0(\mathbf{x}, t)$, where

$$\nabla^2\psi_0 = 0. \tag{11.8.2}$$

We consider particular solutions of the form

$$\psi_1(\mathbf{x}, t) = Uy - \frac{\omega_0}{2}\left(\frac{Ax^2 + By^2}{A+B}\right), \tag{11.8.3}$$

where A and B are arbitrary constants satisfying the restriction $A + B \neq 0$. For instance, setting $A = B = 1$, the stream function ψ_1 reduces to that for solid-body rotation about a point C_R located at $y_C = 2U/\omega_0$ (Figure 11.11*a*). Choosing $A = 0$ and $B = 1$ yields a uniform shear flow with shearing rate ω_0 (Figure 11.11*b*). Once the external flow stream function ψ_1 is selected, the problem reduces to solving for the harmonic function ψ_0 subject to appropriate boundary conditions. For instance, for flow past a fixed body with the no-penetration condition applied on its outer

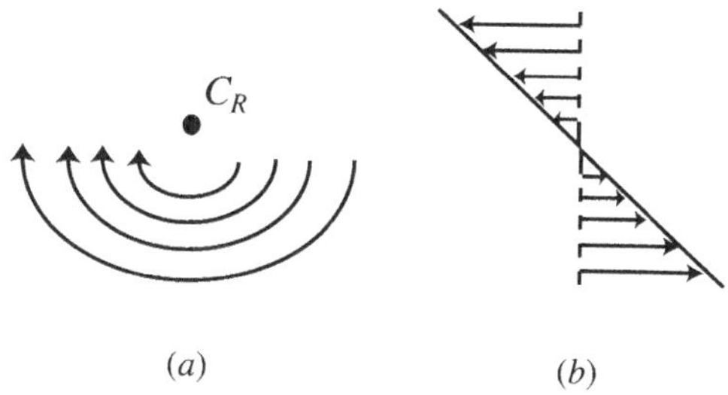

Figure 11.11 External flows generated from the expression (11.8.3) with appropriate choices for the constants A and B: (a) solid-body rotation ($A = B = 1$); (b) uniform shear ($A = 0$, $B = 1$).

boundary C, the boundary condition for ψ_0 becomes $\psi_0|_C = \text{const} - \psi_1|_C$. The solution for ψ_0 can be obtained using some of the same methods described in Chapter 9 for potential flow fields, although care must be taken to ensure satisfaction of the correct boundary conditions.

For steady flow, the pressure is given by (8.4.7) as

$$\frac{dB}{d\psi} = -\omega_0, \tag{11.8.4}$$

where $B \equiv p/\rho + \kappa$ is the Bernoulli coefficient. Integrating (11.8.4) over stream function gives

$$B = B_0 - \omega_0 \psi, \tag{11.8.5}$$

where B_0 is a constant of integration. If p_0 denotes pressure at a stagnation point, at which the stream function vanishes, the pressure field is obtained as

$$p = p_0 - \rho\kappa - \rho\omega_0\psi, \tag{11.8.6}$$

which is the same as the Bernoulli equation with the addition of a term that is linear in ψ. The boundary condition on ψ requires this additional term to be constant on the surface of a fixed body with no-penetration boundary condition.

Example 11.8.1 Uniform Shear Flow Past a Circular Cylinder. We consider a circular cylinder of radius a centered at the coordinate origin and immersed in a uniform shear flow (Figure 11.11b), whose stream function in polar coordinates is

$$\psi_1 = Ur\sin\theta - \tfrac{1}{2}\omega_0 r^2 \sin^2\theta. \tag{11.8.7}$$

Setting $\psi_0 + \psi_1 = C = \text{const}$ on the cylinder surface, the boundary condition for ψ_0 becomes

$$\psi_0\big|_{r=a} = C - Ua\sin\theta + \tfrac{1}{2}\omega_0 a^2 \sin^2\theta. \tag{11.8.8}$$

Solving the Laplace equation for ψ_0 using a classical separation of variables approach (Kreyszig, 1999), with $\psi_0(r,\theta) = R(r)T(\theta)$, we obtain

$$\frac{r}{R}\frac{d}{dr}\left(r\frac{dR}{dr}\right) = -\frac{1}{T}\frac{d^2T}{d\theta^2} = k^2, \tag{11.8.9}$$

where k is a real-valued constant. Equation (11.8.9) yields two ordinary differential equations for $R(r)$ and $T(\theta)$ as

$$r^2\frac{d^2R}{dr^2} + r\frac{dR}{dr} - k^2R = 0, \qquad \frac{d^2T}{d\theta^2} + k^2T = 0. \tag{11.8.10}$$

The solution for $T(\theta)$ is of the form

$$T(\theta) = C_{1n}\sin(k_n\theta) + C_{2n}\cos(k_n\theta), \tag{11.8.11}$$

where k_n is set equal to some integer n in order to satisfy the condition that ψ_0 be periodic over an angle 2π. Equation (11.8.10) for $R(r)$ is of the Euler-Cauchy form, with solution

$$R(r) = \frac{C_{3n}}{r^n} + C_{4n}r^n. \tag{11.8.12}$$

The boundary condition (11.8.8) for ψ_0 contains terms involving both $\sin\theta$ and $\sin^2\theta$. Recalling that $\sin^2\theta = (1 - \cos 2\theta)/2$, we concentrate on the terms in (11.8.11) proportional to $\sin\theta$ and $\cos 2\theta$. At infinity, ψ_0 must approach zero or a constant so that the velocity field approaches the uniform shear flow given by the particular solution (11.8.7). This restriction implies that the C_{4n} coefficient in (11.8.12) vanishes for $n > 0$. We thus consider a solution of the form

$$\psi_0(r,\theta) = \left(\frac{B}{r}\right)\sin\theta + \left(\frac{A}{r^2}\right)\cos 2\theta, \tag{11.8.13}$$

where A and B are constants. Evaluating (11.8.13) at $r = a$ and substituting $\cos 2\theta = 1 - 2\sin^2\theta$ gives

$$\psi_0\big|_{r=a} = \frac{A}{a^2} + \left(\frac{B}{a}\right)\sin\theta - \left(\frac{2A}{a^2}\right)\sin^2\theta. \tag{11.8.14}$$

Equating (11.8.14) to (11.8.8) gives values for these coefficients as

$$A = -\omega_0 a^4/4, \qquad B = -Ua^2. \tag{11.8.15}$$

The final solution for stream function is given by

$$\psi = U\left(r - \frac{a^2}{r}\right)\sin\theta - \frac{\omega_0 r^2}{2}\sin^2\theta - \frac{\omega_0 a^4}{4r^2}\cos(2\theta). \tag{11.8.16}$$

The azimuthal velocity component on the cylinder surface is

$$u_\theta\big|_{r=a} = -\frac{\partial\psi}{\partial r}\Big|_{r=a} = -2U\sin\theta + 2\omega_0 a\sin^2\theta - \omega_0 a/2. \qquad (11.8.17)$$

The cylinder surface pressure can be obtained by substituting (11.8.17) into the equation (11.8.6) for pressure, with $\kappa = u_\theta^2/2$ and $\psi = C$ on the cylinder surface. Symmetry of u_θ over the y-axis implies that there is no drag force acting on the cylinder. The lift force per unit length of the cylinder is given by

$$L = -\int_0^{2\pi} p(a,\theta)\sin\theta\, a\, d\theta = -2\pi a^2\omega_0\rho U. \qquad (11.8.18)$$

BIBLIOGRAPHY

Almgren, A.S., T. Buttke, and P. Colella (1994). "A fast adaptive vortex method in three dimensions," *Journal of Computational Physics* **113**, 177–200.

Anderson, C.R. (1986). "A method of local corrections for computing the velocity field due to a distribution of vortex blobs," *Journal of Computational Physics* **62**, 111–123.

Aref, H. (1983). "Integrable, chaotic and turbulent vortex motions in two-dimensional flows," *Annual Review of Fluid Mechanics* **15**, 345–389.

Aref, H., and N. Pomphrey (1980). "Integrable and chaotic motions of four vortices," *Physics Letters* **78A** (4), 297–300.

Baker, G.R. (1990). "A study of the numerical stability of the method of contour dynamics," *Philosophical Transactions of the Royal Society of London A* **333**, 391–400.

Beale, J.T. (1986). "On the accuracy of vortex methods," in *Proceedings of the Workshop on Computational Fluid Dynamics & Reacting Gas Flows*, I.M.A., University of Minnesota, Minneapolis, MN.

Birkhoff, G. (1962). "Helmholtz and Taylor instability," in *Proceedings of Symposium in Applied Mathematics*, Vol. 13, American Mathematical Society, Providence, RI.

Chen, H., and J.S. Marshall (1999). "A Lagrangian vorticity method for two-phase particulate flows with two-way phase coupling," *Journal of Computational Physics* **148**, 169–198.

Christiansen, J.P. (1973). "Numerical simulation of hydrodynamics by the method of point vortices," *Journal of Computational Physics* **13**, 363–379.

Cottet, G.-H. and P. Koumoutsakos (2000). *Vortex Methods: Theory and Practice* Cambridge University Press: New York.

Dritchel, D.G. (1988). "Contour surgery: a topological reconnection scheme for extended interactions using contour dynamics," *Journal of Computational Physics* **77**, 240–266.

Gharakhani, A., and A.F. Ghoniem (1996). "Massively parallel implementation of a 3D vortex-boundary element method," *European Series in Applied and Industrial Mathematics: Proceedings* **1**, 213–223.

Kaden, H. (1931). "Aufwicklung einer unstabilen Unstetigkeitsfläche," *Ingenieur Archiv* **2**, 140–168.

Krasny, R. (1986). "A study of singularity formation in a vortex sheet by the point-vortex approximation," *Journal of Fluid Mechanics* **167**, 65–93.

Kreyszig, E. (1999). *Advanced Engineering Mathematics*, 8th ed. John Wiley & Sons: New York.

Lamb, H. (1932). *Hydrodynamics*. Cambridge University Press: Cambridge (reprinted Dover Publications, New York, 1945).

Leonard, A.T. (1985). "Computing three-dimensional incompressible flows with vortex elements," *Annual Reviews of Fluid Mechanics* **17**, 523–559.

Marshall, J.S., and J.R. Grant (1996). "Penetration of a blade into a vortex core: vorticity response and unsteady blade forces," *Journal of Fluid Mechanics* **306**, 83–109.

Moore, D.W. (1979). "The spontaneous appearance of a singularity in the shape of an evolving vortex sheet," *Proceedings of the Royal Society of London A* **365**, 105–119.

Moore, D.W., and P.G. Saffman (1971). "Structure of a line vortex in an imposed strain," in *Aircraft Wake Turbulence and Its Detection*, edited by J.H. Olsen, A. Goldberg, and M. Rogers, Plenum Press, New York, pp. 339–354.

Ozorio de Almeida, A.M. (1988). *Hamiltonian Systems: Chaos and Quantization*. Cambridge University Press: New York.

Pullin, D.I. (1992). "Contour dynamics methods," *Annual Review of Fluid Mechanics* **24**, 89–115.

Rott, N. (1956). "Diffraction of a weak shock with vortex generation," *Journal of Fluid Mechanics* **1**, 111–128.

Saffman, P.G. (1992). *Vortex Dynamics*. Cambridge University Press: Cambridge.

Salmon, J.K., and M.S. Warren (1994). "Skeletons from the treecode closet," *Journal of Computational Physics* **111**, 136–155.

Strickland, J.H., and R.S. Baty (1996). "A pragmatic overview of fast multipole methods," *Lectures in Applied Mathematics* **32**, 807–830.

Vanden-Broeck, J.-M. (1993). "Rosenhead's point vortex approximation revisited," *Physics of Fluids A* **5** (11), 2786–2791.

Waugh, D.W. (1992). "Efficiency of symmetric vortex merger," *Physics of Fluids A* **4**(8), 1745–1758.

Whittaker, E.T. (1937). *A Treatise on the Analytical Dynamics of Particles & Rigid Bodies*. Cambridge University Press: Cambridge.

Winckelmans, G.S., J.K. Salmon, A. Leonard, and M.S. Warren (1995). "Three-dimensional vortex particle and panel methods: fast tree-code solvers with active error control for arbitrary distributions/geometries," paper presented at the Forum on Vortex Methods for Engineering Applications, Sandia Nat. Lab.: Albuquerque, NM, Feb. 22–24.

Zabusky, N.J., and E.A. Overman II (1983). "Regularization of contour dynamical algorithms. I. Tangential regularization," *Journal of Computational Physics* **52**, 351–373.

Zabusky, N.J., M.H. Hughes, and K.V. Roberts (1979). "Contour dynamics for the Euler equations in two dimensions," *Journal of Computational Physics* **30**, 96–106.

PROBLEMS

1. Determine the rotation rate of a system initially composed of three Rankine vortices of equal strength Γ and core radius a, placed at the vertices of an equilateral triangle with side length b, as shown in Figure 11.12. Assume first that

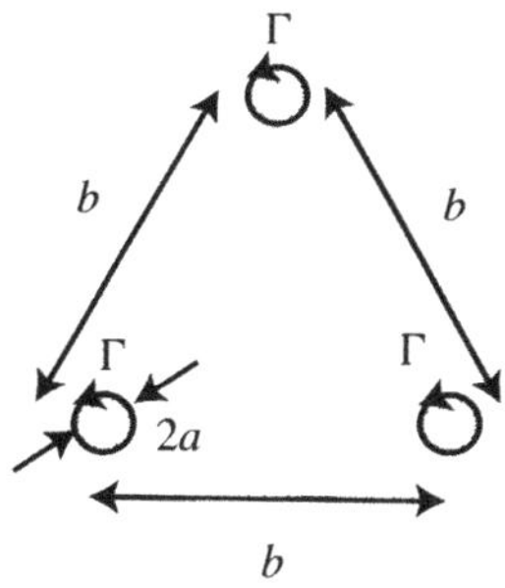

Figure 11.12 System of three equal strength vortices with core radius a arranged in an equilateral triangle with side length b.

$\varepsilon = a/b \ll 1$ and obtain your solution to leading order in ε, such that the vortices can be treated as point vortices. What would be the (qualitative) effects, both with respect to the vortex core and the rotation rate of the system, of allowing finite vortex core radius?

2. Assuming that the vortices shown in Figure 11.12 are point vortices, derive expressions for the interaction energy, linear impulse, angular impulse, vorticity centroid, and radius of gyration of this system.

3. Substitute the similarity solution (11.6.26) into the Birkhoff-Rott equation to obtain the integro-differential equation (11.6.27) for the problem of roll-up of a vortex sheet.

4. Determine the doublet strength $\boldsymbol{\mu}$ for the doublet sheet equivalent to a two-dimensional, flat vortex sheet with constant strength $\boldsymbol{\gamma}$.

5. Derive an expression for the displacement $y_B(x)$ for the cambered airfoil obtained by the conformal mapping in Example 10.5.3. Solve for the lift coefficient of the airfoil using the solution (11.7.8) of small-deflection theory and plot your result as a function of the ratio of maximum deflection h to chord length ℓ. Compare your result to the exact solution obtained from conformal mapping (Section 10.5).

6. Consider a circle of radius a centered at the origin and immersed in a uniformly rotating, two-dimensional flow field, with stream function given by

$$\psi_1 = -\frac{\omega_0}{4}(x^2 + y^2) + Uy,$$

where $2U/\omega_0 > a$.

(a) Find the solution for the stream function $\psi = \psi_0 + \psi_1$ that satisfies the no-penetration boundary condition on the surface of the circle.

(b) Solve for the pressure on the surface of the circle.

(c) Integrate to obtain the force acting on the circle.

COMPUTATIONAL PROJECTS

1. Write a computer code to follow the motion of an initially elliptical vortex patch with uniform vorticity ω_0 internal to the patch boundary using the contour dynamics method.
   (a) Validate your code by examining the self-induced rotation of the vortex patch and comparing you result for rotation rate to that derived for the Kirchhoff elliptical vortex in Section 11.2.
   (b) Add an irrotational straining flow with velocity components $u = ex$ and $v = -ey$, where e is the straining rate. Examine the vortex behavior for cases with $e/\omega_0 = 0.05, 0.10, 0.14$, and 0.16. Compare your results to the analytical solution of Moore and Saffman (1971) (Section 16.2).
2. Write a computer code for one or more initially circular vortex patches with uniform vorticity ω_0 using the discrete vortex method. Apply your code to the problem of merger of two initially circular vortex patches with unit radius and equal strength, separated by an initial distance d, as shown in Figure 11.3.
   (a) Validate your code by computing the velocity field for a single Rankine vortex and comparing with the exact solution (Section 11.2).
   (b) Compute the interaction of the vortex pair for cases with $d = 5.0, 3.5, 2.5$, and compare your observations of vortex merger with the results of Waugh (1992) (Section 11.3).
3. Write a code for computation of an initially circular vortex patch of radius a with uniform vorticity ω_0 internal to the patch using the vortex-in-cell method and the basic discrete vortex method.
   (a) Validate your code by computing the velocity field of a Rankine vortex of radius a and comparing to the exact solution (Section 11.2), for both the vortex-in-cell and basic vortex methods.
   (b) Compare the time required to compute the velocity field for a Rankine vortex for the basic vortex method and vortex-in-cell method as a function of the number of Lagrangian points N. Plot the CPU time versus N on a log-log plot and examine the slope of the resulting curve.
   (c) Examine the Rankine vortex evolution over time with the vortex-in-cell method using grid sizes d of $d/a = 0.1, 0.25, 0.5$. Plot the velocity profile at the end of five vortex rotations for each case. Repeat the same computations with the basic vortex method and comment on the differences between the two cases.

CHAPTER 12

THREE-DIMENSIONAL POTENTIAL FLOWS

Solution of three-dimensional potential flows is more difficult than solution of two-dimensional flows, both because of the complications introduced by an additional spatial dimension and because the analogy between analytic complex functions and potential flows no longer holds. Most of the solution methods employed in Chapter 9 for two-dimensional potential flows cannot be used for three-dimensional flows. Certain three-dimensional flows can be reduced in difficulty by the assumption of axisymmetry, which applies whenever the velocity components expressed in spherical or cylindrical polar coordinates are independent of angle around a symmetry axis. This assumption is commonly employed to model the flow about axisymmetric bodies, such as falling parachutes, missiles, the nose region of submarines, and so forth. The axisymmetry assumption is also frequently employed to model certain aspects of the core dynamics of vortex structures, such as wave propagation with variable core area. Axisymmetric flows without swirl (i.e., with zero azimuthal velocity around the symmetry axis) admit further simplifications deriving from the fact that there is only one nonzero component of vorticity and two nonzero components of velocity. The current chapter describes various solution methods for both axisymmetric and fully three-dimensional potential flows. Axisymmetric flows with nonzero vorticity are discussed in Chapter 13 and fully three-dimensional vortex flows are discussed in Chapter 14.

12.1 GOVERNING EQUATIONS

The governing equation for a three-dimensional potential flow is the Laplace equation $\nabla^2\phi = 0$ for the velocity potential ϕ, where velocity is given by $\mathbf{u} = \nabla\phi$. It

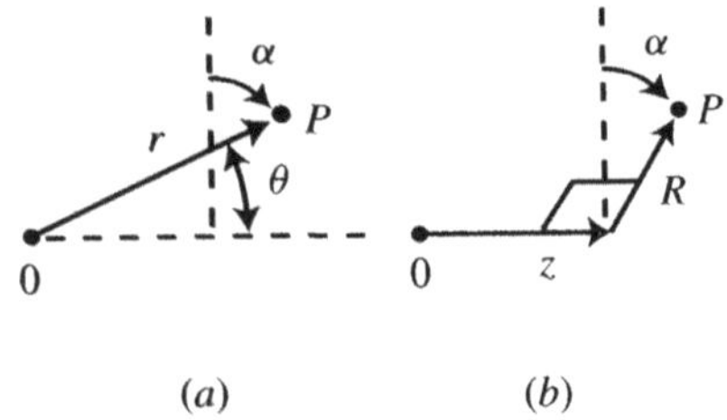

Figure 12.1 Coordinate systems used for axisymmetric flow: (a) spherical coordinates (r, θ, α) and (b) cylindrical polar coordinates (R, α, z).

is not in general possible to define a stream function for a fully three-dimensional flow, unless the problem is such that some special coordinate system can be chosen in which the continuity equation reduces to only two terms.

Axisymmetric flows are conveniently described in either spherical coordinates (r, θ, α) or cylindrical polar coordinates (R, α, z), as illustrated in Figure 12.1. The two systems are related by $R = r\sin\theta$ and $z = r\cos\theta$. For axisymmetric flow with no swirl, the vorticity is oriented in the α-direction and the velocity has nonzero components in the r and θ directions such that

$$\boldsymbol{\omega} = \omega \mathbf{e}_\alpha, \qquad \mathbf{u} = u_r\mathbf{e}_r + u_\theta\mathbf{e}_\theta. \tag{12.1.1}$$

In spherical coordinates, the incompressibility condition becomes

$$\frac{1}{r^2}\frac{\partial}{\partial r}(r^2 u_r) + \frac{1}{r\sin\theta}\frac{\partial}{\partial\theta}(\sin\theta\, u_\theta) = 0, \tag{12.1.2}$$

and the condition of irrotationality is

$$\frac{\partial u_r}{\partial\theta} - \frac{\partial}{\partial r}(r u_\theta) = 0. \tag{12.1.3}$$

Equation (12.1.3) can be identically satisfied by writing the equation in terms of a potential ϕ, such that in spherical coordinates velocity has the components

$$u_r = \frac{\partial\phi}{\partial r}, \qquad u_\theta = \frac{1}{r}\frac{\partial\phi}{\partial\theta}. \tag{12.1.4}$$

Substituting (12.1.4) into the incompressibility condition yields the Laplace equation, which is written in spherical coordinates as

$$\frac{1}{r^2}\frac{\partial}{\partial r}\left(r^2\frac{\partial\phi}{\partial r}\right) + \frac{1}{r^2\sin\theta}\frac{\partial}{\partial\theta}\left(\sin\theta\frac{\partial\phi}{\partial\theta}\right) = 0. \tag{12.1.5}$$

Alternatively, the incompressibility condition can be identically satisfied by introducing the *Stokes stream function* ψ, defined by

$$u_r = \frac{1}{r^2 \sin\theta} \frac{\partial \psi}{\partial \theta}, \qquad u_\theta = -\frac{1}{r \sin\theta} \frac{\partial \psi}{\partial r}. \tag{12.1.6}$$

The Stokes stream function is related to the vector potential $\boldsymbol{\beta}$ by $\boldsymbol{\beta} = [\psi/(r \sin\theta)]\mathbf{e}_\alpha$, where $\mathbf{u} = \nabla \times \boldsymbol{\beta}$. Substituting (12.1.6) into the irrotationality condition (12.1.3) gives the governing equation for stream function as

$$\frac{\partial}{\partial r}\left(\frac{1}{\sin\theta} \frac{\partial \psi}{\partial r}\right) + \frac{1}{r^2} \frac{\partial}{\partial \theta}\left(\frac{1}{\sin\theta} \frac{\partial \psi}{\partial \theta}\right) = 0. \tag{12.1.7}$$

Since (12.1.7) is not the Laplace equation, the stream function is not a harmonic function in axisymmetric potential flows. One consequence of this observation is that the analogy between two-dimensional potential flows and analytic functions of a complex variable no longer holds for axisymmetric flows. Many of the solution techniques developed for two-dimensional flows in Chapters 9 and 10, such as conformal mapping and the Blasius integral laws, cannot be used for axisymmetric flows.

In cylindrical polar coordinates, the expressions for velocity potential and Stokes stream function in terms of the velocity components u_R and u_z become

$$u_R = \frac{\partial \phi}{\partial R} = -\frac{1}{R} \frac{\partial \psi}{\partial z}, \qquad u_z = \frac{\partial \phi}{\partial z} = \frac{1}{R} \frac{\partial \psi}{\partial R}. \tag{12.1.8}$$

The velocity components in the two coordinate systems are related by

$$u_R = u_r \sin\theta + u_\theta \cos\theta, \qquad u_z = u_r \cos\theta - u_\theta \sin\theta. \tag{12.1.9}$$

12.2 BASIC POTENTIAL FLOWS

Irrotational flows (with no vorticity anywhere in the flow field) are generated either from the boundary conditions or from a distribution of dilatation rate $\Delta(\mathbf{x}, t)$, so the velocity potential in a three-dimensional irrotational flow is given by

$$\phi(\mathbf{x}, t) = \phi_0(\mathbf{x}, t) - \frac{1}{4\pi} \int_V \frac{\Delta(\mathbf{x}', t)}{|\mathbf{x} - \mathbf{x}'|} \, dv', \tag{12.2.1}$$

where the harmonic function ϕ_0 for an unbounded flow is determined from boundary conditions at infinity. Since the governing equation (12.1.5) for ϕ is linear, any two solutions for ϕ can be superposed to generate a new solution. The solution of a complicated flow can often be obtained by adding together in an appropriate manner a series of more basic "building-block" solutions. Several examples of building-block solutions of this type are given below.

12.2.1 Uniform Flow

The velocity components for uniform flow with speed U along the axis of symmetry are

$$u_r = U\cos\theta, \qquad u_\theta = -U\sin\theta. \tag{12.2.2}$$

Integration of definitions (12.1.4) and (12.1.6) for potential function and stream function yields

$$\phi = Ur\cos\theta, \qquad \psi = \tfrac{1}{2}Ur^2\sin^2\theta. \tag{12.2.3}$$

In cylindrical polar coordinates, this result has the form

$$\phi = Uz, \qquad \psi = \tfrac{1}{2}UR^2. \tag{12.2.4}$$

12.2.2 Straining Flow

Axisymmetric straining flow is often used to represent the effect of fluid stretching on a small structure immersed in the flow, such as a particle, droplet, or bubble in a two-phase flow, a long-chain molecule in a polymeric fluid, or a coherent vortical structure in a turbulent flow. This flow field is also representative of the inviscid flow near a stagnation point, such as the nose region of a submarine. The velocity components for axisymmetric straining flow with strain rate c are given in cylindrical polar coordinates by

$$u_z = 2cz, \qquad u_R = -cR. \tag{12.2.5}$$

Substituting these into the definition (12.1.8) of velocity potential and stream function in cylindrical polar coordinates, and integrating over R or z as necessary, gives

$$\phi = c\left(z^2 - \tfrac{1}{2}R^2\right), \qquad \psi = cR^2z. \tag{12.2.6}$$

In spherical coordinates, the velocity potential and stream function become

$$\phi = \tfrac{1}{2}cr^2(2\cos^2\theta - \sin^2\theta), \qquad \psi = cr^3\sin^2\theta\cos\theta. \tag{12.2.7}$$

12.2.3 Point Source

A point source or sink is generated from a singularity of the rate of dilatation field of the form $\Delta = m\delta(\mathbf{x})$, where the source strength m is equal to the volumetric flow rate emitted from the source. Substituting the dilatation rate distribution into (12.2.1), with $\phi_0 = 0$, yields the potential function as

$$\phi = -\frac{m}{4\pi r}. \tag{12.2.8}$$

The corresponding velocity components are

$$u_r = \frac{m}{4\pi r^2}, \qquad u_\theta = 0. \tag{12.2.9}$$

Substituting these velocity components into the expression (12.1.6) for stream function and integrating gives

$$\psi = -\frac{m}{4\pi}(1 + \cos\theta). \tag{12.2.10}$$

In cylindrical polar coordinates, ϕ and ψ for a point source are given by

$$\phi = -\frac{m}{4\pi(R^2 + z^2)^{1/2}}, \qquad \psi = -\frac{m}{4\pi}\left[1 + \frac{z}{(R^2 + z^2)^{1/2}}\right]. \tag{12.2.11}$$

12.2.4 Point Doublet

A point doublet can be formed by placing a source and a sink of equal strength $\pm m$ at a distance δz from each other, and then letting $m\delta z$ approach a constant as $\delta z \to 0$. The source is placed at the origin of a spherical coordinate system and the sink is placed at a position $z = \delta z$ along the symmetry axis, as shown in Figure 12.2. The value of the velocity potential at a point P, located at (r, θ), is given by the sum of the potential functions from the source and sink, giving

$$\phi = -\frac{m}{4\pi r} + \frac{m}{4\pi(r - \delta r)} = -\frac{m}{4\pi r}\left(1 - \frac{1}{1 - \delta r/r}\right), \tag{12.2.12}$$

where δr is the difference between the radius from the source center to point P and that from the sink center to point P. For $\delta r/r \ll 1$, (12.2.12) becomes

$$\phi \cong \frac{m\delta r}{4\pi r^2}. \tag{12.2.13}$$

The law of cosines can be used to show that for $\delta r/r \ll 1$, $\delta r \cong \delta z \cos\theta$, so that to leading order in $\delta r/r$ the velocity potential becomes

$$\phi \cong \frac{m\delta z \cos\theta}{4\pi r^2}. \tag{12.2.14}$$

Letting the sink approach the source ($\delta z \to 0$) as the strength m increases to infinity, such that the product $m\delta z$ approaches the nonzero doublet strength μ, the velocity

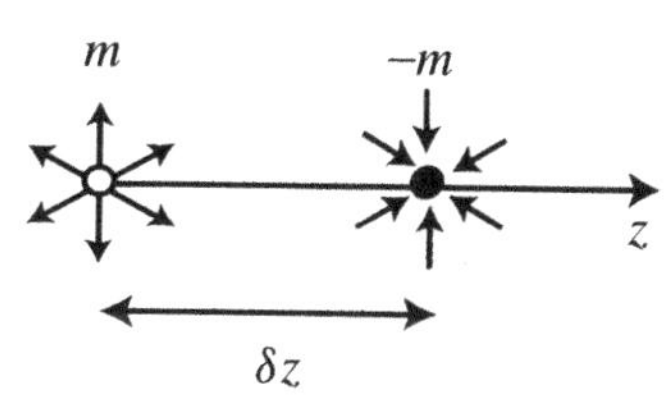

Figure 12.2 Source and sink system used to generate a point doublet as $\delta z \to 0$.

potential for a doublet is obtained as

$$\phi = \frac{\mu}{4\pi r^2} \cos\theta. \tag{12.2.15}$$

The velocity components for a doublet are

$$u_r = -\frac{\mu}{2\pi r^3} \cos\theta, \qquad u_\theta = -\frac{\mu}{4\pi r^3} \sin\theta. \tag{12.2.16}$$

Substituting the expression (12.2.16) for the velocity components into (12.1.6) and integrating give the stream function for a doublet as

$$\psi = -\frac{\mu}{4\pi r} \sin^2\theta. \tag{12.2.17}$$

12.2.5 Line Source of Finite Length

A line source of finite length (Figure 12.3) may be formed by superposing point sources along a section (z_1, z_2) of the z-axis. The strength of the line source per unit length is denoted by $q(z)$ such that the volumetric flow rate emanating from a length dz of the line source is given simply by $q\,dz$. The line source is discretized into N line segments, with length $\Delta\xi = (z_2 - z_1)/N$ and center points at $z = \xi_n$, $n = 1, \ldots, N$. A point source is placed at the center point of each line segment with strength $q(\xi_n)\Delta\xi$. The stream function induced by this set of point sources, which is most conveniently expressed in cylindrical polar coordinates, is obtained using (12.2.11) as

$$\psi(R, z) = -\sum_{n=1}^{N} \frac{q(\xi_n)\Delta\xi}{4\pi} \left(1 + \frac{z - \xi_n}{[R^2 + (z - \xi_n)^2]^{1/2}}\right). \tag{12.2.18}$$

Letting $N \to \infty$, the definition of a Riemann integral can be used to write the stream function for the continuous line source as

$$\psi(R, z) = -\int_{z_1}^{z_2} \frac{q(\xi)}{4\pi} \left(1 + \frac{z - \xi}{[R^2 + (z - \xi)^2]^{1/2}}\right) d\xi. \tag{12.2.19}$$

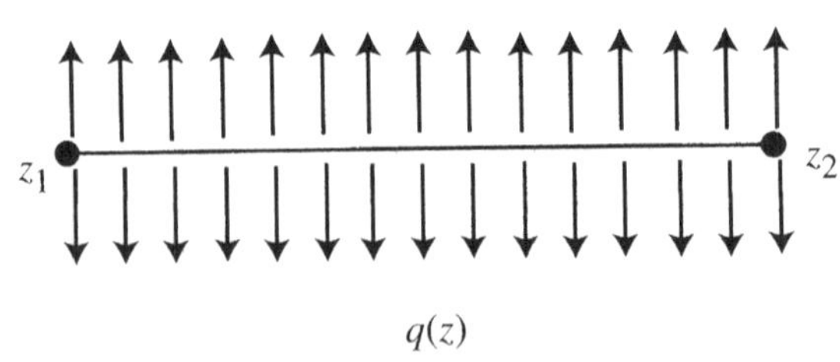

Figure 12.3 Line source spanning interval (z_1, z_2).

For the special case of uniform source strength (q = const), (12.2.19) can be integrated to yield

$$\psi(R, z) = -\frac{q}{4\pi}\left(z_2 - z_1 - [R^2 + (z - z_2)^2]^{1/2} + [R^2 + (z - z_1)^2]^{1/2}\right). \tag{12.2.20}$$

Denoting the length of the line source by $L \equiv z_2 - z_1$ and defining $r_2 \equiv [R^2 + (z - z_2)^2]^{1/2}$ and $r_1 \equiv [R^2 + (z - z_1)^2]^{1/2}$ as the distances from the end points at z_2 and z_1 to the point (R, z), the above result can be rewritten as

$$\psi(R, z) = -\frac{q}{4\pi}(L - r_2 + r_1). \tag{12.2.21}$$

12.3 SOME AXISYMMETRIC FLOWS WITH IMMERSED BODIES

In this section, we describe several examples of axisymmetric flows with immersed bodies that can be generated by superposition of the basic flow solutions derived in the previous section. Numerous other examples exist that are not discussed here. A particularly rich collection of solutions for axisymmetric and fully three-dimensional potential flows with single or multiple bodies of various shapes can be found in the text by Milne-Thompson (1968). The examples included in the present text are fairly simple; more complex problems might best be handled using one of the computational methods discussed in Sections 12.6–12.7 or (for slender bodies) the approximate method described in Section 12.5.

12.3.1 Uniform Flow Past a Sphere

Uniform flow past a sphere of radius a can be generated by superposing uniform flow along the symmetry axis with a point doublet, giving the velocity potential as

$$\phi = Ur\cos\theta + \frac{\mu}{4\pi r}\cos\theta. \tag{12.3.1}$$

The radial velocity component on the sphere surface $r = a$ is

$$u_r\big|_{r=a} = \left(U - \frac{\mu}{2\pi a^3}\right)\cos\theta, \tag{12.3.2}$$

so that the no-penetration condition is satisfied by setting the doublet strength as $\mu = 2\pi a^3 U$. Substituting this result into the expression for velocity potential gives

$$\phi = U\left(r + \frac{a^3}{2r^2}\right)\cos\theta, \tag{12.3.3}$$

and the corresponding velocity components are

$$u_r = U\left(1 - \frac{a^3}{r^3}\right)\cos\theta, \qquad u_\theta = -U\left(1 + \frac{a^3}{2r^3}\right)\sin\theta. \tag{12.3.4}$$

Substituting the velocity components into (12.1.6) and integrating gives the stream function as

$$\psi = \frac{1}{2}U\left(r^2 - \frac{a^3}{r}\right)\sin^2\theta. \tag{12.3.5}$$

The velocity on the sphere surface is obtained from (12.3.5) as $\mathbf{u}(a,\theta) = \frac{3}{2}U\sin\theta\mathbf{e}_\theta$ such that the maximum velocity (at $\theta = 90°$) has magnitude equal to $\frac{3}{2}$ of the free-stream velocity. The pressure field $p_s(\theta)$ on the surface of a fixed sphere in a steady uniform flow is obtained by substituting the surface velocity field into the Bernoulli equation (8.2.3), giving

$$\frac{p_s(\theta) - p_\infty}{\rho} = \frac{1}{2}U^2\left(1 - \frac{9}{4}\sin^2\theta\right). \tag{12.3.6}$$

The pressure coefficient $c_p(\theta) \equiv (p_s(\theta) - p_\infty)/\frac{1}{2}\rho U^2 = 1 - \frac{9}{4}\sin^2\theta$ for the sphere, varying from a value of 1.0 at the front and back stagnation points ($\theta = 0, \pi$) to a value of 1.25 at the shoulder ($\theta = \pi/2$). The symmetry of the pressure field suggests that there is no net force acting on the sphere.

12.3.2 Rankine Half-Body

Superposition of uniform flow and a point source generates a nonclosed body (or half-body), as shown in Figure 12.4, which mimics the flow past the leading edge of a blunt body. The velocity potential and stream function for this flow are

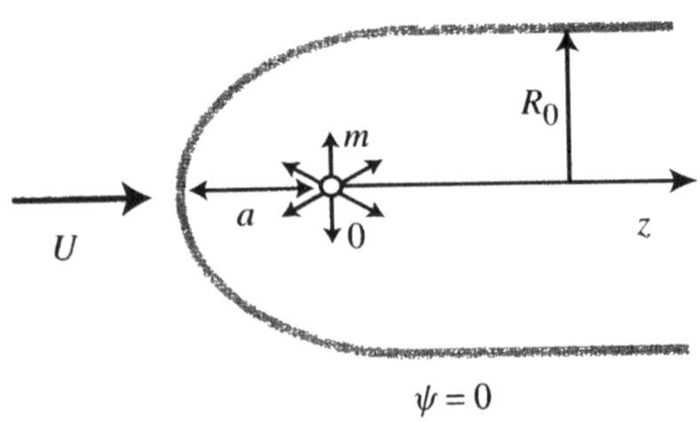

Figure 12.4 Rankine half-body formed from a point source and uniform flow.

$$\phi = Ur\cos\theta - \frac{m}{4\pi r}, \qquad \psi = \frac{1}{2}Ur^2\sin^2\theta - \frac{m}{4\pi}(1+\cos\theta), \tag{12.3.7}$$

and the corresponding velocity components are

$$u_r = U\cos\theta + \frac{m}{4\pi r^2}, \qquad u_\theta = -U\sin\theta. \tag{12.3.8}$$

The position a of the stagnation point upstream of the point source can be obtained by setting $u_r(a, \pi) = 0$, which yields

$$a = \left(\frac{m}{4\pi U}\right)^{1/2}. \tag{12.3.9}$$

The dividing streamline between the interior and exterior of the half-body corresponds to $r = r_0(\theta)$, which coincides with the streamline that passes through the stagnation point at $z = -a$. From (12.3.7), we see that $\psi = 0$ whenever $\theta = \pi$ such that the dividing streamline for this problem corresponds to $\psi = 0$. [This is in fact the reason that the constant 1 is usually added to $\cos\theta$ in the expression (12.2.10) for the stream function of a point source.] Solving for $r_0(\theta)$ from (12.3.7) with $\psi = 0$ yields

$$r_0(\theta) = \left(\frac{m}{4\pi U}\right)^{1/2} \frac{1}{\sin(\theta/2)}. \tag{12.3.10}$$

The cylindrical polar radius $R_0 = r_0 \sin\theta$ of the half-body is given by

$$R_0(\theta) = \left(\frac{m}{4\pi U}\right)^{1/2} \frac{\sin\theta}{\sin(\theta/2)}. \tag{12.3.11}$$

Far downstream of the point source (as $\theta \to 0$), the half-body approaches the limiting radius $R_0 \to (m/\pi U)^{1/2} = 2a$.

12.3.3 Rankine Solids

Flow past a fixed, closed body can be generated by superposing uniform flow with a set of sources and sinks having zero net mass flow rate. One example of such a flow is given by superposing uniform flow with a single point source at $z = -a$ and point sink at $z = a$ (Figure 12.5), known as a Rankine solid. Letting the angle between the source and a point P be denoted by θ_1 and that between the sink and point P be denoted by θ_2, the stream function for this flow can be written as

$$\psi = \frac{1}{2}Ur^2\sin^2\theta - \frac{m}{4\pi}(\cos\theta_1 - \cos\theta_2). \tag{12.3.12}$$

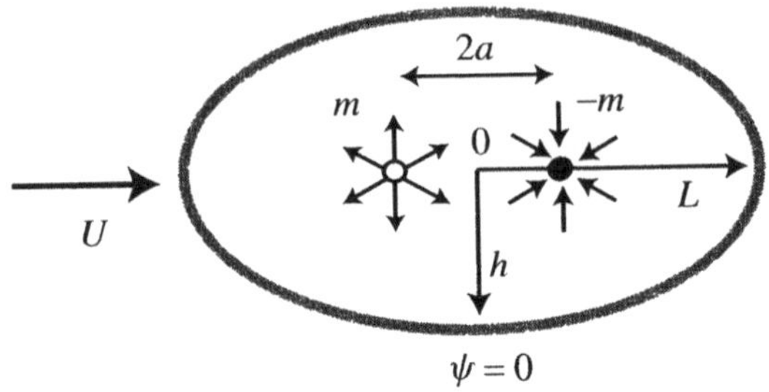

Figure 12.5 Rankine solid formed from a point source, a point sink, and uniform flow.

The dividing streamline corresponds to $\psi = 0$, so the body radius $r_0(\theta)$ is obtained from (12.3.12) as

$$r_0^2(\theta) = \frac{m}{2\pi U \sin^2\theta}(\cos\theta_1 - \cos\theta_2). \tag{12.3.13}$$

The Rankine solid can be characterized by its half-length L and half-width h. The half-length can be determined by setting u_r equal to zero at the stagnation point at $(r,\theta) = (L, 0)$, or

$$u_r(L, 0) = U + \frac{m}{4\pi(L+a)^2} - \frac{m}{4\pi(L-a)^2} = 0. \tag{12.3.14}$$

Rearranging gives a nonlinear equation for L of the form

$$\left[\left(\frac{L}{a}\right)^2 - 1\right]^2 = \frac{m}{\pi U a^2}\frac{L}{a}. \tag{12.3.15}$$

Setting $r_0(\pi/2) = h$, an equation for the half-width can be obtained from (12.3.13) as

$$\left(\frac{h}{a}\right)^2\left[\left(\frac{h}{a}\right)^2 + 1\right]^{1/2} = \frac{m}{\pi U a^2}. \tag{12.3.16}$$

The result (12.3.16) is derived by noting that at the angle $\theta = \pi/2$, θ_1 and θ_2 satisfy $\tan\theta_1 = -\tan\theta_2 = h/a$, so that $\cos\theta_1 = -\cos\theta_2 = 1/[(h/a)^2 + 1]^{1/2}$. A plot of h/a and L/a versus the dimensionless parameter $m/\pi U a^2$ is given in Figure 12.6.

12.3.4 Tapered Bodies

Flow past a closed body with a tapered (or streamlined) back end can be generated by superposing uniform flow with a point source of strength m at the origin and a line sink extending from the origin to point $z = a$ with strength $-m/a$ per unit length

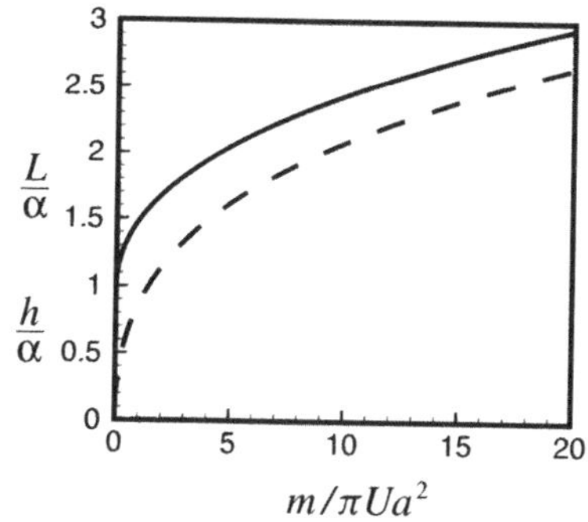

Figure 12.6 Variation of the half-length L (solid line) and half-thickness h (dashed line) of a Rankine solid as a function of the dimensionless parameter $m/\pi U a^2$.

(Figure 12.7). The stream function is given by

$$\psi(r,\theta) = \frac{1}{2} U r^2 \sin^2\theta - \frac{m}{4\pi}(1+\cos\theta) + \frac{m}{4\pi a}(a + r - r_2), \qquad (12.3.17)$$

where $r_2 = [r^2 + a^2 - 2ra\cos\theta]^{1/2}$ is the distance between the end point $z = a$ of the line sink and the point (r, θ). The dividing streamline (representing the body surface) corresponds to $\psi = 0$, since $r_2 \to r + a$ and $\cos\theta \to -1$ as $\theta \to \pi$ and $r_2 \to r - a$ and $\cos\theta \to 1$ as $\theta \to 0$. The body radius $r_0(\theta)$ may therefore be obtained by solving for r in (12.3.17) with $\psi = 0$.

12.3.5 Oscillating Bubble

Oscillation of air bubbles entrained in the wake of marine vehicles are a major noise source, which is important for applications such as acoustic ship detection. The flow past an oscillating small bubble, which is assumed to have approximately spherical shape, can be modeled by superposing uniform flow and a point doublet and point source located at the origin. The velocity potential for the resulting flow is

$$\phi = U r\cos\theta + \frac{\mu}{4\pi r^2}\cos\theta - \frac{m}{4\pi r}. \qquad (12.3.18)$$

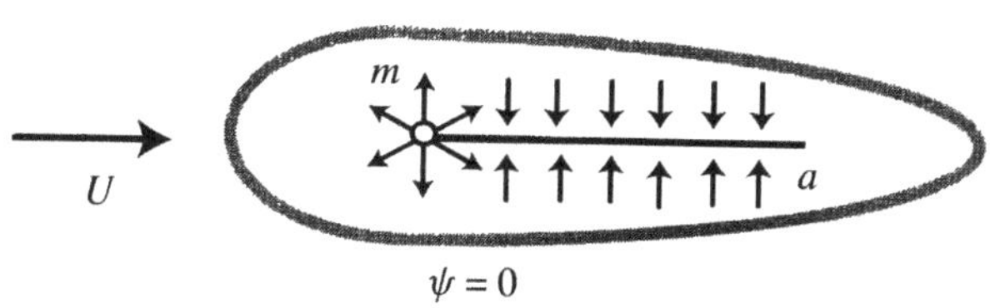

Figure 12.7 Flow past a closed tapered body generated by superposing uniform flow, a point source, and a line sink of length a.

Setting the radial velocity evaluated from (12.3.18) at the surface of the bubble equal to the rate of change of the bubble radius $a(t)$, as required by the no-penetration condition, gives

$$u_r(a,\theta,t) = \left(U - \frac{\mu}{2\pi a^3}\right)\cos\theta + \frac{m}{4\pi a^2} = \dot{a}. \qquad (12.3.19)$$

The requirement that (12.3.19) be satisfied for all angles θ yields expressions for the source and doublet strength as

$$\mu = 2\pi a^3 U, \qquad m = 4\pi a^2 \dot{a}. \qquad (12.3.20)$$

Considering now the special case $U = 0$, the unsteady Bernoulli equation (8.2.2) with zero body force gives

$$\frac{p - p_\infty}{\rho} + \kappa + \frac{\partial \phi}{\partial t} = 0, \qquad (12.3.21)$$

where $p_\infty(t)$ is the pressure far away from the bubble. The Laplace formula (5.3.12) gives the pressure in the liquid at the bubble surface as

$$p(a,\theta,t) = p_B(t) - \frac{2\gamma}{a}, \qquad (12.3.22)$$

where γ is surface tension and $p_B(t)$ is the pressure inside the bubble. The Bernoulli equation can be evaluated at the surface of the bubble using (12.3.19)–(12.3.22) to obtain an ordinary differential equation for the bubble radius $a(t)$ as

$$\frac{3}{2}(\dot{a})^2 + a\ddot{a} + \frac{2\gamma}{\rho a} = \frac{p_B - p_\infty}{\rho}, \qquad (12.3.23)$$

which is known as the *Rayleigh equation*.

Solution for $a(t)$ from the Rayleigh equation depends on the variation of the external pressure and the pressure within the bubble. Assuming that the gas inside the bubble behaves as an isentropic ideal gas, the interior pressure varies with bubble radius as $p_B(t) = p_{B0}(a_0/a)^{3k}$, where $k = c_p/c_v$ is the ratio of the specific heats. If p_∞ is constant, so that the bubble oscillates in the absence of forcing, then (12.3.23) has the form of an autonomous equation, such that all of the terms depend only on a and its derivatives and not explicitly on time. In this case, (12.3.23) may be reduced to a first-order equation by the substitution $f(a) \equiv \dot{a}$, giving

$$\frac{a}{2}\frac{d(f^2)}{da} + \frac{3}{2}f^2 + \frac{2\gamma}{\rho a} = \frac{p_B - p_\infty}{\rho}. \qquad (12.3.24)$$

Equation (12.3.24) is linear in the variable $f^2(a)$ and can be solved by standard methods.

12.4 SPHERE THEOREMS

Analogous to the circle theorem for two-dimensional potential flows (Section 9.4), sphere theorems allow one to easily insert a spherical body in an arbitrary external flow, subject to certain restrictions. Two different sphere theorems are described in this section—the Butler (1953) theorem, which deals with stream function and applies only to axisymmetric flows, and the Weiss (1945) theorem, which deals with velocity potential and can be used for arbitrary three-dimensional potential flows.

12.4.1 Butler Sphere Theorem

The Butler sphere theorem can be used to insert a fixed sphere of radius a at the origin of any axisymmetric potential flow in which there are no singularities within a distance a of the origin and there are no rigid surfaces.

Theorem 12.4.1 (Butler Sphere Theorem). Consider an incompressible, irrotational, axisymmetric flow with no rigid boundaries and no singularities within a distance a of the origin, which has stream function $\psi_0(r, \theta)$ such that $\psi_0 = O(r^2)$ near the origin. If a rigid sphere of radius a is introduced in the flow, the stream function becomes

$$\psi(r, \theta) = \psi_0(r, \theta) - \frac{r}{a}\psi_0\left(\frac{a^2}{r}, \theta\right). \qquad (12.4.1)$$

To prove this theorem, we must show that the stream function given by (12.4.1) (i) satisfies the irrotationality condition, (ii) is constant on the sphere surface $r = a$, (iii) is nonsingular outside the sphere ($r \geq a$), and (iv) approaches the same flow as $\psi_0(r, \theta)$ at infinity with zero net mass flow rate. Condition (ii) is clearly satisfied since (12.4.1) gives $\psi(a, \theta) = 0$. Satisfaction of condition (iii) follows from the realization that r and a^2/r are inverse points with respect to a sphere of radius a, such that if $r > a$ then $a^2/r < a$ and vice versa. We require in the statement of the theorem that all singularities of $\psi_0(r, \theta)$ are *outside* of the sphere, so correspondingly all singularities of $\psi_0(a^2/r, \theta)$ must be *inside* the sphere. To prove condition (iv), we again use the idea of inverse points to note that if $\psi_0(r, \theta) = O(r^2)$ as $r \to 0$, then $\psi_0(a^2/r, \theta) = O(1/r^2)$ as $r \to \infty$. Using the definition (12.1.6) of the Stokes stream function, the velocity induced by the second term on the right-hand side of (12.4.1), which arises from the presence of the sphere, must vary as $O(1/r^3)$ as $r \to \infty$.

Satisfaction of condition (i) can be demonstrated by substituting the second term on the right-hand side of (12.4.1) into the form (12.1.7) of the irrotationality condition expressed in terms of stream function, which gives

$$r^2 \frac{\partial^2}{\partial r^2}\left[\frac{r}{a}\psi_0\left(\frac{a^2}{r}, \theta\right)\right] + \sin\theta \frac{\partial}{\partial \theta}\left(\frac{1}{\sin\theta}\frac{\partial}{\partial \theta}\right)\left[\frac{r}{a}\psi_0\left(\frac{a^2}{r}, \theta\right)\right] = 0. \qquad (12.4.2)$$

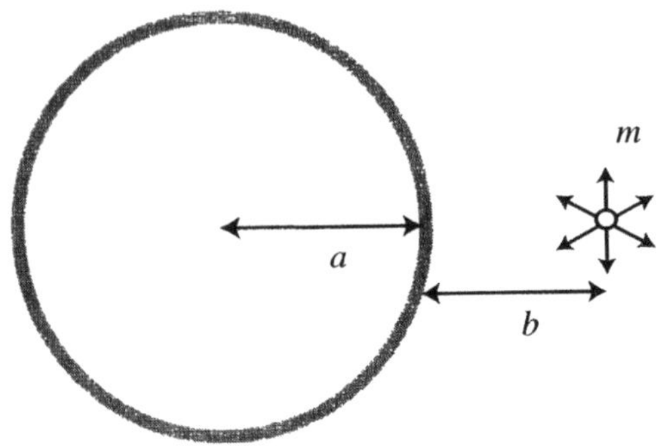

Figure 12.8 Point source at position $z = b$ outside of a sphere of radius a.

Making the change of variables $\xi = a^2/r$, (12.4.2) becomes

$$\xi^2 \frac{\partial^2}{\partial \xi^2} \psi_0(\xi, \theta) + \sin\theta \frac{\partial}{\partial \theta} \left(\frac{1}{\sin\theta} \frac{\partial}{\partial \theta} \right) \psi_0(\xi, \theta) = 0, \tag{12.4.3}$$

which has the same form as the original irrotationality condition (12.1.7) acting on ψ_0. The modified flow given by (12.4.1) is therefore irrotational if the original flow is irrotational.

Example 12.4.1 Point Source in the Vicinity of a Sphere. We consider the flow induced by a point source placed at a point B with position $z = b$ outside of a sphere of radius a, where $b > a$, as illustrated in Figure 12.8. The stream function for the source alone is given by

$$\psi_0(r, \theta) = -\frac{m}{4\pi}(1 + \cos\theta_1), \tag{12.4.4}$$

where θ_1 is the angle of point P, with position (r, θ), about the source at B. Using trigonometric identities, we can write $\cos\theta_1$ in terms of r and θ, such that (12.4.4) becomes

$$\psi_0(r, \theta) = -\frac{m}{4\pi} \left(1 + \frac{r\cos\theta - b}{[r^2 + b^2 - 2rb\cos\theta]^{1/2}} \right). \tag{12.4.5}$$

Applying the sphere theorem (12.4.1), the stream function for the flow with the sphere present is given by

$$\psi(r, \theta) = -\frac{m}{4\pi} \left(1 + \frac{r\cos\theta - b}{[r^2 + b^2 - 2rb\cos\theta]^{1/2}} \right) + \frac{m}{4\pi}\frac{r}{a} \left(1 + \frac{\left(\frac{a^2}{r}\right)\cos\theta - b}{\left[\frac{a^4}{r^2} + b^2 - \left(\frac{2ba^2}{r}\right)\cos\theta\right]^{1/2}} \right). \tag{12.4.6}$$

This result can be simplified by noting that the coordinates (r_2, θ_2) of point P relative to the inverse point B' of the source with respect to the sphere, located at $z = b' \equiv a^2/b$, satisfy

$$r_2^2 = r^2 + b'^2 - 2b'r\cos\theta, \qquad \cos\theta_2 = \frac{r\cos\theta - b'}{[r^2 + b'^2 - 2b'r\cos\theta]^{1/2}}, \tag{12.4.7}$$

so that the stream function becomes

$$\psi(r,\theta) = -\frac{m}{4\pi}(1 + \cos\theta_1) - \frac{ma}{4\pi b}\cos\theta_2 + \frac{m}{4\pi a}(r - r_2). \tag{12.4.8}$$

The first term on the right-hand side of (12.4.8) is the original source, the second term is an image source of strength ma/b located at B', and the third term is an image line sink of strength m/a per unit length spanning a distance a^2/b from B' to the origin. Since the sphere is rigid, the image field satisfies the restriction of zero total mass flow rate. A schematic of the image field is given in Figure 12.9.

12.4.2 Weiss Sphere Theorem

The Weiss sphere theorem can be used to insert a sphere of radius a at the origin of an arbitrary three-dimensional potential flow, subject to the usual restrictions of no singularities within a distance a of the origin and no rigid surfaces in the original flow. The Weiss theorem is more cumbersome to use than the Butler theorem, since it requires evaluation of an integral over the radius, so for axisymmetric flows the Butler theorem is usually preferable.

Theorem 12.4.2 (Weiss Sphere Theorem). Consider an incompressible, irrotational flow with no rigid boundaries and no singularities within a distance a of the

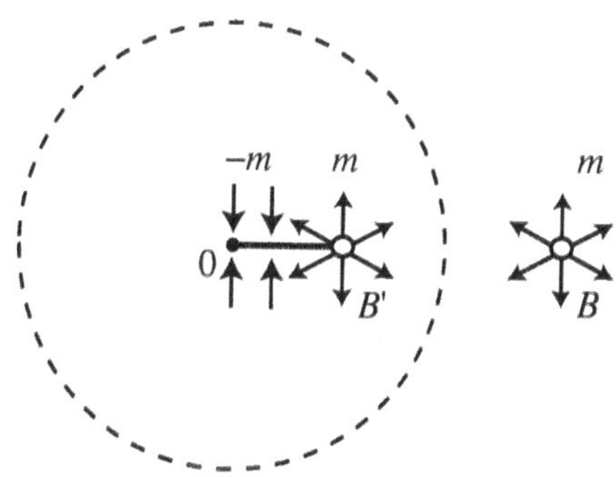

Figure 12.9 Image field of a point source at B outside of a sphere, consisting of a point source at the inverse point B' and a line sink stretching from B' to the center of the sphere.

origin, which has velocity potential $\phi_0(r, \theta, \alpha)$. If a rigid sphere of radius a is introduced in the flow, the velocity potential becomes

$$\phi(r, \theta, \alpha) = \phi_0(r, \theta, \alpha) + \frac{1}{a} \int_0^{a^2/r} \xi \frac{\partial \phi_0(\xi, \theta, \alpha)}{\partial \xi} \, d\xi. \tag{12.4.9}$$

Denoting the last term in (12.4.9) by $\chi(r, \theta, \alpha)$, proof of this theorem requires that (i) χ is irrotational ($\nabla^2 \chi = 0$), (ii) the no-penetration condition $\partial \phi / \partial r = 0$ is satisfied on the surface $r = a$ of the sphere, (iii) there are no singularities of χ outside the sphere (in $r \geq a$), and (iv) the velocity field associated with χ vanishes at infinity such that χ exhibits no net mass injection into the flow. Satisfaction of condition (ii) can be verified by directly differentiating (12.4.9) with respect to r, giving

$$\frac{\partial \phi}{\partial r} = \frac{\partial \phi_0}{\partial r} + \frac{1}{a} \left(-\frac{a^2}{r^2} \right) \left(\xi \frac{\partial \phi_0}{\partial \xi} \right) \Bigg|_{\xi = a^2/r}. \tag{12.4.10}$$

Setting $r = a = \xi$, the right-hand side of (12.4.10) reduces to zero. Satisfaction of condition (iii) follows from the fact that a^2/r is the inverse point of r over the sphere surface. Expanding χ as

$$\chi = \frac{1}{a} [\xi \phi_0(\xi, \theta, \alpha)]_{\xi = a^2/r} - \frac{1}{a} \int_0^{a^2/r} \phi_0 \, d\xi, \tag{12.4.11}$$

it is clear that if $\phi_0(r, \theta, \alpha)$ has no singularities inside the sphere (in $r \leq a$), then $\phi_0(r, \theta, \alpha)$ has no singularities outside the sphere (in $r \geq a$). Satisfaction of condition (iv) can be assured by assuming that ϕ_0 has a power series expansion of the form $\phi_0 \sim A_0 + A_1 r + \cdots$ near the origin. Substituting this expansion into the expression for χ and performing the integration over ξ gives $\chi \sim a^3 A_1 / 2r^2$ as $r \to \infty$, where we recall that $r \to \infty$ is the inverse point of $r = 0$. The radial velocity induced by χ must decrease at infinity as $O(1/r^3)$, such that no net mass source is generated by χ.

Satisfaction of condition (i) can be verified by noting that the change of variables $\xi = a^2/r$ yields

$$\begin{aligned}
\frac{\partial}{\partial r} \left(r^2 \frac{\partial \chi}{\partial r} \right) &= \frac{\xi^2}{a} \frac{\partial}{\partial \xi} \left(\xi \frac{\partial \phi_0}{\partial \xi} \right) \\
&= \frac{1}{a} \int_0^{a^2/r} \frac{\partial}{\partial \xi} \left[\xi^2 \frac{\partial}{\partial \xi} \left(\xi \frac{\partial \phi_0}{\partial \xi} \right) \right] d\xi \\
&= \frac{1}{a} \int_0^{a^2/r} \xi \frac{\partial}{\partial \xi} \left[\frac{\partial}{\partial \xi} \left(\xi^2 \frac{\partial \phi_0}{\partial \xi} \right) \right] d\xi.
\end{aligned} \tag{12.4.12}$$

We can thus write

$$\nabla^2\chi = \frac{\xi^2}{a^5}\int_0^{a^2/r} \xi\frac{\partial}{\partial\xi}\left[\frac{\partial}{\partial\xi}\left(\xi^2\frac{\partial\phi_0}{\partial\xi}\right) + \frac{1}{\sin\theta}\frac{\partial}{\partial\theta}\left(\sin\theta\frac{\partial\phi_0}{\partial\theta}\right) + \frac{1}{\sin^2\theta}\frac{\partial^2\phi_0}{\partial\alpha^2}\right]d\xi. \tag{12.4.13}$$

The term in brackets in (12.4.13) is equal to ξ^2 times the Laplacian of $\phi_0(\xi, \theta, \alpha)$. Since the initial flow is assumed to be irrotational, $\nabla^2\phi_0$ must vanish, so (12.4.13) yields $\nabla^2\chi = 0$.

12.5 SLENDER-BODY THEORY

The superposition method discussed in Section 12.3 can be used to obtain the velocity field for flow past a body or family of bodies of a given shape. In many applications, however, one is instead given a body of some arbitrary shape, or else asked to design a body of optimal shape, for which it may be difficult to obtain a solution by such a procedure. We examine two approaches that can be employed in such situations. The first approach (slender-body theory) is applicable only to bodies satisfying certain "slenderness" conditions and utilizes an asymptotic procedure to obtain an approximate solution for the flow field. The other approach (discussed in the next two sections) uses a computational method that yields a numerical solution of the problem to any desired degree of accuracy, restricted only by the limitations of the computing system. Our discussion of slender-body theory is presented in two parts: the first dealing with the relatively simple case of purely axisymmetric flow and the second dealing with axisymmetric bodies immersed in fully three-dimensional flows.

12.5.1 Axisymmetric Flows

The basic idea of slender-body theory is to replace the body by a line source along the axis of symmetry, running from one end of the body to the other, and then to set the source distribution along the line source to approximately satisfy the no-penetration condition at the body surface. The basis for this approximation can be motivated by reexamining the solution for a line source distribution given in Section 12.2. Expressed in cylindrical polar coordinates, the velocity field induced by a line source of uniform strength q extending over the interval $(0, L)$ on the symmetry axis is given by

$$u_R(R, z) = \frac{q}{4\pi R}\left\{\frac{z}{[R^2 + z^2]^{1/2}} - \frac{z - L}{[R^2 + (z - L)^2]^{1/2}}\right\}, \tag{12.5.1a}$$

$$u_z(R, z) = -\frac{q}{4\pi R}\left\{\frac{R}{[R^2 + z^2]^{1/2}} - \frac{R}{[R^2 + (z - L)^2]^{1/2}}\right\}. \tag{12.5.1b}$$

We evaluate the velocity at a point (R, z) located close to the line source $(R/L \ll 1)$ but not close to either of its end points $[z/R = O(L/R), (z-L)/R = O(L/R)]$. To leading order in R/L, the velocity components become

$$u_R = \frac{q}{2\pi R}\left[1 + O\left(\frac{R}{L}\right)^2\right], \qquad u_z = O\left(\frac{u_R R}{L}\right). \tag{12.5.2}$$

The radial velocity has the same form as that induced by a two-dimensional point source, and the axial velocity induced by the line source is small compared to the radial velocity. We therefore conclude that close to a line source, the induced flow field is similar to that induced by the cross section of the line source in a two-dimensional flow. This observation is the basis for the asymptotic approximation used to specify the line source strength in slender-body theory.

We consider uniform flow along the axis of an axisymmetric body extending along the interval (z_1, z_2) on the symmetry axis (Figure 12.10a). The body length is $L = z_2 - z_1$ and the body radius is $R_0(z)$, where the maximum radius is denoted by $R_{\max} \equiv \max_{z_1<z<z_2} R_0(z)$. A small parameter ε is defined by

$$\varepsilon \equiv \frac{R_{\max}}{L}. \tag{12.5.3}$$

The body is required to satisfy the "slenderness" conditions $\varepsilon \ll 1$ and $dR_0/dz = O(\varepsilon)$, which imply that the body is much longer than it is wide and that the slope is everywhere small.

To solve for the flow field, the body is replaced by a line source placed along the symmetry axis over the interval (z_1, z_2). The strength of the line source is assumed to vary slowly with axial distance. More specifically, the variation of $q(z)$ is assumed to satisfy

$$\frac{dq}{dz} = O(\varepsilon^2 U). \tag{12.5.4}$$

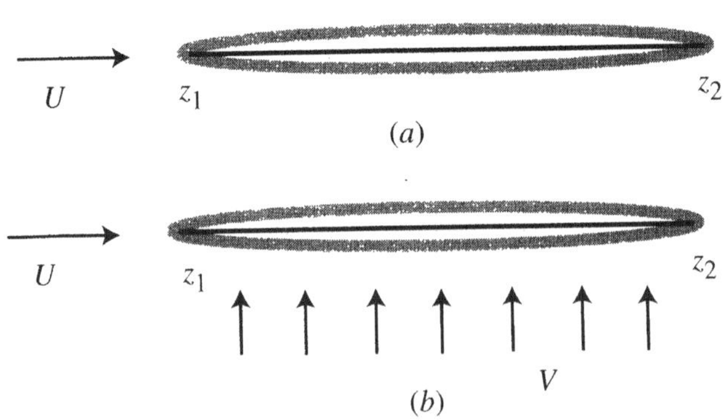

Figure 12.10 Slender axisymmetric body with (a) head-on flow only and (b) both head-on flow and cross flow.

The no-penetration condition at the body surface requires that

$$u_z \frac{dR_0}{dz} = u_R \qquad \text{on} \qquad R = R_0(z). \tag{12.5.5}$$

Based on the argument leading to (12.5.2) and the assumption that q changes gradually with z, the radial and axial velocity components in the boundary condition (12.5.5) can be approximated as

$$u_R(R_0, z) = \frac{q(z)}{2\pi R_0}[1 + O(\varepsilon^2)], \qquad u_z(R_0, z) = U[1 + O(\varepsilon)]. \tag{12.5.6}$$

Substituting (12.5.6) into the boundary condition (12.5.5) and solving for $q(z)$ gives the source strength distribution to leading order in ε as

$$q(z) = 2\pi U R_0 \frac{dR_0}{dz}. \tag{12.5.7}$$

Since $dR_0/dz = O(\varepsilon)$, this solution satisfies the assumption (12.5.4). Integrating (12.5.7) over the length of the body yields

$$\int_{z_1}^{z_2} q(z)\, dz = \pi U \int_{z_1}^{z_2} d(R_0^2) = \pi U[R_0^2(z_2) - R_0^2(z_1)] = 0, \tag{12.5.8}$$

such that the total volumetric source of the line source vanishes, as required for the body to be closed. Substituting (12.5.7) into the expression (12.2.19) for the stream function for a line source and adding the uniform-flow stream function (12.2.4) yields the total stream function for flow past the body.

12.5.2 Axisymmetric Body with Cross Flow

Slender-body theory can be generalized for the case of an axisymmetric body with cross flow $V\mathbf{e}_y$, as shown in Figure 12.10*b*, by placing both a line source and a line doublet on the symmetry axis over an interval (z_1, z_2). The velocity potential induced by a line doublet with vector strength $\boldsymbol{\beta}(z)$ per unit length is given by

$$\phi(x, y, z) = \frac{1}{4\pi} \int_{z_1}^{z_2} \frac{\boldsymbol{\beta}(\xi) \cdot \mathbf{x}}{[x^2 + y^2 + (z - \xi)^2]^{3/2}}\, d\xi. \tag{12.5.9}$$

For the present application, the line doublet is aligned in the direction parallel to the cross flow, such that $\boldsymbol{\beta} = \beta \mathbf{e}_y$. For a line doublet with uniform strength over the interval $(0, L)$, (12.5.9) can be integrated to yield

$$\phi(x, y, z) = \frac{\beta y}{4\pi(x^2 + y^2)} \left\{ \frac{z}{[x^2 + y^2 + z^2]^{1/2}} - \frac{z - L}{[x^2 + y^2 + (z - L)^2]^{1/2}} \right\}. \tag{12.5.10}$$

For a point close to the line doublet but not close to the end points [satisfying the conditions stated prior to (12.5.2)], (12.5.10) can be approximated to leading order in R/L as

$$\phi(x, y, z) = \frac{\beta y}{2\pi(x^2 + y^2)}\left[1 + O\left(\frac{R}{L}\right)^2\right]. \tag{12.5.11}$$

The result (12.5.11) has the same form as the velocity potential of a two-dimensional point doublet (9.3.4) oriented in the y-direction.

The radial velocity is decomposed into two parts, u_{R1} and u_{R2}. The part u_{R1} is induced by the line source and is chosen to satisfy the boundary condition (12.5.5) from the axial flow. The resulting expression for the source strength distribution is the same as given in (12.5.7). The part u_{R2} is induced by the line doublet and is required to vanish on the body surface. Based on the result (9.3.15) for the doublet strength for two-dimensional flow past a circular cylinder, the doublet strength is obtained to leading order in the small parameter ε as

$$\beta(z) = 2\pi R_0^2 V. \tag{12.5.12}$$

The total velocity potential induced by the body is obtained by adding that induced by the line source and the line doublet, or

$$\phi(x, y, z) = \frac{1}{4\pi}\int_{z_1}^{z_2}\left\{-\frac{q(\xi)}{[x^2 + y^2 + (z-\xi)^2]^{1/2}} + \frac{\beta(\xi)y}{[x^2 + y^2 + (z-\xi)^2]^{3/2}}\right\} d\xi, \tag{12.5.13}$$

where q and β are given by (12.5.7) and (12.5.12), respectively.

12.6 SOURCE AND DOUBLET SHEET BOUNDARY-INTEGRAL METHODS

In this section and the next, we develop the theory behind boundary-integral methods, which are commonly used for solution of potential flow past bodies of arbitrary shape in either two or three dimensions. These methods are standard tools in the aerodynamics and marine industries for design of lifting surfaces, such as airfoil sections and propellers. Boundary-integral methods can also be used in flows with nonzero vorticity to determine the contribution to the flow field induced by the presence of the body.

We consider flow past a closed surface S, which encloses a region V and has outward unit normal $\mathbf{n}$. Green's second identity (2.7.8) for any two functions $f(\mathbf{x})$ and $g(\mathbf{x})$ states that

$$\int_V (f\nabla^2 g - g\nabla^2 f)\, dv = \int_S \left(f\frac{\partial g}{\partial n} - g\frac{\partial f}{\partial n}\right) da. \tag{12.6.1}$$

Choosing f as the velocity potential $\phi(\mathbf{x}, t)$ and g as the three-dimensional Green's function $G(\mathbf{x} - \mathbf{x}')$, and making use of the property (6.3.5) of the Green's function and the fact that ϕ is a harmonic function, yields

$$\beta\phi(\mathbf{x}, t) = \int_S \left(\phi' \frac{\partial G}{\partial n'} - G \frac{\partial' \phi}{\partial n'}\right) da', \tag{12.6.2}$$

where $\beta = 0$ if $\mathbf{x} \notin V$, $\beta = \frac{1}{2}$ if $\mathbf{x} \in S$, and $\beta = 1$ if $\mathbf{x} \in V$. A prime is used to denote dependence on $\mathbf{x}'$. Hence, if the value of ϕ and its normal derivative are known for all points on the surface S, then ϕ at any point within V can be obtained by evaluation of the integral in (12.6.2).

Equation (12.6.2) is applied to flow past a body immersed in a prescribed irrotational flow field $\mathbf{u}_0(\mathbf{x}, t) = \nabla\phi_0$, in which the body may be moving with velocity $\mathbf{u}_B(\mathbf{x}, t)$. The body surface is denoted by S_B, the region inside the body is V_I, and the region outside the body is V_E, as shown in Figure 12.11. The unit normal $\mathbf{n}$ of S_B is defined to point into V_E. Boundary-integral methods replace the actual body by a sheet of either sources or doublets fixed to the body surface, where the strength of the source or doublet sheet is set so as to enforce the no-penetration condition on S_B. In this representation, the interior of the body is treated as a fluid, even if in actuality it is solid, in order to get the correct flow field in the body exterior for external flow problems, and vice versa for interior flow problems. Applying (12.6.2) to the interior region V_I gives

$$\beta_I\phi_I(\mathbf{x}, t) = \int_{S_B} \left(\phi_I' \frac{\partial G}{\partial n'} - G \frac{\partial \phi_I'}{\partial n'}\right) da', \tag{12.6.3}$$

where $\phi_I(\mathbf{x}, t)$ denotes the velocity potential interior to the body surface and $\beta_I = 0$ if $\mathbf{x} \notin V_I$, $\beta_I = \frac{1}{2}$ if $\mathbf{x} \in S_B$, and $\beta_I = 1$ if $\mathbf{x} \in V_I$. Next, applying (12.6.2) to the exterior region such that the unit normal $\mathbf{n}$ in (12.6.2) is replaced by $-\mathbf{n}$ gives

$$\beta_E\phi_E(\mathbf{x}, t) = -\int_{S_B} \left(\phi_E' \frac{\partial G}{\partial n'} - G \frac{\partial \phi_E'}{\partial n'}\right) da' + \int_{S_0} \left(\phi' \frac{\partial G}{\partial n'} - G \frac{\partial \phi'}{\partial n'}\right) da', \tag{12.6.4}$$

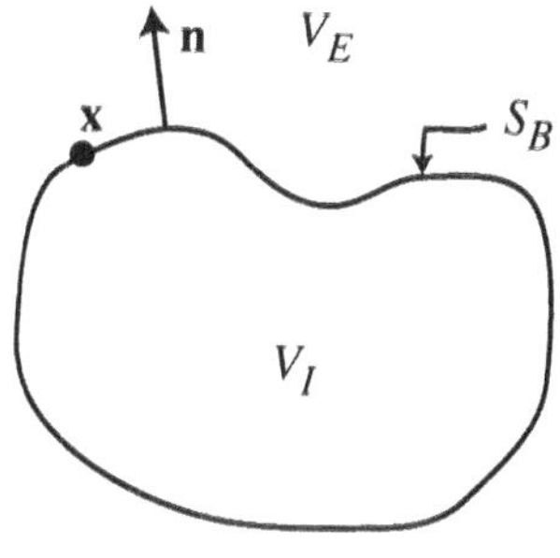

Figure 12.11 Sketch showing a point $\mathbf{x}$ on a closed surface S_B containing an internal volume V_I with outward unit normal $\mathbf{n}$ pointing into an external volume V_E.

where $\beta_E = 1 - \beta_I$, $\phi_E(\mathbf{x}, t)$ denotes the velocity potential exterior to S_B, and S_0 is a closed surface enclosing the body. The velocity field induced by the body decays sufficiently rapidly with distance away from the body (see Section 6.6) that in the limit as S_0 moves outward to infinity, the velocity potential in the last integral in (12.6.4) can be replaced by the potential ϕ_0 associated with the prescribed irrotational flow. Applying (12.6.2) to the region V_0 bounded by the closed surface S_0 yields for any $\mathbf{x} \in V_0$ the result

$$\phi_0(\mathbf{x}, t) = \int_{S_0} \left(\phi_0' \frac{\partial G}{\partial n'} - G \frac{\partial \phi_0'}{\partial n'} \right) da', \tag{12.6.5}$$

so that (12.6.4) becomes

$$\beta_E \phi_E(\mathbf{x}, t) = -\int_{S_B} \left(\phi_E' \frac{\partial G}{\partial n'} - G \frac{\partial \phi_E'}{\partial n'} \right) da' + \phi_0(\mathbf{x}, t). \tag{12.6.6}$$

Adding (12.6.3) and (12.6.6) yields

$$\phi(\mathbf{x}, t) - \phi_0(\mathbf{x}, t) = \int_{S_B} \mu' \frac{\partial G}{\partial n'} da' + \int_{S_B} q' G \, da', \tag{12.6.7}$$

where $q(\mathbf{x}, t) \equiv \partial(\phi_E - \phi_I)/\partial n$ and $\mu(\mathbf{x}, t) \equiv -(\phi_E - \phi_I)$ are the source and doublet sheet strengths, respectively. The function ϕ is equal to ϕ_E for $\mathbf{x} \in V_E$, ϕ_I for $\mathbf{x} \in V_I$, and $(\phi_E + \phi_I)/2$ for $\mathbf{x} \in S_B$.

12.6.1 Source Sheet Method

In the source sheet method (Hess and Smith, 1966), the boundary conditions of the potential fields ϕ_E and ϕ_I are set such that $\mu = 0$ for all $\mathbf{x} \in S_B$. A boundary-integral equation for the source sheet strength $q(\mathbf{x}, t)$ is obtained by evaluating (12.6.7) on the body surface and taking the normal derivative, which gives

$$\frac{1}{2} \left(\frac{\partial \phi_E}{\partial n} + \frac{\partial \phi_I}{\partial n} \right) - \mathbf{u}_0 \cdot \mathbf{n} = \int_{S_B} q(\mathbf{x}', t) \frac{\partial G(\mathbf{x} - \mathbf{x}')}{\partial n} da', \tag{12.6.8}$$

where we note that the unit normal within the integral on the right-hand side is evaluated at $\mathbf{x}$ rather than at $\mathbf{x}'$.

For external flow problems (such that the fluid occupies the region V_E), we set $\partial \phi_E / \partial n = \mathbf{u}_B \cdot \mathbf{n}$ on the body surface, where $\mathbf{u}_B$ is the prescribed velocity of the body such that

$$\frac{1}{2} \left(\frac{\partial \phi_E}{\partial n} + \frac{\partial \phi_I}{\partial n} \right) = -\frac{q}{2} + \mathbf{u}_B \cdot \mathbf{n} \quad \text{for all } \mathbf{x} \in S_B. \tag{12.6.9}$$

Substituting (12.6.9) into (12.6.8) yields an integral equation for q as

$$(\mathbf{u}_B - \mathbf{u}_0) \cdot \mathbf{n} = \frac{1}{2} q(\mathbf{x}, t) + \int_{S_B} q(\mathbf{x}', t) \frac{\partial G(\mathbf{x} - \mathbf{x}')}{\partial n} \, da'. \tag{12.6.10}$$

Taking the gradient of (12.6.7) gives the velocity field external to the body as

$$\mathbf{u}(\mathbf{x}, t) - \mathbf{u}_0(\mathbf{x}, t) = \int_{S_B} q(\mathbf{x}', t) \nabla G(\mathbf{x} - \mathbf{x}') \, da'. \tag{12.6.11}$$

Once the source sheet strength is obtained by solution of (12.6.8), the velocity at any point in the flow field can be obtained by performing the integration over the source sheet indicated in (12.6.11).

For internal flow problems (such that the fluid occupies the region V_I), we set $\partial \phi_I / \partial n = \mathbf{u}_B \cdot \mathbf{n}$ on the body surface such that

$$\frac{1}{2} \left(\frac{\partial \phi_E}{\partial n} + \frac{\partial \phi_I}{\partial n} \right) = \frac{q}{2} + \mathbf{u}_B \cdot \mathbf{n} \quad \text{for all } \mathbf{x} \in S_B. \tag{12.6.12}$$

Setting $\mathbf{u}_0 = 0$, the boundary integral equation (12.6.8) reduces for internal flow problems to

$$\mathbf{u}_B \cdot \mathbf{n} = -\frac{1}{2} q(\mathbf{x}, t) + \int_{S_B} q(\mathbf{x}', t) \frac{\partial G(\mathbf{x} - \mathbf{x}')}{\partial n} \, da'. \tag{12.6.13}$$

The velocity expression (12.6.11) holds for both internal and external flows.

The Green's functions for two- and three-dimensional flows are given by (6.5.5). For three-dimensional flows, the boundary-integral equation and velocity expression become

$$(\mathbf{u}_B - \mathbf{u}_0) \cdot \mathbf{n} = \pm \frac{1}{2} q(\mathbf{x}, t) + \frac{1}{4\pi} \int_{S_B} q(\mathbf{x}', t) \frac{\mathbf{n} \cdot \mathbf{r}}{r^3} \, da' \tag{12.6.14a}$$

and

$$\mathbf{u}(\mathbf{x}, t) = \mathbf{u}_0(\mathbf{x}, t) + \frac{1}{4\pi} \int_{S_B} \frac{q(\mathbf{x}', t) \mathbf{r}}{r^3} \, da', \tag{12.6.14b}$$

where $\mathbf{r} \equiv \mathbf{x} - \mathbf{x}'$, $r \equiv |\mathbf{r}|$, and the plus and minus signs in front of the first term in (12.6.14a) correspond to external and internal flows, respectively. For two-dimensional flows, after replacing the body surface S_B with element of area da by the body contour C_B with element of length $d\ell$, the equivalent expressions are

$$(\mathbf{u}_B - \mathbf{u}_0) \cdot \mathbf{n} = \pm \frac{1}{2} q(\mathbf{x}, t) + \frac{1}{2\pi} \int_{C_B} q(\mathbf{x}', t) \frac{\mathbf{n} \cdot \mathbf{r}}{r^2} \, d\ell' \tag{12.6.15a}$$

and

$$\mathbf{u}(\mathbf{x},t) = \mathbf{u}_0(\mathbf{x},t) + \frac{1}{2\pi}\int_{C_B} \frac{q(\mathbf{x}',t)\mathbf{r}}{r^2}\, d\ell'. \tag{12.6.15b}$$

We note that the source sheet method applies only to bodies without lift, since by construction the circulation must vanish about any circuit enclosing the body. The boundary-integral equation (12.6.10) is of the form of a Fredholm equation of the second kind, which has well-known uniqueness and existence properties (Courant and Hilbert, 1953). Numerical solution of the boundary-integral equation (12.6.14a), or (12.6.15a) for two-dimensional flows, can be obtained by discretizing the body surface into a series of N *panels*, which are usually chosen to be small flat surfaces, of either quadrilateral or triangular shape, over which the sheet strength is constant. The integration can be performed analytically over each panel (Hess and Smith, 1966; Newman, 1986), yielding a $N \times N$ matrix equation for the source panel sheet strengths. This solution procedure can be both computer time and memory intensive for large values of N, and several alternatives exist for acceleration of this procedure. For instance, the time required for construction of the matrix system can be reduced by evaluating the integral in (12.6.14a) using an appropriate numerical quadrature, rather than exact integration, particularly for distant panels. The time required for solution of the matrix equation can be reduced by use of any of several iterative techniques. The memory usage, required to store the interaction matrix resulting from the discretized integral in (12.6.14a), as well as computation time can be reduced using the multipole expansion method or other methods that cluster distant panels (Rokhlin, 1983), similar to those discussed in Section 11.5 with reference to vortex methods in two dimensions. A thorough discussion of different solution approaches for second-kind boundary-integral equations, including iteration and numerical integration techniques, is given in the monograph by Atkinson (1997).

12.6.2 Doublet Sheet Method

In the doublet sheet method (Maskew, 1982), the boundary conditions of the potential fields ϕ_E and ϕ_I are set such that $q = 0$ for all $\mathbf{x} \in S_B$, which implies that the fluid normal velocity is continuous over S_B. A boundary-integral equation for the doublet sheet strength $\mu(\mathbf{x},t)$ is obtained by evaluating (12.6.7) on the body surface and using the definition of μ to write $(\phi_I + \phi_E)/2 = -\mu/2 + \phi_I = \mu/2 + \phi_E$, giving

$$\phi_I(\mathbf{x},t) - \phi_0(\mathbf{x},t) = \frac{1}{2}\mu(\mathbf{x},t) + \int_{S_B} \mu(\mathbf{x}',t)\frac{\partial G(\mathbf{x}-\mathbf{x}')}{\partial n'}\, da' \tag{12.6.16}$$

for external flows and

$$\phi_E(\mathbf{x},t) = -\frac{1}{2}\mu(\mathbf{x},t) + \int_{S_B} \mu(\mathbf{x}',t)\frac{\partial G(\mathbf{x}-\mathbf{x}')}{\partial n'}\, da' \tag{12.6.17}$$

for internal flows. The boundary conditions at the body surface are specified by prescribing the value of ϕ_I or ϕ_E for external or internal flow problems, respectively.

For instance, for an external flow problem involving uniform flow $\mathbf{u}_0(t)$ past a body, where the body is translated with velocity $\mathbf{u}_B(t)$, we choose $\phi_I = \mathbf{u}_B \cdot \mathbf{x}$ and $\phi_0 = \mathbf{u}_0 \cdot \mathbf{x}$, so that the boundary-integral equation (12.6.16) becomes

$$(\mathbf{u}_B - \mathbf{u}_0) \cdot \mathbf{x} = \frac{1}{2}\mu(\mathbf{x}, t) + \int_{S_B} \mu(\mathbf{x}', t)\frac{\partial G(\mathbf{x} - \mathbf{x}')}{\partial n'}\, da'. \tag{12.6.18}$$

Discretizing the doublet sheet by a set of flat panels, (12.6.18) can be converted into a matrix equation, solution of which yields the doublet sheet strength on each panel. Once the doublet sheet strength is known, the potential function ϕ can be obtained at any point in the flow field by the integration indicated in (12.6.7) with q set equal to zero.

12.7 VORTEX SHEET BOUNDARY-INTEGRAL METHOD

It is proved in Section 11.6 that a doublet sheet of strength μ is equivalent to a vortex sheet with strength $\boldsymbol{\gamma} = \nabla \times (\mu \mathbf{n})$. The doublet sheet method developed in the previous section can be used to obtain an expression for the velocity potential ϕ, provided that the value of ϕ is specified on the side of the body surface opposite to that on which the fluid lies (e.g., ϕ_I for an external flow). The doublet sheet formulation is convenient for some applications, such as wave propagation (Section 15.6), in which the velocity potential is directly involved in the dynamic boundary condition on the interface S_B. In other applications, however, it is more convenient to express the boundary conditions in terms of the velocity on S_B. The doublet sheet method can be used for such cases by taking the gradient of (12.6.7), which, as shown in Section 11.6, can be rearranged in the form of the velocity induced by a vortex sheet. In such cases, the boundary-integral formulation is most appropriately called a vortex sheet method.

The velocity induced at a point $\mathbf{x}$ by a vortex sheet of strength $\boldsymbol{\gamma}$ lying on the body surface S_B is given by

$$\mathbf{u}(\mathbf{x}, t) = \mathbf{u}_0(\mathbf{x}, t) + \int_{S_B} \boldsymbol{\gamma}' \times \nabla G\, da'. \tag{12.7.1}$$

The boundary-integral equation for the vortex sheet strength in an external flow can be obtained by taking the gradient of the integral equation (12.6.16) for the doublet sheet strength, giving

$$\mathbf{u}_I(\mathbf{x}, t) - \mathbf{u}_0(\mathbf{x}, t) = \frac{1}{2}\nabla\mu(\mathbf{x}, t) + \nabla \int_{S_B} \mu'\mathbf{n}' \cdot \nabla' G\, da'. \tag{12.7.2}$$

It is shown in Section 11.6 that the last term in (12.7.2) is equivalent to the integral on the right-hand side of (12.7.1), provided that the vortex sheet strength is set as indicated in (11.6.13). The boundary condition on the body surface S_B requires that the projection of the interior velocity tangent to the body surface be equal to the tangent projection of the prescribed body velocity, or $(\mathbf{n} \times \mathbf{u}_I) \times \mathbf{n} = (\mathbf{n} \times \mathbf{u}_B) \times \mathbf{n}$ for all $\mathbf{x} \in S_B$. Taking the projection of (12.7.2) tangent to S_B and applying this boundary condition give

$$[\mathbf{n} \times (\mathbf{u}_B - \mathbf{u}_0)] \times \mathbf{n} = \frac{1}{2}(\mathbf{n} \times \nabla\mu) \times \mathbf{n} + \int_{S_B} [\mathbf{n} \times (\boldsymbol{\gamma}' \times \nabla G)] \times \mathbf{n}\, da'. \quad (12.7.3)$$

The first term on the right-hand side of (12.7.3) is shown in Section 11.6 to be equal to $\mathbf{n} \times \boldsymbol{\gamma}$, so that the boundary-integral equation for the vortex sheet strength becomes

$$[\mathbf{n} \times (\mathbf{u}_B - \mathbf{u}_0)] \times \mathbf{n} = \frac{1}{2}\mathbf{n} \times \boldsymbol{\gamma} + \int_{S_B} [\mathbf{n} \times (\boldsymbol{\gamma}' \times \nabla G)] \times \mathbf{n}\, da'. \quad (12.7.4)$$

One issue that stands out from the above derivation concerns the manner in which the tangent velocity in the internal region is used to set the boundary condition on the body surface, as stated just prior to (12.7.3). This approach essentially collapses the boundary layer that forms about the body in any real, viscous flow, into an infinitely thin vorticity sheet. The tangent velocity condition for the vortex sheet method is related to the requirement of no slip at the wall in a viscous flow (i.e., as the body surface S_B is approached from the internal region V_I). This method mimics the process by which vorticity is produced in real fluid flows, and it also guarantees satisfaction of the no-penetration condition on the body. The latter fact is assured by the following theorem, due to Martensen (1959).

Theorem 12.7.1 (Martensen Theorem). Vanishing of the tangent velocity projection everywhere on a closed, fixed surface S with no vorticity in the region V_I interior to S implies that the velocity component normal to S must also vanish.

Martensen's theorem assures that if the no-slip condition is satisfied everywhere along the internal side of the vortex sheet covering the surface of an immersed, fixed body, then the no-penetration condition will also be everywhere satisfied on the surface. This theorem can be proved by contradiction. Let us suppose that the no-penetration condition is not satisfied, such that there exists a streamline that penetrates the body surface, denoted by C_1 in Figure 12.12. The two end points, a and b, of the curve C_1 are connected by a curve C_2 lying on the surface S, such that the union of C_1 and C_2 forms a closed circuit C. Since the flow interior to S is irrotational, the circulation about the closed circuit C must vanish, such that

$$\int_C \mathbf{u} \cdot d\mathbf{x} = \int_{C_1} \mathbf{u} \cdot d\mathbf{x} + \int_{C_2} \mathbf{u} \cdot d\mathbf{x} = 0. \quad (12.7.5)$$

However, the tangent velocity to C_2 must vanish, as stipulated in the theorem, so (12.7.5) reduces to

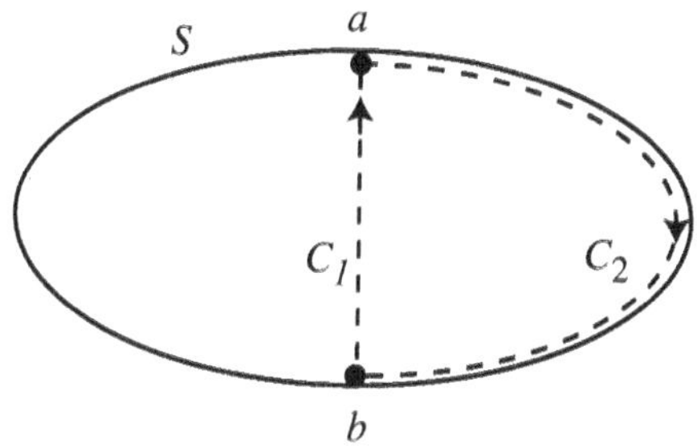

Figure 12.12 Sketch used in the proof of Martensen's theorem.

$$\int_{C_1} \mathbf{u} \cdot d\mathbf{x} = 0. \tag{12.7.6}$$

Since the velocity is unidirectional along the streamline C_1, (12.7.6) requires that the velocity must vanish everywhere on C_1, which implies that a streamline cannot penetrate the body surface.

Using the expressions for Green's function given in (6.5.5), the boundary-integral equation (12.7.4) for the vortex sheet strength $\boldsymbol{\gamma}$ can be written in a three-dimensional flow as

$$[\mathbf{n} \times (\mathbf{u}_B - \mathbf{u}_0)] \times \mathbf{n} = \frac{1}{2}\mathbf{n} \times \boldsymbol{\gamma} + \frac{1}{4\pi} \int_{S_B} \left[\mathbf{n} \times \left(\frac{\boldsymbol{\gamma}' \times \mathbf{r}}{r^3}\right)\right] \times \mathbf{n}\, da'. \tag{12.7.7}$$

Once the integral equation (12.7.7) is solved for the two components of $\boldsymbol{\gamma}$ tangent to S, the velocity at any point in the flow field can be obtained from the vortex sheet form (12.7.1) of the Biot-Savart integral as

$$\mathbf{u}(\mathbf{x}, t) = \mathbf{u}_0(\mathbf{x}, t) - \frac{1}{4\pi} \int_{S_B} \frac{\mathbf{r} \times \boldsymbol{\gamma}'}{r^3}\, da'. \tag{12.7.8}$$

In a two-dimensional flow, the boundary-integral equation for the vortex sheet strength is

$$[\mathbf{n} \times (\mathbf{u}_B - \mathbf{u}_0)] \times \mathbf{n} = \frac{1}{2}\mathbf{n} \times \boldsymbol{\gamma} + \frac{1}{2\pi} \int_{C_B} \left[\mathbf{n} \times \left(\frac{\boldsymbol{\gamma}' \times \mathbf{r}}{r^2}\right)\right] \times \mathbf{n}\, d\ell', \tag{12.7.9}$$

and the Biot-Savart integral is

$$\mathbf{u}(\mathbf{x}, t) = \mathbf{u}_0(\mathbf{x}, t) - \frac{1}{2\pi} \int_{C_B} \frac{\mathbf{r} \times \boldsymbol{\gamma}'}{r^2}\, d\ell'. \tag{12.7.10}$$

In two dimensions, the only nonzero component of the vortex sheet strength is oriented normal to the plane of the flow. Details about the numerical implementation of these boundary-integral equations can be found in references such as Lewis (1991).

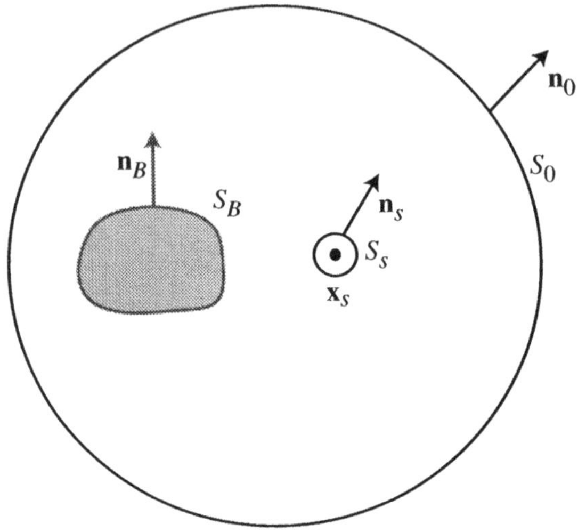

Figure 12.13 Control surfaces used to determine the force on an immersed body with surface S in the presence of a singularity at $\mathbf{x}_s$.

12.8 FORCES INDUCED BY SINGULARITIES

In this section, we examine the force exerted on an immersed body of arbitrary shape in the presence of steady flow formed by the combination of uniform flow at infinity and some set of singularities of the dilatation rate, such as point sources and point doublets. For definiteness, consider the body shown in Figure 12.13 with surface S_B having outward unit normal $\mathbf{n}_B$. A singularity is placed at a point $\mathbf{x}_s$ outside of the body, and the singularity is enclosed by a small closed surface S_s with outward unit normal $\mathbf{n}_s$. We select as a control volume the region enclosed by the two surfaces S_B and S_s and another surface S_0, with outward unit normal $\mathbf{n}_0$, far from the body. Applying the conservation of momentum in control-volume form (4.2.12) to this region and imposing the restriction of steady flow gives an expression for the fluid force $\mathbf{F}$ acting on the body as

$$\mathbf{F} = -\int_{S_B} p\mathbf{n}_B \, da = -\int_{S_0} [p\mathbf{n}_0 + \rho\mathbf{u}(\mathbf{u}\cdot\mathbf{n}_0)] \, da + \int_{S_s} [p\mathbf{n}_s + \mathbf{u}(\mathbf{u}\cdot\mathbf{n}_s)] \, da. \quad (12.8.1)$$

The Bernoulli equation (8.2.3) for steady, irrotational flow with no gravity gives

$$p = p_\infty + \tfrac{1}{2}\rho(\mathbf{U}\cdot\mathbf{U} - \mathbf{u}\cdot\mathbf{u}), \quad (12.8.2)$$

Noting that the integral of a constant over a closed surface vanishes and substituting (12.8.2) into (12.8.1) yields

$$\mathbf{F} = -\rho \int_{S_0} \left[-\frac{1}{2}(\mathbf{u} \cdot \mathbf{u})\mathbf{n}_0 + \mathbf{u}(\mathbf{u} \cdot \mathbf{n}_0) \right] da$$
$$+ \rho \int_{S_s} \left[-\frac{1}{2}(\mathbf{u} \cdot \mathbf{u})\mathbf{n}_s + \mathbf{u}(\mathbf{u} \cdot \mathbf{n}_s) \right] da. \qquad (12.8.3)$$

We consider the first integral on the right-hand side of (12.8.3) as the surface S_0 is taken far away from the body. Letting $\mathbf{u} = \mathbf{U} + \mathbf{u}'$, where $\mathbf{u}'$ decays at least as fast as $O(1/r^2)$ as $r \to \infty$, the first integral in (12.8.3) can be rewritten as

$$\int_{S_0} \left[-\frac{1}{2}(\mathbf{u} \cdot \mathbf{u})\mathbf{n}_0 + \mathbf{u}(\mathbf{u} \cdot \mathbf{n}_0) \right] da$$
$$= \int_{S_0} \left[-\frac{1}{2}(\mathbf{u}' \cdot \mathbf{u}')\mathbf{n}_0 + \mathbf{u}'(\mathbf{u}' \cdot \mathbf{n}_0) + \mathbf{U} \times (\mathbf{u}' \times \mathbf{n}_0) \right] da. \qquad (12.8.4)$$

The surface S_0 is chosen to be a sphere of radius r, where we consider the limit $r \to \infty$. The velocity perturbation far from the body has the asymptotic form $\mathbf{u}' = (m/4\pi r^2)\mathbf{e}_r + O(1/r^3)$, where m is the total volumetric output from the singularity, so the integrals over the first two terms on the right-hand side of (12.8.4) approach zero as $O(1/r^2)$. Since $\mathbf{e}_r \times \mathbf{e}_r = 0$, the product $\mathbf{u}' \times \mathbf{n}_0$ varies as $O(1/r^3)$ far from the body. The integral over the last term in (12.8.4) therefore approaches zero in proportion to $O(1/r)$ as $r \to \infty$, such that the first integral in (12.8.3) vanishes when S_0 is taken to infinity. With no singularity present this result gives $\mathbf{F} = 0$, which is the three-dimensional manifestation of D'Alembert's paradox.

For flows with a singularity present, (12.8.3) becomes

$$\mathbf{F} = \rho \int_{S_s} \left[-\frac{1}{2}(\mathbf{u} \cdot \mathbf{u})\mathbf{n}_s + \mathbf{u}(\mathbf{u} \cdot \mathbf{n}_s) \right] da. \qquad (12.8.5)$$

The force on the body is therefore entirely determined by the flow field in the neighborhood of the singularity. Derivation of subsequent expressions for $\mathbf{F}$ depend on the type of singularity under consideration. The cases of a single point source and a single doublet are discussed in the following two subsections.

12.8.1 Force Induced by a Point Source

The surface S_s surrounding the singularity is taken to be a sphere with small radius ε. When the singularity at $\mathbf{x}_s$ is a point source of strength m, the velocity at S_s for small values of ε approaches

$$\mathbf{u} \cong \frac{m}{4\pi\varepsilon^2}\mathbf{n}_s + \mathbf{u}_I, \qquad (12.8.6)$$

where $\mathbf{u}_I$ denotes the velocity induced at the singular point by all other contributions to the flow (i.e., the uniform flow and the image set within the body). Substituting

(12.8.6) into (12.8.5) gives

$$\mathbf{F} = \rho \int_{S_s} \left[\frac{m^2}{32\pi^2\varepsilon^4}\mathbf{n}_s - \frac{1}{2}(\mathbf{u}_I \cdot \mathbf{u}_I)\mathbf{n}_s + \frac{m}{4\pi\varepsilon^2}\mathbf{u}_I + (\mathbf{u}_I \cdot \mathbf{n}_s)\mathbf{u}_I \right] da. \quad (12.8.7)$$

The generalized Green's theorem (2.7.1) can be used to show that the first and second terms in (12.8.7), which have the form of a constant times a unit normal, and the last term all vanish when integrated over a closed surface, where we assume that ε is sufficiently small that $\mathbf{u}_I$ can be regarded as a constant in (12.8.7). Since the integrand of the third term is constant and $\int_{S_s} da = 4\pi\varepsilon^2$ is the surface area of S_s, the force on the body is given by (12.8.7) as

$$\mathbf{F} = \rho m \mathbf{u}_I. \quad (12.8.8)$$

The force exerted on a solid body in the presence of a point source is therefore equal to the mass flow rate ρm generated by the source times the induced velocity evaluated at the point source. This result is the three-dimensional version of Lagally's theorem, stated for two-dimensional flow in Section 10.3.

Example 12.8.1 Force Induced on a Sphere by a Nearby Point Source. The flow field induced by a point source outside of a sphere is solved in Example 12.4.1. Differentiating the second term of the stream function given in (12.4.6) with respect to θ and setting $(r, \theta) = (b, 0)$, the induced velocity at the point source location is obtained in spherical coordinates as

$$\mathbf{u}_I = \frac{ma^3}{4\pi b(b^2 - a^2)^2}\mathbf{e}_z. \quad (12.8.9)$$

The force acting on the sphere is given by Lagally's theorem as

$$\mathbf{F} = \frac{\rho m^2 a^3}{4\pi b(b^2 - a^2)^2}\mathbf{e}_z. \quad (12.8.10)$$

The force on the sphere is directed toward the source, as would be expected from the fact that the velocity magnitude is higher on the side with the source than on the opposite side.

12.8.2 Force Induced by a Point Doublet

The force on a body in the presence of a point doublet can be derived by treating the doublet as the superposition of a source and a sink and using the result of the previous subsection. Suppose that a source of strength m is located at a point $\mathbf{x}_s$ and a sink of strength $-m$ is located at $\mathbf{x}_s + \Delta\mathbf{x}$. We let $\mathbf{u}_I$ denote the "image" velocity at the source location, induced by the image of the singularity within the body. The velocity at the source location $\mathbf{x}_s$ (not counting that induced by the source itself) is

$$\mathbf{u}_{\text{source}} = \frac{m}{4\pi\,\Delta x^2}\mathbf{a} + \mathbf{u}_I, \tag{12.8.11}$$

where $\Delta x = |\Delta\mathbf{x}|$ and $\mathbf{a}$ is a unit vector pointing from the source to the sink. The first term in (12.8.11) is the velocity induced by the sink. The velocity at the sink location $\mathbf{x}_s + \Delta\mathbf{x}$ (not counting that induced by the sink itself) is

$$\mathbf{u}_{\text{sink}} = \frac{m}{4\pi\,\Delta x^2}\mathbf{a} + \mathbf{u}_I + \Delta\mathbf{x}\cdot\nabla\mathbf{u}_I, \tag{12.8.12}$$

where a Taylor expansion is used to approximate the value of $\mathbf{u}_I$ at the sink location. The net force acting on the body is the sum of the force $\rho m\mathbf{u}_{\text{source}}$ due to the source and the force $-\rho m\mathbf{u}_{\text{sink}}$ due to the sink. Using (12.8.11) and (12.8.12), the total force acting on the body is obtained as

$$\mathbf{F} = -\rho m\Delta\mathbf{x}\cdot\nabla\mathbf{u}_I. \tag{12.8.13}$$

Taking the limit in which the source and sink approach a doublet, where $\nabla\mathbf{x} \to 0$ and $m \to \infty$ such that $m\Delta\mathbf{x} \to \boldsymbol{\mu}$ (the vector doublet strength $\boldsymbol{\mu}$ is related to the scalar strength μ by $\boldsymbol{\mu} = \mu\mathbf{a}$), the formula for the force on the body in the presence of a point doublet reduces to

$$\mathbf{F} = -\rho\boldsymbol{\mu}\cdot\nabla\mathbf{u}_I. \tag{12.8.14}$$

12.9 ADDED MASS AND BUOYANCY FORCES

The D'Alembert paradox states that a fixed three-dimensional body exposed to a steady uniform flow, with no singularities outside the body, experiences no force due to the flow. The present section is concerned with forces that arise in unsteady flows, for instance, when a body accelerates through a fluid or when the uniform velocity at infinity changes in time. It might seem strange to the reader that a nonzero force on the body should arise in unsteady irrotational flows, since the Laplace equation for ϕ is independent of time and the velocity field for accelerating flow past a fixed body is no different than for a steady flow with the same instantaneous far-field velocity. The symmetry of the velocity field on the front and back sides of a sphere immersed in a uniform flow is therefore retained even if the uniform-flow speed changes in time. It should be recalled, however, that the unsteady Bernoulli equation (8.2.2) contains a term proportional to $\partial\phi/\partial t$ that explicitly depends on time. For this reason, even if the velocity field retains the front-back symmetry for unsteady flow past a sphere, the pressure field will not be symmetric, resulting in a net force on the sphere along the direction of the free-stream velocity.

There are two physical effects that give rise to unsteady fluid forces on an immersed body. The first of these is related to the fact that when a body accelerates relative to the ambient speed of the surrounding fluid, the net kinetic energy of the fluid must change. In order to bring about this change in kinetic energy, it is necessary for the body to exert work on the fluid. A similar circumstance occurs for a container of liquid traveling through empty space, where in this simple example the container plays the role of the body and the liquid within the container plays the role of the fluid surrounding the body. If the speed of the container changes, the container must perform work on the liquid to change the liquid speed accordingly. The net amount of force required to change the container speed is equal to the sum of the container mass and the liquid mass times the acceleration of the container. Similarly in the problem of a body traveling through a fluid, the surrounding fluid can be treated as an "added mass" to the body, and the force that the body must exert to accelerate the fluid is called the *added mass force*.

The second physical effect that is responsible for force on a body in an unsteady flow is related to the fact that acceleration of the uniform flow at infinity gives rise to an ambient pressure gradient spanning the flow field. It is well known that a body placed in a pressure gradient experiences a force directed toward the low-pressure side of the body. For instance, a body placed in a fluid with a hydrostatic pressure gradient (caused by gravity) experiences an upward buoyancy force, as any child holding a beach ball under water will readily attest. A similar *buoyancy force* acts on a body immersed in an unsteady uniform flow, although in this case the pressure gradient is caused by the fluid acceleration rather than by gravity.

12.9.1 Added Mass Force

The kinetic energy of the flow about a body immersed in a nonzero uniform flow is infinite. For this reason, it is most convenient in discussing added mass force to consider a reference frame in which the velocity vanishes at infinity. The added mass force depends only on the difference between the uniform-flow speed and the body advection speed, so this translation of reference frame does not affect the added mass force. We consider a body of arbitrary shape with bounding surface S and outward unit normal $\mathbf{n}$, as shown in Figure 12.14. Far away from the body, a sphere S_0 is positioned with outward unit normal $\mathbf{n}_0$. The kinetic energy in the region V bounded by the surfaces S and S_0 is given by

$$T = \frac{1}{2}\rho \int_V \mathbf{u} \cdot \mathbf{u}\, dv = \frac{1}{2}\rho \int_V \nabla\phi \cdot \nabla\phi\, dv. \tag{12.9.1}$$

Since $\nabla^2\phi = 0$, (12.9.1) can be rewritten as

$$T = \frac{1}{2}\rho \int_V \nabla \cdot (\phi\nabla\phi)\, dv. \tag{12.9.2}$$

Using the divergence theorem, we obtain

$$T = -\frac{1}{2}\rho \int_S \phi(\mathbf{n} \cdot \nabla\phi)\, da + \frac{1}{2}\rho \int_{S_0} \phi(\mathbf{n}_0 \cdot \nabla\phi)\, da, \tag{12.9.3}$$

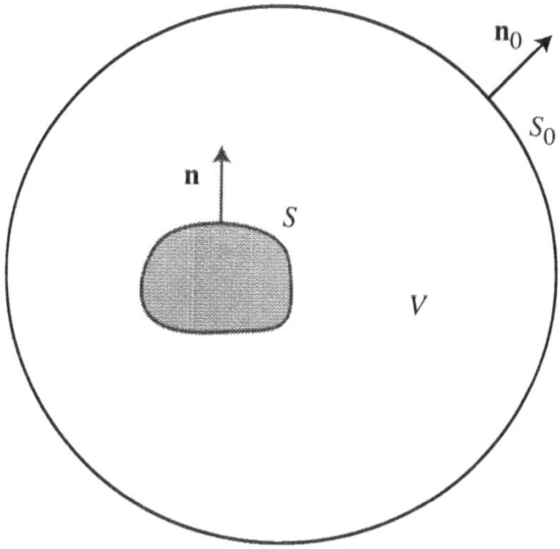

Figure 12.14 Control volume used to determine the kinetic energy in the fluid surrounding a moving body.

where the negative sign in front of the first term is due to the fact that the normal vector **n** of S points into the region V. Since the reference frame is chosen such that the velocity vanishes far from the body, ϕ must approach a constant and $\nabla\phi$ must decrease in proportion to $O(1/r^3)$ (for a body of fixed shape) on S_0 as the radius r of S_0 is taken to infinity. The second integral in (12.9.3) therefore approaches zero in proportion to $O(1/r)$ as $r \to \infty$. The resulting kinetic energy of a body moving in an unbounded space, with zero velocity at infinity, can be expressed as an integral over the surface of the body as

$$T = -\frac{1}{2}\rho \int_S \phi \frac{\partial \phi}{\partial n}\, da. \tag{12.9.4}$$

The *added mass* of a body moving in a fluid is defined as that mass of fluid that, if it were advected with the same translation velocity as the body, would have the same kinetic energy as the entire fluid. This definition is illustrated in Figure 12.15, where the total kinetic energy contained in the fluid volume V in Figure 12.15*a* is the same as the kinetic energy of a mass M of fluid transported with the body in Figure 12.15*b*. For the case with no fluid velocity at infinity, the result (12.9.4) can be used to express this definition mathematically as

$$\frac{1}{2} M U_B^2 = -\frac{1}{2}\rho \int_S \phi \frac{\partial \phi}{\partial n}\, da, \tag{12.9.5}$$

where U_B is the translation speed of the body. Solving for the added mass M from

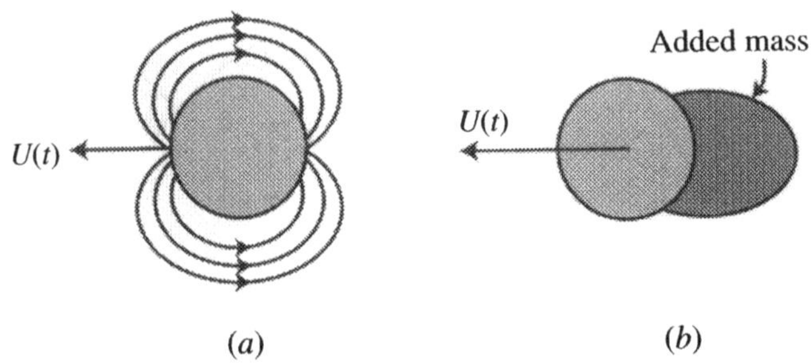

Figure 12.15 Sketches illustrating the concept of fluid added mass: (*a*) actual flow with distributed kinetic energy; (*b*) model of body with added mass.

(12.9.5) gives

$$M = -\frac{\rho}{U_B^2}\int_S \phi \frac{\partial \phi}{\partial n}\, da. \tag{12.9.6}$$

Example 12.9.1 Added Mass of a Sphere. We consider a sphere of radius a traveling with speed U_B through an otherwise stationary fluid. Subtracting the uniform flow from the solution (12.3.3), the velocity potential for this flow is given by

$$\phi = U_B \frac{a^3}{2r^2}\cos\theta. \tag{12.9.7}$$

Substituting this result into (12.9.6) gives the added mass as

$$M = -\frac{\rho}{U_B^2}\int_0^{2\pi} d\alpha \int_0^{\pi}\left(-\frac{1}{2}aU_B^2\cos^2\theta\right)a^2\sin\theta\, d\theta, \tag{12.9.8}$$

which after performing the indicated integration yields

$$M = \frac{2}{3}\pi a^3 \rho. \tag{12.9.9}$$

For this case, the added mass is equal to one-half of the mass of fluid displaced by the body. The added mass of a translated rigid body is in general proportional to the mass of displaced fluid, but the coefficient of proportionality depends on the shape of the body.

The *added mass force* is the force exerted on a body accelerating in an otherwise still fluid due to the work that is performed by the body in changing the kinetic energy of the surrounding fluid. It is straightforward to determine the added mass force using the model illustrated in Figure 12.15. If the kinetic energy due to the added mass is $\frac{1}{2}MU_B^2$, then the corresponding momentum is $M\mathbf{U}_B$. Newton's second law states that force is equal to the rate of change of momentum. Assuming the body to be rigid (so that M is constant), the force exerted by the body on the fluid is $M d\mathbf{U}_B/dt$. The added mass force $\mathbf{F}_A$ exerted on the body is opposite to that which the body exerts on the fluid, or

$$\mathbf{F}_A = -M\frac{d\mathbf{U}_B}{dt}. \tag{12.9.10}$$

We note that the added mass M is a constant for a rigid body with translation velocity in a fixed direction, and it is independent of the speed of translation of the body through the fluid. The equation of motion of a body of mass m translating with velocity $\mathbf{U}_B$ through an otherwise stationary fluid of density ρ and subjected to some external force $\mathbf{P}$ has the form

$$(m + M)\frac{d\mathbf{U}_B}{dt} = \mathbf{P}, \tag{12.9.11}$$

where M is given by (12.9.6). The body behaves as if its total mass were $m + M$, rather than its actual mass m.

For the case where the fluid is not at rest at infinity but approaches some constant value $\mathbf{U}_\infty$, the added mass force depends only on the velocity of the body relative to that of the uniform stream. For this case, the expression (12.9.10) for added mass force is generalized as

$$\mathbf{F}_A = -M\frac{d(\mathbf{U}_B - \mathbf{U}_\infty)}{dt}. \tag{12.9.12}$$

12.9.2 Buoyancy Force

A body of volume V immersed in a fluid with constant ambient pressure gradient ∇p experiences a buoyancy force $\mathbf{F}_B$ given by

$$\mathbf{F}_B = -V\nabla p. \tag{12.9.13}$$

This result can be derived by the following argument. Consider a body of volume V with bounding surface S and outward unit normal $\mathbf{n}$. The surrounding fluid exerts a force $\mathbf{F}_B = -\int_S p\mathbf{n}\,da$ on the body surface. This integral can be converted into a volume integral using the generalized Green's theorem (2.7.1) such that the force on the body is given by

$$\mathbf{F}_B = -\int_V \nabla p\,dv. \tag{12.9.14}$$

If the pressure gradient is constant, it can be taken out of the integral in (12.9.14), which after integration over the volume yields the result (12.9.13).

Application of this result to the common buoyancy force due to a body subjected to a hydrostatic pressure gradient is examined in the following example.

Example 12.9.2 Gravitational Buoyancy Force. We consider a stationary body of volume V immersed in a fluid of density ρ subject to a hydrostatic pressure gradient $\nabla p = \rho \mathbf{g}$, where $\mathbf{g}$ is the gravitational acceleration vector. Application of (12.9.13) gives the buoyancy force acting on the body as $\mathbf{F}_B = -\rho V \mathbf{g}$. This result is equivalent to Archimedes' principle, stating that the force exerted on the body by the fluid is oriented opposite to the direction of gravity and has magnitude equal to the weight of the fluid displaced by the body.

This same approach can be used to obtain an expression for the *inertial buoyancy force* acting on a body immersed in an accelerating uniform stream. The unsteady Bernoulli equation (8.2.2) gives the pressure variation in the fluid as

$$\frac{p - p_0}{\rho} + \frac{1}{2}(\mathbf{u} \cdot \mathbf{u} - \mathbf{u}_0 \cdot \mathbf{u}_0) + \frac{\partial(\phi - \phi_0)}{\partial t} = 0, \tag{12.9.15}$$

where p_0 is the pressure at some arbitrary reference point far from the body. The kinetic energy term in (12.9.15) produces no net force on the body, since it is the same for steady and unsteady flow with the same far-field boundary conditions. The added mass force can be associated with the $\partial(\phi - \phi_0)/\partial t$ term in (12.9.15) and the buoyancy force can be associated with spatial variation of the far-field pressure p_0. That these two effects are additive is guaranteed by (12.9.15).

The Euler equation for the far-field flow, for which the velocity $\mathbf{u}_0$ approaches $\mathbf{U}_\infty(t)$, gives the gradient of p_0 as

$$\nabla p_0 = -\rho \frac{d\mathbf{U}_\infty}{dt}. \tag{12.9.16}$$

Substituting this result into (12.9.13) gives the buoyancy force on the body as

$$\mathbf{F}_B = \rho V \frac{d\mathbf{U}_\infty}{dt}. \tag{12.9.17}$$

The buoyancy force depends only on the far-field velocity and is independent of the motion of the body in the fluid.

Example 12.9.3 Force on a Small Particle in the Flow Field of a Line Vortex. We consider a small particle with volume V that is placed in the flow field of a line vortex of strength Γ, as shown in Figure 12.16. The maximum particle length scale a is assumed to be much smaller than the distance R between the vortex axis of the particle centroid. The particle has density ρ_p and mass $m_p = \rho_p V$, and the fluid has density ρ_f. The particle is assumed to initially travel in the azimuthal direction around the vortex with the same speed $v = \Gamma/2\pi R$ as the surrounding fluid particles. The centrifugal force exerted on the particle, given by

$$\mathbf{F}_C = \frac{m_p v^2}{R}\mathbf{e}_R = \frac{\rho_p V \Gamma^2}{4\pi^2 R^3}\mathbf{e}_R, \tag{12.9.18}$$

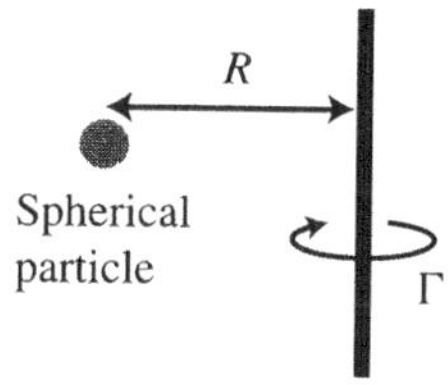

Figure 12.16 A small particle located at a distance R from the axis of a line vortex.

acts to throw the particle away from the vortex axis. The radial pressure gradient associated with the line vortex is given by

$$\nabla p = \frac{\rho_f \Gamma^2}{4\pi^2 R^3} \mathbf{e}_R. \tag{12.9.19}$$

This pressure gradient induces a buoyancy force on the particle, obtained from (12.9.13) as

$$\mathbf{F}_B = -V \nabla p = -\frac{\rho_f V \Gamma^2}{4\pi^2 R^3} \mathbf{e}_R, \tag{12.9.20}$$

which acts to pull the particle toward the vortex axis. The net radial force on the particle is given by

$$\mathbf{F}_C + \mathbf{F}_B = \frac{(\rho_p - \rho_f) V \Gamma^2}{4\pi^2 R^3} \mathbf{e}_R. \tag{12.9.21}$$

If the particle density is greater than the fluid density, the net force is oriented in the positive radial direction and the particle is thrown away from the vortex axis. If the particle density is less than the fluid density, the net force is oriented in the negative radial direction and the particle is pulled toward the vortex core. For this reason, air bubbles in water are pulled into nearby vortex structures, and in the presence of a sufficient number of bubbles the vortex cores can become completely filled by air. On the other hand, particles of sand and other heavy solids immersed in water are centrifuged away from the vortex cores. This effect forms interesting streaky patterns in the near-wall region of turbulent boundary layers as the particles collect in the interstitial regions between the cores of coherent vortex structures (Kaftori et al., 1995).

BIBLIOGRAPHY

Atkinson, K.E. (1997). *The Numerical Solution of Integral Equations of the Second Kind*, Cambridge University Press: New York.

Butler, S.F.J. (1953). "A note on Stokes's stream function for motion with a spherical boundary," *Proceedings of the Cambridge Philosophical Society* **49**, 169–174.

Courant, R., and D. Hilbert (1953). *Methods of Mathematical Physics*, Vol. I, John Wiley & Sons: New York (reprinted in 1989).

Hess, J.L. and A.M.O. Smith (1966). "Calculation of potential flow about arbitrary bodies," *Progress in Aeronautical Sciences* **8**, 1–138.

Kaftori, D., G. Hetsroni, and S. Banerjee (1995). "Particle behavior in the turbulent boundary layer. I. Motion, deposition, and entrainment," *Physics of Fluids*, **7**(5), 1095–1106.

Lewis, R.I. (1991). *Vortex Element Methods for Fluid Dynamic Analysis of Engineering Systems*, Cambridge University Press: Cambridge.

Martensen, E. (1959). "Berechnung der Druckverteilung an Gitterprofilen in ebener Potentialstromung mit einer Fredholmschen Integralgleichung," *Archive for Rational Mechanics and Analysis* **3**, 235–270.

Maskew, B. (1982). "Prediction of subsonic aerodynamic characteristics: a case for low-order panel methods," *Journal of Aircraft* **19**(2), 157–163.

Milne-Thompson, L.M. (1968). *Theoretical Hydrodynamics*, 5th ed., Macmillan & Company: London.

Morino, L. (1993). "Boundary integral equations in aerodynamics," *Applied Mechanics Reviews* **46**(8), 445–466.

Newman, J.N. (1986). "Distributions of sources and normal dipoles over a quadrilateral panel," *Journal of Engineering Mathematics* **20**, 13–126.

Porter, D., and D.S.G. Stirling (1990). *Integral Equations*, Cambridge University Press: Cambridge.

Rokhlin, V. (1983). "Rapid solution of integral equations of classical potential theory," *Journal of Computational Physics* **60**, 187–207.

Weiss, P. (1945). "On hydrodynamical images: arbitrary irrotational flow disturbed by a sphere," *Proceedings of the Cambridge Philosophical Society* **40**, 259–261.

PROBLEMS

1. Determine the natural oscillation frequency of a spherical gas bubble immersed in an otherwise stationary liquid. Assume that the change in bubble radius is small compared to the equilibrium radius, so that the equations governing the bubble oscillations can be linearized.

2. Use the Butler sphere theorem to solve for the flow field generated by a sphere of radius a immersed in an axisymmetric straining field with straining rate c.

3. Consider the problem of uniform flow $U = 1$ in the z-direction past a thin, axisymmetric body with polar radius $R_0(z) = \varepsilon(1 - z^2)$, which lies along the z-axis in the interval $z = (-1, 1)$. The maximum body half-width is given by $\varepsilon = 0.03$. Ignoring the difficulties at the end points $z = -1$ and $z = 1$, use slender body theory to plot the slip velocity on the body surface as a function of z. Your answer should be valid through $O(\varepsilon)$.

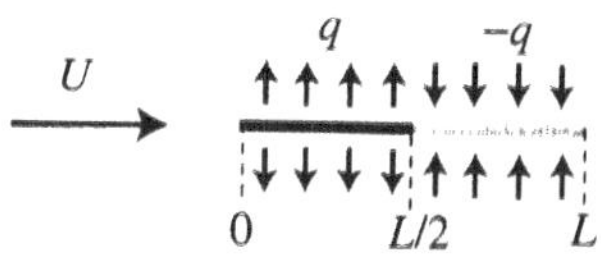

Figure 12.17 Uniform flow past a line source of length L, with strength q in the interval $(0, L/2)$ and strength $-q$ in the interval $(L/2, L)$.

4. Consider the problem of uniform flow past a line source of length L with constant strength q over the interval $(0, L/2)$ and constant strength $-q$ over the interval $(L/2, L)$, as shown in Figure 12.17.
   (a) Use slender-body theory to estimate the body shape.
   (b) Compute the exact shape of the body surface by determining the location of the stagnation point upstream of the body and setting the stream function equal to a constant on the body surface.
   (c) Make a plot comparing the exact body shape with that obtained by slender body theory for a case with $q/LU = 0.01$. Are the assumptions made in development of slender-body theory satisfied for this problem?
5. Reconsider the problem described in Example 12.9.3, in which a small particle of volume V is located at a distance R from a line vortex with strength Γ. The particle is a sphere with radius a and density ρ_p, and the surrounding fluid has density ρ_f. Assume that in a real fluid the particle is sufficiently small that its viscous drag force is given by the linear Stokes formula $\mathbf{F}_D = 6\pi a\mu(\mathbf{u} - \mathbf{v})$, where μ is the viscosity, $\mathbf{v}$ is the particle velocity and $\mathbf{u}$ is the fluid velocity measured at the particle location. All other forces (added mass, buoyancy, and centrifugal force) are the same as in inviscid flow.
   (a) Write the equation of motion for the particle and perform a dimensional analysis to determine the important dimensionless parameters affecting the particle motion.
   (b) Write a computer program to solve for the particle motion over time for cases with $\rho_f < \rho_p$ and $\rho_f > \rho_p$. Plot R versus time and the particle pathline for each case.
6. A neutrally buoyant, spherical, acoustic transmitting device is suspended in a fixed position by a thin cable in a uniform current with constant speed U (Figure 12.18). The spherical device transmits sound to the fluid by changing its radius a periodically in time as $a(t) = a_0 + A\sin(\sigma t)$, where a_0, A, and σ are constants and $a_0 > A$.
   (a) State the boundary condition on the sphere surface $r = a(t)$ at time t. Using superposition of axisymmetric potential flow solutions (e.g., uniform flow, point source, point doublet), solve for the flow field about the pulsating sphere (ignoring the presence of the cable).
   (b) Determine the drag D on the spherical device as a function of time, assuming the fluid to be inviscid and nearly incompressible.

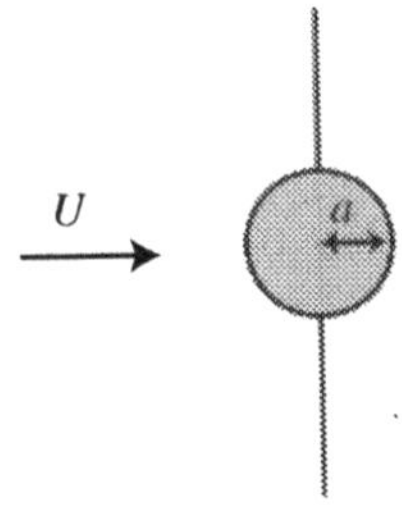

Figure 12.18 A spherical transmitting device with radius $a(t)$ suspended on a cable and immersed in a uniform stream.

7. Consider the problem of a doublet of strength $\boldsymbol{\mu}$ located at a distance b from a sphere of radius a, where the doublet strength vector points toward the sphere center.
   (a) Use the Butler sphere theorem to solve for the velocity field in this flow.
   (b) Use Lagally's theorem to determine the force on the sphere.

COMPUTATIONAL PROJECTS

1. Write a computer code that implements either the source sheet or vortex sheet boundary integral method to compute uniform flow with speed U past a pair of circular cylinders of radius a and separated by a distance b (Figure 12.19).

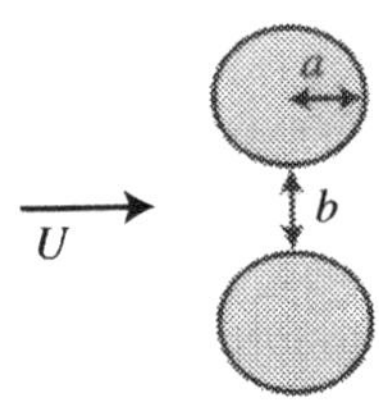

Figure 12.19 Uniform flow past a pair of circular cylinders of radius a with separation distance b.

   (a) Test the computer code by computation of flow past a single cylinder and comparison to the exact solution obtained in Chapter 9.
   (b) Apply the code to compute the force on one cylinder of the pair for cases with $a/b = 1, 3, 10$. For the case with $a/b = 10$, plot the cylinder force versus the number of source or vortex panels in order to determine the body resolution necessary for the flow solution to be nearly independent of the number of panels.
2. Consider the problem of a two-dimensional point vortex initially located at a distance b upstream of the leading edge of a Joukowski airfoil of thickness t, as

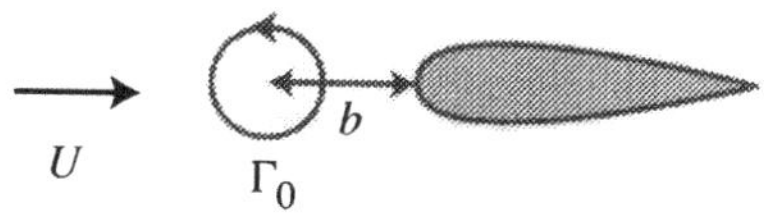

Figure 12.20 A two-dimensional point vortex located initially at a distance b upstream of a Joukowski airfoil immersed in a uniform stream.

shown in Figure 12.20. The vortex circulation is Γ_0, and a uniform flow is present with speed U in the x-direction. The airfoil chord length ℓ is chosen such that $\Gamma_0/U\ell = 0.1$, $t/\ell = 0.12$, and $b/\ell = 1$.

(a) Write a code that implements the vortex sheet boundary-integral method, with the Kutta condition applied at the trailing edge of the plate. Test your code by computation of steady flow about the airfoil at different angles of attack. Compare the lift force on the airfoil with the analytic solution (10.5.8) for angle of attack of 10°. Plot the lift as a function of number of vorticity panels to determine the number of panels necessary for resolution of the vortex sheet.

(b) Introduce the point vortex and compute the force on the airfoil with zero angle of attack, assuming the vortex to be fixed in space. Plot the flow streamlines. Compare your result to an analytic solution obtained using the Joukowski transformation to map the airfoil into the circle indicated in Example 10.5.2.

(c) Use your code to compute the motion of the vortex around the airfoil, as determined by the uniform flow and the image vorticity from the foil. Plot the vortex path.

CHAPTER 13

AXISYMMETRIC VORTEX FLOWS

The assumption of axisymmetry provides a useful simplification for obtaining solutions that illustrate a number of important vortex dynamics phenomena. Axisymmetric vortex flows with vanishing swirl about the symmetry axis include a family of ringlike vortices that induce upon themselves a steady propagation velocity. Axisymmetric flows with nonzero swirl are used to study propagation of waves with variable core area on columnar vortices. Several of the computational methods for two-dimensional vortex dynamics possess axisymmetric counterparts, which are useful for study of nonlinear wave propagation and instabilities on vortex cores.

13.1 GENERAL THEORY

13.1.1 Transport Equations

Axisymmetric vortex flows are most conveniently described in terms of cylindrical polar coordinates (R, α, z). The velocity vector is written in component form as $\mathbf{u} = u\mathbf{e}_R + v\mathbf{e}_\alpha + w\mathbf{e}_z$, where the assumption of axisymmetry implies that the velocity components are functions only of R and z. The current chapter uses only cylindrical polar coordinates, so the subscripts attached to the velocity components in Chapter 12 are dropped. The continuity equation for axisymmetric flow is

$$\frac{1}{R}\frac{\partial}{\partial R}(Ru) + \frac{\partial w}{\partial z} = 0. \tag{13.1.1}$$

As described in Section 12.1, the Stokes stream function is defined such that the continuity equation is identically satisfied, which gives

$$u = -\frac{1}{R}\frac{\partial \psi}{\partial z}, \qquad w = \frac{1}{R}\frac{\partial \psi}{\partial R}. \tag{13.1.2}$$

The Stokes stream function is related only to the radial and axial velocity components and is independent of the azimuthal velocity component.

The vorticity vector in axisymmetric flow is given by

$$\boldsymbol{\omega} = -\frac{\partial v}{\partial z}\mathbf{e}_R + \left(\frac{\partial u}{\partial z} - \frac{\partial w}{\partial R}\right)\mathbf{e}_\alpha + \frac{1}{R}\frac{\partial}{\partial R}(Rv)\mathbf{e}_z. \tag{13.1.3}$$

The vorticity transport equation for a three-dimensional, inviscid flow is

$$\frac{D\boldsymbol{\omega}}{Dt} = \frac{\partial \boldsymbol{\omega}}{\partial t} + (\mathbf{u}\cdot\nabla)\boldsymbol{\omega} = (\boldsymbol{\omega}\cdot\nabla)\mathbf{u}. \tag{13.1.4}$$

In expressing (13.1.4) in component form, it is important to note that the unit vectors $\mathbf{e}_R$ and $\mathbf{e}_\alpha$ have nonzero gradient with respect to α such that

$$\frac{\partial \mathbf{e}_R}{\partial \alpha} = \mathbf{e}_\alpha, \qquad \frac{\partial \mathbf{e}_\alpha}{\partial \alpha} = -\mathbf{e}_R. \tag{13.1.5}$$

Using (13.1.5), as well as (13.1.3) for the α-component of vorticity, the component form of the vorticity transport equation for axisymmetric flow is obtained as

$$\frac{D\omega_R}{Dt} = \omega_R\frac{\partial u}{\partial R} + \omega_z\frac{\partial u}{\partial z}, \tag{13.1.6a}$$

$$\frac{D\omega_\alpha}{Dt} = \frac{u\omega_\alpha - 2v\omega_R}{R}, \tag{13.1.6b}$$

$$\frac{D\omega_z}{Dt} = \omega_R\frac{\partial w}{\partial R} + \omega_z\frac{\partial w}{\partial z}. \tag{13.1.6c}$$

The transport equation for ω_α can be written alternatively as

$$\frac{D}{Dt}\left(\frac{\omega_\alpha}{R}\right) = -\frac{2v\omega_R}{R^2}. \tag{13.1.7}$$

For the special case of axisymmetric flow with no swirl ($v = 0$), the vorticity is oriented entirely in the azimuthal direction, or $\boldsymbol{\omega} = \omega(R, z)\mathbf{e}_\alpha$. For this case, the vorticity transport equations in the R- and z-directions vanish identically and (13.1.7) becomes

$$\frac{D}{Dt}\left(\frac{\omega}{R}\right) = 0, \tag{13.1.8}$$

which implies that the ratio ω/R is conserved on any material point. This result, of course, follows directly from the relationship (7.2.7), which states that the ratio of the vorticity magnitude at the current time to that at an initial time is proportional to the stretch of the vortex line in any inviscid flow.

The physical mechanisms responsible for generation of azimuthal vorticity are more clearly seen by rewriting the azimuthal vorticity equation (13.1.6b) using the vorticity definition (13.1.3) as

$$\frac{D\omega_\alpha}{Dt} = \frac{u\omega_\alpha}{R} + R\left[\omega_R \frac{\partial}{\partial R}\left(\frac{v}{R}\right) + \omega_z \frac{\partial}{\partial z}\left(\frac{v}{R}\right)\right]. \qquad (13.1.9)$$

The first term on the right-hand side of (13.1.9) represents stretching of existing azimuthal vorticity by the radial velocity and the second term [which is equal to the right-hand side of (13.1.7)] represents generation of new azimuthal vorticity by tilting of the radial and axial vorticity components in the azimuthal direction due to radial and axial gradients, respectively, of the angular rotation rate v/R about the symmetry axis. Equation (13.1.7) indicates that positive azimuthal vorticity is generated on material points with negative radial vorticity and vice versa (assuming swirl velocity v to be positive).

The R- and z-components of the vorticity transport equation (13.1.6) reduce to the flow invariant

$$\frac{D(Rv)}{Dt} = 0, \qquad (13.1.10)$$

which states that the circulation is constant about a circle enclosing the symmetry axis when the circle moves with the surrounding fluid particles. This result, of course, follows immediately from Kelvin's circulation theorem (Theorem 7.3.1). This invariant can also be derived from the azimuthal component of the momentum equation, which can be written in the various forms

$$\frac{1}{R}\frac{D(Rv)}{Dt} = \frac{Dv}{Dt} + \frac{uv}{R} = \frac{\partial v}{\partial t} + u\omega_z - w\omega_R = 0. \qquad (13.1.11)$$

The R- and z-components of the vorticity transport equation are obtained by taking $-\partial/\partial z$ of (13.1.11) and $(1/R)(\partial/\partial R)$ of R times (13.1.11), respectively.

13.1.2 Green's Function Solution for Stream Function

The velocity field for any incompressible flow can be written in terms of the vector potential $\boldsymbol{\beta}$ as $\mathbf{u} = \nabla \times \boldsymbol{\beta}$, where for axisymmetric flows the azimuthal component of $\boldsymbol{\beta}$ is related to the Stokes stream function by $\beta_\alpha = \psi/R$. Taking the azimuthal component of the Green's function solution (6.5.4) for $\boldsymbol{\beta}$ in a three-dimensional flow, a solution for the Stokes stream function is obtained as

$$\psi(R, z, t) = R\boldsymbol{\beta}\cdot\mathbf{e}_\alpha = \frac{R}{4\pi}\int_V \frac{\omega(R', z', t)\mathbf{e}_\alpha \cdot \mathbf{e}'_\alpha}{s}\, dv' \qquad (13.1.12)$$

with $s^2 \equiv |\mathbf{x} - \mathbf{x}'|^2 = (z - z')^2 + R^2 + R'^2 - 2RR'\cos\phi$ and $\phi \equiv \alpha - \alpha'$. Noting that $\mathbf{e}_\alpha \cdot \mathbf{e}'_\alpha = \cos\phi$, the stream function can be rewritten in terms of integration over cylindrical polar coordinates as

$$\psi(R, z, t) = \int_{-\infty}^{\infty} \int_{0}^{\infty} \omega(R', z', t) G(R, R', z - z') \, dR' \, dz', \tag{13.1.13}$$

where $G(R, R', z - z')$ plays the role of a Green's function for axisymmetric flows and is defined by

$$G(R, R', z - z') \equiv \frac{RR'}{4\pi} \int_{0}^{2\pi} \frac{\cos\phi}{[(z - z')^2 + R^2 + R'^2 - 2RR'\cos\phi]^{1/2}} \, d\phi. \tag{13.1.14}$$

The integration over ϕ can be performed in terms of the complete elliptic integrals $K(k)$ and $E(k)$ of the first and second kinds, respectively, which are defined by (Abramowitz and Stegun, 1964)

$$K(k) \equiv \int_{0}^{\pi/2} \frac{d\phi}{(1 - k^2 \sin^2\phi)^{1/2}},$$
$$E(k) \equiv \int_{0}^{\pi/2} (1 - k^2 \sin^2\phi)^{1/2} \, d\phi. \tag{13.1.15}$$

Defining the dimensionless variable k by

$$k^2 \equiv \frac{4R'R}{(z - z')^2 + (R + R')^2}, \tag{13.1.16}$$

the Green's function can be written using standard transformations (Helmholtz, 1858) as

$$G(R, R', z - z') = \frac{(R'R)^{1/2}}{2\pi} \left[\left(\frac{2}{k} - k \right) K(k) - \frac{2}{k} E(k) \right]. \tag{13.1.17}$$

13.1.3 Flow Invariants

It is of interest to examine the form taken for axisymmetric swirling flows of various quantities, such as linear and angular impulse, kinetic energy, and helicity, which are known to be invariant for unbounded flows with vanishing velocity at infinity (Section 7.4). In deriving these expressions, it is useful to recall that the position of any point P in the R-z plane can be expressed in cylindrical polar coordinates as $\mathbf{x} = R\mathbf{e}_R + z\mathbf{e}_z$, where the base vectors vary with angle α as

$$\mathbf{e}_R = \cos\alpha \, \mathbf{e}_x + \sin\alpha \, \mathbf{e}_y, \qquad \mathbf{e}_\alpha = -\sin\alpha \, \mathbf{e}_x + \cos\alpha \, \mathbf{e}_y. \tag{13.1.18}$$

The integral of $\mathbf{e}_R$ or $\mathbf{e}_\alpha$ over the interval $(0, 2\pi)$ in α vanishes. The linear impulse $\mathbf{P}$, given by (7.4.1), is one-half of $\mathbf{x} \times \boldsymbol{\omega}$ integrated over the flow volume. The integrals over the $\mathbf{e}_R$ and $\mathbf{e}_\alpha$ components of this cross product vanish, such that $P_R = P_\alpha = 0$

and

$$P_z = \pi \int_{-\infty}^{\infty} \int_0^{\infty} \omega_\alpha R^2 \, dR \, dz. \tag{13.1.19}$$

For the angular impulse **L**, the $\mathbf{e}_R$- and $\mathbf{e}_\alpha$-components of the expression in (7.4.1) again vanish after integration over α (such that $L_R = L_\alpha = 0$) and the axial component becomes

$$L_z = -\pi \int_{-\infty}^{\infty} \int_0^{\infty} \omega_z R^3 \, dR \, dz. \tag{13.1.20}$$

Using the expression (13.1.3) for the axial vorticity and integrating by parts, the angular impulse can alternatively be written as

$$L_z = 2\pi \int_{-\infty}^{\infty} \int_0^{\infty} v R^2 \, dR \, dz. \tag{13.1.21}$$

The kinetic energy is expressed in (7.4.20) as one-half of the product, $\boldsymbol{\beta} \cdot \boldsymbol{\omega}$, of vector potential and vorticity integrated over the fluid volume. The vector potential in an axisymmetric flow with nonzero swirl can be written in terms of the Stokes stream function as $\boldsymbol{\beta} = \beta_R \mathbf{e}_R + (\psi/R)\mathbf{e}_\alpha + \beta_z \mathbf{e}_z$, so that

$$\boldsymbol{\beta} \cdot \boldsymbol{\omega} = \frac{1}{R} \omega_\alpha \psi + \beta_R \omega_R + \beta_z \omega_z. \tag{13.1.22}$$

Using the definition (13.1.3) of vorticity in an axisymmetric flow and the fact that $v = \partial \beta_R / \partial z - \partial \beta_z / \partial R$, we can write

$$\begin{aligned} R(\beta_R \omega_R + \beta_z \omega_z) &= -R\beta_R \frac{\partial v}{\partial z} + \beta_z \frac{\partial}{\partial R}(Rv) \\ &= -\frac{\partial}{\partial z}(R\beta_R v) + \frac{\partial}{\partial R}(R\beta_z v) + Rv^2. \end{aligned} \tag{13.1.23}$$

The first two terms on the right-hand side of (13.1.23) vanish upon integration over z and R, respectively. The resulting expression for kinetic energy becomes

$$T = \pi \int_{-\infty}^{\infty} \int_0^{\infty} (\psi \omega_\alpha + R v^2) \, dR \, dz. \tag{13.1.24}$$

The helicity is defined in (7.4.23) as the integral of the product $\mathbf{u} \cdot \boldsymbol{\omega}$ over the fluid volume. Using the definition (13.1.3) of the vorticity components, we can write

$$\begin{aligned} R\mathbf{u} \cdot \boldsymbol{\omega} &= R\left[-u \frac{\partial v}{\partial z} + v\omega_\alpha + \frac{w}{R} \frac{\partial}{\partial R}(Rv)\right] \\ &= -\frac{\partial}{\partial z}(Ruv) + \frac{\partial}{\partial R}(Rwv) + 2v\omega_\alpha. \end{aligned} \tag{13.1.25}$$

The first two terms on the right-hand side of (13.1.25) vanish upon integrating over z and R, respectively, so that the resulting expression for helicity becomes

$$J = 4\pi \int_{-\infty}^{\infty} \int_{0}^{\infty} v\omega_\alpha R\, dR\, dz. \tag{13.1.26}$$

The vorticity centroid (R_C, z_C) is defined by

$$R_C^2 \equiv \frac{\int_{-\infty}^{\infty} \int_0^{\infty} \omega_\alpha R^2\, dR\, dz}{\int_{-\infty}^{\infty} \int_0^{\infty} \omega_\alpha\, dR\, dz}, \qquad z_C \equiv \frac{\int_{-\infty}^{\infty} \int_0^{\infty} \omega_\alpha R^2 z\, dR\, dz}{\int_{-\infty}^{\infty} \int_0^{\infty} \omega_\alpha R^2\, dR\, dz}. \tag{13.1.27}$$

The expression for R_C^2 can be rewritten in terms of the linear impulse P_z and the total circulation $\Gamma_\infty \equiv \int_{-\infty}^{\infty} \int_0^{\infty} \omega_\alpha\, dR\, dz$ as

$$R_C^2 = \frac{P_z}{\pi \Gamma_\infty}. \tag{13.1.28}$$

Since P_z and Γ_∞ are both invariants of motion for a flow with no solid boundaries, the centroid radius R_C must also be invariant. The axial position z_C of the vorticity centroid is not invariant and typically advects in time in response to the self-induced velocity of the vorticity field.

13.2 THIN-CORE VORTEX RINGS

We consider a vortex ring with ring radius b and circular core of radius a (Figure 13.1). The centerline of the core lies on a circle C enclosing the symmetry axis. At the instant $t = 0$, C is located at $(R, z) = (b, 0)$. The vorticity is everywhere oriented in the azimuthal direction, so that at $t = 0$ the vorticity is given by $\boldsymbol{\omega} = \omega(R, z)\mathbf{e}_\alpha$.

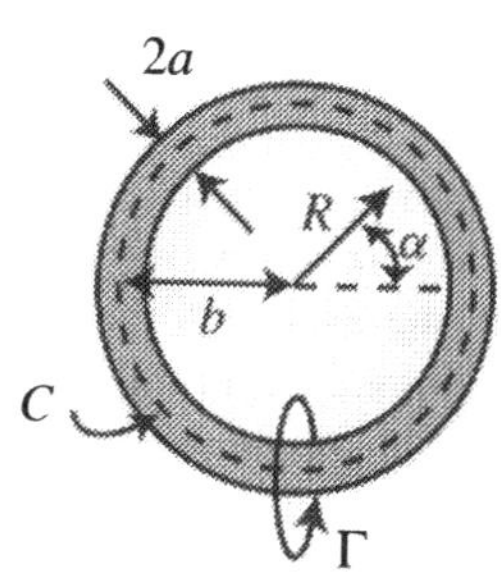

Figure 13.1 Thin-core vortex ring with ring radius b and core radius a.

13.2.1 External Flow Field

In a thin-core vortex ring, it is assumed that the core radius is much smaller than the ring radius, or $\varepsilon \equiv a/b \ll 1$. To leading order in ε, the vorticity field can be approximated at $t = 0$ as being contained within a single vortex filament lying on the core centerline C, such that $\omega(R, z) = \Gamma\delta(R - b, z)$. In this case, the expression (13.1.13) for stream function can be integrated to yield

$$\psi(R, z) = \Gamma G(R, b, z). \tag{13.2.1}$$

Using the expression (13.1.17) for the Green's function, the solution for stream function for the circular vortex filament is obtained as

$$\psi(R, z) = \frac{(bR)^{1/2}\Gamma}{2\pi}\left[\left(\frac{2}{k} - k\right)K(k) - \frac{2}{k}E(k)\right], \tag{13.2.2}$$

where the dimensionless variable k is given by

$$k^2 \equiv \frac{4bR}{z^2 + (R + b)^2}. \tag{13.2.3}$$

The velocity field induced by the vortex ring is obtained by differentiating the stream function, as indicated in (13.1.2). The fluid impulse is similarly obtained by the thin-core approximation as

$$P_z = \pi \int_{-\infty}^{\infty}\int_0^{\infty} R^2\omega\, dR\, dz \cong \pi b^2\Gamma. \tag{13.2.4}$$

The function $G(R, b, z)$ can alternatively be written in terms of Bessel functions as (Lamb, 1932)

$$G(R, b, z) = 2\pi \int_0^{\infty} e^{-\lambda z} J_1(\lambda b) J_1(\lambda R)\, d\lambda. \tag{13.2.5}$$

This form is used, for instance, by Strickland and Amos (1992) in deriving a multipole acceleration method for axisymmetric discrete-vortex methods.

13.2.2 Self-Induced Propagation Velocity

One characteristic feature of vortex rings is that they induce a propagation velocity upon themselves such that an isolated vortex ring translates steadily with time along its symmetry axis. However, if one evaluates the velocity given by the expression (13.2.2) for stream function at a point (R, z) lying on the curve C, the resulting value of the vortex self-induced velocity is infinite. Clearly, something must be wrong! The error lies in the fact that in making the assumption that all vorticity lies on a single filament C, it is implicitly assumed that the distance between the point (R, z) where the velocity is evaluated and any point on C is much greater than the vortex core radius a. As the point (R, z) is moved close to the core centerline C, this assumption is violated, leading to the spurious result of infinite ring propagation velocity.

For accurate computation of the vortex ring propagation speed, it is necessary to account for the finite size of the vortex core and the vorticity variation over the core cross section. The solution for propagation speed of a thin-core ring with uniform-vorticity profile was given without proof by Kelvin as a postscript to Tait's (1867) translation of Helmholtz's (1858) paper on vortex rings. Derivation of this result was later provided by Hicks (1885) and subsequently by several other researchers. Fraenkel (1970) presents a systematic asymptotic method in the small parameter ε for computation of vortex ring speed with arbitrary vorticity profile.

Lamb (1932) presents an alternative method for finding the ring propagation speed, which we shall follow, using the relationship between the fluid axial impulse and the kinetic energy. We recall that the definitions of invariant quantities in Section 13.1 require that the velocity vanishes at infinity. By contrast, the solution for the vortex ring is most conveniently obtained in a frame moving with the ring (the so-called ring-fixed frame), for which the flow field is steady. We let $\mathbf{u}$ denote velocity in the frame with zero velocity at infinity and $\hat{\mathbf{u}}$ denote velocity in the ring-fixed frame such that

$$\mathbf{u} = \mathbf{U} + \hat{\mathbf{u}}, \tag{13.2.6}$$

where $\mathbf{U} = U\mathbf{e}_z$ is the vortex ring self-induced propagation velocity. Substituting (13.2.6) into the expression (7.4.17) for kinetic energy in a three-dimensional flow and using the definition (7.4.1) for impulse yield

$$T = 2UP_z + \int_V \hat{\mathbf{u}} \cdot (\mathbf{x} \times \boldsymbol{\omega})\, dv. \tag{13.2.7}$$

The impulse for a vortex ring is given by (13.2.4), so it remains only to develop suitable approximations for T and the volume integral $\int_V \hat{\mathbf{u}} \cdot (\mathbf{x} \times \boldsymbol{\omega})\, dv$ and to substitute into (13.2.7) to solve for U.

An expression for kinetic energy in axisymmetric flows with no swirl is obtained from (13.1.24) as

$$T = \pi \int_{-\infty}^{\infty} \int_0^{\infty} \psi\omega\, dR\, dz. \tag{13.2.8}$$

The volume integral in (13.2.7) can be written for axisymmetric flows as

$$\int_V \hat{\mathbf{u}} \cdot (\mathbf{x} \times \boldsymbol{\omega})\, dv = 2\pi \int_{-\infty}^{\infty} \int_0^{\infty} R\omega(-z\hat{u} + R\hat{w})\, dR\, dz. \tag{13.2.9}$$

The vorticity, stream function, and velocity components for a thin-core vortex ring scale as

$$\omega = O\left(\frac{\Gamma}{a^2}\right), \qquad \hat{u} = O\left(\frac{\Gamma}{a}\right),$$
$$\hat{w} = O\left(\frac{\Gamma}{a}\right), \qquad \psi = O(b\Gamma). \tag{13.2.10}$$

From the formula given by Kelvin (1867), the vortex propagation velocity is known to scale as $U = O(\Gamma/b)$, which is smaller than the leading-order velocity swirling around the vortex core by a factor $\varepsilon = a/b$. Using these scalings, the first two terms in (13.2.7) are found to scale as $O(b\Gamma^2)$, whereas both parts of the last term [given in (13.2.9)] scale as $O(b\Gamma^2/\varepsilon)$. In order to develop a consistent asymptotic expression for U, it is important that all three terms in (13.2.7) are of the same order of magnitude.

This problem can be resolved using the following formula (Lamb, 1932):

$$R\omega(R\hat{w} - z\hat{u}) = -3zR\hat{u}\omega + \frac{\partial}{\partial z}(zR^2\hat{w}\omega) + \frac{\partial}{\partial R}(zR^2\hat{u}\omega). \tag{13.2.11}$$

Substituting (13.2.11) into (13.2.9), the integrals of the derivatives over R and z vanish, giving

$$\int_V \hat{\mathbf{u}} \cdot (\mathbf{x} \times \boldsymbol{\omega})\, dv = -6\pi \int_{-\infty}^{\infty} \int_0^{\infty} zR\hat{u}\omega\, dR\, dz, \tag{13.2.12}$$

where the right-hand side of (13.2.12) scales as $O(b\Gamma^2)$. The formula (13.2.11) can be derived by noting that

$$\frac{\partial}{\partial z}(zR^2\hat{w}\omega) = R^2\hat{w}\omega + zR^2\left[\frac{\partial \hat{w}}{\partial z}\omega + \hat{w}\frac{\partial \omega}{\partial z}\right], \tag{13.2.13a}$$

$$\frac{\partial}{\partial R}(zR^2\hat{u}\omega) = zR\hat{u}\omega + zR\left[\frac{\partial (R\hat{u})}{\partial R}\omega + R\hat{u}\frac{\partial \omega}{\partial R}\right]. \tag{13.2.13b}$$

Adding these two expressions gives

$$\frac{\partial}{\partial R}(zR^2\hat{u}\omega) + \frac{\partial}{\partial z}(zR^2\hat{w}\omega) = zR\hat{u}\omega + R^2\hat{w}\omega + zR^2\omega\left[\frac{1}{R}\frac{\partial (R\hat{u})}{\partial R} + \frac{\partial \hat{w}}{\partial z}\right]$$
$$+ zR^2\left[\hat{u}\frac{\partial \omega}{\partial R} + \hat{w}\frac{\partial \omega}{\partial z}\right]. \tag{13.2.14}$$

The continuity condition (13.1.1) requires that the first term in brackets vanishes in (13.2.14). The steady-state form of the vorticity transport equation (13.1.7) implies that the second term in brackets in (13.2.14) can be replaced by $\hat{u}\omega/R$, so that (13.2.14) becomes

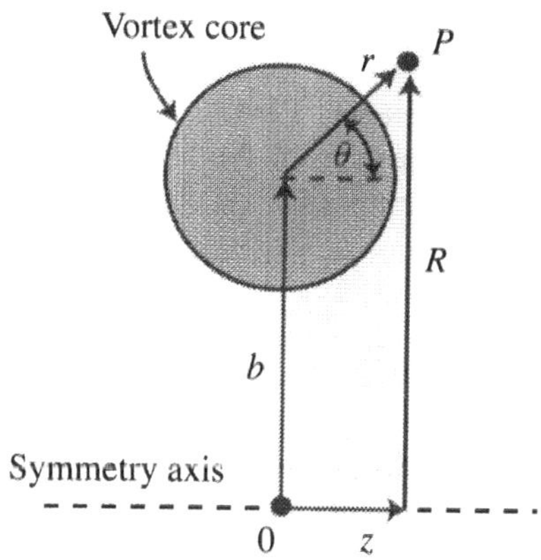

Figure 13.2 Core-centered coordinate system (r, θ) used for determining the propagation speed of a vortex ring.

$$\frac{\partial}{\partial R}(zR^2\hat{u}\omega) + \frac{\partial}{\partial z}(zR^2\hat{w}\omega) = 2zR\hat{u}\omega + R^2\hat{w}\omega. \tag{13.2.15}$$

Subtracting $3zR\hat{u}\omega$ from both sides of (13.2.15) gives formula (13.2.11).

In evaluating the different terms in (13.2.7), it is convenient to employ a core-centered coordinate system (r, θ), shown in Figure 13.2, defined by

$$R = b + r\sin\theta, \qquad z = r\cos\theta. \tag{13.2.16}$$

We assume that the vortex core cross section is circular and let $\Gamma_0(r)$ denote the circulation contained within a distance r of the core center in the R-z plane. The azimuthal vorticity within the core can be approximated by $\omega \cong \omega_0(r) = (1/2\pi r)(d\Gamma_0/dr)$. Making the substitution (13.2.16) on the right-hand side of (13.2.12) and using the additional approximations $da \equiv dRdz \cong rdrd\theta$ and $\hat{u} \cong [\Gamma_0(r)/2\pi r]\cos\theta$ give

$$\begin{aligned}\int_V \hat{\mathbf{u}} \cdot (\mathbf{x} \times \boldsymbol{\omega})\, dv &= -3b\left[\int_0^{2\pi} \cos^2\theta d\theta\right]\left[\int_0^a \Gamma_0(r)\omega_0(r) r\, dr\, d\theta\right] \\ &= -\frac{3b}{2}\int_0^a \Gamma_0 \frac{d\Gamma_0}{dr}\, dr = -\frac{3b}{4}\Gamma^2. \end{aligned} \tag{13.2.17}$$

In order to evaluate the kinetic energy using the expression (13.2.8), it is necessary to first obtain an approximation for the stream function close to the vortex core. For general axisymmetric flows, the stream function is given by the integral (13.1.13) over the Green's function $G(R, R', z - z')$, where the Green's function can be expressed in terms of the complete elliptic integrals $K(k)$ and $E(k)$ as given in (13.1.17). As the point (R, z) approaches the vortex core, the modulus k of the elliptic integrals approaches unity. Near $k = 1$, the elliptic integrals have the limiting values $K(k) \sim \ln(4/k')$ and $E(k) \sim 1$, where $k' \equiv (1-k^2)^{1/2}$ is the complementary modulus (Abramowitz and Stegun, 1964, p. 591). These results can be used in (13.1.17) to obtain an asymptotic expansion for the Green's function $G(R, R', z-z')$ close to the vortex core. When expressed in the core-centered coordinates (r, θ), this asymptotic expansion is given by

$$G(r, r', \theta-\theta') \sim \frac{b}{2\pi}\left[\ln\left(\frac{8b}{[r^2+r'^2-2rr'\cos(\theta-\theta')]^{1/2}}\right)-2\right], \qquad (13.2.18)$$

with error of $O(\varepsilon)$. Substituting this expansion into (13.1.13) gives an approximation for stream function near the vortex core as $\psi \sim \psi_0(r)$, where

$$\psi_0(r) = \frac{b}{2\pi}\int_0^a \omega_0(r')\left(\int_0^{2\pi}\left[\ln(8b)-2\right.\right.$$
$$\left.\left.-\frac{1}{2}\ln\left(r^2+r'^2-2rr'\cos(\theta-\theta')\right)\right]d\theta'\right)r'\,dr'. \qquad (13.2.19)$$

In evaluating the integral over θ', it is noted that

$$\int_0^{2\pi}\ln[r^2+r'^2-2rr'\cos(\theta-\theta')]\,d\theta' = \begin{cases} 4\pi\ln r' & \text{for } r' > r, \\ 4\pi\ln r & \text{for } r > r'. \end{cases} \qquad (13.2.20)$$

The stream function can then be written as

$$\psi_0(r) = b\left[\ln\left(\frac{8b}{r}\right)-2\right]\int_0^r r'\omega_0(r')\,dr'$$
$$+\,b\int_r^a r'\omega_0(r')\left[\ln\left(\frac{8b}{r'}\right)-2\right]dr'. \qquad (13.2.21)$$

Using $\omega_0(r) = (1/2\pi r)(d\Gamma_0/dr)$ and integrating by parts in the second integral give

$$\psi_0(r) = \frac{b\Gamma}{2\pi}\left[\ln\left(\frac{8b}{a}\right)-2\right]+\frac{b}{2\pi}\int_r^a\frac{\Gamma_0(r')}{r'}\,dr', \qquad (13.2.22)$$

where $\Gamma = \Gamma_0(a)$.

The kinetic energy is obtained from (13.2.8) as

$$T = \pi\int_0^a \psi_0(r)\omega_0(r)2\pi r\,dr$$
$$= \frac{b\Gamma}{2}\left[\ln\left(\frac{8b}{a}\right)-2\right]\int_0^a\frac{d\Gamma_0}{dr}\,dr+\frac{b}{2}\int_0^a\frac{d\Gamma_0}{dr}\left[\int_r^a\frac{\Gamma_0(r')}{r'}\,dr'\right]dr. \qquad (13.2.23)$$

Integrating by parts in the last integral gives the kinetic energy for a thin-core vortex ring as

$$T = \frac{b\Gamma^2}{2}\left[\ln\left(\frac{8b}{a}\right) - 2\right] + \frac{b}{2}\int_0^a \frac{\Gamma_0^2(r)}{r}\,dr. \tag{13.2.24}$$

Substituting the expression (13.2.24) for kinetic energy, (13.2.17) for the volumetric integral, and (13.2.4) for impulse into (13.2.7) and solving for U gives the propagation velocity of the vortex ring as

$$U = \frac{\Gamma}{4\pi b}\left[\ln\left(\frac{8b}{a}\right) - \frac{1}{2} + \frac{1}{\Gamma^2}\int_0^a \frac{\Gamma_0^2(r)}{r}\,dr\right]. \tag{13.2.25}$$

The ring propagation velocity thus depends on both the core radius and the vorticity profile within the core. For the special case of a vortex with uniform vorticity distribution, $\Gamma_0(r)$ is proportional to the area contained within a disk of radius r, or $\Gamma_0(r) = r^2\Gamma/a^2$, so the integral in (13.2.25) has the value $\frac{1}{4}$ and the vortex propagation velocity becomes

$$U = \frac{\Gamma}{4\pi b}\left[\ln\left(\frac{8b}{a}\right) - \frac{1}{4}\right], \tag{13.2.26}$$

which is the formula given by Kelvin (1867).

13.3 HILL'S SPHERICAL VORTEX

An exact solution for vorticity transport in axisymmetric flow is given by Hill (1894) for a vortex occupying the interior of a sphere of radius a. It is most convenient to develop this solution for steady-state flow, so we assume that the vortex self-induced propagation velocity is balanced by a uniform flow with speed U oriented in the $-z$-direction, as shown in Figure 13.3. The value of the vortex propagation speed U is obtained as part of the solution.

The azimuthal vorticity field for the spherical vortex is given by

$$\omega = \begin{cases} AR & \text{for } R^2 + z^2 \le a^2, \\ 0 & \text{for } R^2 + z^2 > a^2, \end{cases} \tag{13.3.1}$$

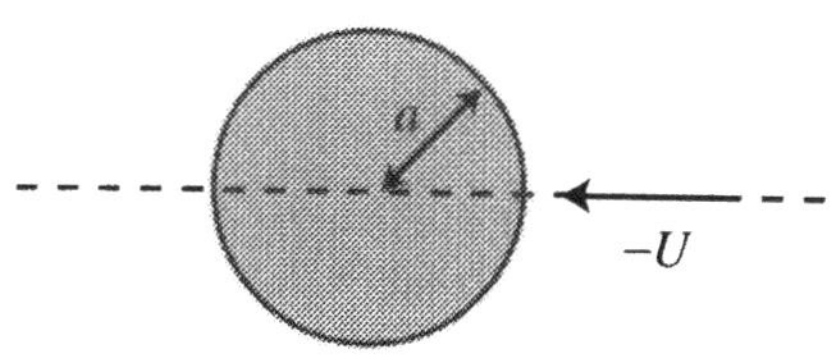

Figure 13.3 Schematic diagram of Hill's spherical vortex immersed in a uniform flow of speed U oriented in the $-z$ direction.

where the constant A is called the vortex strength. This form for vorticity clearly satisfies the vorticity transport equation in the form (13.1.8), since ω/R is constant on material points both within and external to the sphere $R^2 + z^2 = a^2$ provided that there is no exchange of fluid between the sphere interior and exterior.

The governing equation for Stokes stream function is obtained from (13.1.2)–(13.1.3) as

$$\frac{\partial^2 \psi}{\partial R^2} - \frac{1}{R}\frac{\partial \psi}{\partial R} + \frac{\partial^2 \psi}{\partial z^2} = -R\omega. \tag{13.3.2}$$

The stream function is obtained by solving (13.3.2) both inside and outside the spherical boundary of the vortex and then matching the two solutions. Outside the sphere, the stream function must approach $-UR^2/2$ as $R \to \infty$ or $z \to \pm\infty$, corresponding to uniform flow far from the vortex. Inside the sphere, the stream function and its first derivatives are everywhere bounded. Velocity continuity at the vortex boundary requires that ψ and $\partial\psi/\partial z$ be continuous on $R^2 + z^2 = a^2$, from which it follows that $\partial\psi/\partial R$ will be continuous as well. A solution for stream function can be obtained using standard methods (e.g., by transforming into spherical coordinates and using the separation of variables method), and is found to have the form

$$\psi = \begin{cases} \dfrac{A}{10} R^2(a^2 - R^2 - z^2) & \text{in } R^2 + z^2 \le a^2 \\ -\dfrac{1}{2} U R^2 \left[1 - \dfrac{a^3}{(R^2 + z^2)^{3/2}}\right] & \text{in } R^2 + z^2 > a^2. \end{cases} \tag{13.3.3}$$

For both the interior and exterior solutions, the stream function vanishes on the sphere surface, so that the no-penetration condition and the condition of continuity of ψ are satisfied by construction. The exterior solution is simply the stream function solution for uniform flow past a sphere obtained previously in Section 12.3, expressed in cylindrical polar coordinates (with U replaced by $-U$), and hence satisfies the boundary condition at infinity. The remaining matching condition requires that $\partial\psi/\partial z$ be continuous on the sphere surface. Equating the value of this derivative for the interior and exterior solutions in (13.3.3) with $z^2 = a^2 - R^2$ gives

$$\left.\frac{\partial \psi}{\partial z}\right|_{z^2 = a^2 - R^2} = -BR^2(a^2 - R^2)^{1/2} = -\frac{3}{2}\frac{U}{a^2} R^2 (a^2 - R^2)^{1/2}, \tag{13.3.4}$$

which yields a solution for the vortex propagation speed as

$$U = \tfrac{2}{15} a^2 A. \tag{13.3.5}$$

The impulse and kinetic energy of Hill's vortex can be obtained by direct integration from the expressions given in (13.1.19) and (13.1.24), using the solutions for the vorticity and stream function given in (13.3.1) and (13.3.3), as

$$P_z = \tfrac{4}{15}\pi A a^5 = 2\pi U a^3 \tag{13.3.6}$$

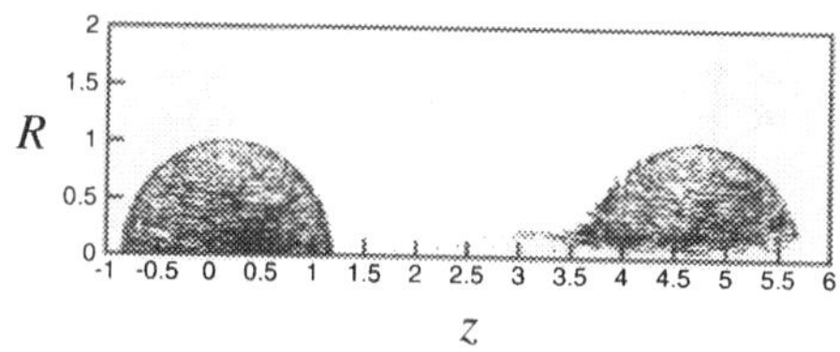

Figure 13.4 Computational result showing the development of a Hill's spherical vortex, with the first image near the initial time and the second image exhibiting a tail of vorticity shed from the rear due to axisymmetric instability of the vortex. This result was obtained using an axisymmetric discrete-vortex method, and the plots show velocity vectors plotted at each Lagrangian computational point.

and

$$T = \frac{4}{525}\pi A^2 a^7 = \frac{1}{7}\pi U^2 a^3. \tag{13.3.7}$$

Moffat and Moore (1978) show that Hill's vortex is unstable to axisymmetric disturbances. As the vortex propagates along the symmetry axis, this instability causes a trail of vorticity to be shed from the rear of the vortex, as shown in the computational result in Figure 13.4, which was obtained using an axisymmetric discrete-vortex method. Shedding of vorticity in the rear of the vortex has the effect of removing vorticity from the region near the symmetry axis such that the vortex evolves into a donut-type shape. Norbury (1973) describes a family of vortex rings with azimuthal vorticity proportional to the distance R from the symmetry axis, ranging from Hill's spherical vortex at one end to thin-core vortex rings at the other end.

13.4 AXISYMMETRIC CONTOUR DYNAMICS

In this section, an extension of the contour dynamics method (Section 11.3) is developed for axisymmetric flows in which vorticity varies as $\boldsymbol{\omega} = cR\mathbf{e}_\alpha$, where c is a constant, within a surface of revolution that intersects a closed curve C in the R-z plane (Figure 13.5a). The region enclosed by C in the R-z plane is denoted by A, where the increment of area on A is $da = dR\,dz$. The derivation presented in this section follows that of Pozrikidis (1986). For reasons that will become apparent presently, two different approaches must be used to obtain the contour integrals for the radial and axial velocity components.

For the radial velocity component, we begin by substituting the Green's function solution (13.1.13) for stream function into the (13.1.2) to obtain

$$u(R, z, t) = -\frac{1}{R}\int_A \omega(R')\frac{\partial}{\partial z}[G(R, R', z - z')]\,da', \tag{13.4.1}$$

where we use the assumption that vorticity depends only on R within the region A. Since G is a function of the difference $z - z'$, we can write $\partial G/\partial z = -\partial G/\partial z'$, so

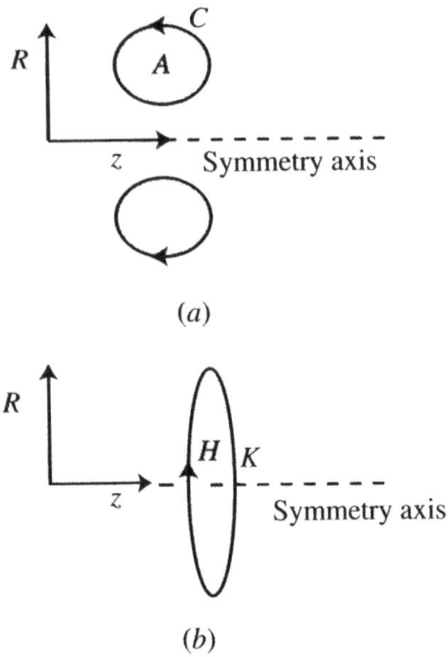

Figure 13.5 Diagram showing (*a*) a contour C in the R-z plane bounding a volume of rotation and (*b*) a circle K enclosing the symmetry axis.

that (13.4.1) becomes

$$u(R, z, t) = \frac{1}{R} \int_A \frac{\partial}{\partial z'} [\omega(R') G(R, R', z - z')] \, da'. \tag{13.4.2}$$

Performing the integration over z' gives a contour integral for u as

$$u(R, z, t) = \frac{1}{R} \int_C \omega(R') G(R, R', z - z') \, dR'. \tag{13.4.3}$$

It is necessary for this derivation that vorticity ω remains independent of z not only at the initial time but also for all later times. This condition is guaranteed if the vorticity varies linearly with R, for which case $\omega / R = \text{const}$ and the form (13.1.8) of the vorticity transport equation is identically satisfied.

A similar approach unfortunately cannot be used to obtain a contour integral for the axial velocity component due to the fact that the Green's function does not depend on the difference $R - R'$, but rather on R and R' independently. Instead, we note that the fluid outside of the vortex boundary is irrotational, and therefore admits a velocity potential ϕ such that $\mathbf{u} = \nabla \phi$. Since w is given by the z-derivative of ϕ, we can use a similar approach as employed in obtaining (13.4.3) in writing w as a contour integral.

It is first necessary to derive an expression for velocity potential. The velocity of an incompressible flow can be written as $\mathbf{u} = \nabla \times \boldsymbol{\beta}$ such that substitution of the

Green's function solution (6.5.4) for $\boldsymbol{\beta}$ gives

$$\mathbf{u}(\mathbf{x}, t) = \nabla \times \left(\frac{1}{4\pi} \int_V \frac{\boldsymbol{\omega}(\mathbf{x}', t)}{s} \, dv' \right), \tag{13.4.4}$$

where $s \equiv |\mathbf{x} - \mathbf{x}'|$. Substituting $\boldsymbol{\omega} = \omega(R, z, t)\mathbf{e}_\alpha$ for nonswirling flow and writing $dv' = R'\,dR'\,d\alpha'\,dz'$ give

$$\mathbf{u}(\mathbf{x}, t) = \frac{1}{4\pi} \int_{-\infty}^{\infty} \int_0^{\infty} \omega(R', z', t) \left(\nabla \times \int_0^{2\pi} \frac{1}{s} R' \mathbf{e}'_\alpha \, d\alpha' \right) dR'\,dz'. \tag{13.4.5}$$

Letting K denote a circle around the symmetry axis passing through point (R', z') as shown in Figure 13.5b, (13.4.5) can be written as

$$\mathbf{u}(\mathbf{x}, t) = \frac{1}{4\pi} \int_{-\infty}^{\infty} \int_0^{\infty} \omega(R', z', t) \left(\nabla \times \int_K \frac{1}{s} d\mathbf{x}' \right) dR'\,dz', \tag{13.4.6}$$

where $d\mathbf{x}' = R'\mathbf{e}'_\alpha d\alpha'$ is an element of length directed tangent to K. The integral identity (2.7.13) is used to rewrite the integral over K in (13.4.6) as an integral over the open surface H with unit normal $\mathbf{n}$ bounded by K as

$$\int_K \frac{1}{s} d\mathbf{x}' = \int_H \mathbf{n}' \times \nabla' \left(\frac{1}{s} \right) da'. \tag{13.4.7}$$

For the current application of this identity, H is a disk of radius R' and $\mathbf{n} = \mathbf{e}_z$. Taking the curl of the integral in (13.4.7) and using the vector identity (2.6.8), where we note that $\mathbf{n}$ is constant, give

$$\nabla \times \int_K \frac{1}{s} d\mathbf{x}' = \int_H \left\{ \mathbf{n}' \left[\nabla \cdot \nabla' \left(\frac{1}{s} \right) \right] - (\mathbf{n}' \cdot \nabla) \nabla' \left(\frac{1}{s} \right) \right\} da'. \tag{13.4.8}$$

Since s is a function of the difference $\mathbf{x} - \mathbf{x}'$ and $-1/4\pi s$ is the Green's function of the Poisson equation in three dimensions (Section 6.5), we can write

$$\nabla \cdot \nabla' \left(\frac{1}{s} \right) = -\nabla^2 \left(\frac{1}{s} \right) = 4\pi \delta(\mathbf{x} - \mathbf{x}'). \tag{13.4.9}$$

After substitution into the solution (13.4.6) for $\mathbf{u}$ and integration over the Dirac delta, the first term in the integral over H in (13.4.8) gives no contribution to the velocity field for points $\mathbf{x}$ at which the vorticity vanishes (i.e., at points outside of the vortex boundary).

The solution (13.4.6) for $\mathbf{u}$ thus becomes

$$\mathbf{u}(\mathbf{x}, t) = -\frac{1}{4\pi} \int_{-\infty}^{\infty} \int_0^{\infty} \omega(R', z', t) \left(\int_H (\mathbf{n}' \cdot \nabla) \nabla' \left(\frac{1}{s} \right) da' \right) dR'\,dz'. \tag{13.4.10}$$

Since $\mathbf{n}$ is constant and s depends only on the difference $\mathbf{x} - \mathbf{x}'$, we can write $(\mathbf{n}' \cdot \nabla)\nabla'(1/s) = \nabla[\mathbf{n}' \cdot \nabla'(1/s)]$, such that (13.4.10) can be written in the form $\mathbf{u} = \nabla\phi$, where

$$\phi(R, z, t) = \int_{-\infty}^{\infty} \int_{0}^{\infty} \omega(R', z', t) F(R, R', z - z')\, dR'\, dz' \tag{13.4.11}$$

and the function F is defined, using $\nabla'(1/s) = \mathbf{s}/s^3$, by

$$F(R, R', z - z') \equiv -\frac{\Omega}{4\pi} \equiv -\frac{1}{4\pi} \int_H \frac{\mathbf{n}' \cdot \mathbf{s}}{s^3}\, da'. \tag{13.4.12}$$

The quantity Ω can be interpreted as the solid angle subtended by the surface H at point $\mathbf{x}$. While the above definition for Ω is multiple valued, it can be made single valued by treating the disk H as a branch cut such that Ω increases by 4π with each passage through H.

The axial velocity component is obtained by differentiating (13.4.11) as

$$w(R, z, t) = \int_{-\infty}^{\infty} \int_{0}^{\infty} \omega(R', z', t) \frac{\partial}{\partial z} F(R, R', z - z')\, dR'\, dz'. \tag{13.4.13}$$

Using $\partial F/\partial z = -\partial F/\partial z'$ and the assumption that vorticity is independent of z within the region A gives

$$w(R, z, t) = -\int_A \frac{\partial}{\partial z'}[\omega(R') F(R, R', z - z')]\, da', \tag{13.4.14}$$

which upon integration over z yields the contour integral

$$w(R, z, t) = -\int_C \omega(R') F(R, R', z - z')\, dR'. \tag{13.4.15}$$

Numerical computations with this method are performed in a manner similar to two-dimensional contour dynamics. A set of points is placed on the vortex contour C, which are advected as material points in the R-z plane. The velocity components at each point are obtained by numerical solution of the contour integrals (13.4.3) and (13.4.15) using an appropriate numerical quadrature method. An expression for the Green's function $G(R, R', z - z')$ in terms of the complete elliptic integrals of the first and second kind is given by (13.1.17). The function $F(R, R', z - z')$ is given by (13.4.12), where the solid angle Ω can be evaluated in terms of complete elliptic integrals of the third kind $\Pi(a \backslash k^2)$ (Abramowitz and Stegun, 1964, p. 589) as

$$\Omega = -2\pi \frac{z - z'}{|z - z'|} - \frac{1}{D} \frac{z - z'}{[(z - z')^2 + (R + R')^2]^{1/2}} \times \left[(E + 1) n_1 \Pi(-n_1 \backslash k^2) - (E - 1) n_2 \Pi(n_2 \backslash k^2)\right], \tag{13.4.16}$$

where k^2 is defined by (13.1.16) and

$$D \equiv \frac{R}{[(z-z')^2 + R^2]^{1/2}}, \qquad E \equiv \frac{R'}{[(z-z')^2 + R^2]^{1/2}},$$

$$n_1 \equiv \frac{2D}{1-D}, \qquad n_2 \equiv \frac{2D}{1+D}.$$

As noted in Section 13.2, the Green's function $G(R, R', z - z')$ has a singularity as $(R', z') \to (R, z)$, with limiting value of the form $G \sim (R/2\pi)\ln(4/k')$, where $k' \equiv (1 - k^2)^{1/2}$ is the complementary modulus. This singularity can be added and subtracted from the contour integral such that the integral evaluated numerically is nonsingular.

13.5 STEADY AXISYMMETRIC FLOWS

The azimuthal vorticity component can be expressed in any axisymmetric flow in terms of the Stokes stream function using (13.1.2)–(13.1.3) as

$$\omega_\alpha = -\frac{1}{R} D^2 \psi, \tag{13.5.1}$$

where the differential operator D^2 is defined by

$$D^2 \equiv \frac{\partial^2}{\partial R^2} - \frac{1}{R}\frac{\partial}{\partial R} + \frac{\partial^2}{\partial z^2}. \tag{13.5.2}$$

The azimuthal component of the vorticity transport equation (13.1.7) can be rewritten using the definition of vorticity in (13.1.3) as

$$\frac{D}{Dt}\left(\frac{\omega_\alpha}{R}\right) = -\frac{2v\omega_R}{R^2} = \frac{1}{R^2}\frac{\partial (v^2)}{\partial z}, \tag{13.5.3}$$

where D/Dt is our usual notation for material derivative.

The objective of the present section is to show how the assumption of steady-state flow can be employed to obtain an integral of (13.5.3), giving a second relationship between azimuthal vorticity and stream function that when substituted into (13.5.1) yields a nonlinear equation for stream function. The starting point of this derivation is the fact that the product Rv is invariant on a material point in any axisymmetric flow, as indicated by (13.1.10). In a *steady* axisymmetric flow, the fluid particles initially lying on a given streamline will advect over time on a surface of revolution formed by rotating the streamline about the symmetry axis. It follows that $D\psi/Dt = 0$ in a steady flow, which implies that

$$Rv = F(\psi). \tag{13.5.4}$$

Taking the gradient of (13.5.4) and using the definition (13.1.2) of the Stokes stream function give

$$\nabla(Rv) = R\frac{dF}{d\psi}(w\mathbf{e}_R - u\mathbf{e}_z). \tag{13.5.5}$$

Using (13.5.4)–(13.5.5), the azimuthal vorticity equation (13.5.3) can be written as

$$\frac{D}{Dt}\left(\frac{\omega_\alpha}{R}\right) = \frac{2F}{R^4}\frac{\partial F}{\partial z} = -\frac{2uF}{R^2}\frac{dF}{d\psi} = \frac{D}{Dt}\left(\frac{F}{R^2}\frac{dF}{d\psi}\right). \tag{13.5.6}$$

In writing the last expression in (13.5.6), we make use of the fact that any function of stream function can pass through the material derivative in a steady flow (since $D\psi/Dt = 0$). Integrating (13.5.6) over time gives

$$\frac{\omega_\alpha}{R} = \frac{F}{R^2}\frac{dF}{d\psi} + C(\psi), \tag{13.5.7}$$

where $C(\psi)$ is an arbitrary function of ψ that arises as a constant of integration along the streamline. Substituting (13.5.7) into (13.5.1) yields an equation for stream function as

$$D^2\psi = -F\frac{dF}{d\psi} - R^2C. \tag{13.5.8}$$

It remains to interpret the coefficient of integration $C(\psi)$. We note that the vector product of velocity and vorticity can be written using the vorticity definition (13.1.3) as

$$\mathbf{u}\times\boldsymbol{\omega} = \left[\frac{v}{R}\frac{\partial}{\partial R}(Rv) - w\omega_\alpha\right]\mathbf{e}_R + (w\omega_R - u\omega_z)\mathbf{e}_\alpha + \left(v\frac{\partial v}{\partial z} + u\omega_\alpha\right)\mathbf{e}_z. \tag{13.5.9}$$

From the azimuthal momentum equation (13.1.11), the azimuthal component of $\mathbf{u}\times\boldsymbol{\omega}$ is found to vanish in any steady axisymmetric flow. Substituting the expression (13.5.4) for v into the remaining terms of (13.5.9) and using (13.5.5) to evaluate the derivatives give

$$\mathbf{u}\times\boldsymbol{\omega} = \left[\frac{F}{R}\frac{dF}{d\psi} - \omega_\alpha\right](w\mathbf{e}_R - u\mathbf{e}_z). \tag{13.5.10}$$

Substituting the solution (13.5.7) for ω_α into (13.5.10) gives

$$\mathbf{u}\times\boldsymbol{\omega} = -RC(\psi)(w\mathbf{e}_R - u\mathbf{e}_z). \tag{13.5.11}$$

For any steady flow with conservative body force, the momentum conservation equation can be expressed as

$$\mathbf{u}\times\boldsymbol{\omega} = \nabla B, \tag{13.5.12}$$

where B is the Bernoulli coefficient defined in (8.1.3). Since B is invariant on a fluid particle in any steady flow (Section 8.2), B must depend only on stream function, so the definition (13.1.2) can be used to write

$$\nabla B = R\frac{dB}{d\psi}(w\mathbf{e}_R - u\mathbf{e}_z). \tag{13.5.13}$$

Equating (13.5.11) and (13.5.13) gives an expression for the coefficient $C(\psi)$ in terms of the Bernoulli coefficient B as

$$C(\psi) = -\frac{dB}{d\psi}. \tag{13.5.14}$$

Substituting the expression (13.5.14) for C into (13.5.8) yields a nonlinear equation for stream function in a steady axisymmetric flow as

$$D^2\psi = -F\frac{dF}{d\psi} + R^2\frac{dB}{d\psi}. \tag{13.5.15}$$

This equation appears to have been first derived in a paper by Bragg and Hawthorne (1950) and is generally known as the *Bragg-Hawthorne equation*. It has been used as the starting point for numerous studies of axisymmetric rotational flows seeking to find either exact or approximate solutions. For instance, the Hill's spherical vortex (Section 13.3) can be obtained as a solution of the Bragg-Hawthorne equation with $F = 0$ everywhere (i.e., no swirl about the symmetry axis) and

$$B = \begin{cases} -A\psi & \text{for } R^2 + z^2 \leq a^2, \\ 0 & \text{for } R^2 + z^2 > a^2. \end{cases} \tag{13.5.16}$$

The Bragg-Hawthorne equation plays a central role in development of the theory for solitary waves of variable core area on a columnar vortex, as described by Benjamin (1967) and Leibovich and Kribus (1990).

13.6 WAVES OF VARIABLE CORE AREA

The vorticity transport equation in the form (13.1.7) indicates that vorticity oriented in the radial direction in a flow with nonzero swirl velocity leads to generation of vorticity in the azimuthal direction. Physically, this occurs by tilting of the radial vorticity in the azimuthal direction by radial variation of the angular rotation rate v/R. Radial vorticity in a columnar vortex is associated with variation in vortex core area, which can arise from a variety of causes. For instance, if a vortex is stretched such that the stretching rate is nonuniform along the vortex axis, the core will grow thinnest at positions of greatest stretching rate. The azimuthal vorticity generated by tilting of radial vorticity acts to inhibit core area variation, such as to attempt to make the core everywhere of uniform thickness. Variation of vortex core area can alternatively be described in terms of the pressure force on the lateral surface of the core. As the core area increases, the lateral pressure also increases, thus providing a radial restoring force on the core. Area-varying waves on vortex cores arise from the joint action of the fluid inertia and the pressure force on the lateral surface.

In this section, we derive the solution for Kelvin waves (Kelvin, 1880), which are small-amplitude perturbations on a columnar vortex with strength Γ, equilibrium core radius a, and ambient swirl velocity profile

$$V(R) = \begin{cases} \dfrac{\Gamma R}{2\pi a^2} & \text{for } R \leq a, \\ \dfrac{\Gamma}{2\pi R} & \text{for } R > a. \end{cases} \tag{13.6.1}$$

The velocity field is assumed to exhibit axisymmetric perturbations such that

$$\begin{aligned} u(R, z, t) &= \hat{u}(R)e^{i(\omega t - kz)}, \\ v(R, z, t) &= V(R) + \hat{v}(R)e^{i(\omega t - kz)}, \\ w(R, z, t) &= \hat{w}(R)e^{i(\omega t - kz)}, \end{aligned} \tag{13.6.2}$$

where ω is the perturbation frequency and k is the axial wavenumber. Substituting (13.6.2) into the continuity equation (13.1.1) yields

$$\frac{1}{R}\frac{\partial}{\partial R}(R\hat{u}) = ik\hat{w}. \tag{13.6.3}$$

Substituting (13.6.2) into (13.1.3) gives the vorticity components as

$$\begin{aligned} \omega_R(R, z, t) &= ik\hat{v}e^{i(\omega t - kz)}, \\ \omega_\alpha(R, z, t) &= -\left(ik\hat{u} + \frac{d\hat{w}}{dR}\right)e^{i(\omega t - kz)}, \\ \omega_z(R, z, t) &= \frac{1}{R}\frac{d(RV)}{dR} + \frac{1}{R}\frac{d(R\hat{v})}{dR}e^{i(\omega t - kz)}. \end{aligned} \tag{13.6.4}$$

Under the assumption that the radial displacement D of the vortex core is small compared to the core radius ($D/a \ll 1$), the terms that are quadratic in the perturbation quantities upon substitution of (13.6.2) and (13.6.4) into the vorticity transport equations (13.1.6) can be neglected to yield a set of linear ordinary differential equations for $(\hat{u}, \hat{v}, \hat{w})$ as

$$i\omega\hat{v} = -\frac{1}{R}\frac{d(RV)}{dR}\hat{u}, \tag{13.6.5a}$$

$$\omega\left(ik\hat{u} + \frac{d\hat{w}}{dR}\right) = \frac{2kV\hat{v}}{R}, \tag{13.6.5b}$$

$$\frac{i\omega}{R}\frac{d(R\hat{v})}{dR} + \frac{d}{dR}\left[\frac{1}{R}\frac{d(RV)}{dR}\right]\hat{u} = -\frac{ik}{R}\frac{d(RV)}{dR}\hat{w}. \tag{13.6.5c}$$

Multiplying (13.6.5a) by R and then taking the derivative $\partial/\partial R$ and using the continuity equation (13.6.3) yield the vorticity equation (13.6.5c) in the z-direction. Thus, between (13.6.3) and (13.6.5a)–(13.6.5c) there are only three independent equations for the velocity perturbations $(\hat{u}, \hat{v}, \hat{w})$. Since the swirl velocity $V(R)$ has a slope discontinuity at $R = a$, we solve for the perturbation velocity amplitudes $(\hat{u}, \hat{v}, \hat{w})$ independently in the vortex core interior $(R \leq a)$ and exterior $(R > a)$, and then match the solutions at $R = a$.

In the interior region, the perturbation equations (13.6.5) reduce to

$$-i\omega\hat{v} = 2C\hat{u}, \tag{13.6.6a}$$

$$\omega\left(ik\hat{u} + \frac{d\hat{w}}{dR}\right) = 2kC\hat{v}, \tag{13.6.6b}$$

$$\omega\left(\frac{d\hat{v}}{dR} + \frac{\hat{v}}{R}\right) = -2kC\hat{w}, \tag{13.6.6c}$$

where $C \equiv \Gamma/2\pi a^2$. Eliminating $\hat{u}$ and $\hat{v}$ from (13.6.6) gives an ordinary differential equation for $\hat{w}$ as

$$\frac{d^2\hat{w}}{dR^2} + \frac{1}{R}\frac{d\hat{w}}{dR} + \left(\frac{q}{a}\right)^2 \hat{w} = 0, \tag{13.6.7}$$

where $q \equiv ka[(\Gamma^2/\pi^2 a^4\omega^2) - 1]^{1/2}$ is a constant dimensionless parameter. Equation (13.6.7) is a Bessel equation that, after imposing the condition that $\hat{w}$ remains bounded as $R \to 0$, yields a solution in terms of the Bessel function of the first kind as

$$\hat{w}(R) = C_1 J_0\left(\frac{qR}{a}\right). \tag{13.6.8}$$

Exterior to the vortex core $(R > a)$, the perturbation equations (13.6.5) reduce to $\hat{v} = 0$ and

$$ik\hat{u} + \frac{d\hat{w}}{dR} = 0. \tag{13.6.9}$$

Combining (13.6.9) with the result (13.6.3) of the continuity equation yields an equation for $\hat{w}$ as

$$\frac{d^2\hat{w}}{dR^2} + \frac{1}{R}\frac{d\hat{w}}{dR} - k^2\hat{w} = 0. \tag{13.6.10}$$

Equation (13.6.10) is a modified Bessel equation, which after imposing the condition that $\hat{w}$ vanishes as $R \to \infty$ yields a solution in terms of the modified Bessel function of the second kind as

$$\hat{w}(R) = C_2 K_0(kR). \tag{13.6.11}$$

The matching condition requires that the three velocity components are continuous at the vortex core lateral boundary. In the linear theory, it is sufficient to perform the matching at $R = a$ rather than at the actual boundary of the vortex core. It suffices to require continuity of $\hat{u}$ and $\hat{w}$, from which continuity of $\hat{v}$ follows. Continuity of $\hat{w}$ is enforced by equating (13.6.8) and (13.6.11) at $R = a$, yielding

$$C_1 J_0(q) = C_2 K_0(ka). \tag{13.6.12}$$

The radial velocity is obtained from the solution for $\hat{w}$ and the perturbation equations (13.6.6) or (13.6.9) such that continuity of $\hat{u}$ at $R = a$ yields

$$\left(\frac{ka}{q}\right) C_1 J_1(q) = -C_2 K_1(ka). \tag{13.6.13}$$

Equations (13.6.12) and (13.6.13) form an eigenvalue problem, for which the solvability condition of vanishing determinant yields

$$\frac{ka}{q}\frac{J_1(q)}{J_0(q)} = -\frac{K_1(ka)}{K_0(ka)}. \tag{13.6.14}$$

The dimensionless frequency $f \equiv \omega\pi a^2/\Gamma$ is obtained from the definition of q as

$$f \equiv \frac{\pi a^2 \omega}{\Gamma} = \frac{ka}{(q^2 + k^2a^2)^{1/2}}. \tag{13.6.15}$$

The dimensionless frequency is equal to the wave frequency ω divided by the vorticity magnitude $\Gamma/\pi a^2$ within the undisturbed vortex, and its value can be obtained for a given ka by solving (13.6.14) for q and then substituting into (13.6.15). A plot of f as a function of ka is given in Figure 13.6.

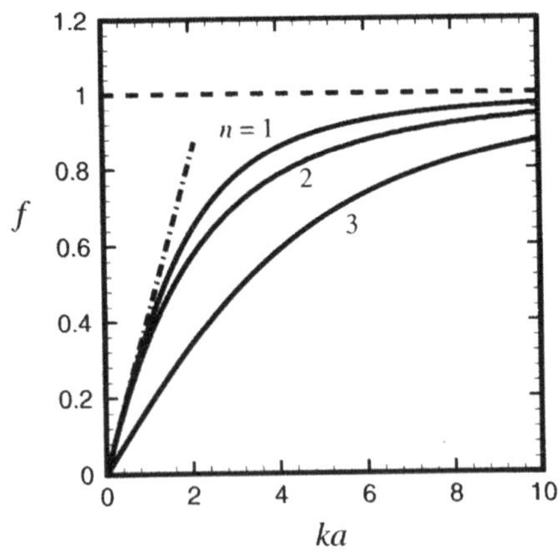

Figure 13.6 Dimensionless frequency as a function of dimensionless wavenumber ka for axisymmetric Kelvin waves for the first three roots of equation (13.6.14) (solid curves) and for the long-wave solution (13.6.16) (dashed line).

The above analysis requires that the wave amplitude and slope are small ($D/a \ll 1$ and $kD \ll 1$), but it places no restriction on the wavelength relative to the vortex core radius. As such, the result (13.6.14) is valid for arbitrary values of ka. Recalling that $K_0(z) \sim -\ln z$ and $K_1(z) \sim 1/z$ as $z \to 0$, we observe that the right-hand side of (13.6.14) becomes unbounded in the long wave limit $ka \to 0$. Since $J_1(z)$ is bounded for finite z, it follows from (13.6.14) that the solution for q for long waves must correspond to a root of $J_0(q)$. With $q \gg ka$ in the denominator of (13.6.15), the dimensionless frequency for long axisymmetric waves reduces to

$$f \cong \frac{ka}{q} \qquad \text{where } J_0(q) = 0. \tag{13.6.16}$$

13.7 PLUG-FLOW MODEL

The Kelvin wave analysis is limited to waves of small amplitude that are periodic along the vortex axis. There exist numerous situations that involve waves of finite amplitude or nonperiodic axial flows in which the Kelvin wave analysis does not apply, but for which we nevertheless desire to model the variation of the vortex axial flow and core radius. Such situations can be handled using the so-called plug-flow model (Lundgren and Ashurst, 1989), in which the axial flow is assumed to be uniform over the vortex core. In its simplest form, this model is analogous to shallow water theory used in engineering hydraulics and, in fact, results in a similar set of equations, although higher-order versions of the model exist (Leonard, 1994; Marshall, 1991). The plug-flow model is valid only for long wave motion ($ka \lll 1$), but it does not require small wave amplitude or periodic boundary conditions.

In the plug-flow model, the velocity within the vortex core is written in terms of the core radius $\sigma(z, t)$ and is assumed to have the form

$$u = \left(\frac{R}{\sigma}\right) u_0(z, t), \qquad v = \frac{\Gamma R}{2\pi\sigma^2}, \qquad w = w(z, t). \tag{13.7.1}$$

The function $u_0(z, t)$ can be obtained by the kinematic boundary condition at the core lateral boundary (i.e., the requirement that the core boundary must be a material surface) as

$$u(\sigma, z, t) = u_0(z, t) = \frac{\partial \sigma}{\partial t} + w\frac{\partial \sigma}{\partial z}. \tag{13.7.2}$$

Substitution of (13.7.1) into the incompressibility condition (13.1.1) gives the vortex continuity equation

$$\frac{2}{\sigma}\left(\frac{\partial \sigma}{\partial t} + w\frac{\partial \sigma}{\partial z}\right) + \frac{\partial w}{\partial z} = 0. \tag{13.7.3}$$

Defining a one-dimensional material derivative of any function $f = f(z, t)$ by

$$\frac{Df}{Dt} \equiv \frac{\partial f}{\partial t} + w\frac{\partial f}{\partial z}, \tag{13.7.4}$$

the continuity equation can be written as

$$2\frac{D\sigma}{Dt} + \sigma\frac{\partial w}{\partial z} = 0. \tag{13.7.5}$$

A second equation relating the core radius σ and the axial velocity w is obtained by integrating the axial momentum equation over the core radius. We recall that the radial and axial momentum equations for an axisymmetric flow are given by

$$\frac{\partial u}{\partial t} + u\frac{\partial u}{\partial R} + w\frac{\partial u}{\partial z} - \frac{v^2}{R} = -\frac{\partial (p/\rho)}{\partial R}, \tag{13.7.6a}$$

$$\frac{\partial w}{\partial t} + u\frac{\partial w}{\partial R} + w\frac{\partial w}{\partial z} = -\frac{\partial (p/\rho)}{\partial z}, \tag{13.7.6b}$$

and the azimuthal momentum equation reduces to the condition that $vR = \text{const}$. Integrating the axial momentum equation (13.7.6b) over the core cross section gives

$$\frac{\sigma^2}{2}\frac{Dw}{Dt} = -\int_0^{\sigma} R\frac{\partial (p/\rho)}{\partial z}\, dR. \tag{13.7.7}$$

An approximation for the pressure field is obtained by solving the radial momentum equation (13.7.6a) for the leading-order terms in the small parameter ka, where a is a typical value of the core radius and $2\pi/k$ is a typical axial length scale (e.g., the wavelength of an area-varying wave). The centrifugal term v^2/R in (13.7.6a) scales as $O(\Gamma^2/a^3)$, whereas the first three terms in (13.7.6a), which we will refer to collectively as the radial momentum terms, scale as $O(a\omega^2)$, where ω denotes a typical frequency of the motion. Equating ω to the frequency of long Kelvin waves of wavenumber k, given by (13.6.16), yields $\omega = O(\Gamma k/a)$. The ratio of the radial momentum terms to the centrifugal term in the radial momentum equation thus scales as $O(ka)^2$, which indicates that the radial momentum terms are negligible in the long-wave limit. Using the expression (13.7.1) for the azimuthal velocity, the radial momentum equation can be integrated inside the vortex core to yield the radial pressure variation as

$$\frac{p(R) - p(\sigma)}{\rho} = -\frac{\Gamma^2}{8\pi^2\sigma^4}(\sigma^2 - R^2), \tag{13.7.8}$$

where $p(\sigma)$ is the pressure at the core lateral surface. Using the expression $v = \Gamma/2\pi R$ for the azimuthal velocity external to the vortex core yields the pressure at the core lateral surface as

$$\frac{p(\sigma) - p_\infty}{\rho} = -\frac{\Gamma^2}{8\pi^2\sigma^2}, \tag{13.7.9}$$

where p_∞ is the pressure as $R \to \infty$. Substituting (13.7.8) and (13.7.9) into (13.7.7) and performing the integration over radius, the integrated axial momentum equation becomes

$$\frac{\sigma^2}{2}\frac{Dw}{Dt} = -\frac{\Gamma^2}{8\pi^2\sigma}\frac{\partial\sigma}{\partial z} - \frac{1}{\rho}\frac{\partial p_\infty}{\partial z}. \tag{13.7.10}$$

The continuity equation (13.7.5) and the axial momentum equation (13.7.10) together constitute the governing equations for $\sigma(z, t)$ and $w(z, t)$ in the plug-flow model. These equations form a one-dimensional hyperbolic system and are analogous to the one-dimensional gas dynamics equations or the shallow-water equations. The equations can be solved numerically using standard algorithms for one-dimensional hyperbolic systems (e.g., Peyret and Taylor, 1983).

It is of interest to reexamine the problem of small-amplitude waves using the plug-flow model, where in this case we admit the presence of an ambient axial flow W. The perturbed core radius and axial velocity fields are given by

$$\sigma(z, t) = a + Ae^{i(\omega t - kz)}, \qquad w(z, t) = W + \hat{w}e^{i(\omega t - kz)}, \tag{13.7.11}$$

where (a, W) are the equilibrium values and $(A, \hat{w})$ are the perturbation amplitudes. Substituting (13.7.11) into the continuity and axial momentum equations (13.7.5) and (13.7.10) and linearizing for small perturbations yield

$$2(\omega - Wk)A - ka\hat{w} = 0, \tag{13.7.12a}$$

$$\frac{a^2}{2}(\omega - kW)\hat{w} + \frac{\Gamma^2}{8\pi^2 a}A = 0. \tag{13.7.12b}$$

The solvability condition of the eigenvalue problem (13.7.12) gives the dispersion relationship for the frequency ω as

$$\omega = kW \pm \frac{k\Gamma}{2\sqrt{2}\pi a}. \tag{13.7.13}$$

The phase speed $c \equiv \omega/k$ is given by

$$c = W \pm \frac{\Gamma}{2\sqrt{2}\pi a}. \tag{13.7.14}$$

Waves corresponding to the plus and minus signs in (13.7.14) are typically called right-running and left-running waves, respectively. The solution for phase speed is independent of wavelength, implying that long area-varying vortex waves are nondispersive.

With zero ambient axial flow ($W = 0$), an expression for dimensionless frequency $f \equiv \pi a^2 \omega/\Gamma$ is obtained from (13.7.13) as

$$f = \frac{ka}{2\sqrt{2}}. \tag{13.7.15}$$

The long-wave limit of the Kelvin wave solution is given by (13.6.16) as $f \cong ka/q$ where $J_0(q) = 0$. The first root of the Bessel function $J_0(q)$ occurs at $q \cong 2.405$, from which we find that the solution (13.7.15) of the plug-flow model differs from the exact Kelvin wave solution by 15%. This difference, which is typical of the error associated with the plug-flow model, occurs due to the fact that the velocity perturbations in the plug-flow model are restricted to be of the form (13.7.1).

Another feature of the solution (13.7.14) for phase velocity of small-amplitude waves is that for cases with sufficiently strong axial flow ($W > \Gamma/2\sqrt{2}\pi a$ in the plug-flow model), both left-running and right-running waves are swept downstream. In this condition, the vortex is said to be *supercritical*. The situation has obvious analogies to supersonic gas flows and to supercritical water layers. Vortices with axial flow W less than the critical value are said to be *subcritical*. The transition from supercritical to subcritical vortex flow is believed to be related to the phenomenon of *vortex breakdown*. A vortex breakdown is an abrupt transition of a vortex flow, which typically takes the form of either a sudden swelling of the core radius or an abrupt kinking of the core, as shown in Figure 13.7. These two types of breakdown are called bubble-type and spiral-type, respectively. Other vortex breakdown types also exist, as discussed in the flow visualization studies by Sarpkaya (1971) and Faler and Leibovich (1977). Further, the so-called bubble-type breakdown appears, at least in some cases, to be more of a tightly wound spiral, rather like a ball of yarn, than an axisymmetric swelling of the vortex. Stationary vortex breakdowns are observed in a variety of applications, such as in the separation vortex above a delta wing and in a vortex along the axis of swirling flow through a diverging tube, in which the pressure gradient external to the vortex is oriented in the direction opposite the vortex axial flow.

One explanation for vortex breakdown (first put forward by Benjamin, 1962) is that it is a transition from supercritical to subcritical vortex flow (in a frame traveling

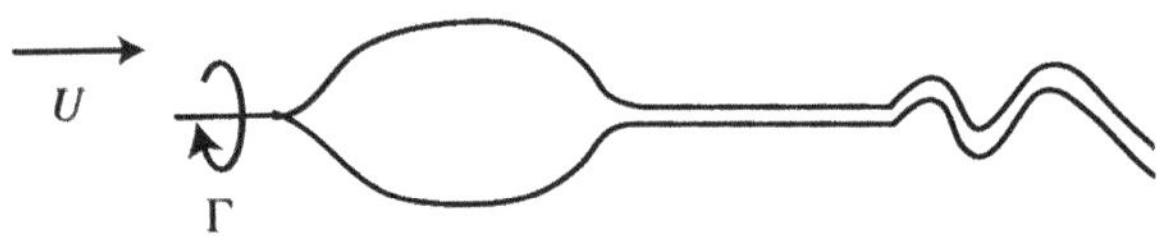

(*a*)

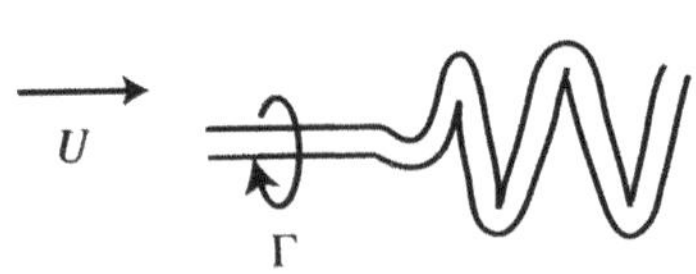

(*b*)

Figure 13.7 Schematic diagrams showing two common types of vortex breakdown: (*a*) bubble type with downstream spiral and (*b*) spiral type. Based on flow visualization studies of Sarpkaya (1971), Faler and Leibovich (1977), and Escudier and Zehnder (1982).

with the breakdown) and is therefore analogous to a hydraulic jump of a shallow liquid layer or to a shock in gas dynamics. We recall that in hydraulics, while the shallow-water model is not valid within the hydraulic jump itself, it can nevertheless be used to predict the necessary conditions for occurrence of the hydraulic jump and to estimate the change in fluid depth over the jump and the rate at which energy must be dissipated within the jump. The one-dimensional inviscid gas dynamics equations can similarly be used to predict the occurrence of a shock wave and the jump conditions across the shock, even though viscosity is important within the shock itself. Like shallow-water flow and gas dynamics, the plug-flow vortex model is governed by a nonlinear hyperbolic system of differential equations, which admit a shocklike solution (called a *vortex shock*) characterized by a discontinuity in vortex core radius and axial velocity. Jump conditions across a vortex shock can be derived using methods similar to those introduced in Chapter 5 (Lundgren and Ashurst, 1989; Marshall, 1991). For a vortex shock with translation velocity W_S along the vortex axis, the jump conditions in mass and momentum require that

$$\left\langle \sigma^2 (W_S - w) \right\rangle = 0, \tag{13.7.16a}$$

$$\left\langle \sigma^2 w (w - W_S) \right\rangle = -\frac{\Gamma^2}{4\pi^2} \langle \ln \sigma \rangle, \tag{13.7.16b}$$

which can be used to solve for the jumps in σ and w if W_S is prescribed or for W_S and the jump in σ if the jump in axial velocity w is prescribed. The notation $\langle f \rangle \equiv f_2 - f_1$ denotes the jump in some quantity f over the discontinuity. An energy jump condition can also be derived (Marshall, 1991), which gives the total energy dissipation rate Φ within the vortex breakdown as

$$\frac{\Phi}{\pi \rho} = -\left\langle \frac{1}{2} \sigma^2 (w - W_S) \left(w^2 + \frac{\Gamma^2}{8\pi^2 \sigma^2} \right) - \frac{3\Gamma^2 w}{16\pi^2} + \frac{\Gamma^2 W_S}{4\pi^2} \ln \sigma \right\rangle. \tag{13.7.17}$$

The *swirl ratio* is defined by $\Omega \equiv \Gamma / 2\pi \sigma w$, where subcritical and supercritical flow conditions correspond to $\Omega > \Omega_{\text{crit}}$ and $\Omega < \Omega_{\text{crit}}$, respectively, and the plug-flow model gives the critical condition as $\Omega_{\text{crit}} = \sqrt{2}$. The jump conditions (13.7.16) and (13.7.17) yield a nonlinear equation for the ratio $A \equiv \sigma_2 / \sigma_1$ of downstream to upstream core radius and an expression for the dimensionless dissipation rate $\Phi^* \equiv \sigma_1 \Phi / \rho \Gamma^3$ for a stationary vortex shock ($W_S = 0$) as

$$A^2 (1 - \Omega_1^2 \ln A) = 1, \tag{13.7.18a}$$

and

$$\Phi^* = \frac{1}{16\pi^2 \Omega_1} \left(\frac{1 - A^2}{A^2} \right) \left[1 - \frac{1}{\Omega_1^2} \left(\frac{1 + A^2}{A^2} \right) \right]. \tag{13.7.18b}$$

For supercritical vortices ($\Omega < \Omega_{\text{crit}}$), the solutions from (13.7.18) give $A > 1$ and $\Phi^* > 0$, whereas for subcritical vortices ($\Omega > \Omega_{\text{crit}}$) they yield $A < 1$ and $\Phi^* < 0$. Since the rate of energy dissipation must be positive, stationary vortex shocks are prohibited for subcritical flows.

The available experimental evidence appears to be consistent with the result that stationary vortex breakdowns cannot occur for subcritical flows. Our argument that a shocklike discontinuity in the plug-flow model would manifest itself physically as a vortex breakdown is also supported by recent direct comparisons between both experiments and direct computations and solutions of the plug-flow model (Krishnamoorthy and Marshall, 1994; Marshall and Krishnamoorthy, 1997), although more work along this line is needed. A more complete account of experimental results, along with description of alternative theories for vortex breakdown, is given in the review by Leibovich (1984).

13.8 AXISYMMETRIC DISCRETE-VORTEX METHOD

The discrete-vortex method, described for two-dimensional flow in Section 11.4, is extended in the present section for axisymmetric flow with or without swirl. The axisymmetric discrete-vortex method discretizes the vorticity field by a set of vortex rings with finite core radius. The centerline of each ring, identified for ring n by (R_n, z_n), is advected in the R-z plane as

$$\frac{dR_n}{dt} = u(R_n, z_n, t), \qquad \frac{dz_n}{dt} = w(R_n, z_n, t) + W_n, \tag{13.8.1}$$

where u and w are components of the velocity field induced by the other vortex elements and W_n is the element self-induced velocity.

The vorticity field is discretized by the sum of N ring elements as

$$\boldsymbol{\omega}(\mathbf{x}, t) = \sum_{n=1}^{N} \boldsymbol{\Gamma}_n(t) f_n(\mathbf{x} - \mathbf{x}_n), \tag{13.8.2}$$

where $\boldsymbol{\Gamma}_n(t)$ is the element amplitude and the core function $f_n(\mathbf{x} - \mathbf{x}_n)$ is normalized such that

$$\int_{-\infty}^{\infty} \int_{0}^{\infty} f_n(R - R_n, z - z_n) R\, dR\, dz = R_n. \tag{13.8.3}$$

For example, if the vorticity is assumed to vary as a Gaussian within the element core with core radius δ_n, such that $f_n = C_n \exp(-r^2/\delta_n^2)$, where $r^2 \equiv (R - R_n)^2 + (z - z_n)^2$, the normalization (13.8.3) gives the coefficient C_n as

$$C_n = \frac{2R_n}{\pi^{1/2}\delta_n^3 + 2\pi R_n \delta_n^2}. \tag{13.8.4}$$

The element amplitude $\boldsymbol{\Gamma}_n(t)$ is obtained from the vorticity $\boldsymbol{\omega}_n = \boldsymbol{\omega}(R_n, z_z, t)$ on element n using one of the two methods discussed in Section 11.4. In the traditional approach, we set $\boldsymbol{\Gamma}_n = \boldsymbol{\omega}_n h_n^2$, where the incompressibility condition for axisymmetric flows requires that $h_n^2 R_n = \text{const}$. Alternatively, the iterative procedure (11.4.10), with ω_n and Γ_n replaced by the corresponding vector-valued quantities, can be used to refit the amplitudes to the computational points at each time step.

The velocity components u and w can be obtained from the definition (13.1.2) of the Stokes stream function ψ, which is given in (13.1.13) in terms of an integral over the Green's function $G(R, R', z - z')$ times the vorticity field. Substituting the vorticity discretization (13.8.2) into (13.1.13) gives the stream function as

$$\psi(R, z, t) = \sum_{n=1}^{N} \Gamma_{\alpha,n}(t) \int_{-\infty}^{\infty} \int_{0}^{\infty} f_n(R' - R_n, z' - z_n) G(R, R', z - z')\, dR'\, dz', \tag{13.8.5}$$

where $\Gamma_{\alpha,n}$ is the azimuthal component of the amplitude of element n. For elements with $\delta_n / R_n \ll 1$, (13.8.5) has the form of a Laplace integral when f_n is a Gaussian, which may be approximated using standard asymptotic methods (Bender and Orszag, 1978) by the product

$$\psi(R, z, t) = \sum_{n=1}^{N} \xi_n(r) \psi_n(R, R_n, z - z_n), \tag{13.8.6}$$

where $r \equiv [(R - R_n)^2 + (z - z_n)^2]^{1/2}$. Here $\psi_n(R, R_n, z - z_n)$ is the stream function induced by a singular vortex filament of strength $\Gamma_{\alpha,n}$ located at (R_n, z_n), given by (13.2.2) with b replaced by R_n and z replaced by $z - z_n$. The function $\xi_n(r)$, which regularizes the singularity at the element centerline, is given by

$$\xi_n(r) = 2\pi \int_0^r f_n(r') r'\, dr'. \tag{13.8.7}$$

When f_n has a Gaussian form, (13.8.7) gives the regularization function as $\xi_n(r) = 1 - \exp(-r^2/\delta_n^2)$. The assumption $\delta_n / R_n \ll 1$ breaks down close to the symmetry axis. However, since the azimuthal vorticity component is small for small R, and must vanish on $R = 0$, vorticity elements with centroid within a distance of order δ_n from the symmetry axis do not contribute significantly to the flow field in the R-z plane.

The direct computation of the velocity components from (13.8.6) requires $O(N^2)$ computations per time step, which can be unmanageable for very large N. A multipole acceleration procedure for axisymmetric flows is described by Strickland and Amos (1992) that reduces the number of computations to $O(N \ln N)$ for large N. The basic approach is analogous to that described for two-dimensional flows in Section 11.5, and the interested reader is referred to the original paper for details.

The easiest way to obtain the swirl velocity is by using the fact that the product vR is invariant on any material point. Hence, if the initial swirl velocity is known on computational point n, the swirl velocity at any later time t can be obtained as

$$v_n(t) = \frac{v_n(0)R_n(0)}{R_n(t)}. \tag{13.8.8}$$

This approach does not extend to viscous flows, for which the product vR is no longer invariant. An alternative axisymmetric discrete vortex method that applies to both inviscid and viscous flows is given by Marshall and Grant (1997), in which the swirl velocity is obtained from the axisymmetric form of the Biot-Savart integral as an integral over the axial and radial vorticity components. When using this alternative approach, special care must be taken to resolve the axial component of vorticity near the symmetry axis for proper determination of the swirl velocity.

The self-induced velocity for a vortex ring has the form (Section 13.2)

$$W_n = \frac{\Gamma_{\alpha,n}}{4\pi R_n}\left[\ln\left(\frac{8R_n}{\delta_n}\right) + C\right], \tag{13.8.9}$$

where C is an $O(1)$ constant that depends on the form of core function f_n. For instance, for a Gaussian core function, (13.2.25) gives $C = 0.583$. The importance of the element self-induced velocity, in comparison to the total velocity field, depends on the resolution of the flow field. For instance, suppose that the core of an isolated vortex ring of ring radius b, core radius a, and strength Γ is discretized using N vortex ring elements of equal strength $\mathbf{\Gamma}_n = (\Gamma/N)\mathbf{e}_\alpha$ and element radius $\delta_n = B(\pi a^2/N)^{1/2}$, where B is an $O(1)$ constant. If $b \gg a$, so that the thin-core assumption applies, we can let $R_n \cong b$ to write the element self-induced velocity as $W_n \cong (\Gamma/4\pi bN)[\ln(N) + \ln(8b/a) - \ln(B\pi^{1/2}) + C]$. The ring self-induced propagation velocity as a whole is approximately $W = (\Gamma/4\pi b)\,[\ln(8b/a) + D]$, where D is another $O(1)$ constant. For large values of N, these two results give

$$\frac{W_n}{W} = O\left(\frac{1}{N}\ln N\right), \tag{13.8.10}$$

which vanishes as $N \to \infty$. For well-resolved flow fields (i.e., for large N), the element self-induced velocity is small compared to that induced by the other vortex elements, and for this reason it is often neglected.

The vorticity components are evolved using the vorticity transport equations (13.1.6). The velocity components u and w in the R-z plane are determined entirely from the azimuthal vorticity component ω_α, and the swirl velocity v can be determined using (13.8.8) from the element radial displacement. For flows with swirl, the radial vorticity component ω_R must also be computed since it appears on the right-hand side of the transport equation for ω_α.

There are several options available for computation of the vorticity stretching terms on the right-hand side of the vorticity transport equations (13.1.6), which involve differentiation in the R- and z-directions. One approach is to analytically differentiate the sums for u and w obtained from the expression (13.8.6) for stream function. This approach increases the computation time (which for large N is nearly entirely devoted to the velocity summation) by a factor of three, and hence is not very efficient. Alternatively, the directional derivative $\omega_R(\partial/\partial R) + \omega_z(\partial/\partial z)$ of the velocity components may be approximated by a finite difference, using the difference in velocity value evaluated at a distance ε slightly in front of point n and a distance $-\varepsilon$ slightly behind point n along a line oriented in the direction of the radial and axial vorticity components. This approach requires two velocity computations for each element n, assuming that the velocity at n is taken as the average of the two values evaluated on either side of n. A still more efficient approach is to approximate the velocity field by a least-squares polynomial fit to the velocity at nearby computational points, and then compute the derivatives by differentiation of the polynomial fit (Marshall and Grant, 1997). This approach requires only one velocity computation at each element per time step, and the least-square procedure has the added benefit of smoothing any undesirable fluctuations in the computed velocity field that might introduce noise in the stretching terms.

An example of a computation performed using an axisymmetric discrete vortex method is shown in Figure 13.8 (Marshall, 1997). The computation examines the interaction of two vortex rings of strength $\pm\Gamma_S$ encircling a columnar vortex of strength Γ_C. The rings are initially placed at a radius b and are separated by an axial distance $L/2$, where for both cases shown $b/L = \frac{1}{3}$. The flow is periodic in the axial direction over a length L. The plots show velocity vectors attached to each of the Lagrangian computational points used to discretize the vorticity field. In Figure 13.8a, the rings are sufficiently weak compared to the columnar vortex ($\Gamma_S/\Gamma_C = 0.2$) that the azimuthal vorticity generated within the columnar vortex by the mechanism of tilting of radial and axial vorticity, discussed in Section 13.1 with reference to (13.1.9), is able to counter the outward radial velocity induced by the rings. The columnar vortex responds to the forcing from the rings by formation of a standing wave, in which the core radius bulges outward and compresses inward in an oscillatory manner. The dashed line in Figure 13.8 indicates the initial boundary of the vortex core. In Fig-

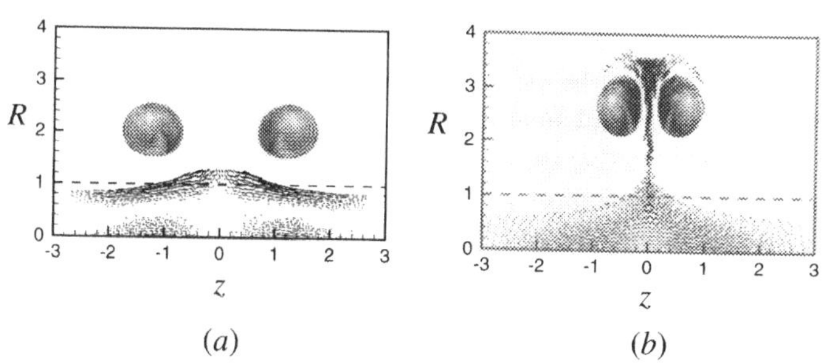

Figure 13.8 Numerical computation of the interaction of a columnar vortex and a pair of vortex rings of opposite sign, obtained using an axisymmetric discrete-vortex method. In (a), the ring strength is sufficiently weak ($\Gamma_S/\Gamma_C = 0.2$) that a standing wave exists on the columnar vortex. In (b), the ring strength is strong enough ($\Gamma_S/\Gamma_C = 1$) to strip vorticity from the columnar vortex core. The dashed line indicates the initial core boundary, and both (a) and (b) are plotted at the same time.

ure 13.8*b*, the rings are sufficiently strong ($\Gamma_S/\Gamma_C = 1.0$) that a sheet of vorticity is ejected, or stripped, from the columnar vortex. Vorticity stripping, which is sometimes described in the literature as "bursting," is one of the principal mechanisms responsible for decay of high Reynolds number vortices, such as aircraft trailing vortices, in the presence of external turbulence (Liu, 1992; Sarpkaya and Daly, 1987).

BIBLIOGRAPHY

Abramowitz, M., and I.A. Stegun (1964). *Handbook of Mathematical Functions.* National Bureau of Standards, Washington, DC (reprinted Dover Publications, New York, 1965).

Bender, C.M. and S.A. Orszag (1978). *Advanced Mathematical Methods for Scientists and Engineers.* McGraw-Hill, New York.

Benjamin, T.B. (1962). "Theory of the vortex breakdown phenomenon," *Journal of Fluid Mechanics* **14**, 593–629.

Benjamin, T.B. (1967). "Some developments in the theory of vortex breakdown," *Journal of Fluid Mechanics* **28**, 65–84.

Bragg, S.L. and W.R. Hawthorne (1950). "Some exact solutions of the flow through annular cascade actuator discs," *Journal of the Aeronautical Sciences* **17**, 243–249.

Escudier, M.P. and N. Zehnder (1982). "Vortex-flow regimes," *Journal of Fluid Mechanics* **115**, 105–121.

Faler, J.H., and S. Leibovich (1977). "Disrupted states of vortex flow and vortex breakdown," *Physics of Fluids* **20**(9), 1385–1400.

Fraenkel, L.E. (1970). "On steady vortex rings of small cross-section in an ideal fluid," *Proceedings of the Royal Society of London A* **316**, 29–62.

Helmholtz, H. (1858). "Uber Integrale der hydrodynamischen Gleichungen welche den Wirbelbewegungen entsprechen," *Crelles Journal* **55**, 25–55.

Hicks, W.M. (1885). "Researches on the theory of vortex rings—part II," *Philosophical Transactions of the Royal Society of London A* **176**, 725–780.

Hill, M.J.M. (1894). "On a spherical vortex," *Philosophical Transactions of the Royal Society of London A* **185**, 213–245.

Kelvin, Lord (1867). "The translatory velocity of a circular vortex ring," *Philosophical Magazine* **33**, 511–512 (reprinted *Mathematical and Physical Papers*, Vol. 4, Cambridge University Press, Cambridge, 1910).

Kelvin, Lord (1880). "Vibrations of a columnar vortex," *Philosophical Magazine* **10**, 155–168 (reprinted *Mathematical and Physical Papers* Vol. 4, Cambridge University Press, Cambridge, 1910).

Krishnamoorthy, S., and J.S. Marshall (1994). "An experimental investigation of 'vortex shocks,' " *Physics of Fluids* **6**(11), 3737–3741.

Lamb, H. (1932). *Hydrodynamics*, Cambridge University Press, Cambridge (reprinted Dover Publications, New York, 1945).

Leibovich, S. (1984). "Vortex stability and breakdown: survey and extension," *AIAA Journal* **22**(9), 1192–1206.

Leibovich, S. and A. Kribus (1990). "Large-amplitude wavetrains and solitary waves in vortices," *Journal of Fluid Mechanics* **216**, 459–504.

Leonard, A. (1994). "Nonlocal theory of area-varying waves on axisymmetric vortex tubes," *Physics of Fluids A* **6**(2), 765–777.

Liu, H.T. (1992). "Effects of ambient turbulence on the decay of a trailing vortex wake," *Journal of Aircraft* **29**, 255–263.

Lundgren, T.S. and W.T. Ashurst (1989). "Area-varying waves on curved vortex tubes with application to vortex breakdown," *Journal of Fluid Mechanics* **200**, 283–307.

Marshall, J.S. (1991). "A general theory of curved vortices with circular cross-section and variable core area," *Journal of Fluid Mechanics* **229**, 311–338.

Marshall, J.S. (1997). "The flow induced by periodic vortex rings wrapped around a columnar vortex core," *Journal of Fluid Mechanics* **345**, 1–30.

Marshall, J.S. and J.R. Grant (1997). "A Lagrangian vorticity collocation method for viscous, axisymmetric flows with and without swirl," *Journal of Computational Physics* **138**, 302–330.

Marshall, J.S. and S. Krishnamoorthy (1997). "On the instantaneous cutting of a columnar vortex with non-zero axial flow," *Journal of Fluid Mechanics* **351**, 41–74.

Moffat, H.K., and D.W. Moore (1978). "The response of Hill's spherical vortex to a small axisymmetric disturbance," *Journal of Fluid Mechanics* **87**, 749–760.

Norbury, J. (1973). "A family of steady vortex rings," *Journal of Fluid Mechanics* **57**, 417–431.

Peyret, R. and T.D. Taylor (1983). *Computational Methods for Fluid Flow*, Springer-Verlag, New York.

Pozrikidis, C. (1986). "The nonlinear instability of Hill's vortex," *Journal of Fluid Mechanics* **168**, 337–367.

Saffman, P. G. (1992). *Vortex Dynamics*, Cambridge University Press, Cambridge.

Sarpkaya, T. (1971). "On stationary and traveling vortex breakdown," *Journal of Fluid Mechanics* **45**(3), 545–559.

Sarpkaya, T. and J.J. Daly (1987). "Effect of ambient turbulence on trailing vortices," *Journal of Aircraft* **24**, 399–404.

Strickland, J.H. and D.E. Amos (1992). "Fast solver for systems of axisymmetric ring vortices," *AIAA Journal* **30**(3), 737–746.

Tait, P.G. (1867). "Translation of 'On integrals of the hydrodynamical equations which express vortex-motion' by H. Helmholtz," *Philosophical Magazine* **33**, 485–512.

Yamada, H. and T. Matsui (1978). "Preliminary study of mutual slip-through of a pair of vortices," *Physics of Fluids* **21**, 292–294.

PROBLEMS

1. It is often observed in numerical computations of turbulent flows that in regions of intense vorticity, the vorticity vector is approximately aligned with the middle eigenvector of the rate of deformation tensor **D**, such that the eigenvalue $\lambda^{(2)}$ associated with the middle eigenvector is intermediate between the other two eigenvalues. Consider a Lamb vortex with vorticity $\boldsymbol{\omega} = \Gamma \exp(-R^2/a^2)\mathbf{e}_z$ and azimuthal velocity $v = (\Gamma/2\pi R)[1 - \exp(-R^2/a^2)]$, where a is a parameter that

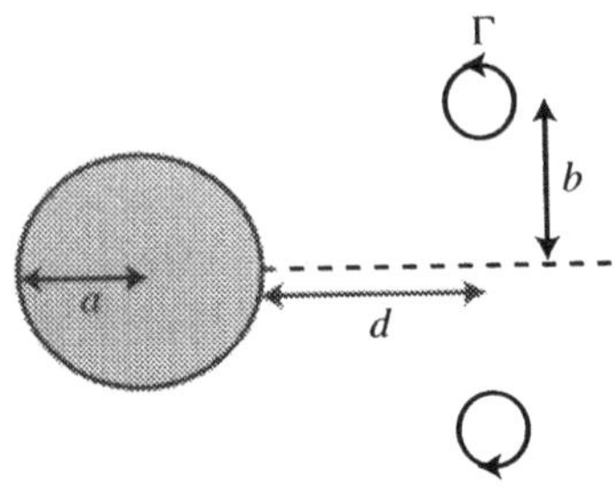

Figure 13.9 A vortex ring of radius b located at a distance d from a sphere of radius a.

determines the vortex core size. Suppose that the vortex is stretched by an axisymmetric stretching flow with stretching rate c.

(a) Solve for the variation of a with time.

(b) Determine the three eigenvectors and associated eigenvalues of the rate of deformation tensor as a function of R. Under what conditions does the vorticity vector align with the middle eigenvector of $\mathbf{D}$?

2. A "vortex gun" is developed to shoot vortex rings in water in order to set off undersea mines around naval vessels and in harbors. The vortex rings have ring radius $b = 10$ cm, core radius $a = 0.5$ cm, strength $\Gamma = 50$ cm^2 /s, and uniform vorticity profile within the core.

   (a) Determine the ring propagation speed.

   (b) Estimate the maximum impact (force integrated over time) of the ring on a mine that it hits head-on. Assuming that a characteristic time scale for the ring-mine interaction is the time required for the ring to propagate over a distance equal to the ring radius, estimate the order of magnitude of the maximum force induced on the mine from the ring.

   (c) If one ring is produced by the gun every second, estimate the minimum power requirements for operation of the vortex gun (i.e., assuming that all the gun energy is transmitted to the ring).

   (d) Would this be an effective device for naval mine detonation? Why or why not?

3. Use the Butler sphere theorem to derive the velocity field due to a vortex ring of ring radius b at a distance d from the center of a sphere of radius a, as shown in Figure 13.9.

4. Prove that the velocity field induced by a vortex ring of strength Γ, ring radius b, and core radius a approaches that induced by a point doublet at large distances from the ring. What is the strength and orientation of the effective point doublet?

5. Derive expressions for the impulse, kinetic energy, helicity, and vorticity centroid for Hill's spherical vortex.

6. Show that expression (13.5.16) yields the stream function for Hill's spherical vortex when substituted into the Bragg-Hawthorne equation. Determine the pressure at the center of Hill's vortex relative to that at infinity.

7. Use the plug-flow model to derive a solution for axisymmetric standing waves on a columnar vortex with equilibrium core radius a and perturbation axial velocity of the form $w(z, t) = \hat{w} \sin(kz) \sin(\omega t)$. Derive an expression for wave frequency ω as a function of wavenumber k.

COMPUTATIONAL PROJECTS

1. Write a computer code implementing the axisymmetric vortex method described in Section 13.8. Use your code to compute the motion of a vortex ring with vorticity distribution proportional to the radius R of a cylindrical polar coordinate system. The core of the ring should be formed using at least 100 vortex elements. Evaluate the propagation velocity of a vortex ring with ratio of ring radius b to core radius a of $b/a = 10, 5, 3, 2, 1.5$. Plot the computed propagation velocity versus b/a and compare to Kelvin's thin-core vortex ring prediction (13.2.26).
2. Use an axisymmetric vortex method to compute the "leap-frogging" motion of two vortex rings, illustrated in Figure 13.10. Initialize the rings with ring radius to core radius ratio $b/a = 4$ and with axial separation of $2a$. Plot the location of the vortex elements over a time series during which the rings exchange places and then switch back again. Compare the computed oscillation frequency of the rings with an estimate obtained using the Kelvin expression (13.2.26) for propagation velocity of a thin-core ring.

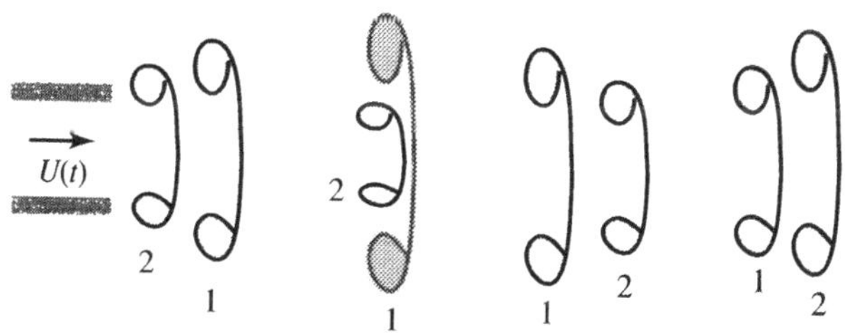

Figure 13.10 Leap-frogging motion of two vortex rings ejected from an orifice. Based on flow visualizations of Yamada and Matsui (1978).

CHAPTER 14

VORTEX TUBES

Vortex structures in three-dimensional flows often take the form of thin tubes within which vorticity is aligned approximately along the tube. Vortex tubes may be isolated, interacting only with themselves, or they may interact with other tubes or solid bodies in their vicinity. Tubular vortices are most noticeable in systems consisting of only a few strong vortices, such as the trailing vortices behind the wings of an aircraft, the helical cavitation vortex behind a propeller blade, or the vortices shed from a ship hull or the rear of an automobile into the body wake. Tornadoes are a dramatic example of a tubular vortex, for which bending motion of the tornado core is often visible. On a more fundamental level, large-scale tubular coherent vortices control a wide range of features of turbulent flows, such as mixing rates of mass, heat, and momentum, which ultimately determine such quantities as the skin friction, heat flux, and separation point of a turbulent boundary layer and the mixing of chemical species in a combustion chamber.

The dynamics of tubular vortices can be classified into two categories, loosely termed *filament dynamics* and *core dynamics*. The current chapter concentrates on filament dynamics, which treats the tubular vortex in bulk using the assumption that all vorticity is concentrated on a single curve along the tube center, although this restriction must be relaxed somewhat to determine the self-induced motion of the vortex. Filament models are typically used for problems such as propagation and growth of bending waves on the vortex core and displacement of the vortex during interaction with a solid body. Core dynamics, on the other hand, deals with the evolution of vorticity within the vortex core on the scale of the core radius. Some examples of core dynamics are discussed in other parts of this book, including propagation of waves of variable core area (Chapter 13) and the self-induced rotation of vortex cores of elliptical shape (Chapters 11 and 16). The time and spatial scales of core dynamics problems are often much smaller than for filament dynamics problems, which in many cases allows the two to be treated separately.

14.1 VELOCITY FIELD INDUCED BY A CURVED VORTEX FILAMENT

A vortex filament is a space curve endowed with certain dynamical properties. Before introducing the dynamics of vortex filaments, it is necessary to review some aspects of the geometry of space curves. Consider a space curve C with position vector $\mathbf{r} = \mathbf{r}(\xi, t)$, where the scalar variable ξ identifies fluid particles on C. It is convenient to choose ξ to be a Lagrangian variable such that the value of ξ is constant on a fluid particle. Three orthogonal unit vectors, denoted by $\boldsymbol{\lambda}_1$, $\boldsymbol{\lambda}_2$, and $\boldsymbol{\lambda}_3$, used to indicate direction relative to C are defined by

$$\boldsymbol{\lambda}_3 \equiv \frac{\partial \mathbf{r}/\partial \xi}{|\partial \mathbf{r}/\partial \xi|}, \qquad \boldsymbol{\lambda}_1 \equiv \frac{\partial \boldsymbol{\lambda}_3/\partial \xi}{|\partial \boldsymbol{\lambda}_3/\partial \xi|}, \qquad \boldsymbol{\lambda}_2 \equiv \boldsymbol{\lambda}_3 \times \boldsymbol{\lambda}_1. \tag{14.1.1}$$

The *unit tangent* vector $\boldsymbol{\lambda}_3$ is proportional to the change in position vector between two nearby points on C. The vectors $\boldsymbol{\lambda}_1$ and $\boldsymbol{\lambda}_2$, which are both oriented orthogonal to C, are called the *principal normal* and *binormal* vectors, respectively. The principal normal is proportional to the change in the tangent vector between two nearby points on C, and the binormal is the vector orthogonal to both the tangent vector and the principal normal vector. Two examples showing the orientation of these three vectors are given in Figure 14.1 for a ring and a helix. For a ring, the tangent vector is oriented in the azimuthal direction, the principal normal vector is oriented radially inward, and the binormal vector is oriented along the ring symmetry axis. The principal normal vector for a helix also points radially inward. The binormal and tangent vectors for a helix have components in both the azimuthal and axial directions and are obtained by rotating the tangent and binormal vectors for a ring about the $\boldsymbol{\lambda}_1$-axis.

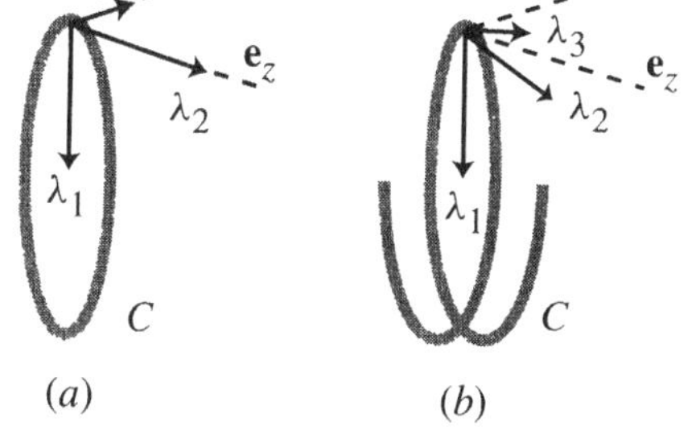

Figure 14.1 Examples showing the orientation of the tangent vector $\boldsymbol{\lambda}_3$, the principal normal vector $\boldsymbol{\lambda}_1$, and the binormal vector $\boldsymbol{\lambda}_2$ for (a) a ring and (b) a helix.

The increment in arc length ds along C is given by

$$ds = \left|\frac{\partial \mathbf{r}}{\partial \xi}\right| d\xi. \tag{14.1.2}$$

The definition of the unit vectors in (14.1.1) can equivalently be written using arc length in place of the Lagrangian variable ξ. The shape of the curve can be described by the curvature κ and the torsion τ, which are functions of time and position along C. The curvature is proportional to the rate of change of the tangent vector with distance along C, whereas the torsion is proportional to the rate of change of the binormal vector. Definitions of these two variables are given by the *Serret-Frenet equations*

$$\frac{\partial \boldsymbol{\lambda}_3}{\partial s} = \kappa \boldsymbol{\lambda}_1, \qquad \frac{\partial \boldsymbol{\lambda}_2}{\partial s} = -\tau \boldsymbol{\lambda}_1, \qquad \frac{\partial \boldsymbol{\lambda}_1}{\partial s} = \tau \boldsymbol{\lambda}_2 - \kappa \boldsymbol{\lambda}_3. \tag{14.1.3}$$

The first two equations in (14.1.3) serve as definitions of κ and τ, whereas the last equation follows from differentiation of the identity $\boldsymbol{\lambda}_1 = \boldsymbol{\lambda}_2 \times \boldsymbol{\lambda}_3$. The curvature of the ring in Figure 14.1*a* is equal to the reciprocal of the ring radius, whereas the ring has zero torsion because the binormal vector is constant. The helix in Figure 14.1*b* has the same curvature as the ring, but it has a constant torsion that is proportional to the axial distance that the helix advances with each turn.

A filament-based curvilinear coordinate system (n_1, n_2, s), with basis vectors $(\boldsymbol{\lambda}_1, \boldsymbol{\lambda}_2, \boldsymbol{\lambda}_3)$ respectively, is introduced such that C corresponds to $n_1 = n_2 = 0$. The Biot-Savart equation for a vortex filament can be obtained by writing the vorticity field as $\boldsymbol{\omega} = \Gamma \delta(n_1, n_2)\boldsymbol{\lambda}_3$, so that substitution into the general Biot-Savart equation (6.5.7b) and integration over the directions normal to the filament gives

$$\mathbf{u}(\mathbf{x}, t) = -\frac{\Gamma}{4\pi} \int_C \frac{(\mathbf{x} - \mathbf{x}') \times \boldsymbol{\lambda}_3'}{|\mathbf{x} - \mathbf{x}'|^3}\, ds'. \tag{14.1.4}$$

Using the definitions (14.1.1) and (14.1.2) and the fact that $\mathbf{x}' = \mathbf{r}(\xi', t)$ on C, the Biot-Savart equation for a filament can be rewritten in the alternative form

$$\mathbf{u}(\mathbf{x}, t) = -\frac{\Gamma}{4\pi} \int_C \frac{\mathbf{x} - \mathbf{r}(\xi', t)}{|\mathbf{x} - \mathbf{r}(\xi', t)|^3} \times \frac{\partial \mathbf{r}'}{\partial \xi'}\, d\xi'. \tag{14.1.5}$$

14.2 CUT-OFF MODEL FOR A VORTEX TUBE

The filament Biot-Savart integral (14.1.5) is singular if the point $\mathbf{x}$ at which the velocity is evaluated lies on the curve C, giving an infinite self-induced velocity of the filament. To see this singular behavior more clearly, we use a Taylor expansion to examine the integrand of (14.1.4) near the singular point $\mathbf{x} = \mathbf{r}(s_0)$. Letting $\zeta \equiv s - s_0$, the Taylor series expansion gives

$$\mathbf{r}(s) - \mathbf{r}(s_0) = \zeta \left.\frac{\partial \mathbf{r}}{\partial s}\right|_{s_0} + \frac{\zeta^2}{2} \left.\frac{\partial^2 \mathbf{r}}{\partial s^2}\right|_{s_0} + O(\zeta^3). \tag{14.2.1}$$

Differentiating (14.2.1) with respect to s gives the unit tangent vector evaluated at s as

$$\boldsymbol{\lambda}_3(s) = \frac{\partial \mathbf{r}}{\partial s} = \left.\frac{\partial \mathbf{r}}{\partial s}\right|_{s_0} + \zeta \left.\frac{\partial^2 \mathbf{r}}{\partial s^2}\right|_{s_0} + O(\zeta^2). \tag{14.2.2}$$

Using this expansion, the integrand of (14.1.4) near the point $\mathbf{x} = \mathbf{r}(s_0)$ becomes

$$\begin{aligned} \frac{\mathbf{x} - \mathbf{r}(s)}{|\mathbf{x} - \mathbf{r}(s)|^3} \times \boldsymbol{\lambda}_3 &= -\frac{1}{|\zeta|^3} \left(\zeta \left.\frac{\partial \mathbf{r}}{\partial s}\right|_{s_0} + \frac{\zeta^2}{2} \left.\frac{\partial^2 \mathbf{r}}{\partial s^2}\right|_{s_0} \right) \times \left(\left.\frac{\partial \mathbf{r}}{\partial s}\right|_{s_0} + \zeta \left.\frac{\partial^2 \mathbf{r}}{\partial s^2}\right|_{s_0} \right) + O(1) \\ &= -\frac{1}{2|\zeta|} \left(\left.\frac{\partial \mathbf{r}}{\partial s}\right|_{s_0} \times \left.\frac{\partial^2 \mathbf{r}}{\partial s^2}\right|_{s_0} \right) + O(1). \end{aligned} \tag{14.2.3}$$

This result can be expressed in terms of the filament unit vectors using (14.1.1)–(14.1.3) as

$$\frac{\mathbf{x} - \mathbf{r}(s)}{|\mathbf{x} - \mathbf{r}(s)|^3} \times \boldsymbol{\lambda}_3 = -\frac{\kappa}{2|\zeta|} \boldsymbol{\lambda}_2 + O(1), \tag{14.2.4}$$

where κ is the vortex curvature and $\boldsymbol{\lambda}_2$ is the binormal vector at s_0. The integrand thus exhibits a nonintegrable $1/\zeta$ singularity at the point s_0.

This singularity arises from the nonphysical nature of the filament model for evaluating the contribution to the velocity at a point $\mathbf{x}$ on the filament centerline C due to vorticity at points near $\mathbf{x}$. In particular, in clumping all of the vorticity within the vortex tube on the filament curve C, it is required that the distance between a point $\mathbf{x}$ where the velocity is evaluated and any point $\mathbf{r}(\xi)$ on the filament is much greater than the core radius a. This assumption is clearly violated when $\mathbf{x}$ lies on the filament for points ξ such that $|\mathbf{x} - \mathbf{r}(\xi)| \leq O(a)$. In considering how to treat this singularity, it is useful to consider the relative contribution of different parts of a vortex tube to the self-induced velocity of the tube. There are two length scales associated with a vortex tube of circular cross section—the core radius a and the radius of curvature $1/\kappa$. In evaluating the induced velocity $\mathbf{u}(\mathbf{x})$ at a point $\mathbf{x}$ on the centerline C of a vortex tube, it is useful to divide the vortex into three parts, labeled I, II, and III in Figure 14.2. Region I spans a length $\delta = O(a)$ along the vortex axis on either side of point $\mathbf{x}$. Assuming that the curvature κ of C satisfies $\kappa a \ll 1$, the centerline curve of the vortex tube in region I is nearly a straight line. Since a straight vortex exhibits no self-induced velocity at its centerline, region I does not contribute significantly to $\mathbf{u}(\mathbf{x})$. This observation is in stark contrast to the prediction of filament theory, for which the region of the vortex nearest to $\mathbf{x}$ provides the greatest contribution to $\mathbf{u}(\mathbf{x})$.

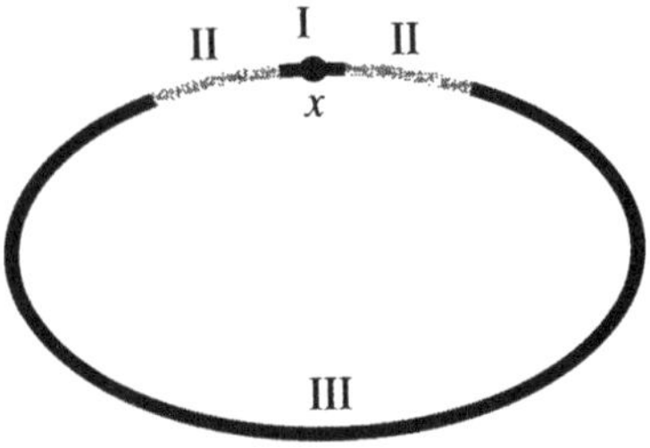

Figure 14.2 Three regions of a curved vortex tube with respect to a point **x**: (I) near region with points within a distance of $O(a)$ from **x**, (II) middle region with points within a distance of $O(1/\kappa)$ from **x**, and (III) distant region.

Region III, lying at a distance greater than $\ell = O(1/\kappa)$ from **x**, is sufficiently far away that the contribution to **u**(**x**) of any section of a given length of $O(1/\kappa)$ in region III is weak compared to that of a section of similar length in region II. Nevertheless, the vortex in region III may often be much longer than that in region II, and the contribution of the distant portions of the vortex plays a critical role in many vortex dynamics problems.

In the *cut-off model* for a thin vortex tube, a small region with length $\delta = \delta_C a$ is removed from the curve C in the filament integral (14.1.4) on both sides of the singular point, which is equivalent to neglecting region I in the model presented in Figure 14.2. The cut-off is indicated by writing the region of integration in (14.1.4) as $C[\delta]$, rather than simply C. It is important to accurately determine the cut-off coefficient δ_C (sometimes called the Crow cut-off constant), which has an $O(1)$ effect on the filament velocity. The utility of the cut-off model lies in the observation that the cut-off length is independent of the form of the curve C, which follows from the restriction $\kappa a \ll 1$ and the fact that the cut-off length is of $O(a)$, such that the vortex is nearly straight over the cut-off region. This being so, the self-induced velocity of a vortex tube of arbitrary shape with curvature κ at a point **x** will be the same as a vortex ring of radius $1/\kappa$ locally tangent to the vortex filament at **x**. The validity of the cut-off approximation for small κa has been justified empirically by Crow (1970) and using asymptotic methods by Moore and Saffman (1972) and Fukumoto and Miyazaki (1991).

The cut-off coefficient δ_C can be determined by comparing the self-induced velocity predicted by the filament Biot-Savart equation (14.1.4) with cut-off of length $\delta = \delta_C a$ for a circular vortex ring to the result (13.2.25) obtained using the Lamb transformation. For a thin vortex tube in the shape of a ring of radius b, the self-induced velocity U is determined from (14.1.4) as

$$U = -\frac{\Gamma}{4\pi}\mathbf{e}_z \cdot \int_{C[\delta]} \frac{\mathbf{r}(s_0) - \mathbf{r}(s')}{|\mathbf{r}(s_0) - \mathbf{r}(s')|^3} \times \boldsymbol{\lambda}_3'\, ds' \tag{14.2.5}$$

for any point s_0 on the filament. Since the ring tangent vector is oriented in the azimuthal direction, we can write $\boldsymbol{\lambda}_3'\, ds' = b\mathbf{e}_\alpha'\, d\alpha'$, so that (14.2.5) becomes

$$U = -\frac{\Gamma}{4\pi}\mathbf{e}_z \cdot \int_{\delta_C a/b}^{2\pi-\delta_C a/b} \frac{(\mathbf{r}_0 - \mathbf{r}') \times \mathbf{e}'_\alpha}{|\mathbf{r}_0 - \mathbf{r}'|^3} b\, d\alpha', \tag{14.2.6}$$

where for convenience $\mathbf{r}_0$ is taken to coincide with the point $(R, \alpha) = (b, 0)$. Elementary trigonometry can be used to show that $\mathbf{e}_z \cdot [(\mathbf{r}_0 - \mathbf{r}') \times \mathbf{e}'_\alpha] = -b(1 - \cos\alpha')$ and $|\mathbf{r}_0 - \mathbf{r}'|^3 = 2\sqrt{2}b^3(1 - \cos\alpha')^{3/2}$, so (14.2.6) becomes

$$U = \frac{\Gamma}{8\sqrt{2}\pi b} \int_{\delta_C a/b}^{2\pi-\delta_C a/b} \frac{d\alpha'}{(1 - \cos\alpha')^{1/2}} = \frac{\Gamma}{4\pi b} \ln\left(\frac{4b}{\delta_C a}\right). \tag{14.2.7}$$

The self-induced velocity computed for a thin-core vortex ring in Section 13.2 is given by

$$U = \frac{\Gamma}{4\pi b}\left[\ln\left(\frac{8b}{a}\right) - \frac{1}{2} + \frac{1}{\Gamma^2}\int_0^a \frac{\Gamma_0^2(r)}{r}\, dr\right], \tag{14.2.8}$$

where r is the core-centered coordinate and $\Gamma_0(r)$ is the core circulation profile. Equating the last two results yields an expression for the cut-off coefficient δ_C as

$$\ln(2\delta_C) = \frac{1}{2} - \frac{1}{\Gamma^2}\int_0^a \frac{\Gamma_0^2(r)}{r}\, dr. \tag{14.2.9}$$

For uniform vorticity within the core, (14.2.9) reduces to $\delta_C = 0.5\exp(\frac{1}{4}) \cong 0.642$.

An alternative procedure for regularization of the singularity in the filament Biot-Savart equation, proposed by Rosenhead (1930), involves maintaining the integration over the full curve C in (14.1.4) but adding a small constant to the denominator, giving

$$\mathbf{u}(\mathbf{x}, t) = -\frac{\Gamma}{4\pi}\int_C \frac{(\mathbf{x} - \mathbf{x}') \times \boldsymbol{\lambda}_3}{[|\mathbf{x} - \mathbf{x}'|^2 + \mu^2]^{3/2}}\, ds'. \tag{14.2.10}$$

The constant μ is assumed to be proportional to the core radius a, such that $\mu = \delta_R a$ where δ_R is called the Rosenhead cut-off constant. This regularization procedure is more convenient for numerical computation since the integration is performed over the entire curve C. The Rosenhead regularization procedure yields equivalent results to the cut-off procedure described previously if δ_R is set as

$$\delta_R = \frac{\delta_C}{e}. \tag{14.2.11}$$

Extension of the cut-off theory for uniform axial flow within the vortex core is given by Widnall and Bliss (1971) and Moore and Saffman (1972). The latter paper notes that the cut-off model may experience difficulties for large axial flow rates, for which case an alternative model given by Moore and Saffman (1972) is available.

The cut-off approximation is used extensively in the vortex dynamics literature and has generally been found to provide a good approximation for the bulk motion of vortex tubes provided that κa is small and that the vortex filaments do not come too close to each other. Even for cases with $\kappa a = O(1)$ the predictions of the cut-off model are typically at least qualitatively correct, although in such cases the small-scale vorticity deformation within the vortex core may become significant. One interesting example of use of the cut-off model is given by the motion of an elliptical vortex ring (Dhanak and De Bernardinis, 1981). As shown in Figure 14.3, the ring is initialized as an ellipse with aspect ratio $A > 1$ lying on a planar surface. For values of A close to unity, the ring performs a nearly periodic "dance-type" motion in time (which vortex dynamicists are fond of replicating with their fingers). In the first step, the ring deforms from a planar shape as the two sides with the largest curvature move forward faster than the sides with smaller curvature. (It is shown in Section 13.2 that the self-induced velocity of a vortex ring increases with increase in curvature.) This deformation introduces a curvature along the mean propagation direction of the ring, causing the sides with smaller curvature to propagate outward and the sides with higher curvature to propagate inward toward the ring center axis. At the end of this second step, the axes of the ring have switched. The third and fourth steps of the dance are essentially the same as the first two, but rotated by 90° due to the axis switching, such that at the final time the ring returns to a nearly planar elliptical shape, after which the motion starts again. For larger values of aspect ratio A, the ring motion becomes increasingly nonlinear and it does not return exactly to its initial state at the end of each sequence of the motion. For cases with A greater than about five, the middle sections of the ring collapse into each other. At this point the computations with the cut-off model must be stopped, since the motion induced

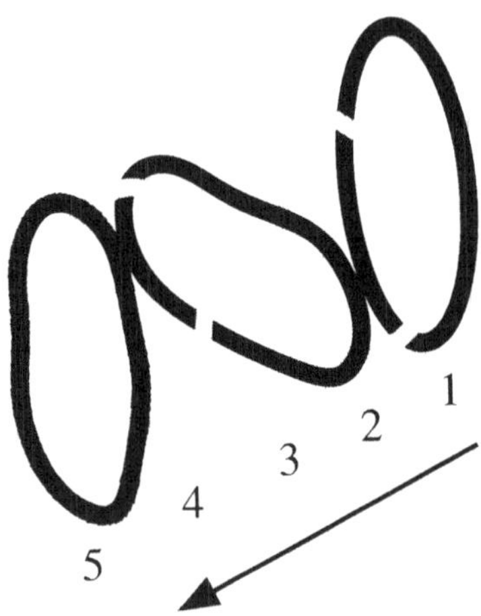

Figure 14.3 Periodic motion of an elliptical vortex ring with initial aspect ratio $A = 2$, computed for a single vortex filament using the cut-off model. The ring is shown in perspective view at five times during one oscillation period, with the initial, half-period and full-period views drawn in black and the quarter-period and three-quarter-period views drawn in gray.

by one part of the vortex yields an infinite propagation velocity on another part of the vortex at the touching point, although one might speculate that the ring will pinch-off into two rings at a later time.

14.3 LOCAL-INDUCTION APPROXIMATION

The local-induction approximation assumes that the contribution of the intermediate region II of the vortex tube in Figure 14.2 so dominates that of the far-field region III that the far field can be neglected. The integration in the Biot-Savart equation (14.1.4) is cut off both over a small interval of length $\delta = \delta_C a$ on either side of the singularity at $\mathbf{x}$ (the usual cut-off approximation) and at distances greater than some length ℓ from the singularity. So long as $\ell \gg a$, it does not matter exactly how the length ℓ is chosen, but for dimensional consistency in the current development we set $\ell = c/\kappa$, where c is an $O(1)$ constant. The Biot-Savart integeral (14.1.4) can be approximated using the expansion (14.2.4) as

$$\mathbf{u}(\mathbf{x}, t) = \frac{\Gamma\kappa}{4\pi}\left[\int_{\delta}^{\ell}\frac{d\zeta}{\zeta} + O(1)\right]\boldsymbol{\lambda}_2. \tag{14.3.1}$$

Performing the integration in (14.3.1) yields the local-induction approximation for the vortex velocity as

$$\mathbf{u}(\mathbf{x}, t) = \frac{\Gamma\kappa}{4\pi}\left[\ln\left(\frac{1}{\kappa a}\right) + O(1)\right]\boldsymbol{\lambda}_2. \tag{14.3.2}$$

The local-induction approximation assumes that the product κa is so small that $\ln(1/\kappa a) \gg 1$, such that the $O(1)$ term in brackets in (14.3.2) can be neglected in comparison to the logarithmic term. It is rare in practical vortex flows that κa is sufficiently small for this approximation to be valid. For instance, for a vortex with core radius of 1 mm, the radius of curvature of the vortex centerline measures 22 m when $\ln(1/\kappa a) = 10$ and it exceeds 10^{43} m when $\ln(1/\kappa a) = 100$! In more typical applications, where the radius of curvature is perhaps an order of magnitude larger than the core radius, the logarithmic term $\ln(1/\kappa a)$ in (14.3.2) has values in the range 2–3. The local-induction approximation can nevertheless be used to obtain a qualitative feel for the vortex behavior for some types of phenomena, although one must always be mindful of the consequences of neglect of the far-field contributions, which can in some situations lead to major qualitative changes in the vortex dynamics. For those cases where it is appropriate, the local-induction equation (14.3.2) has the advantage that it yields a differential equation for filament position, which can be written as

$$\frac{d\mathbf{r}}{dt}(\xi, t) = \frac{\Gamma\kappa}{4\pi}\ln\left(\frac{1}{\kappa a}\right)\boldsymbol{\lambda}_2, \tag{14.3.3}$$

where the curvature and binormal vector can be expressed in terms of $\mathbf{r}(\xi, t)$ using (14.1.1)–(14.1.3). Since the logarithmic term in (14.3.3) varies slowly, it is common to make the further approximation that this term is constant, so that the filament motion is governed by an equation of the form

$$\frac{d\mathbf{r}}{dt}(\xi, t) \cong A\kappa \boldsymbol{\lambda}_2, \tag{14.3.4}$$

where A is a constant. The result (14.3.4) is much more amenable to analytical solution than the integro-differential equation given by the full filament Biot-Savart equation with only local cut-off.

Several properties of vortex motion under the local-induction approximation are derived by Arms and Hama (1965). The first of these properties concerns the length of the vortex arc lying between points identified by the Lagrangian variables ξ_1 and ξ_2, defined by

$$\ell(\xi_1, \xi_2) = \int_{\xi_1}^{\xi_2} \left| \frac{\partial \mathbf{r}}{\partial \xi} \right| d\xi. \tag{14.3.5}$$

Differentiating the integrand of the length integral with respect to time and using the definition (14.1.1) of the tangent vector gives

$$\frac{d}{dt}\left| \frac{\partial \mathbf{r}}{\partial \xi} \right| = \boldsymbol{\lambda}_3 \cdot \frac{\partial}{\partial \xi}\left(\frac{d\mathbf{r}}{dt} \right). \tag{14.3.6}$$

Substituting the local-induction approximation in the form (14.3.4) into (14.3.6) and using the Serret-Frenet equation (14.1.3) yields

$$\frac{d}{dt}\left| \frac{\partial \mathbf{r}}{\partial \xi} \right| = \boldsymbol{\lambda}_3 \cdot \left[\frac{\partial (\kappa A)}{\partial \xi} \boldsymbol{\lambda}_2 - \tau \kappa A \left| \frac{\partial \mathbf{r}}{\partial \xi} \right| \boldsymbol{\lambda}_1 \right] = 0. \tag{14.3.7}$$

The result (14.3.7) implies that $|\partial \mathbf{r}/\partial \xi| = \text{const}$, so from (14.3.5) we conclude that the self-induced velocity of a vortex filament whose motion is governed by the local-induction approximation preserves the arc length between any two points on the filament.

A second property of the local-induction approximation is that it satisfies the requirement of invariance of the vortex impulse **P** and angular impulse **L**. This property may be shown by direct computation, where the impulse **P** is defined for a filament by

$$\mathbf{P} \equiv \frac{\Gamma}{2} \int_C \mathbf{r} \times \boldsymbol{\lambda}_3 \, ds. \tag{14.3.8}$$

For convenience, we consider equivalently the impulse per unit strength $\mathbf{p} \equiv \mathbf{P}/\Gamma$. It is noted that the vector **p** has a geometrical interpretation. The magnitude of **p** is equal to the maximum projected area traced by a line extending from the origin O to any point ξ on the vortex filament, as ξ is traversed along the length of C (Figure 14.4). The direction of **p** is orthogonal to the plane in which this projected

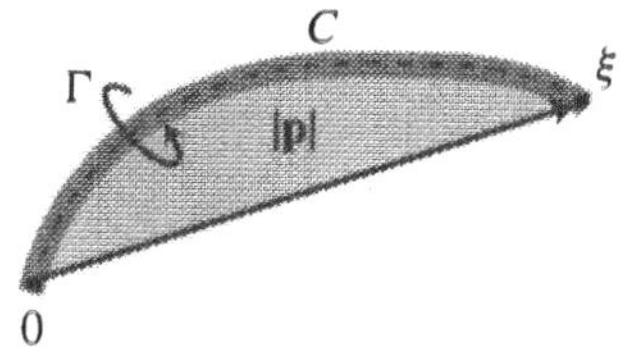

Figure 14.4 Area traced out by a line segment joining the origin O to a point ξ on the vortex filament C.

area is a maximum. Using the result that the length of C is constant, so that the arc length s is equivalent to the Lagrangian variable ξ, differentiation of $\mathbf{p}$ with respect to time yields

$$\frac{d\mathbf{p}}{dt} = \frac{1}{2}\int_C \left[\frac{d\mathbf{r}}{dt} \times \boldsymbol{\lambda}_3 + \mathbf{r} \times \frac{\partial}{\partial s}\left(\frac{d\mathbf{r}}{dt}\right)\right] ds. \tag{14.3.9}$$

Performing integration by parts for the second term and using the definition of $\boldsymbol{\lambda}_3$, we obtain

$$\frac{d\mathbf{p}}{dt} = \int_C \frac{d\mathbf{r}}{dt} \times \boldsymbol{\lambda}_3 \, ds. \tag{14.3.10}$$

Substituting the governing equation (14.3.4) of the local-induction approximation into (14.3.10) gives

$$\frac{d\mathbf{p}}{dt} = \int_C A\kappa \boldsymbol{\lambda}_2 \times \boldsymbol{\lambda}_3 \, ds. \tag{14.3.11}$$

Orthogonality of the unit vectors requires that $\boldsymbol{\lambda}_1 = \boldsymbol{\lambda}_2 \times \boldsymbol{\lambda}_3$, which together with the Serret-Frenet equation (14.1.3) allows us to write (14.3.11) as

$$\frac{d\mathbf{p}}{dt} = \int_C A \frac{\partial \boldsymbol{\lambda}_3}{\partial s} \, ds. \tag{14.3.12}$$

Taking A to be a constant and performing the integration in (14.3.12) over the closed loop C gives the result $d\mathbf{p}/dt = 0$, thus proving invariance of the impulse. Invariance of angular impulse can be proved in a similar manner.

Hasimoto (1972) demonstrated by use of some clever transformations that the governing equations of the local-induction approximation can be reduced to the cubic Schrödinger equation. In particular, defining a scaled time $T \equiv t/A$ using the constant A in (14.3.4) and introducing a new complex-valued variable ψ, defined in terms of the filament curvature κ and torsion τ as

$$\psi(s, T) \equiv \kappa \exp\left(i \int_0^s \tau \, ds\right), \tag{14.3.13}$$

the governing equation of the vortex can be reduced to the nonlinear equation

$$\frac{\partial \psi}{\partial T} = i\left[\frac{\partial^2 \psi}{\partial s^2} + \frac{1}{2}(|\psi|^2 + B)\psi\right], \tag{14.3.14}$$

where $B = B(T)$.

Equation (14.3.14), referred to as the cubic or nonlinear Schrödinger equation, is known to exhibit solitary-wave-type solutions (Infeld and Rowlands, 1990). Defining a variable $\eta \equiv s - cT$ that translates with the wave, where c is a constant wave translation velocity along the vortex axis, and assuming that the wave is of permanent form, such that κ and τ can be expressed as functions of η alone, substitution of (14.3.13) into (14.3.14) and separation into real and imaginary parts give

$$-c\kappa[\tau(\eta) - \tau(-cT)] = \kappa'' - \kappa\tau^2 + \tfrac{1}{2}(\kappa^2 + B)\kappa, \tag{14.3.15a}$$

$$c\kappa' = 2\kappa'\tau + \kappa\tau'. \tag{14.3.15b}$$

The prime in (14.3.15) denotes differentiation with respect to η. Integrating (14.3.15b) gives

$$(c - 2\tau)\kappa^2 = 0. \tag{14.3.16}$$

If the vortex curvature is nonzero, it follows that the torsion must be constant and equal to half of the wave propagation speed, or

$$\tau = \tau_0 = c/2. \tag{14.3.17}$$

Integrating (14.3.15a) gives a solution for the vortex curvature as

$$\kappa = 2\nu\,\mathrm{sech}(\nu\eta), \tag{14.3.18}$$

where we choose $B = 2(\tau_0^2 - \nu^2)$. The parameter ν can be eliminated by rescaling the arc length as $\eta^* \equiv \nu\eta$ and similarly rescaling the curvature and torsion as $\kappa^* \equiv \kappa/\nu$ and $\tau^* = \tau/\nu$. Variation of τ^* yields qualitatively different solitary waves, as shown in Figure 14.5.

14.4 BENDING WAVES ON A VORTEX TUBE

In this section, we develop solutions for helical bending waves of amplitude D and axial wavenumber k on a vortex tube of circular cross section with core radius a. Two different approaches are employed. For the long-wave case, in which $ka \ll 1$, the cut-off method developed in Section 14.2 can be used to solve for the wave propagation. For the special case of very long waves ($ka \lll 1$), the local-induction approximation can be used to further simplify this solution. Alternatively, for small-

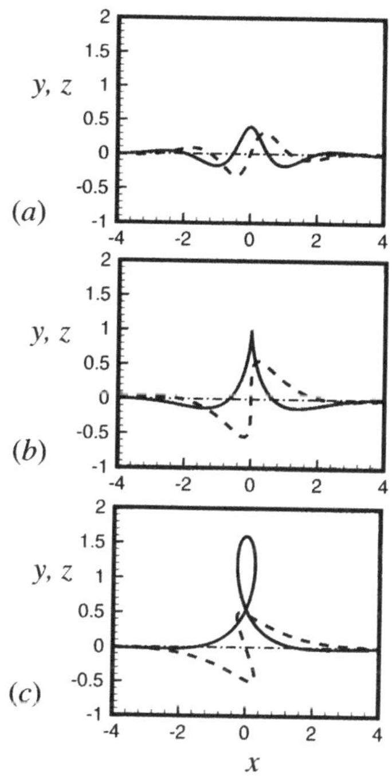

Figure 14.5 Solitary helical waves of a vortex filament, showing projections of the wave form in the x-y plane (solid line) and the x-z plane (dashed line) for cases with (a) $\tau^* = 2.0$, (b) $\tau^* = 1.0$, and (c) $\tau^* = 0.5$.

amplitude waves (with $D/a \ll 1$), a solution by Kelvin (1880) for a Rankine vortex with small perturbations gives the wave frequency for all values of ka. Results of these approaches are compared for the case of long helical waves of small amplitude.

14.4.1 Helical Wave Solution from Cut-off Model

The motion of the vortex tube under the cut-off model is obtained from (14.1.14) as

$$\frac{d\mathbf{r}}{dt}(s,t) = -\frac{\Gamma}{4\pi}\int_{C[\delta]} \frac{\mathbf{r}(s,t)-\mathbf{r}(s',t)}{|\mathbf{r}(s,t)-\mathbf{r}(s',t)|^3} \times \boldsymbol{\lambda}_3(s',t)\,ds', \tag{14.4.1}$$

where s is the arc length along the centerline curve C. The position vector to C is given in Cartesian coordinates by

$$\mathbf{r} = X(\zeta,t)\mathbf{e}_x + Y(\zeta,t)\mathbf{e}_y + \zeta\mathbf{e}_z, \tag{14.4.2}$$

where ζ denotes distance along the central axis of the helix (coinciding with the z-coordinate axis) and X and Y are perturbations of C in the lateral directions. A directed element of length along C can be written as

$$\boldsymbol{\lambda}_3\, ds = \left(\frac{\partial X}{\partial \zeta} \mathbf{e}_x + \frac{\partial Y}{\partial \zeta} \mathbf{e}_y + \mathbf{e}_z \right) d\zeta. \tag{14.4.3}$$

While the use of the cut-off model requires only that $ka \ll 1$, the helical vortex solution obtained in the current section makes the further approximation that the helix amplitude is much smaller than the wave length, or $kD \ll 1$, in order that the filament Biot-Savart equation (14.4.1) may be linearized. Substituting (14.4.2) and (14.4.3) into (14.4.1) and linearizing gives the equation for tube motion as

$$\frac{d\mathbf{r}}{dt} = -\frac{\Gamma}{4\pi} \int_{C[\delta]} \frac{\left[Y - Y' - (\zeta - \zeta') \left(\frac{\partial Y'}{\partial \zeta'} \right) \right] \mathbf{e}_x - \left[X - X' - (\zeta - \zeta') \left(\frac{\partial X'}{\partial \zeta'} \right) \right] \mathbf{e}_y}{|\zeta - \zeta'|^3} d\zeta'. \tag{14.4.4}$$

We now consider solutions for the perturbation quantities X and Y in the form of helical waves, so that

$$X(\zeta, t) = \hat{X} e^{i(\omega t - k\zeta)}, \qquad Y(\zeta, t) = \hat{Y} e^{i(\omega t - k\zeta)}, \tag{14.4.5}$$

where $\hat{X}$ and $\hat{Y}$ are complex constants and ω is the wave frequency. The equation of motion (14.4.4) reduces to a system of algebraic equations for $\hat{X}$ and $\hat{Y}$ of the form

$$i\omega \hat{X} = \frac{\Gamma k^2}{2\pi} I(k\delta) \hat{Y}, \qquad i\omega \hat{Y} = -\frac{\Gamma k^2}{2\pi} I(k\delta) \hat{X}, \tag{14.4.6}$$

where $I(k\delta)$ is the self-induction integral, defined by

$$I(k\delta) = \int_{\delta k}^{\infty} \frac{\cos\tau + \tau \sin\tau - 1}{\tau^3} d\tau, \tag{14.4.7}$$

and $\delta = \delta_C a$ is the cut-off distance. This expression for the self-induction integral can be derived by substituting the perturbation (14.4.5) into (14.4.4), dividing by $\exp[i(\omega t - k\zeta)]$, and taking the real part of the remaining integral on the right-hand side. The dimensionless wave frequency $f \equiv \pi a^2 \omega / \Gamma$ is obtained from the solvability condition of the eigenvalue problem (14.4.6) as

$$f = \frac{(ka)^2}{2} I(\delta_C ka). \tag{14.4.8}$$

The self-induction integral can be expressed in terms of the cosine-integral $Ci(z) \equiv -\int_z^{\infty} (\cos\tau)/\tau\, d\tau$ as

$$I(z) = \frac{1}{2} \left[\frac{\cos z - 1}{z^2} + \frac{\sin z}{z} - Ci(z) \right]. \tag{14.4.9}$$

The above expression can be further simplified based on the fact that $ka \ll 1$ is required for validity of the cut-off approximation. Using the series expansion for the cosine integral with small arguments of the form

$$Ci(z) = \gamma + \ln z + O(z^2), \tag{14.4.10}$$

where $\gamma = 0.5772\ldots$ is the Euler number, together with the standard Taylor expansions for the sine and cosine functions, the leading-order solution for the dimensionless wave frequency is obtained as

$$f - \frac{(ka)^2}{4}\left[\ln\left(\frac{1}{\delta_C ka}\right) + \frac{1}{2} - \gamma\right]. \tag{14.4.11}$$

The solution of this problem from the local-induction approximation can be obtained by neglecting all terms of $O(1)$ in the solution (14.4.11) in favor of the term proportional to $\ln(1/ka)$, which gives the very long wavelength limit for the dimensionless wave frequency as

$$f \sim \frac{(ka)^2}{4}\ln\left(\frac{1}{ka}\right). \tag{14.4.12}$$

14.4.2 Kelvin Waves

An exact solution for small-amplitude vortex bending waves ($D/a \ll 1$) is given by Kelvin (1880) for a Rankine vortex with helical wave velocity perturbations of the form

$$\begin{aligned} u(R,\alpha,z,t) &= \hat{u}(R)e^{i(\omega t - m\alpha - kz)}, \\ v(R,\alpha,z,t) &= V(R) + \hat{v}(R)e^{i(\omega t - m\alpha - kz)}, \\ w(R,\alpha,z,t) &= \hat{w}(R)e^{i(\omega t - m\alpha - kz)}, \end{aligned} \tag{14.4.13}$$

where (R, α, z) are cylindrical polar coordinates, $V(R)$ is the azimuthal velocity component of the Rankine vortex, given in (13.6.1), and $(\hat{u}, \hat{v}, \hat{w})$ are the perturbation amplitudes. The wave frequency ω must be determined as a function of the azimuthal and axial wavenumbers, m and k, respectively. The case with $m = 0$ corresponds to an axisymmetric wave of variable core area, which is discussed in Section 13.6. Cases with $m \geq 2$ correspond to waves of deformation of core shape, perhaps the most interesting of which is the elliptical deformation for $m = 2$. The current section is concerned with the case of bending waves, corresponding to $m = 1$, in which the core remains circular (to leading order) but the vortex centerline is displaced into a helical shape. The analysis for small-amplitude helical waves follows a similar approach to that described for axisymmetric waves in Section 13.6, with the difference that the full three-dimensional vorticity transport equations must be employed rather than the axisymmetric equations. Details can be found in the original paper of Kelvin (1880).

Letting $\Omega \equiv \Gamma/2\pi a^2$, where Γ is the vortex strength, the linear analysis yields a dispersion relationship for waves of the form (14.4.13) on a Rankine vortex as

$$\frac{(\omega - m\Omega)^2}{4\Omega^2 - g^2}\left[\frac{\nu a J'_{|m|}(\nu a)}{J_{|m|}(\nu a)} - \frac{2\Omega m}{g}\right] = -a|k|\frac{K'_{|m|}(|k|a)}{K_{|m|}(|k|a)}, \qquad (14.4.14)$$

where $\nu^2 \equiv k^2(4\Omega^2 - g^2)/g^2$, $g \equiv \omega - m\Omega$, and $J_{|m|}(\cdot)$ and $K_{|m|}(\cdot)$ are the Bessel function of the first kind and the modified Bessel function of the second kind of order $|m|$. This analysis was extended by Krishnamoorthy (1966) for the case of a Rankine vortex with a uniform axial velocity W within the core, for which the dispersion relationship has the same form as that given above with the parameter g replaced by $g = \omega - m\Omega - kW$.

Equation (14.4.14) can be solved for the dimensionless frequency $f \equiv \pi a^2\omega/\Gamma = \omega/2\Omega$ of bending waves ($m = 1$) for any given value of the product ka. This expression yields an infinite number of solutions for f corresponding to the different oscillations of the Bessel function. Each of the solutions f_n is called a *mode* and $n \geq 1$ denotes the mode number. The dimensionless frequency may have either a positive or negative sign, depending on the mode number and the value of ka. The wave propagates both along the vortex axis with phase velocity $c = \omega/k$ and in the azimuthal direction around the vortex core with angular phase velocity $C = \pm\omega$. The angular velocity of a bending wave relative to the swirling fluid particles within the Rankine vortex core is $\omega - \Omega$, which is equal to the parameter g defined above for the case $W = 0$. Waves with $g > 0$ are called *cograde*, since to a rotating observer the waves appear to be propagating in the direction of vortex rotation. Conversely, waves with $g < 0$ are called *retrograde*. A plot of the dimensionless frequencies for

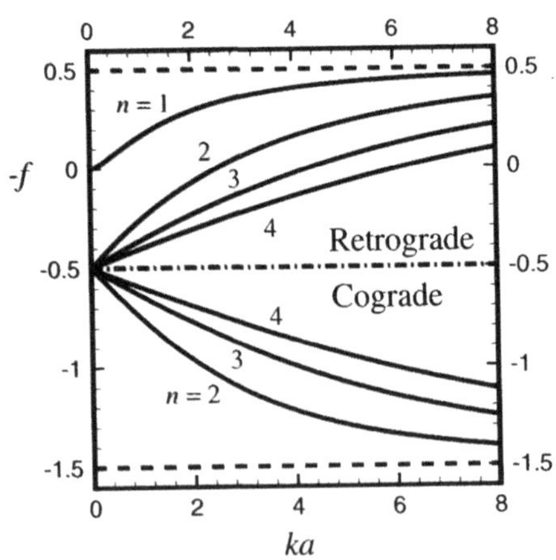

Figure 14.6 Dimensionless frequency $f \equiv \pi a^2\omega/\Gamma$ for small-amplitude bending waves on a Rankine vortex for the first four modes ($n = 1, \ldots, 4$), as predicted from the Kelvin dispersion relationship. Short-wave asymptotic limits are indicated by dashed lines.

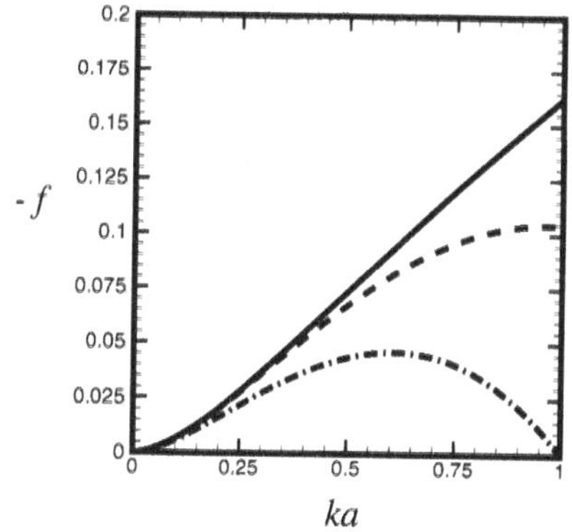

Figure 14.7 Comparison of the predictions for dimensionless frequency of first-mode bending waves from the small-amplitude Kelvin theory (solid line), the vortex filament model with cut-off (dashed line), and the local-induction approximation (dashed-dotted line).

cograde and retrograde bending waves of the first four modes as a function of ka is given in Figure 14.6. For the first mode, only the retrograde solution exists, but both cograde and retrograde solutions exist for all higher modes. All solutions asymptote to $|f - \frac{1}{2}| = 1$ as $ka \to \infty$, which is indicated by dashed lines in Figure 14.6.

For the case of small-amplitude bending waves with long wavelength, such that both $D/a \ll 1$ and $ka \ll 1$, the condition $kD \ll 1$ required for linearization of the Biot-Savart equation is also satisfied. For this case, both the Kelvin wave theory and the cut-off model prediction (14.4.8) are valid. A comparison of the frequency predictions from the Kelvin wave solution (14.4.14) for retrograde waves of the first mode, the linearized cut-off model solution (14.4.8), and the solution (14.4.12) from the local-induction approximation are given in Figure 14.7 as a function of ka. All three solutions collapse to the same curve in the limit $ka \to 0$. The local-induction approximation is observed to deviate from the Kelvin wave solutions quite early, with differences of more than 10% for $ka > 0.035$. The cut-off model solution, on the other hand, remains valid for waves of moderate length, and does not exhibit deviations of more than 10% of the Kelvin wave solution until $ka > 0.5$.

BIBLIOGRAPHY

Arms, R.J. and F.R. Hama (1965). "Localized-induction concept on a curved vortex and motion of an elliptic vortex ring," *Physics of Fluids* **8**(4), 553–559.

Crow, S.C. (1970). "Stability theory for a pair of trailing vortices," *AIAA Journal* **8**, 2172–2179.

Dhanak, M.R., and B. de Bernardinis (1981). "The evolution of an elliptic vortex ring," *Journal of Fluid Mechanics* **109**, 189–216.

Fukumoto, Y., and T. Miyazaki (1991). "Three-dimensional distortions of a vortex filament with axial velocity," *Journal of Fluid Mechanics* **222**, 369–416.

Hasimoto, H. (1972). "A soliton on a vortex filament," *Journal of Fluid Mechanics* **51**, 477–485.

Infeld, E. and G. Rowlands (1990). *Nonlinear Waves, Solitons and Chaos*, Cambridge University Press, Cambridge.

Kelvin, Lord (1880). "Vibrations of a columnar vortex," *Philosophical Magazine* **10**, 155–168 (reprinted *Mathematical and Physical Papers*, Vol. 4, Cambridge University Press, Cambridge, 1910).

Krishnamoorthy, V. (1966). "Vortex breakdown and measurements of pressure fluctuation over slender wings," Ph.D. dissertation, University of Southampton, Southampton.

Lundgren, T.S. and W.T. Ashurst (1989). "Area-varying waves on curved vortex tubes with application to vortex breakdown," *Journal of Fluid Mechanics* **200**, 283–307.

Marshall, J.S. (1991). "A general theory of curved vortices with circular cross-section and variable core area," *Journal of Fluid Mechanics* **229**, 311–338.

Marshall, J.S. and J.R. Grant (1994). "Evolution and break-up of vortex rings in straining and shearing flows," *Journal of Fluid Mechanics* **273**, 285–312.

Moore, D.W. and P.G. Saffman (1972). "The motion of a vortex filament with axial flow," *Philosophical Transactions of the Royal Society of London A* **272**, 403–429.

Rosenhead, L. (1930). "The speed of vorticity in the wake behind a cylinder," *Proceedings of the Royal Society of London A* **127**, 590–612.

Widnall, S.E. and D.B. Bliss (1971). "Slender body analysis of the motion and stability of a vortex filament containing an axial flow," *Journal of Fluid Mechanics* **50**, 335–353.

PROBLEMS

1. Use the orthogonality relationship $\boldsymbol{\lambda}_1 = \boldsymbol{\lambda}_2 \times \boldsymbol{\lambda}_3$ together with the first two Serret-Frenet equations $(14.1.3)_{1,2}$ to derive the third Serret-Frenet equation $(14.1.3)_3$.
2. Substitute the periodic form (14.4.5) for the vortex perturbations X and Y of a helical wave into the linearized filament Biot-Savart equation (14.4.4) to obtain expressions for the perturbation amplitudes $\hat{X}$ and $\hat{Y}$.
3. Plot the ratio of the dimensionless frequency f for axisymmetric waves and bending waves on a columnar Rankine vortex with core radius a as a function of dimensionless wavenumber ka for $ka < 0.5$. Use the first-mode Kelvin wave solution for axisymmetric waves and the cut-off approximation for bending waves. Comment on the relative speed of axisymmetric waves and bending waves on a vortex core.
4. Compute the propagation speed for vortex rings for the cases listed in the table below. For each case consider (a) uniform vorticity profile, (b) Gaussian vorticity profile [proportional to $\exp(-r^2/a^2)$], and (c) local induction approximation. Examine your results and discuss the relative importance of vorticity profile, ring and core radii, and nonlocal vorticity in determining the self-induced velocity of vortex tubes.

Case	Ring Radius, b (cm)	Core Radius, a (cm)	Strength, Γ (cm²/s)
A	5	1	20
B	10	1	20
C	10	2	20

COMPUTATIONAL PROJECTS

Write a computer program for evolution of a vortex ring using the Rosenhead cut-off model and uniform vorticity profile within the core. Test the code by evaluating the self-induced propagation velocity of a circular ring with ratio of ring radius to core radius of $b/a = 4$, 10. Compare your results to the analytical solution of Kelvin.

1. Apply the vortex filament code to compute the motion of an initially elliptical vortex ring. The ring aspect ratio A is defined as the ratio of the semimajor to semiminor axes of the ellipse. Examine the motion of the ellipse for cases with $A = 1.5, 3, 6$. Compare your results with those obtained by Dhanak and de Bernardinis (1981).
2. Apply the vortex filament code to examine evolution of an initially circular vortex ring immersed in a plane straining flow. Both the initial vortex ring and the nonzero velocity components of the straining flow are in the x-y plane, as shown in Figure 14.8. Examine cases with $2\pi a^2 c/\Gamma = 0.01, 0.03, 0.05$, where c is the strain rate, a is the vortex core radius, and Γ is the vortex circulation. Set the initial ring radius b such that $b/a = 10$. Compare your results to those of Marshall and Grant (1994).

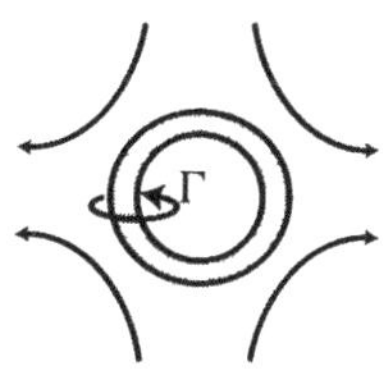

Figure 14.8 Circular vortex ring placed in a plane straining flow.

CHAPTER 15

INTERFACIAL WAVE MOTION

Waves at the interface between two immiscible fluid masses are observed at the surface of oceans and lakes subject to wind forcing or on a water stream flowing over an uneven surface. In cases where one fluid has much greater density than the other fluid, such as water and air, the interfacial waves can often be idealized as *free-surface waves*, where the inertial force and shear stress imposed by the lighter fluid are neglected. Waves occurring between two fluids with similar density, such as water and oil, are called *internal waves*, and in such cases it is necessary to account for the inertia of both fluid masses. In oceans or large lakes, surface waves can exert large forces and moments on marine structures, such as ships and coastal structures. Currents generated by wave motion are responsible for most coastal sediment transport. Internal waves are important for regulating turbulence mixing in the upper layers of oceans and lakes. The equations of motion for interfacial waves can also be used to develop the theory for a number of fluid flow instabilities, which are discussed in Chapter 16.

15.1 INTERNAL WAVES IN LAYERED MEDIA

In this section, equations are derived that govern the motion of internal waves between two immiscible fluids of density ρ_1 and ρ_2, which in the equilibrium state occupy layers with thickness h_1 and h_2, respectively, as shown in Figure 15.1. There exists a current in each layer of strength U_1 and U_2. For simplicity, the flow is assumed to be two dimensional such that the interface position is given by $y = \eta(x, t)$, where $\eta = 0$ in the equilibrium state. The flow in each fluid region is assumed to be irrotational such that the potential functions, ϕ_1 and ϕ_2, satisfy the Laplace equation

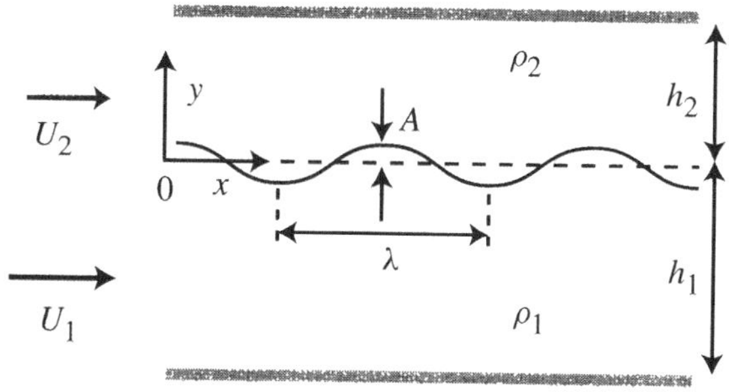

Figure 15.1 Schematic showing different parameters used to describe a two-dimensional internal wave system.

$$\nabla^2\phi_1 = 0, \qquad \nabla^2\phi_2 = 0. \tag{15.1.1}$$

At the top and bottom boundaries, the potential functions satisfy the no-penetration condition

$$\left.\frac{\partial \phi_1}{\partial y}\right|_{y=-h_1} = 0, \qquad \left.\frac{\partial \phi_2}{\partial y}\right|_{y=h_2} = 0. \tag{15.1.2}$$

The interface $y = \eta(x, t)$ is assumed to be a material surface, consisting of the same set of fluid particles for all time. The interface can be defined by $f(x, y, t) \equiv y - \eta(x, t) = 0$, so the requirement that the interface is a material surface can be expressed by setting the material derivative of $f(x, y, t)$ equal to zero, or

$$\frac{Df}{Dt} = v - \frac{D\eta}{Dt} = 0 \qquad \text{on } y = \eta. \tag{15.1.3}$$

Expanding the material derivative of η and applying the condition (15.1.3) to both sides of the fluid interface yield the *kinematic interface matching condition*

$$\frac{\partial \eta}{\partial t} + u_1 \frac{\partial \eta}{\partial x} = v_1, \qquad \frac{\partial \eta}{\partial t} + u_2 \frac{\partial \eta}{\partial x} = v_2. \tag{15.1.4}$$

Using the definition $\mathbf{n} = \nabla f / |\nabla f|$ of the interface unit normal, the matching condition (15.1.3) can alternatively be expressed in terms of the normal velocity component to the fluid interface as

$$\mathbf{u}_1 \cdot \mathbf{n} = \mathbf{u}_2 \cdot \mathbf{n} = \frac{-\partial f/\partial t}{|\nabla f|} = \frac{\partial \eta/\partial t}{\left[1 + (\partial \eta/\partial x)^2\right]^{1/2}}. \tag{15.1.5}$$

The momentum jump condition, as expressed by the Laplace formula (5.3.11), gives the pressure jump over the interface as

$$p_2 - p_1 = \frac{\gamma}{R}, \tag{15.1.6}$$

where the interface radius of curvature, R, is defined in two dimensions by

$$\frac{1}{R} = -\nabla \cdot \mathbf{n} = \frac{\partial^2 \eta / \partial x^2}{\left[1 + (\partial \eta / \partial x)^2\right]^{3/2}}. \tag{15.1.7}$$

The unsteady Bernoulli equation (8.2.2) can be used to relate the pressure jump to the velocity potential such that (15.1.6) yields the *dynamic interface matching condition*

$$\left(\rho_2 \frac{\partial \phi_2}{\partial t} - \rho_1 \frac{\partial \phi_1}{\partial t}\right) + \frac{1}{2}\left[\rho_2(u_2^2 + v_2^2) - \rho_1(u_1^2 + v_1^2)\right] + (\rho_2 - \rho_1) g \eta + \frac{\gamma}{R} = F(t), \tag{15.1.8}$$

where $F(t)$ is an arbitrary function of time.

15.2 LINEAR WAVE THEORY

The governing equations (15.1.1), together with the boundary conditions (15.1.2) and interface matching conditions (15.1.4) and (15.1.8), can be solved for the velocity potentials ϕ_1 and ϕ_2 and the interface displacement height η once appropriate initial conditions have been specified. Unfortunately, an analytic solution of these equations is generally not possible due to the nonlinearity implicit in the interface matching conditions (15.1.4) and (15.1.8). In the current section, an approximate solution of these equations is developed for the case in which the wave amplitude is sufficiently small that the interface matching conditions can be linearized about the equilibrium state. A numerical method for solution of the full nonlinear equations is described in Section 15.6.

In the linear wave theory, an arbitrary periodic waveform is expressed using a Fourier series as

$$\eta(x, t) = \sum_{n=1}^{N} A_n e^{i(k_n x - \sigma_n t)}, \tag{15.2.1}$$

where we consider only the real part of the function on the right-hand side. Each wave has wavelength λ_n, with associated wavenumber $k_n = 2\pi/\lambda_n$, amplitude A_n, and frequency σ_n. In general, A_n and σ_n are complex numbers, whereas the wavenumber k_n is defined to be real valued. The ratio of the imaginary to the real part of A_n determines the wave phase, and the magnitude of A_n is the wave height. If σ_n is real valued, the wave propagates in the x-direction with constant height. If σ_n is complex valued, the wave height either grows or decays exponentially in time. In practice, complex values of σ_n occur in complex-conjugate pairs, such that one member of the pair quickly decays and the other grows, resulting in flow instability. In the present development, we consider waves with a single wavelength λ and set the phase such that the associated amplitude A is real valued. In the linear theory, each wave evolves independently of the others, so the equations governing the wave motion can be derived for each wavelength individually.

The superposition principle for the linear theory can also be used to decompose the velocity potential into a part associated with the steady current and a part due to the wave field, or

$$\phi_n(x,t) = U_n x + \phi_n'(x,t), \qquad n = 1, 2. \tag{15.2.2}$$

Solutions of the Laplace equation for the potential in the upper and lower layers that satisfy the no-penetration boundary conditions at $y = h_2$ and $y = -h_1$ are given by

$$\phi_1'(x,y,t) = B_1 \cosh[k(y+h_1)]e^{i(kx-\sigma t)}, \tag{15.2.3a}$$

$$\phi_2'(x,y,t) = B_2 \cosh[k(y-h_2)]e^{i(kx-\sigma t)}, \tag{15.2.3b}$$

where B_1 and B_2 are complex constants.

The dimensionless wave amplitude is defined by $\varepsilon = Ak$, such that the small-slope assumption in the linear theory can be written as $0 < \varepsilon \ll 1$. In the linearized version of the interface matching conditions (15.1.4) and (15.1.8), all terms quadratic or higher order in ϕ_1, ϕ_2, or η or their cross products can be neglected. The linear kinematic interface matching condition thus becomes

$$\frac{\partial \eta}{\partial t} + U_1 \frac{\partial \eta}{\partial x} = \frac{\partial \phi_1'}{\partial y}, \qquad \frac{\partial \eta}{\partial t} + U_2 \frac{\partial \eta}{\partial x} = \frac{\partial \phi_2'}{\partial y}, \tag{15.2.4}$$

where the second-order terms proportional to $(\partial \phi_n'/\partial x)(\partial \eta/\partial x)$ are omitted. Choosing the arbitrary function $F(t)$ in (15.1.8) to eliminate the nonzero equilibrium terms, the linearized dynamic interface matching condition becomes

$$\left(\rho_2 \frac{\partial \phi_2'}{\partial t} - \rho_1 \frac{\partial \phi_1'}{\partial t}\right) + \left[\rho_2 U_2 \frac{\partial \phi_2'}{\partial x} - \rho_1 U_1 \frac{\partial \phi_1'}{\partial x}\right] + (\rho_2 - \rho_1) g\eta + \gamma \frac{\partial^2 \eta}{\partial x^2} = 0, \tag{15.2.5}$$

where the linearized interface curvature is obtained from (15.1.7) as $1/R \cong \partial^2 \eta/\partial x^2$. In the nonlinear matching conditions, the velocity potential and its derivatives are evaluated on the moving interface $y = \eta(x,t)$, which introduces another source of nonlinearity. However, in the linear theory a Taylor series expansion can be used to displace the matching conditions to the equilibrium interface surface $y = 0$. In particular, letting $f(x,y,t)$ denote the velocity potential or any of its derivatives, the first two terms of the Taylor series expansion about the equilibrium surface are

$$f(x,\eta,t) = f(x,0,t) + \eta \left.\frac{\partial f}{\partial y}\right|_{y=0} + \cdots. \tag{15.2.6}$$

From the solution (15.2.3), the second term on the right-hand side of (15.2.6) is $O(\varepsilon f)$ and can be neglected to leading order in ε. Consequently, the velocity potential and its derivatives in the linearized matching conditions (15.2.4)–(15.2.5) are evaluated on the equilibrium interface position.

Substituting (5.2.1) and (5.2.3) into the interface matching conditions (15.2.4) and (15.2.5) yields a system of linear equations, which can be expressed in matrix form as

$$\mathbf{M} \cdot \mathbf{a} = \mathbf{0}, \tag{15.2.7}$$

where $\mathbf{M}$ and $\mathbf{a}$ are defined by $\mathbf{a}^{\mathrm{T}} = (A, B_1, B_2)$ and

$$\mathbf{M} = \begin{pmatrix} i(kU_1 - \sigma) & -k\sinh(kh_1) & 0 \\ i(kU_2 - \sigma) & 0 & k\sinh(kh_2) \\ \gamma k^2 + (\rho_1 - \rho_2)g & -i\rho_1(\sigma - kU_1)\cosh(kh_1) & i\rho_2(\sigma - kU_2)\cosh(kh_2) \end{pmatrix}.$$

The solvability condition $\det(\mathbf{M}) = 0$ yields a quadratic equation for σ of the form

$$\frac{\rho_2(kU_2 - \sigma)^2}{k\tanh(kh_2)} + \frac{\rho_1(kU_1 - \sigma)^2}{k\tanh(kh_1)} - \gamma k^2 - (\rho_1 - \rho_2)g = 0. \tag{15.2.8}$$

The solution for σ yields the *dispersion equation* for the wave as

$$\sigma = -\frac{b}{2a} \pm \frac{1}{2}\left[\left(\frac{b}{a}\right)^2 - \frac{4c}{a}\right]^{1/2}, \tag{15.2.9}$$

where

$$\frac{b}{a} = -\frac{2k\rho_2 U_2/\tanh(kh_2) + 2k\rho_1 U_1/\tanh(kh_1)}{\rho_2/\tanh(kh_2) + \rho_1/\tanh(kh_1)},$$

$$\frac{c}{a} = \frac{k^2\rho_2 U_2^2/\tanh(kh_2) + k^2\rho_1 U_1^2/\tanh(kh_1) - \gamma k^3 - (\rho_1 - \rho_2)gk}{\rho_2/\tanh(kh_2) + \rho_1/\tanh(kh_1)}.$$

Two limiting cases of the above analysis are often used in applications. These limits are related to the ratio of layer depth to wavelength, as represented by the product kh_n $(n = 1, 2)$. For *shallow-water waves*, $kh_n \ll 1$ and we can make the approximation $\tanh(kh_n) \cong kh_n$. For *deep-water waves*, $kh_n \gg 1$ such that $\tanh(kh_n) \cong 1$. In cases where the inertia of the fluid in the upper layer is negligible compared to that in the lower layer (i.e., $\rho_1 \gg \rho_2$), we consider the free-surface wave limit $\rho_2/\rho_1 \to 0$. The role of surface tension is examined in Section 15.3, but for the present discussion it is convenient to assume that the waves are sufficiently long that the surface tension term is negligible.

It is instructive to record the resulting expressions for certain of these limiting cases. For instance, in the case of shallow-water free-surface waves, the expression (15.2.9) for wave frequency becomes

$$\sigma = k(U_1 \pm \sqrt{gh_1}). \tag{15.2.10}$$

Since σ is real valued, this solution has the form of a traveling wave of constant form. The *phase velocity* $c \equiv \sigma/k$ is the speed of propagation of the wave crests, which for shallow-water waves is given by

$$c = U_1 \pm \sqrt{gh_1}. \tag{15.2.11}$$

Right-running waves, corresponding to the positive sign in (15.2.11), propagate in the positive x-direction, whereas left-running waves, associated with the negative sign, propagate in the negative x-direction. For all linear, free-surface waves, a steady current advects the wave without influencing the wave propagation relative to an observer moving with the current. This passive role of the current is a consequence of the superposition principle of linear wave theory, although, as shown in Section 16.4, the current difference between the layers plays an important role in stability of internal waves.

One interesting property of the expression (15.2.11) is that the phase velocity of shallow-water waves does not depend on the wavelength. Waves with this property are said to be *nondispersive*, where the origin of this term is related to the observation that a packet of waves of different lengths propagate together in a nondispersive wave field, rather than spreading out (or dispersing) according to the wavelength. Other cases of wave propagation do not possess this property. For instance, for deep-water free-surface waves the wave frequency and phase velocity are given by

$$\sigma = kU_1 \pm \sqrt{gk}, \qquad c = U_1 \pm \sqrt{\frac{g}{k}}. \tag{15.2.12}$$

The phase speed for deep-water waves, relative to the mean current, is faster for long waves than for short waves. In such cases, the wave field is said to be *dispersive*.

15.3 CAPILLARY AND GRAVITY WAVES

The relative roles played by surface tension and gravitational force on a traveling wave field are examined in this section, where for simplicity we specialize to the case of deep-water free-surface waves with no mean current. The wave frequency and phase velocity are obtained from the general solution (15.2.9) as

$$\sigma = \left(gk + \frac{\gamma k^3}{\rho}\right)^{1/2}, \qquad c = \left(\frac{g}{k} + \frac{\gamma k}{\rho}\right)^{1/2}. \tag{15.3.1}$$

A plot of phase velocity versus wavenumber is given in Figure 15.2. For very long waves (called *gravity waves*), $k \to 0$ and the gravitational term dominates the surface tension term in (15.3.1). By contrast, for very short waves (called *capillary waves*), $k \to \infty$ and the surface tension term dominates the gravity term. An estimate of the wavelength for which the surface tension term becomes important is given by that corresponding to the minimum value of c, at which point both terms in the expression for c in (15.3.1) are of a similar order of magnitude. Setting $\partial(c^2)/\partial k = 0$, the minimum phase velocity $c_{\min}$ and its corresponding wavelength $\lambda_{\min}$ are obtained as

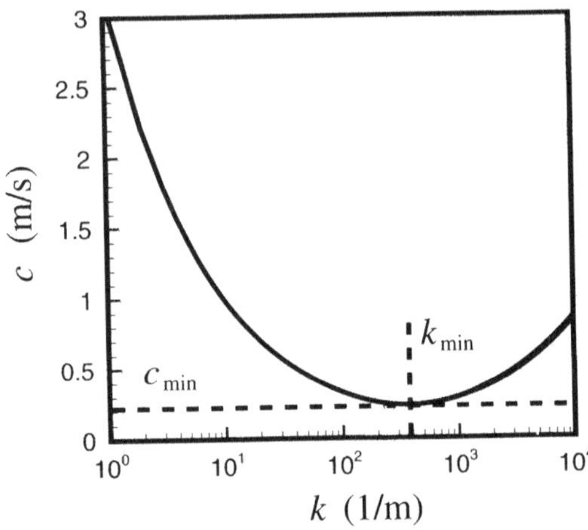

Figure 15.2 Variation of phase velocity with wavenumber for deep-water free-surface waves in the presence of surface tension.

$$c_{\min} = \left(\frac{4\gamma g}{\rho}\right)^{1/4}, \qquad \lambda_{\min} = 2\pi\left(\frac{\gamma}{\rho g}\right)^{1/2}. \tag{15.3.2}$$

For instance, for waves along a clean air-water interface at standard temperature and pressure, $\lambda_{\min}$ is about 1.7 cm, or about the size of a small ripple. Water waves with length much smaller than this value are dominated by surface tension, whereas surface tension is negligible for water waves with length much larger than this value.

15.4 PARTICLE DISPLACEMENT IN A WAVE FIELD

The solution for the fluid velocity field induced by the wave motion, relative to the current, can be obtained by differentiating the solution (15.2.3) for the velocity potential. In the bottom layer, we obtain

$$u_1'(x, y, t) = ikB_1 \cosh[k(y + h_1)]e^{i(kx-\sigma t)}, \tag{15.4.1a}$$

$$v_1'(x, y, t) = kB_1 \sinh[k(y + h_1)]e^{i(kx-\sigma t)}. \tag{15.4.1b}$$

The particle displacements ξ and ζ in the x- and y-directions, respectively, are obtained by integrating (15.4.1) over time as

$$\xi(x, y, t) = -\left(\frac{kB_1}{\sigma}\right) \cosh[k(y + h_1)]e^{i(kx-\sigma t)}, \tag{15.4.2a}$$

$$\zeta(x, y, t) = i\left(\frac{kB_1}{\sigma}\right) \sinh[k(y + h_1)]e^{i(kx-\sigma t)}. \tag{15.4.2b}$$

A solution for B_1 is obtained from the linearized kinematic interface matching condition (15.2.4) as

$$B_1 = iA\frac{kU_1 - \sigma}{k\sinh(kh_1)}. \tag{15.4.3}$$

Substituting (15.4.3) into (15.4.2) and taking the real part of the right-hand side yields

$$\xi(x, y, t) = A\left[\frac{kU_1 - \sigma}{\sigma\sinh(kh_1)}\right]\cosh[k(y + h_1)]\sin(kx - \sigma t), \tag{15.4.4a}$$

$$\zeta(x, y, t) = -A\left[\frac{kU_1 - \sigma}{\sigma\sinh(kh_1)}\right]\sinh[k(y + h_1)]\cos(kx - \sigma t). \tag{15.4.4b}$$

Solving for the sine and cosine terms in (15.4.4) and substituting into the trigonometric identity $\sin^2 z + \cos^2 z = 1$ give

$$\frac{\xi^2}{\cosh^2[k(y + h_1)]} + \frac{\zeta^2}{\sinh^2[k(y + h_1)]} = A^2\left[\frac{kU_1 - \sigma}{\sigma\sinh(kh_1)}\right]^2. \tag{15.4.5}$$

The particle paths, given by (15.4.5), follow the equation of an ellipse with aspect ratio $\tanh[k(y + h_1)]$. Near the bottom boundary, as $y \to -h_1$, the aspect ratio approaches zero, implying that the fluid particles move back and forth on a horizontal line. For shallow-water waves, $\cosh[k(y+h_1)] \cong 1$ and $\sinh[k(y+h_1)] \cong k(y+h_1)$, so that the horizontal particle displacement is approximately independent of depth and the vertical displacement varies linearly with depth. For deep-water waves, the aspect ratio approaches unity, implying that fluid particles move in a circle. With zero current, the radius of oscillation is $A\exp(ky)$, where $y < 0$ within layer 1. An illustration of the particle paths for shallow- and deep-water free-surface waves is given in Figure 15.3.

The equation for particle displacement in layer 2 is the same as (15.4.5) with h_1 replaced by $-h_2$. The aspect ratio of the elliptical particle path approaches zero at the top boundary at $y \to h_2$. For deep-water waves, the particles follow circular paths, which in the absence of current have radii $A\exp(-ky)$, where $y > 0$ in layer 2.

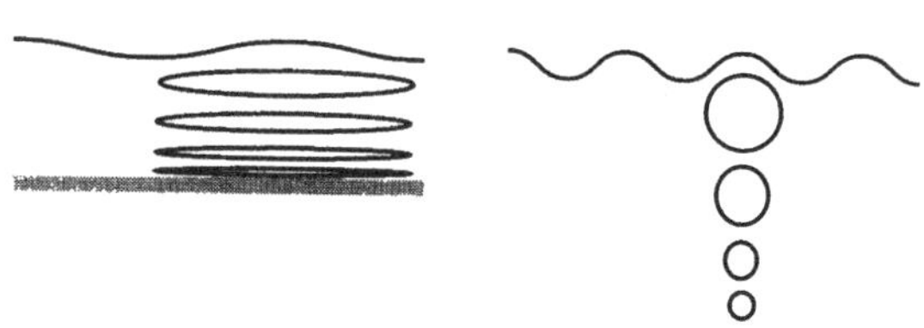

(*a*) (*b*)

Figure 15.3 Particle paths for (*a*) shallow and (*b*) deep water free-surface waves.

15.5 WAVE ENERGY AND GROUP VELOCITY

Although small-amplitude waves do not provide a net transport of mass, since the particle paths relative to an observer moving with the mean current are closed, they do transport energy. Understanding of wave energy transport is important in determining the modification of waves as they propagate onto a beach or the force required to generate wave fields, such as the wave wake of a ship. For simplicity, wave energy and its transport are examined in the current section for free-surface waves with no current, although the results can be readily generalized to internal wave fields. Subscript 1 used to denote layer 1 is dropped in this section.

15.5.1 Wave Energy

Wave energy is composed of two kinds—potential energy P associated with the displacement of the fluid relative to the gravitational field and kinetic energy T associated with motion of the fluid particles. The average potential and kinetic energy over one wavelength are given by

$$P = \frac{1}{\lambda}\int_x^{x+\lambda}\int_{-h}^{\eta} \rho g y \, dy \, dx, \qquad T = \frac{1}{\lambda}\int_x^{x+\lambda}\int_{-h}^{\eta} \frac{\rho}{2}(u^2 + v^2)\, dy \, dx. \quad (15.5.1)$$

With $\eta(x,t) = A\cos(kx - \sigma t)$ and performing the integration indicated above, the potential energy becomes

$$P_{\text{tot}} = \tfrac{1}{2}\rho g h^2 + \tfrac{1}{4}\rho g A^2. \quad (15.5.2)$$

The first term on the right-hand side of (15.5.2) represents the potential energy of the fluid in equilibrium, whereas the second term represents that associated with the wave field, or

$$P = \tfrac{1}{4}\rho g A^2. \quad (15.5.3)$$

The solutions for the velocity components associated with the wave field are

$$u(x,y,t) = \frac{\sigma A}{\sinh(kh)}\cosh[k(y+h)]\cos(kx - \sigma t), \quad (15.5.4a)$$

$$v(x,y,t) = \frac{\sigma A}{\sinh(kh)}\sinh[k(y+h)]\sin(kx - \sigma t). \quad (15.5.4b)$$

Substituting the velocity components into the expression for kinetic energy in (15.5.1) gives

$$T = \frac{\rho}{2\lambda}\left[\frac{\sigma A}{\sinh(kh)}\right]\int_x^{x+\lambda}\int_{-h}^{\eta}\Big\{\cosh^2[k(y+h)]\cos^2(kx - \sigma t) + \sinh^2[k(y+h)]\sin^2(kx - \sigma t)\Big\}\, dy\, dx. \quad (15.5.5)$$

In the linear theory, the free-surface displacement η can be replaced by the equilibrium surface $y = 0$ in the upper limit of integration over y in (15.5.5). The integrand of (15.5.5) can be replaced using standard trigonometric identities by $0.5\{\cosh[2k(y+h)] + \cos[2(kx - \sigma t)]\}$, which can be directly integrated to yield

$$T = \tfrac{1}{4}\rho g A^2. \tag{15.5.6}$$

The mean kinetic energy is equal to the mean potential energy of the wave field, and the total average wave energy E is given by

$$E = \tfrac{1}{2}\rho g A^2. \tag{15.5.7}$$

15.5.2 Energy Transport Rate

The net rate of energy transmission, Q, by the wave field is given by the average rate of working by pressure forces on a vertical surface extending from the bottom boundary to the free surface, or

$$Q = \frac{1}{\tau}\int_t^{t+\tau}\int_{-h}^{\eta} pu\,dy\,dt, \tag{15.5.8}$$

where τ is the wave period. The pressure field due to a free-surface wave can be computed from the linearized Bernoulli equation (8.2.2) as $p = -\rho g y - \rho(\partial\phi/\partial t)$, so that with use of the solution (15.2.3) for the velocity potential we obtain

$$p = -\rho g y + \rho g A\left[\frac{\cosh[k(y+h)]}{\cosh(kh)}\right]\cos(kx - \sigma t). \tag{15.5.9}$$

The hydrostatic term in (15.5.9) produces no net energy transport and can therefore be omitted from the integral (15.5.8). Substituting the second term in (15.5.9), referred to as the "dynamic pressure," and the expression (15.5.4a) for horizontal velocity component into (15.5.8) gives

$$Q = \frac{2\rho g\sigma A^2}{\sinh(2kh)}\left(\int_{-h}^{0}\cosh^2[k(y+h)]\,dy\right)\left(\frac{1}{\tau}\int_t^{t+\tau}\cos^2(kx - \sigma t)\,dt\right), \tag{15.5.10}$$

where the upper limit of integration in the y-direction is set to zero since we wish to retain only second-order terms in the interface displacement. Performing the indicated integrations in (15.5.10) yields the wave transport rate as

$$Q = Ec_g, \tag{15.5.11}$$

where E is the wave energy density, given by (15.5.7), and the constant c_g is given by

$$c_g = \frac{c}{2}\left[1 + \frac{2kh}{\sinh(2kh)}\right]. \tag{15.5.12}$$

The constant c_g represents the rate at which wave energy is transported, and for reasons that will be explained in the last part of this section, it is commonly known as the group velocity.

15.5.3 Group Velocity

To explain the notion of a *wave group*, we consider a case in which two periodic wave trains of the same amplitude propagate in the same direction with slightly different wavenumbers, $k_1 = k - \Delta k/2$ and $k_2 = k + \Delta k/2$, and with associated frequencies $\sigma_1 = \sigma - \Delta\sigma/2$ and $\sigma_2 = \sigma + \Delta\sigma/2$. The superposition of the two wave trains yields the wave displacement as

$$\begin{aligned}
\eta &= A\cos(k_1 x - \sigma_1 t) + A\cos(k_2 x - \sigma_2 t) \\
&= 2A\cos\left\{\tfrac{1}{2}[(k_1 + k_2)x - (\sigma_1 + \sigma_2)t]\right\}\cos\left\{\tfrac{1}{2}[(k_1 - k_2)x - (\sigma_1 - \sigma_2)t]\right\} \\
&= 2A\cos(kx - \sigma t)\cos\left[\frac{\Delta k}{2}\left(x - \frac{\Delta\sigma}{\Delta k}t\right)\right].
\end{aligned} \tag{15.5.13}$$

The superposition of these two wave trains yields a set of waves propagating approximately with speed σ/k, in which the amplitude varies within an envelope that moves with the *group velocity* $c_g = \Delta\sigma/\Delta k$, each oscillation of the envelope forming a wave group as shown in Figure 15.4. The wave crests form at the rear of each group and propagate through the group, with amplitude initially growing, reaching a maximum at the group center, and then decreasing as the wave passes through the group. This phenomenon is readily observable simply by throwing a rock into a pond on a calm day. A portion of the kinetic energy of the rock will be transformed into wave energy when the rock strikes the pond surface, which is manifested as a group of waves propagating in a ring away from the impact point. With careful observation, one notices that the wave crests appear to move more rapidly than the wave group as

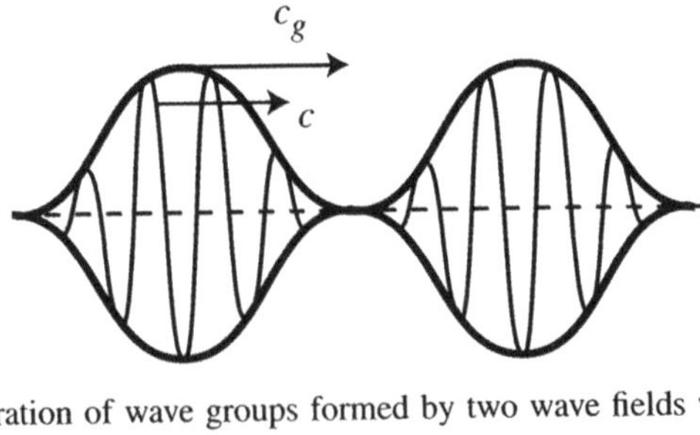

Figure 15.4 Illustration of wave groups formed by two wave fields with slightly different wavenumbers.

a whole. A wave initially at the rear of the group eventually reaches the front of the group and fades away, so that it is no longer detectable, while new waves are formed at the rear of the group.

In the limit $\Delta k \to 0$, the group velocity becomes

$$c_g = \frac{d\sigma}{dk}. \tag{15.5.14}$$

This formula applies to all types of wave phenomena and is generally taken as a definition of group velocity. Upon substitution into (15.5.14) of the dispersion relationship for linear, free-surface water waves (with no mean current or surface tension),

$$\sigma^2 = gk \tanh(kh), \tag{15.5.15}$$

the group velocity reduces to the same expression as given previously in (15.5.12). That the wave energy is transported at the group speed follows from the fact that the instantaneous wave transport rate is proportional to the square of the wave displacement, so the wave energy cannot propagate through the nodal points that separate each group.

15.6 BOUNDARY-INTEGRAL METHOD FOR NONLINEAR INTERFACIAL WAVES

The nonlinear evolution of interfacial waves can be simulated, for both internal and free-surface waves, by replacing the interface with a vortex sheet. This approach is the basis for several different computational methods, such as those described by Longuet-Higgins and Cokelet (1976), Baker et al. (1982), and Tryggvason (1988) These methods differ from each other by the techniques used to speed up the computation and, in the case of the first of these methods, by the use of conformal mapping to transform the interface into a closed contour. In the present discussion, we will not venture into the complexities of the numerical analysis of nonlinear interface dynamics, but instead restrain the discussion to formulation of the basic boundary-integral equation for the vortex sheet motion. For convenience, the theory is developed for two-dimensional internal waves, as shown in Figure 15.1, with the deep-water assumption applied in both the top and bottom layers; however, the analysis is developed such that extension to three-dimensional flows is straightforward. The theory can be generalized to waves of finite depth by introduction of additional vortex sheets along the top and bottom boundaries, where the vortex sheet strength on these boundaries is obtained by solution of the boundary-integral equation for no-slip surfaces derived in Section 12.7.

The velocity induced by a vortex sheet occupying a curve C on the x-y plane with strength $\boldsymbol{\gamma} = \gamma(s)\mathbf{e}_z$ is given by

$$\mathbf{u}(\mathbf{x}, t) = -\frac{1}{2\pi} \int_C \frac{\mathbf{r} \times \boldsymbol{\gamma}'}{r^2}\, ds', \tag{15.6.1}$$

where $\mathbf{r} = \mathbf{x} - \mathbf{x}'$, $r = |\mathbf{r}|$, and s is a measure of arc length on C. The kinematic interface condition, which is a consequence of the requirement that the vortex sheet be a material surface, is satisfied by moving the vortex sheet (representing the interface) with the same normal velocity as the fluid particles on the sheet. The dynamic interface condition, which is related to the jump condition in momentum across the sheet, is then used to derive the boundary-integral equation for variation of the vortex sheet strength along the interface.

We begin by recalling the Euler equation in the form

$$\rho_n \mathbf{a}_n = -\nabla p_n - \rho_n g \mathbf{e}_y, \tag{15.6.2}$$

where $n = \{1, 2\}$ denotes the layer number and $\mathbf{a}_n$ is the acceleration vector. The vortex sheet strength (in two dimensions) is related to the tangential velocity jump across the vortex sheet, or

$$\gamma = (\mathbf{u}_1 - \mathbf{u}_2) \cdot \hat{\mathbf{s}}, \tag{15.6.3}$$

where $\hat{\mathbf{s}}$ is a unit vector oriented tangent to the vortex sheet. Taking the projection of (15.6.2) along C and subtracting the equation across the interface give

$$\frac{1}{2}\left[(\rho_1 + \rho_2)(\mathbf{a}_1 - \mathbf{a}_2) + (\rho_1 - \rho_2)(\mathbf{a}_1 + \mathbf{a}_2)\right] \cdot \hat{\mathbf{s}} = -\frac{\partial}{\partial s}(p_1 - p_2) - (\rho_1 - \rho_2) g \mathbf{e}_y. \tag{15.6.4}$$

The velocity of the interface, denoted by $\mathbf{U}$, has the same normal component as the fluid on either side, since the interface is a material surface. The tangential component of $\mathbf{U}$, which is used to advect the computational points along the vortex sheet, is arbitrary and can thus be specified in any manner that is convenient. For simplicity, in the present discussion we set the tangential component of $\mathbf{U}$ equal to the average tangential fluid velocity on either side of the interface, although other choices have sometimes been used in the literature. The interface velocity $\mathbf{U}$ is thus defined as

$$\mathbf{U} = \tfrac{1}{2}(\mathbf{u}_1 + \mathbf{u}_2), \tag{15.6.5}$$

which is equal to the principal value of the singular integral (15.6.1) when it is evaluated for $\mathbf{x} \in C$. The fluid velocity on either side of the interface can be written as

$$\mathbf{u}_n = \mathbf{U} \pm \tfrac{1}{2}\gamma \hat{\mathbf{s}}, \tag{15.6.6}$$

where in (15.6.6) and in the following equations of this section the top symbol (plus) applies to layer 1 and the bottom symbol (minus) to layer 2. The interface material derivative D_I/Dt is defined such that for any function $f = f(\mathbf{x}, t)$ defined on the interface C,

$$\frac{D_I f}{Dt} \equiv \frac{\partial f}{\partial t} + (\mathbf{U} \cdot \nabla) f. \tag{15.6.7}$$

In terms of this derivative, the fluid acceleration can be written as

$$\mathbf{a}_n = \frac{D_I \mathbf{u}_n}{Dt} \pm \frac{1}{2}\gamma \frac{\partial \mathbf{u}_n}{\partial s}. \tag{15.6.8}$$

The average acceleration, **a**, over the interface is defined by

$$\mathbf{a} \equiv \tfrac{1}{2}(\mathbf{a}_1 + \mathbf{a}_2). \tag{15.6.9}$$

The tangential component of **a** can be written using (15.6.5)–(15.6.9) as

$$\mathbf{a} \cdot \hat{\mathbf{s}} = \frac{D_I \mathbf{U}}{Dt} \cdot \hat{\mathbf{s}} + \frac{1}{8}\frac{\partial(\gamma^2)}{\partial s}. \tag{15.6.10}$$

Dividing (15.6.4) by $\rho_1 + \rho_2$ and using the above definitions yield an equation for the vortex sheet strength as

$$\frac{D_I \gamma}{Dt} + \gamma \frac{\partial \mathbf{U}}{\partial s} \cdot \hat{\mathbf{s}} = -2A\mathbf{a} \cdot \hat{\mathbf{s}} - 2Ag\mathbf{e}_y \cdot \hat{\mathbf{s}} - \frac{2}{\rho_1 + \rho_2}\frac{\partial(p_1 - p_2)}{\partial s}, \tag{15.6.11}$$

where $A \equiv (\rho_1 - \rho_2)/(\rho_1 + \rho_2)$ is the *Atwood ratio*. The pressure difference over the fluid interface is given in terms of the surface tension and the interface radius of curvature by the Laplace formula (15.1.6). Substituting (15.6.10) for the $\mathbf{a}\cdot\hat{\mathbf{s}}$ term into (15.6.11) and finding **U** by evaluating the vortex sheet Biot-Savart equation (15.6.1) on C yields an integro-differential equation for the vortex sheet strength γ.

BIBLIOGRAPHY

Baker, G.R., D.I. Meiron, and S.A. Orszag (1982). "Generalized vortex methods for free-surface flow problems," *Journal of Fluid Mechanics* **123**, 477–501.

Dean, R.G. and R.A. Dalrymple (1984). *Water Wave Mechanics for Engineers and Scientists*, Prentice-Hall, Englewood Cliffs, NJ (reprinted World Scientific Publishing Company, Singapore, 1991).

Lighthill, J. (1978). *Waves in Fluids*, Cambridge University Press, Cambridge.

Longuet-Higgins, M.S. and E.D. Cokelet (1976). "The deformation of steep surface waves on water. I. A numerical method of computation," *Proceedings of the Royal Society of London A* **350**, 1–26.

Tryggvason, G. (1988). "Numerical simulation of the Rayleigh-Taylor instability," *Journal of Computational Physics* **75**, 253–282.

Whitham, G.B. (1974). *Linear and Nonlinear Waves*, John Wiley & Sons, New York.

PROBLEMS

1. For the problem of free-surface waves ($\rho_2 \to 0$) with no surface tension, use the solution for the velocity potential $\phi(x, y, t)$ to obtain an expression for the

stream function $\psi(x, y, t)$ for small-amplitude waves. Use your result to obtain an expression for the complex potential $F(z) = \phi + i\psi$ as a function of the complex variable $z = x + iy$.

2. Find the mean potential energy P per unit surface area for an internal wave with interface displacement $\eta(x, t)$ of the form

$$\eta = A \cos(kx - \sigma t),$$

where A is a constant. The potential energy is given by

$$P = \frac{1}{\lambda} \int_x^{x+\lambda} \int_{-h_1}^{h_2} \rho g y \, dy \, dx - P_0,$$

where $\lambda = 2\pi/k$ is the wavelength and P_0 is the potential energy in the equilibrium state.

3. Determine the phase velocity c and the group velocity c_g for the following cases, and in each case determine whether the wave field is dispersive or nondispersive.
   (a) Free-surface waves in deep water with surface tension, but no mean current.
   (b) Interfacial waves with deep water on both sides of the interface, with no surface tension or mean current.
   (c) Interfacial waves with shallow water on both sides of the interface, with no surface tension or mean current.

4. An axisymmetric wavemaker oscillates periodically with frequency f and generates waves with amplitude A_0 at a distance R_0 from the wavemaker. The wavelength λ is sufficiently small that the waves can be approximated as deep-water free-surface waves. Determine the wave amplitude A at some other distance R from the wavemaker. Assume that both R and R_0 are much larger than the wavelength, so the results obtained for plane waves approximately apply for this problem.

5. Consider a superposition of two sets of free-surface deep-water waves with wavenumber 2.0 and 2.5 m^{-1}. Both sets of waves have the same amplitude. Neglecting surface tension, determine the length of the wave group and the time required for an individual wave crest to propagate through its group.

6. A standing wave can be formed by superposing a left-running and a right-running wave field with the same wavenumber and amplitude. Each standing wave is separated from the others by a node, at which point the interface does not move. Derive expressions for the potential function ϕ and the velocity components in the standing-wave field formed by superposing the right-running wave $\eta_R = A \sin(kx - \sigma t)$ and the left-running wave $\eta_L = A \sin(kx + \sigma t)$, for free-surface waves of finite depth. Determine the positions of the nodes and the oscillation amplitude of the standing wave.

7. Derive an expression for the pressure field under a free-surface wave with small amplitude and a single wavelength λ.

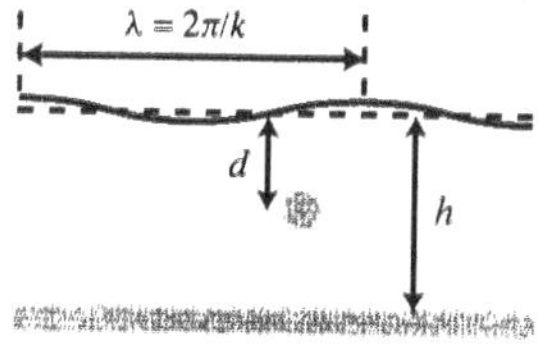

Figure 15.5 A sphere of radius a held at a depth d below the equilibrium surface in a free-surface shallow-water wave field.

8. A sphere of radius a is held fixed at a depth d below the equilibrium surface in a linear free-surface, shallow-water wave field, as shown in Figure 15.5. The waves have wavenumber k, amplitude A, equilibrium depth h, and no mean current. Assume that $Ak \ll 1$, $hk \ll 1$, $ak \ll 1$, and $a/h \ll 1$.
   (a) Determine the time variation of the horizontal velocity and pressure fields at the sphere location induced by the wave when the sphere is not present, making use of the linear wave and shallow-water approximations.
   (b) Use the result of part (a) to estimate the force acting on the sphere in an inviscid flow as a function of time. Carefully explain your reasoning.
9. For a shallow-water free-surface water wave, two pressure sensors are mounted as shown in Figure 15.6. Sensor 1 is placed on the bottom and sensor 2 is placed a distance $b = 0.4$ m above and $d = 4$ m to the right of sensor 1. Assume that there is no surface tension or ambient current. At a given time the gage pressure readings of the sensors are $p_1 = 12{,}000$ N/m^2 and $p_2 = 6{,}733$ N/m^2, and the average fluid depth is $h = 1.1$ m. At this time, the wave crest is located directly above sensor 1 and the distance between the two sensors is less than one wavelength. The water density is 1000 kg/m^3 and the acceleration of gravity is 9.8 m/s^2. Determine the wavelength, the wave amplitude, and the wave frequency. Verify that the shallow-water and linear wave approximations apply.

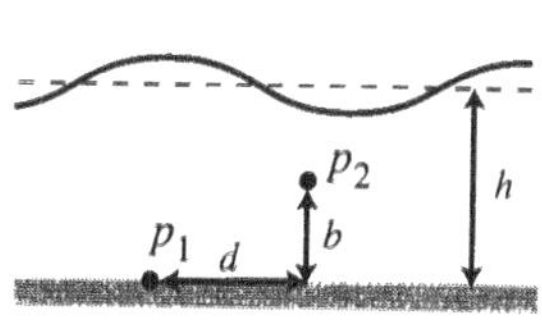

Figure 15.6 Illustration showing locations of two pressure sensors immersed in a progressive wave field.

CHAPTER 16

STABILITY OF FLUID FLOWS

Much of classical fluid mechanics is concerned with finding equilibrium solutions of the Euler or Navier-Stokes equations that are representative of flow situations encountered in various practical problems. Once an appropriate equilibrium solution is found, a critical question to ask is whether or not the solution is stable. In other words, for a fluid system initialized close to the equilibrium solution, but with the various dependent variables slightly perturbed, the stability question concerns whether the system evolves over long time so as to deviate farther from the equilibrium solution, to approach the equilibrium solution, or to remain approximately as far from the equilibrium solution as the initial perturbation. Instabilities observed in nature, occurring as some flow parameter is varied past a critical value, are associated with system transition from an equilibrium state to another equilibrium state, from an equilibrium to a periodic state, from one periodic state to another periodic state with oscillations having different numbers of frequencies, or from a steady or periodic laminar state to a turbulent state. Stability theory, which attempts to predict the occurrence of these bifurcatons in the system state, is an important topic for both inviscid and viscous flows, although inviscid stability theory is more amenable to analytical solution because of the lower order of the resulting differential equations. The current chapter examines some of the basic fluid instabilities for two- and three-dimensional fluid flows. All of the flows examined involve vorticity transport in some part of the flow domain; potential flows are completely determined from the boundary conditions and do not evolve in time, aside from that due to a change in boundary data. The topics are introduced roughly in order of increasing difficulty, progressing from flows where the vorticity is constrained to adopt certain shapes (e.g., ellipses, point singularities, sheets) to flows with unrestricted, distributed vorticity fields.

16.1 GENERAL CONCEPTS

Consider a system having N dependent variables $\{u_1, u_2, \ldots, u_N\}$, which are functions of position and time, and involving M constant dimensionless parameters $\{a_1, a_2, \ldots, a_M\}$. Suppose that an equilibrium solution exists in which the dependent variables attain the values $\{U_1, U_2, \ldots, U_N\}$, which are functions only of position. We introduce some positive-definite measure A that indicates how far the system is from the equilibrium solution. There are many ways in which such a measure can be selected, but a common example is

$$A(t) = \sum_{n=1}^{N} \int_V (u_n - U_n)^2 \, dv. \tag{16.1.1}$$

The equilibrium solution is said to be *stable* if for all positive real numbers ε there can be found some number δ such that the measure $A(t)$ will remain less than ε for all time $t \geq 0$ whenever the measure $A(0)$ of the initial perturbation is less than δ. If in addition $A(t) \to 0$ as $t \to \infty$ for all perturbations such that $A(0) < \delta$, then the equilibrium solution is said to be *asymptotically stable*. Many stable inviscid flow solutions are not asymptotically stable because the perturbations have no dissipating mechanism in the absence of viscosity, so that instead of dying out the perturbations oscillate indefinitely with amplitude of the order of the initial amplitude. If there is any initial perturbation which for some choice ε, no number δ can be found such that $A(t) < \varepsilon$ for all $t \geq 0$ whenever $A(0) < \delta$, then the equilibrium point is said to be *unstable*. It is not necessary that all perturbations of the system have this property; one such perturbation will suffice for the equilibrium point to be unstable.

An equilibrium point is said to be *linearly stable* if the stability condition is satisfied for all ε less than some value ε_0, where ε_0 is chosen to be infinitesimally small (i.e., in the limit $\varepsilon_0 \to 0$). All stable equilibrium solutions are linearly stable, but the converse is not true, since it is possible that the system will become unstable to some perturbation with ε greater than ε_0. On the other hand, an equilibrium solution that is not linearly stable is clearly unstable. It is often convenient for analysis to restrict attention to linear stability theory, for which the equations governing the perturbations can be linearized about the equilibrium solution. Information on finite-amplitude instability theory can be found in books specializing on fluid stability, such as Drazin and Reid (1981).

In order to determine whether a solution is stable or not, it is necessary to consider not just a single perturbation but all "possible" perturbations of the system, i.e., perturbations satisfying the incompressibility constraint and the specified fluid boundary conditions. This task is most easily accomplished by decomposing the dependent variables using a Fourier series, or some other eigenfunction expansion satisfying the fluid boundary conditions, which possesses the property of *completeness*. Complete expansions can be used to construct any arbitrary function satisfying prescribed boundary conditions on a given interval and are typically formed from the eigenfunctions of self-adjoint eigenvalue problems, such as the Sturm-Liouville problem (Kreyszig, 1999; Stakgold, 1979). The fluid stability problems discussed in the current chapter are periodic in at least one direction, so a Fourier series is found to suffice. For instance, if the basic flow $U_n(y)$ depends only on y and the perturbed flow is taken to be periodic in the x-direction, then the perturbation might be expressed as

$$u_n(x, y, t) - U_n(y) = \sum_{m=1}^{\infty} c_{nm}(y) e^{i(k_m x - \sigma_m t)}, \tag{16.1.2}$$

where the coefficients $c_{nm}(y)$ are in general complex valued. The contribution of each m is called a *normal mode*, where this terminology is related to the orthogonality property of Sturm-Liouville eigenfunctions. For linear theory, the perturbations evolve independently of each other such that the frequency σ_m of the mth mode depends only on the corresponding wavenumber k_m. Real values of σ_m correspond to wavenumbers that oscillate in time with constant amplitude. Complex values of σ_m occur in complex conjugate pairs, such that one wave of the pair grows and the other decays in time. The equilibrium solution is stable if σ_m is real for all m, whereas it is unstable if there is any mode m for which σ_m is complex. The stability characteristics of a system are commonly illustrated on a *stability diagram*, in which the regions of stability and instability are indicated on a plot of the wavenumber against one or more of the parameters $\{a_1, a_2, \ldots, a_M\}$.

As the equilibrium solution $\{U_1, U_2, \ldots, U_N\}$ changes for different values of the parameters $\{a_1, a_2, \ldots, a_M\}$, the stability characteristics of the equilibrium point may also change. The value of the set $\{a_1, a_2, \ldots, a_M\}$ at which the nature of the equilibrium solution changes is called a bifurcation point in parameter space. The bifurcations of a system are best illustrated by a bifurcation diagram, in which stable and unstable equilibrium solutions are plotted as functions of one or more of the system parameters $\{a_1, a_2, \ldots, a_M\}$. A few example bifurcation diagrams are shown in Figure 16.1, which, for simplicity, are limited to solutions of ordinary differential equations with a single parameter a. A common type of bifurcation point called a *turning point*, illustrated in Figure 16.1a, occurs when the equilibrium point goes from being stable to unstable, or vice versa, as the bifurcation point is crossed. In a *pitchfork bifurcation* (Figure 16.1b), not only does the stability of the equilibrium solution change as the bifurcation point is crossed, but the number of different equilibrium solutions also changes from one to three. When crossing a *Hopf bifurcation* (Figure 16.1c), a stable equilibrium solution changes into an unstable equilibrium solution plus a *limit cycle*, which is a periodic solution to which systems with nearby initial conditions (in the sense of the measure A) are attracted.

16.2 STABILITY OF AN ELLIPTICAL VORTEX PATCH IN A STRAINING FLOW

The core of a vortex structure may become deformed in shape by straining of the surrounding fluid flow, for instance, resulting from shear in the boundary layer or wake of a solid body or from the velocity induced by other nearby vortices. A simple

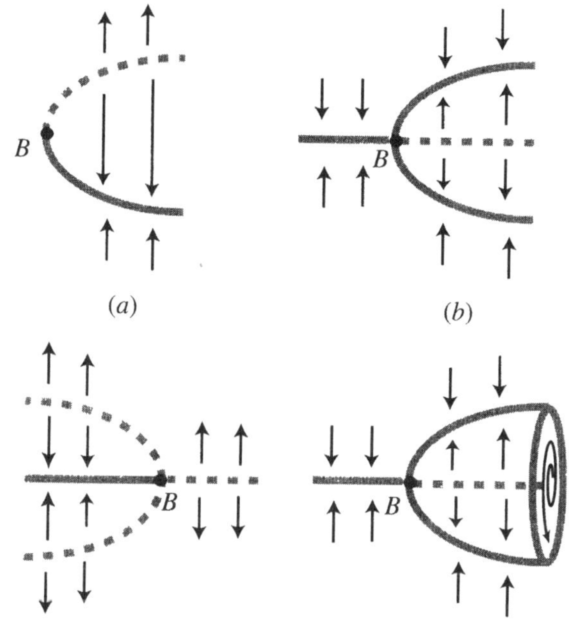

Figure 16.1 Examples of bifurcation diagrams: (*a*) turning point bifurcation, (*b*) supercritical pitchfork bifurcation, (*c*) subcritical pitchfork bifurcation, and (*d*) Hopf bifurcation. Stable and unstable equilibrium solutions are denoted by solid and dashed gray lines, respectively, and the direction of nearby orbits are denoted by arrows.

model of this process (Moore and Saffman, 1971) consists of a two-dimensional patch of uniform vorticity immersed in a uniform straining flow. It is found that the vortex patch deforms in this model into an elliptical shape, whose aspect ratio varies with the ratio of the external straining rate to the vorticity within the vortex core. An equilibrium solution for the strained vortex patch exists for specific values of the vortex orientation angle and aspect ratio. This theory has been extended to a vortex patch in a uniform shear flow by Kida (1981) and to a three-dimenional vortex with axial straining by Neu (1984).

In this section, we derive the equilibrium solution for a strained vortex patch and examine its stability. The position of an elliptical vortex patch with uniform vorticity ω_0 can be specified by its major and minor semiaxes, $a(t)$ and $b(t)$, and by the inclination angle $\theta(t)$ of the semimajor axis to the x-coordinate axis, as illustrated in Figure 16.2. The Kirchhoff solution (Section 11.2) indicates that an isolated elliptical vortex patch rotates without change of form with rotation rate $ab\omega_0/(a+b)^2$. The differential equations governing the three parameters a, b, and θ for an isolated patch are therefore

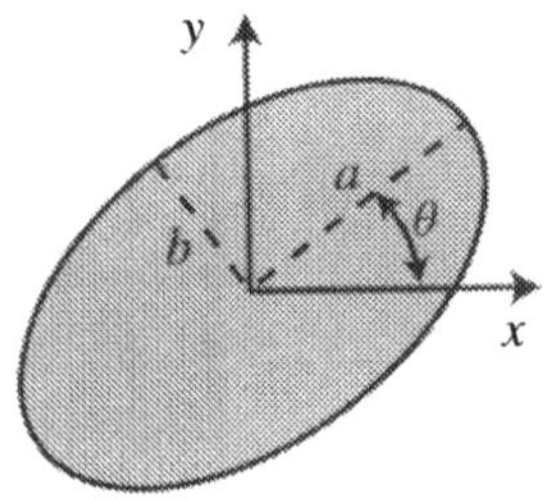

Figure 16.2 An elliptical vortex patch with major and minor semiaxes $a(t)$ and $b(t)$, respectively, and inclination angle $\theta(t)$.

$$\dot{a} = \dot{b} = 0, \qquad \dot{\theta} = \frac{ab\omega_0}{(a+b)^2}, \tag{16.2.1}$$

where an overdot denotes differentiation with respect to time.

An irrotational straining flow, with straining rate e, has velocity components

$$u = ex, \qquad v = -ey. \tag{16.2.2}$$

The equation for the elliptical boundary of the vortex patch can be expressed as

$$f(x, y, t) = \frac{(x\cos\theta + y\sin\theta)^2}{a^2} + \frac{(-x\sin\theta + y\cos\theta)^2}{b^2} - 1 = 0. \tag{16.2.3}$$

Since the boundary of the patch is a material circuit, the function $f(x, y, t)$ must deform under the action of the straining flow in such a way that

$$\frac{Df}{Dt} = \frac{\partial f}{\partial t} + u\frac{\partial f}{\partial x} + v\frac{\partial f}{\partial y} = 0, \tag{16.2.4}$$

if the patch is to maintain an elliptical boundary over time. Substitution of the expression (16.2.3) for f and the velocity components (16.2.2) for a pure straining flow into the kinematic condition (16.2.4) yields an identity provided that the parameters a, b, and θ are evolved according to the equations

$$\dot{a} = ea\cos(2\theta), \qquad \dot{b} = -eb\cos(2\theta),$$
$$\dot{\theta} = -e\left(\frac{a^2 + b^2}{a^2 - b^2}\right)\sin(2\theta). \tag{16.2.5}$$

Equation (16.2.5) describes the evolution of an elliptical material surface immersed in a pure straining flow, not accounting for the induced velocity from the vortex patch.

The Helmholtz decomposition indicates that the velocity field due to a vortex patch in a straining flow is the sum of that due to the straining flow and that due to the vortex patch alone. The resulting governing equations for the three parameters a, b, and θ for a strained vortex patch are obtained by adding the right-hand sides of (16.2.1) and (16.2.5), giving

$$\dot{a} = ea\cos(2\theta), \qquad \dot{b} = -eb\cos(2\theta),$$
$$\dot{\theta} = -e\left(\frac{a^2+b^2}{a^2-b^2}\right)\sin(2\theta) + \frac{ab\omega_0}{(a+b)^2}. \tag{16.2.6}$$

The first two equations in (16.2.6) can be combined to obtain

$$ab = \text{const}, \tag{16.2.7}$$

which is equivalent to the condition that the vortex core area is constant. If time is nondimensionalized using ω_0, (16.2.6) and (16.2.7) can be used to obtain governing equations for the ellipse aspect ratio $R \equiv a/b$ and orientation angle θ in terms of the dimensionless straining rate parameter e/ω_0 as

$$\dot{R} = 2\left(\frac{e}{\omega_0}\right)R\cos(2\theta), \qquad \dot{\theta} = \frac{R}{(R+1)^2} - \left(\frac{e}{\omega_0}\right)\left(\frac{R^2+1}{R^2-1}\right)\sin(2\theta). \tag{16.2.8}$$

An equilibrium solution of the system (16.2.8) exists, denoted by (θ_0, R_0), which satisfies

$$\theta_0 = \frac{\pi}{4}, \qquad \frac{e}{\omega_0} = \frac{R_0(R_0-1)}{(1+R_0)(1+R_0^2)}. \tag{16.2.9}$$

The equilibrium solution for the aspect ratio is plotted versus e/ω_0 in Figure 16.3. There are two solutions for R_0 from (16.2.9) for any e/ω_0 in the interval (0, 0.15), but

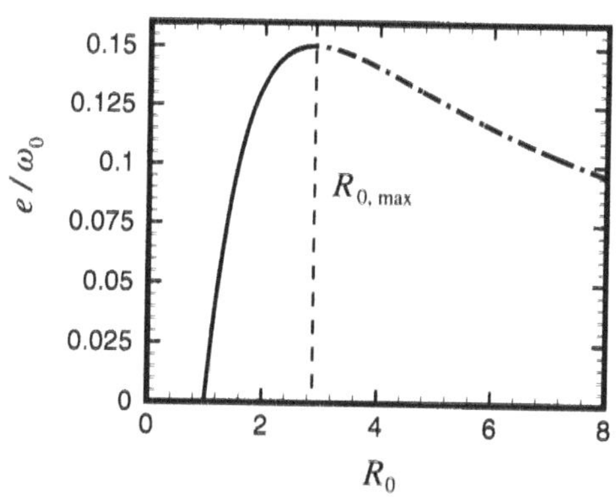

Figure 16.3 Equilibrium solution for aspect ratio R_0 for an elliptical vortex patch with vorticity ω_0 immersed in a uniform straining flow with strain rate e. The solution for $R_0 < R_{0,\max}$ is indicated by a solid line and that for $R_0 > R_{0,\max}$ is indicated by a dashed-dotted line.

no solutions for R_0 when e/ω_0 is greater than 0.15. Numerical computations indicate that the vortex patch stretches out continuously into a thin vortex sheet for cases with $e/\omega_0 > 0.15$. At the maximum value of the straining parameter $e/\omega_0 = 0.15$, the aspect ratio attains a value $R_{0,\max} \cong 2.89$.

We examine the linear stability of the equilibrium solution by initializing R and θ as

$$R(t) = R_0 + R'(t), \qquad \theta(t) = \theta_0 + \theta'(t), \tag{16.2.10}$$

where R' and θ' are small perturbations. Substituting (16.2.10) into (16.2.8), and linearizing by expanding all terms in a Taylor series about the equilibrium values and then neglecting terms that are quadratic and higher order in the primed quantities, yields the perturbation equations as

$$\dot{R}' = -4\left(\frac{e}{\omega_0}\right) R_0\theta', \qquad \dot{\theta}' = BR', \tag{16.2.11}$$

where the constant B is defined by

$$B = \left[\frac{1}{(1+R_0)^2} - \frac{2R_0}{(1+R_0)^3}\right] - \left(\frac{e}{\omega_0}\right)\left[\frac{2R_0}{R_0^2 - 1} - \frac{2R_0(R_0^2+1)}{(R_0^2-1)^2}\right]. \tag{16.2.12}$$

Combining the equations for R' and θ' gives a single second-order equation for R' as

$$\ddot{R}' + 4\left(\frac{e}{\omega_0}\right) R_0 B R' = 0. \tag{16.2.13}$$

Equation (16.2.13) has a general solution of the form

$$R'(t) = Ae^{i\sigma t}, \tag{16.2.14}$$

where the constant σ is obtained by substitution into (16.2.13) as

$$\sigma = \pm\left[4\left(\frac{e}{\omega_0}\right) R_0 B\right]^{1/2}. \tag{16.2.15}$$

For values of R_0 less than $R_{0,\max}$, the constant B in (16.2.12) is positive such that σ is real valued. The solution (16.2.14) then indicates that the aspect ratio and inclination angle will oscillate around the equilibrium value with constant amplitude such that the equilibrium solution is stable. For R_0 greater than $R_{0,\max}$, the constant B is negative and σ is purely imaginary. For this case, the solution (16.2.14) gives one solution that decays exponentially in time and another that increases exponentially, so as to carry the perturbed system progressively farther from the equilibrium solution. The point $R_0 = R_{0,\max}$ is therefore a turning point bifurcation over which the equilibrium solution changes from a stable to an unstable state for increasing R_0.

Figure 16.4 Schematic of the vortex street wake that forms downstream of a circular cylinder. The vortices are indicated by gray patches and the vortex sheets are black lines. Based on flow visualization of S. Taneda, as shown in Van Dyke (1982).

16.3 STABILITY OF TWO-DIMENSIONAL POINT VORTEX ARRAYS

In flows involving many vortices, in which the vortices are sufficiently well separated and close enough in strength that core deformation can be neglected, it is convenient to approximate the vortices by point singularities. Point vortex arrays have been used to model several important fluid flows, such as the spiral vortices that form from the roll-up of a shear layer (Figure 11.8) and the "street" of vortices shed from alternating sides of a cylinder into the body wake (Figure 16.4). Stability of several periodic vortex arrangements, shown in Figure 16.5, is treated in this section, including a

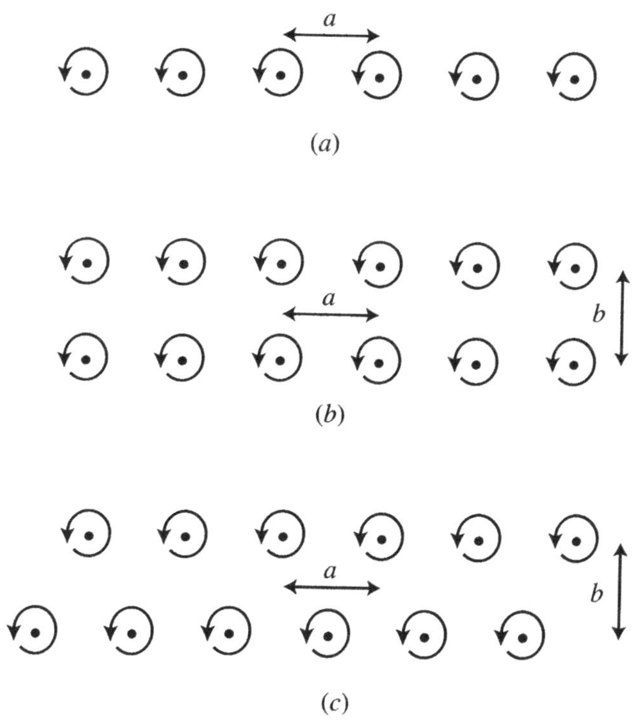

Figure 16.5 Periodic arrays of point vortices: (*a*) single row; (*b*) symmetric double row; (*c*) staggered double row.

single row of evenly spaced vortices, a symmetrical double row of vortices, and a staggered double row of vortices. Further information on point vortex arrays is given by Lamb (1932).

16.3.1 Single Row of Vortices

The complex potential due to a single row of point vortices of strength Γ and spacing distance a, located at $x = \pm na$, $n = 0, 1, 2, 3, \ldots$, is given by

$$F(z) = -\frac{i\Gamma}{2\pi} \sum_{n=-\infty}^{\infty} \ln(z - na). \tag{16.3.1}$$

The infinite sum can be evaluated to give the complex potential $F(z)$ and the associated complex velocity $W(z) = dF/dz$ as

$$F(z) = \frac{i\Gamma}{2\pi} \ln\left(\sin\frac{\pi z}{a}\right), \qquad W(z) = -\frac{i\Gamma}{2a} \cot\frac{\pi z}{a}. \tag{16.3.2}$$

Suppose that the mth vortex is displaced from its equilibrium position ma by a perturbation (x_m, y_m), which varies as a function of time. The equations of motion of the vortex originally located at the origin is given by

$$\frac{dx_0}{dt} = -\frac{\Gamma}{2\pi} \sum_{\substack{m=-\infty \\ m\neq 0}}^{\infty} \frac{y_0 - y_m}{r_m^2}, \qquad \frac{dy_0}{dt} = \frac{\Gamma}{2\pi} \sum_{\substack{m=-\infty \\ m\neq 0}}^{\infty} \frac{x_0 - x_m - ma}{r_m^2}, \tag{16.3.3}$$

where $r_m^2 \equiv (x_0 - x_m - ma)^2 + (y_0 - y_m)^2$. Neglecting second-order terms in the perturbation quantities gives

$$\frac{dx_0}{dt} = -\frac{\Gamma}{2\pi a^2} \sum_{\substack{m=-\infty \\ m\neq 0}}^{\infty} \frac{y_0 - y_m}{m^2},$$

$$\frac{dy_0}{dt} = -\frac{\Gamma}{2\pi a^2} \sum_{\substack{m=-\infty \\ m\neq 0}}^{\infty} \left(\frac{x_0 - x_m}{m^2} + \frac{1}{m}\right), \tag{16.3.4}$$

where we use the fact that $1/r_m^2 \cong (1/m^2a^2)[1 + 2(x_0 - x_m)/ma]$. The sum over the term proportional to $1/m$ in (16.3.4) vanishes from symmetry due to the cancellation of the contributions of vortices with positive and negative values of m.

We consider periodic perturbations of the form

$$x_m(t) = \alpha(t)e^{imk}, \qquad y_m(t) = \beta(t)e^{imk}, \tag{16.3.5}$$

where k is a constant in the interval $(0, 2\pi)$. Substituting these perturbations into (16.3.4) gives

$$\dot{\alpha} = -\lambda\beta, \qquad \dot{\beta} = -\lambda\alpha, \tag{16.3.6}$$

where

$$\lambda \equiv \frac{\Gamma}{\pi a^2} \sum_{m=1}^{\infty} \frac{1-\cos(mk)}{m^2} = \frac{\Gamma}{4\pi a^2} k(2\pi - k) \geq 0. \tag{16.3.7}$$

Combining the two equations in (16.3.6) gives

$$\ddot{\alpha} - \lambda\alpha = 0. \tag{16.3.8}$$

Equation (16.3.8) has solutions of the form $\alpha = Ce^{\pm t\sqrt{\lambda}}$, so that since $\lambda \geq 0$ the flow is unstable for all k. The fastest growing perturbation occurs for $k = \pi$, corresponding to the case where the perturbation of each vortex is opposite that of the two adjacent vortices.

16.3.2 Symmetrical Double Row of Vortices

We now consider a case with two parallel rows of vortices, placed symmetrically on either side of the x-axis, such that all vortices in the upper row have strength Γ and those in the lower row have strength $-\Gamma$ (Figure 16.5b). The vortices in the same row are spaced a distance a and the two rows are separated by a distance b. The system of vortices as a whole advances in the positive x-direction with speed

$$U = \frac{\Gamma}{2a} \coth \frac{\pi b}{a}. \tag{16.3.9}$$

Introducing perturbations (x_m, y_m) for the upper row vortices and (x'_n, y'_n) for the lower row vortices, the vortex positions are given by $(ma + Ut + x_m, \frac{1}{2}b + y_m)$ and $(na + Ut + x'_n, -\frac{1}{2}b + y'_n)$ in the upper and lower rows, respectively. The governing equations for a vortex corresponding to $m = 0$ in the upper row are

$$\frac{dx_0}{dt} = -\frac{\Gamma}{2\pi} \sum_{\substack{m=-\infty \\ m\neq 0}}^{\infty} \frac{y_0 - y_m}{r_m^2} + \frac{\Gamma}{2\pi} \sum_{n=-\infty}^{\infty} \frac{b + y_0 - y'_n}{r'^2_n}, \tag{16.3.10a}$$

$$\frac{dy_0}{dt} = \frac{\Gamma}{2\pi} \sum_{\substack{m=-\infty \\ m\neq 0}}^{\infty} \frac{x_0 - x_m - ma}{r_m^2} - \frac{\Gamma}{2\pi} \sum_{n=-\infty}^{\infty} \frac{x_0 - x'_n - na}{r'^2_n}, \tag{16.3.10b}$$

where $r'^2_n \equiv (x_0 - x'_n - na)^2 + (y_0 - y'_n + b)^2$ and r_m is defined following (16.3.3). The linearized form of the perturbation equations, after elimination of the equilibrium terms, becomes

$$\frac{2\pi}{\Gamma}\frac{dx_0}{dt} = -\sum_{\substack{m=-\infty \\ m\neq 0}}^{\infty} \frac{y_0 - y_m}{m^2 a^2} + \sum_{n=-\infty}^{\infty} \frac{n^2a^2 - b^2}{(n^2a^2 + b^2)^2}(y_0 - y'_n)$$
$$+ \sum_{n=-\infty}^{\infty} \frac{2nab}{(n^2a^2 + b^2)^2}(x_0 - x'_n),$$

$$\frac{2\pi}{\Gamma}\frac{dy_0}{dt} = -\sum_{\substack{m=-\infty \\ m\neq 0}}^{\infty} \frac{x_0 - x_m}{m^2a^2} + \sum_{n=-\infty}^{\infty} \frac{n^2a^2 - b^2}{(n^2a^2 + b^2)^2}(x_0 - x_n')$$
$$- \sum_{n=-\infty}^{\infty} \frac{2nab}{(n^2a^2 + b^2)^2}(y_0 - y_n'). \quad (16.3.11)$$

Considering perturbations of the form

$$x_m(t) = \alpha(t)e^{imk}, \qquad y_m(t) = \beta(t)e^{imk},$$
$$x_n'(t) = \alpha'(t)e^{ink}, \qquad y_n'(t) = \beta'(t)e^{ink}, \quad (16.3.12)$$

where k is a constant in the interval $(0, 2\pi)$, the linear perturbation equations can be written as

$$\frac{d\alpha}{dt^*} = -A\beta - B\alpha' - C\beta', \qquad \frac{d\beta}{dt^*} = -A\alpha - C\alpha' + B\beta',$$
$$\frac{d\alpha'}{dt^*} = A\beta' - B\alpha + C\beta, \qquad \frac{d\beta'}{dt^*} = A\alpha' + C\alpha + B\beta. \quad (16.3.13)$$

The perturbations equations for the lower row (α' and β') are obtained from those for the upper row by exchanging $-\alpha'$ for α and β' for β. Here $t^* = \Gamma t/2\pi a^2$ is the dimensionless time and the constants A, B, and C are defined by

$$A = \sum_{\substack{m=-\infty \\ m\neq 0}}^{\infty} \frac{1 - e^{imk}}{m^2} - \sum_{n=-\infty}^{\infty} \frac{n^2 - \lambda^2}{(n^2 + \lambda^2)^2} = \frac{1}{2}k(2\pi - k) + \frac{\pi^2}{\sinh^2(\lambda\pi)}, \quad (16.3.14a)$$

$$B = \sum_{n=-\infty}^{\infty} \frac{2n\lambda e^{ink}}{(n^2 + \lambda^2)^2} = i\left\{\frac{\pi k \cosh[\lambda(\pi - k)]}{\sinh(\pi\lambda)} - \frac{\pi^2 \sinh(\lambda k)}{\sinh^2(\pi\lambda)}\right\}, \quad (16.3.14b)$$

$$C = \sum_{n=-\infty}^{\infty} \frac{(n^2 - \lambda^2)e^{ink}}{(n^2 + \lambda^2)^2} = -\frac{\pi^2 \cosh(\lambda k)}{\sinh^2(\pi\lambda)} - \frac{\pi k \sinh[\lambda(\pi - k)]}{\sinh(\pi\lambda)}, \quad (16.3.14c)$$

where $\lambda \equiv b/a$ is the spacing ratio.

The solutions of the perturbation equations are of two types. In the first type, $\alpha = \alpha'$ and $\beta = -\beta'$, so that (16.3.13) reduces to

$$\frac{d\alpha}{dt^*} = -B\alpha - (A - C)\beta, \qquad \frac{d\beta}{dt^*} = -(A + C)\alpha - B\beta. \quad (16.3.15)$$

The solutions for α and β are proportional to $\exp(i\sigma_1 t^*)$, where substitution into (16.3.15) gives

$$\sigma_1 = iB \pm i(A^2 - C^2)^{1/2}. \tag{16.3.16}$$

In the second type of solution, $\alpha = -\alpha'$ and $\beta = \beta'$, so that (16.3.13) reduces to

$$\frac{d\alpha}{dt^*} = B\alpha - (A + C)\beta, \qquad \frac{d\beta}{dt^*} = -(A - C)\alpha + B\beta. \tag{16.3.17}$$

The solutions are proportional to $\exp(i\sigma_2 t^*)$, where

$$\sigma_2 = -iB \pm i(A^2 - C^2)^{1/2}. \tag{16.3.18}$$

Since B is purely imaginary, the first term on the right-hand side of both (16.3.16) and (16.3.18) is real valued. The vortex array is therefore stable if $A^2 - C^2 < 0$ and unstable otherwise. For the single-row example, the fastest growing perturbations occur for $k = \pi$, so we consider this perturbation as a special case for the present problem. From (16.3.14) with k set equal to π, we can write

$$A^2 - C^2 = \frac{\pi^4}{4}\tanh^2\left(\frac{\pi\lambda}{2}\right)\coth^2\left(\frac{\pi\lambda}{2}\right). \tag{16.3.19}$$

Since $A^2 - C^2$ is positive for this case, we conclude that the flow is unstable.

16.3.3 Staggered Double Row of Vortices

When the two vortex rows are staggered, such that the x-position of each vortex in the lower row is midway between that of each vortex in the upper row (Figure 16.5c), the propagation speed of the equilibrium system is

$$U = \frac{\Gamma}{2a}\tanh\frac{\pi b}{a}, \tag{16.3.20}$$

and the perturbed vortex positions are $(ma + Ut + x_m, \frac{1}{2}b + y_m)$ in the upper row and $[(n + \frac{1}{2})a + Ut + x'_n, -\frac{1}{2}b + y'_n]$ in the lower row. The perturbed vortex positions are given by the same expression as in (16.3.13), but with the constants A, B, and C defined by

$$\begin{aligned} A &= \sum_{\substack{m=-\infty \\ m\neq 0}}^{\infty} \frac{1 - e^{imk}}{m^2} - \sum_{n=-\infty}^{\infty} \frac{\left(n + \frac{1}{2}\right)^2 - \lambda^2}{\left[\left(n + \frac{1}{2}\right)^2 + \lambda^2\right]^2} \\ &= \frac{1}{2}k(2\pi - k) - \frac{\pi^2}{\cosh^2(\lambda\pi)}, \end{aligned} \tag{16.3.21a}$$

$$B = \sum_{n=-\infty}^{\infty} \frac{(2n+1)\lambda e^{i(n+1/2)k}}{\left[\left(n+\frac{1}{2}\right)^2 + \lambda^2\right]^2}$$

$$= i\left\{\frac{\pi k \sinh[\lambda(\pi - k)]}{\cosh(\pi\lambda)} + \frac{\pi^2 \sinh(\lambda k)}{\cosh^2(\pi\lambda)}\right\}, \qquad (16.3.21b)$$

$$C = \sum_{n=-\infty}^{\infty} \frac{\left[\left(n+\frac{1}{2}\right)^2 - \lambda^2\right] e^{i(n+1/2)k}}{\left[\left(n+\frac{1}{2}\right)^2 + \lambda^2\right]^2}$$

$$= \frac{\pi^2 \cosh(\lambda k)}{\cosh^2(\pi\lambda)} - \frac{\pi k \cosh[\lambda(\pi - k)]}{\cosh(\pi\lambda)}. \qquad (16.3.21c)$$

The results (16.3.15)–(16.3.18) continue to be valid for this case, with the modified definitions of A, B, and C. Thus, the flow is stable if $A^2 - C^2$ is negative for all k, such that the corresponding σ is real valued, and unstable if $A^2 - C^2$ is positive for any k value. We again consider the special case $k = \pi$, which can be shown to correspond to the most unstable perturbation (for details see Lamb, 1932, p. 228). For this case, the constant C vanishes and A is given by

$$A = \frac{\pi^2}{2} - \frac{\pi^2}{\cosh^2(\pi\lambda)}. \qquad (16.3.22)$$

The flow is therefore unstable when $A^2 > 0$ and marginally stable when $A = 0$. The latter case corresponds to values of the spacing ratio such that $\cosh^2(\pi\lambda) = 2$, which upon solving for spacing ratio gives $\lambda = 0.281$.

The staggered vortex double row is linearly unstable unless the spacing ratio b/a has precisely the value 0.281. This result was used by von Kármán and Rubach (1912) in an attempt to model the alternating "vortex street" pattern of vortices shed behind a blunt cylindrical body. This critical spacing ratio value is close to that observed experimentally in the vortex street wake downstream of a circular cylinder in a cross flow, although over long distances (on the order of 20–50 times the body thickness) the spacing ratio is observed to gradually increase to values ranging between 0.4 and 0.6 (Wille, 1960). This growth in spacing ratio has been attributed to effects of the slow viscous growth of the vortex core size, although the issue is still controversial (Hooker, 1936). Even in the case $\lambda = 0.281$, the staggered double point vortex array is unstable to nonlinear disturbances (see references in Wille, 1960). A good overview of the various regimes of vortex shedding in vortex street wakes is given in the review article by Fleischmann and Sallet (1981).

16.4 INTERFACIAL INSTABILITIES

The dynamics of two-dimensional fluid interfaces is discussed in Chapter 15 for both internal and free-surface waves. In the current section, we consider two types of interfacial instabilities that can arise in layered fluids, due either to a difference in current speed between the two layers or to the presence of heavier fluid lying over lighter fluid. The linear analysis for the interface motion is developed in Section 15.2, in which the expression (15.2.9) for the wave frequency σ is obtained for periodic interface displacement of the form (15.2.1). Interfacial instabilities arise whenever σ has a nonzero imaginary part. For convenience, in the following discussion we invoke the deep-water approximation in both layers, which helps simplify the resulting equations.

16.4.1 Rayleigh-Taylor Instability

The Rayleigh-Taylor instability is driven by gravitational force due to the density difference between the layers. In the absence of mean currents, the wave frequency becomes

$$\sigma = \pm\left[\left(\frac{\rho_1 - \rho_2}{\rho_1 + \rho_2}\right) gk + \frac{\gamma k^3}{\rho_1 + \rho_2}\right]^{1/2}. \tag{16.4.1}$$

The flow is unstable whenever the fluid density ρ_2 in the top layer is greater than the density ρ_1 in the bottom layer. The frequency is imaginary, so that the wave grows in time for all wavenumbers less than the critical value

$$k_{\text{crit}} = \left[\frac{(\rho_2 - \rho_1)g}{\gamma}\right]^{1/2}. \tag{16.4.2}$$

Since the instability is driven by the difference in the gravitational force acting on the fluid in the two layers, the critical wavenumber increases with increase in either the density difference $\rho_2 - \rho_1$ or the gravitational acceleration g, thus making progressively shorter waves unstable. For sufficiently short waves, however, the gravitational force becomes of secondary importance compared to the surface tension force (as discussed in Section 15.3), which acts to resist the instability. Consequently, the critical wavenumber decreases with increase in surface tension γ.

While all perturbations with $k < k_{\text{crit}}$ grow in time, it is expected that the wave that eventually dominates is that having the greatest value of the growth rate $|\sigma_I|$, where σ_I is the imaginary part of σ. The fastest growing wave corresponds to a local maximum of σ_I^2. From (16.4.1), σ_I^2 is given by

$$\sigma_I^2 = \left(\frac{\rho_2 - \rho_1}{\rho_1 + \rho_2}\right) gk - \frac{\gamma k^3}{\rho_1 + \rho_2}. \tag{16.4.3}$$

Setting $\partial(\sigma_I^2)/\partial k = 0$ and solving for k yields the fastest growing wavenumber, k_f, as

$$k_f = \left[\frac{(\rho_2 - \rho_1)g}{3\gamma} \right]^{1/2}. \tag{16.4.4}$$

The fastest growing wavenumber is related to the critical wavenumber for this instability by $k_f = k_{\text{crit}}/\sqrt{3}$.

There are many common applications of the Rayleigh-Taylor instability, such as to bubble size selection for film boiling on a flat plate. There is a similar instability that occurs due to unstable density stratification in continuous media, which leads to thermal convection and related overturning in the atmosphere, in the upper layer of oceans and lakes, and (over a much longer time) in the earth's mantle.

16.4.2 Kelvin-Helmholtz Instability

The Kelvin-Helmholtz instability is driven by a current difference between the fluid layers. While we retain surface tension force and density difference between the layers in this analysis, we require that the fluid be stably stratified ($\rho_2 \leq \rho_1$). It also simplifies the equations somewhat to require that the density difference be small compared to the mean density. Hence, if $\rho \equiv (\rho_1 + \rho_2)/2$ and $\Delta\rho \equiv \rho_1 - \rho_2$, we require that $0 \leq \Delta\rho/\rho \ll 1$. To leading order in the small parameter $\Delta\rho/\rho$, the expression (15.2.9) for wave frequency reduces in this case to

$$\sigma = \frac{k}{2}(U_1 + U_2) \pm \left[-\frac{k^2}{4}(U_1 - U_2)^2 + \frac{gk\Delta\rho}{2\rho} + \frac{\gamma k^3}{2\rho} \right]^{1/2}. \tag{16.4.5}$$

The frequency has a nonzero imaginary part for any perturbation with wavenumber k that satisfies the criterion

$$(U_1 - U_2)^2 > \frac{2g}{k}\frac{\Delta\rho}{\rho} + \frac{2\gamma k}{\rho}. \tag{16.4.6}$$

Both the density difference and surface tension are found to resist the instability, thus raising the value of the current difference in (16.4.6) required for the instability to occur.

An example stability diagram for Kelvin-Helmholtz instability is given in Figure 16.6, showing regions of stability and instability on a plot of $|U_1 - U_2|$ versus k. Surface tension force acts to stabilize short waves, and gravitational force associated with $\Delta\rho/\rho > 0$ acts to stabilize long waves. Under the joint action of density difference and surface tension, a critical current difference $|U_1 - U_2|_{\text{crit}}$ exists such that the flow is stable to all perturbations for current differences less than this value. The critical current difference is obtained by taking the local minimum of the right-hand side of (16.4.6) with respect to k as

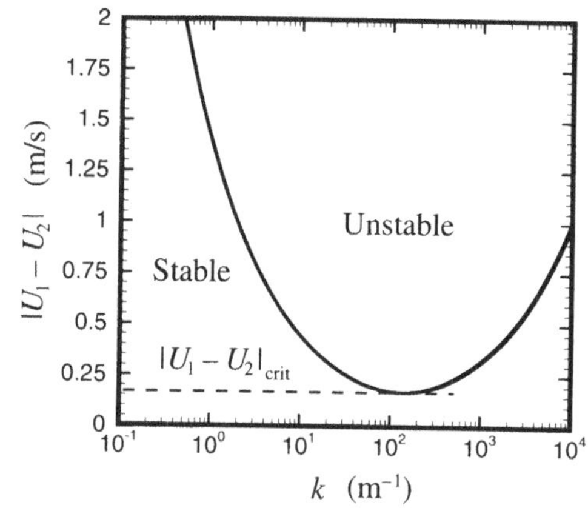

Figure 16.6 Example stability diagram for the Kelvin-Helmholtz instability in the presence of both surface tension and stable density stratification, with $\gamma = 0.05$ N/m, $\rho = 1000$ kg/m^3, and $\Delta\rho = 100$ kg/m^3.

$$|U_1 - U_2|_{\text{crit}} = 2\left[\frac{\gamma g \Delta\rho}{\rho^2}\right]^{1/4}. \tag{16.4.7}$$

The fastest growing wave is obtained by maximizing the imaginary part of the growth rate,

$$\sigma_I^2 = \frac{k^2}{4}(U_1 - U_2)^2 - \frac{gk}{2}\frac{\Delta\rho}{\rho} - \frac{\gamma k^3}{2\rho}, \tag{16.4.8}$$

with respect to k. For cases with nonzero surface tension, the wavenumber of the fastest growing perturbation is

$$k_f = \frac{\rho(U_1 - U_2)^2}{\gamma} \pm \left\{\left[\frac{\rho(U_1 - U_2)^2}{6\gamma}\right]^2 - \frac{g\Delta\rho}{3\gamma}\right\}^{1/2}. \tag{16.4.9}$$

In the absence of surface tension, the flow is unstable for all values of current difference, with the fastest growing wave corresponding to $k \to \infty$, i.e., waves of infinitesimal length. The Kelvin-Helmholtz instability results in roll-up of the vortex sheet to form a series of periodic vortices, which have a form similar to those shown in Figure 11.8.

The observation that the fastest growing perturbations correspond to waves of zero length suggests that there may be something incomplete about the vortex sheet model, and indeed this fact requires that special care be taken in numerical computation of vortex sheet motion (Moore, 1979; Saffman, 1992). A vortex sheet is, of course, a mathematical idealization; vorticity layers of infinitesimal thickness never form in nature due to the diffusive effects of viscosity. Stability calculations with vorticity layers of finite thickness (e.g., Chandrasekhar, 1961, pp. 487–494) indicate stabilization of the short waves (see also Example 16.7.1). A vortex sheet can also be stabilized by stretching along its length (Moore and Griffith-Jones, 1974), which is commonly encountered in flow fields involving multiple vortex structures or turbulence.

16.5 CAPILLARY INSTABILITY OF A LIQUID JET

Another type of interfacial instability occurs for a jet of liquid sprayed into a gaseous atmosphere. Liquid jet flows of this type are important in a variety of industrial applications, such as ink-jet printing and spray deposition processes. In the stability analysis, we again assume that the inertia of the surrounding gas flow is sufficiently small that the pressure within the gas can be taken as constant. Also, gravity is only of secondary importance to the instability under consideration, so for simplicity it is neglected. Liquid jet instabilities differ from the plane interface problem discussed in Section 16.4 mainly due to the transverse curvature of the interface around the circular jet, which has important consequences on the role of the surface tension force on the interface dynamics.

The Laplace equation for velocity potential of an irrotational, incompressible, axisymmetric flow is written in cylindrical polar coordinates (R, α, z) as

$$\frac{1}{R}\frac{\partial}{\partial R}\left(R\frac{\partial \phi}{\partial R}\right) + \frac{\partial^2 \phi}{\partial z^2} = 0. \tag{16.5.1}$$

The free surface of the jet is denoted by $R = a + \eta(z, t)$, where a is the equilibrium radius of the jet and η is the interface displacement. The kinematic condition at the free surface results from the observation that $f(R, z, t) = R - a - \eta(z, t) = 0$ on the surface, so that the material surface condition $Df/Dt = 0$ yields

$$\frac{\partial \eta}{\partial t} + w\frac{\partial \eta}{\partial z} = u. \tag{16.5.2}$$

The dynamic interface condition results from the jump in momentum at the interface, which in the presence of surface tension is given by the Laplace formula as

$$p_\ell - p_{\text{atm}} = \gamma\left(\frac{1}{R_1} + \frac{1}{R_2}\right), \tag{16.5.3}$$

where p_ℓ is the pressure in the liquid at the interface and R_1 and R_2 are the radii of curvature in the streamwise and transverse directions. The atmospheric pressure p_{atm} is set to zero. Substituting (16.5.3) into the unsteady Bernoulli equation (8.2.2) evaluated at the free surface gives

$$\frac{\partial \phi}{\partial t} + \frac{1}{2}(u^2 + w^2) + \frac{\gamma}{\rho}\left(\frac{1}{R_1} + \frac{1}{R_2}\right) = F(t), \tag{16.5.4}$$

where $F(t)$ is an arbitrary function of time.

For small-amplitude perturbations, the interface displacement is assumed to take the form

$$\eta(z, t) = Ae^{i(kz - \sigma t)}, \tag{16.5.5}$$

where A is the wave amplitude, k is the wavenumber, and σ is the frequency. Linear theory is valid provided that the amplitude is small compared to the jet radius and the wavelength, or

$$\frac{A}{a} \ll 1, \qquad Ak \ll 1. \tag{16.5.6}$$

The surface curvatures can be approximated in linear theory by

$$\frac{1}{R_1} \cong \frac{1}{a}\left(1 - \frac{\eta}{a}\right), \qquad \frac{1}{R_2} \cong -\frac{\partial^2 \eta}{\partial z^2}. \tag{16.5.7}$$

The axial jet velocity is W_0 such that the velocity potential can be decomposed as

$$\phi(R, z, t) = W_0 z + \phi'(R, z, t). \tag{16.5.8}$$

The linear forms of the kinematic and dynamic interface conditions are

$$\frac{\partial \eta}{\partial t} + W_0 \frac{\partial \eta}{\partial z} = \frac{\partial \phi'}{\partial R}, \tag{16.5.9}$$

$$\frac{\partial \phi'}{\partial t} + W_0 \frac{\partial \phi'}{\partial z} - \frac{\gamma}{\rho}\left(\frac{\eta}{a^2} + \frac{\partial^2 \eta}{\partial z^2}\right) = 0, \tag{16.5.10}$$

where the arbitrary function $F(t)$ in (16.5.4) is chosen to eliminate the equilibrium terms. In the linear theory, the boundary conditions can be displaced from the actual interface location $R = a + \eta$ to the equilibrium jet surface $R = a$ using a Taylor series expansion similar to that described in Section 15.2.

A solution of the Laplace equation (16.5.1) can be obtained using separation of variables, subject to the conditions of periodic perturbations in the axial direction and zero radial velocity at the symmetry axis. The resulting solution has the form

$$\phi'(R, z, t) = BI_0(kR)e^{i(kz - \sigma t)}, \tag{16.5.11}$$

where B is a complex constant and $I_0(\cdot)$ is the modified Bessel function of the first kind. Substituting (16.5.5) and (16.5.11) into the interface conditions yields an algebraic eigenvalue problem of the form

$$\mathbf{M} \cdot \mathbf{a} = \sigma \mathbf{a}, \tag{16.5.12}$$

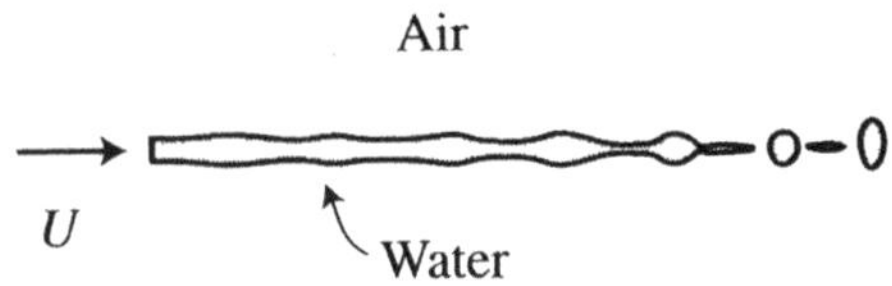

Figure 16.7 Schematic showing the development of capillary instability on a liquid jet, resulting in the formation of droplets.

where $\mathbf{a}^{\mathrm{T}} = (A, B)$ and

$$\mathbf{M} = \begin{pmatrix} kW_0 & ikI_1(ka) \\ \dfrac{i\gamma(1-k^2a^2)}{\rho a^2 I_0(ka)} & kW_0 \end{pmatrix}. \qquad (16.5.13)$$

Solving for the eigenvalue σ by setting the determinant of $\mathbf{M} - \sigma\mathbf{I}$ equal to zero yields an expression for frequency as

$$\sigma = kW_0 \pm \left[\frac{k\gamma(k^2a^2-1)}{\rho a^2}\frac{I_1(ka)}{I_0(ka)}\right]^{1/2}. \qquad (16.5.14)$$

Since the Bessel functions in (16.5.14) are positive for all $ka > 0$, we find that σ is real valued for $ka \geq 1$ and purely imaginary for $ka < 1$, such that instability occurs for wavenumbers less than the critical value $k_{\mathrm{crit}} = 1/a$. Using $\lambda = 2\pi/k$, the liquid jet is found to be unstable for all $\gamma > 0$ to perturbations with wavelength λ greater than the jet circumference $2\pi a$. The fastest growing perturbation is obtained by finding the local maximum of σ_I^2 as $k_f = 0.697/a$.

Surface tension acts both to drive the instability, due to the transverse curvature, and to inhibit the instability for sufficiently short waves, due to the streamwise curvature. This instability results in breakup of the jet into droplets, as illustrated in Figure 16.7, and it plays an important role in atomization processes used to produce fine liquid sprays.

16.6 CENTRIFUGAL INSTABILITY

Rotating flows play an important role in diverse applications, such as centrifugal separation devices, gas bearings, and aircraft wake vortices. In this section, we consider the stability of rotating flows in which the equilibrium solution has azimuthal velocity component $V(R)$ and pressure $P(R)$ given in cylindrical polar coordinates (R, α, z) by

$$V = R\Omega(R), \qquad \frac{dP}{dR} = \rho R\Omega^2(R), \qquad (16.6.1)$$

where $\Omega(R)$ is the angular velocity.

The continuity and momentum equations for axisymmetric flow are

$$\frac{1}{R}\frac{\partial}{\partial R}(Ru) + \frac{\partial w}{\partial z} = 0, \tag{16.6.2a}$$

$$\rho\left(\frac{\partial u}{\partial t} + u\frac{\partial u}{\partial R} + w\frac{\partial u}{\partial z} - \frac{v^2}{R}\right) = -\frac{\partial p}{\partial R}, \tag{16.6.2b}$$

$$\rho\left(\frac{\partial v}{\partial t} + u\frac{\partial v}{\partial R} + w\frac{\partial v}{\partial z} + \frac{uv}{R}\right) = 0, \tag{16.6.2c}$$

$$\rho\left(\frac{\partial w}{\partial t} + u\frac{\partial w}{\partial R} + w\frac{\partial w}{\partial z}\right) = -\frac{\partial p}{\partial z}. \tag{16.6.2d}$$

The velocity and pressure fields are decomposed into mean and perturbation parts as

$$\mathbf{u}(R, z, t) = V(R)\mathbf{e}_\alpha + \mathbf{u}'(R, z, t), \tag{16.6.3a}$$

$$p(R, z, t) = P(R) + p'(R, z, t), \tag{16.6.3b}$$

The perturbations are assumed to be axisymmetric and periodic in the axial direction, so that we can write

$$\mathbf{u}' = [\hat{u}(R)\mathbf{e}_R + \hat{v}(R)\mathbf{e}_\alpha + \hat{w}(R)\mathbf{e}_z]e^{i(kz-\sigma t)}, \qquad p' = \rho\hat{p}(R)e^{i(kz-\sigma t)}, \tag{16.6.4}$$

where the frequency σ is a function of the axial wavenumber k. Substituting the perturbed velocity and pressure fields into (16.6.2) and linearizing yields

$$\frac{d\hat{u}}{dR} + \frac{\hat{u}}{R} + ik\hat{w} = 0, \tag{16.6.5a}$$

$$i\sigma\hat{u} + 2\Omega\hat{v} = \frac{d\hat{p}}{dR}, \tag{16.6.5b}$$

$$i\sigma\hat{v} - \left[\Omega + \frac{d}{dR}(R\Omega)\right]\hat{u} = 0, \tag{16.6.5c}$$

$$i\sigma\hat{w} = ik\hat{p}. \tag{16.6.5d}$$

Eliminating $\hat{v}$, $\hat{w}$, and $\hat{p}$ in the above equations, a second-order differential equation for $\hat{u}$ is obtained as

$$\frac{d}{dR}\left[\frac{1}{R}\frac{d}{dR}(R\hat{u})\right] + k^2\left[\frac{\Phi(R)}{\sigma^2} - 1\right]\hat{u} = 0, \tag{16.6.6}$$

where $\Phi(R) \equiv (2\Omega/R)\,d(R^2\Omega)/dR = (1/R^3)\,d(R^2\Omega)^2/dR$ is the *Rayleigh discriminant.*

Equation (16.6.6) is of the Sturm-Liouville form, and together with homogeneous boundary conditions of the form $\hat{u}(R_1) = \hat{u}(R_2) = 0$ for some R_1 and R_2, it represents a differential eigenvalue problem over the interval $R_1 \leq R \leq R_2$, where $1/\sigma^2$ acts as the eigenvalue. One property of Sturm-Liouville problems is that the eigenvalues are real valued, from which it follows that the wave frequency σ must be either real valued or purely imaginary. An additional property of this system is obtained by multiplying (16.6.6) by $R\hat{u}$ and integrating over the interval (R_1, R_2), giving

$$\int_{R_1}^{R_2} R\hat{u}\frac{d}{dR}\left[\frac{1}{R}\frac{d}{dR}(R\hat{u})\right]dR + \int_{R_1}^{R_2} k^2\frac{\Phi(R)}{\sigma^2}R\hat{u}^2\,dR - \int_{R_1}^{R_2} R\hat{u}^2\,dR = 0. \tag{16.6.7}$$

Using integration by parts in the first term and solving for $1/\sigma^2$ yields

$$\frac{1}{\sigma^2} = \frac{\int_{R_1}^{R_2}\left\{(1/R)\left[(d/dR)(R\hat{u})\right]^2 + R\hat{u}^2\right\}dR}{\int_{R_1}^{R_2} k^2\Phi(R)R\hat{u}^2\,dR}. \tag{16.6.8}$$

The integral in the numerator of (16.6.8) is always positive. If the function $\Phi(R) > 0$ everywhere within the flow, then σ must be real valued and the basic rotating flow solution is stable. If any interval (R_1, R_2) exists in which $\Phi(R) < 0$, then σ must be purely imaginary for perturbations vanishing on the end points of this interval, and the basic rotating flow solution is unstable. Recalling the definition of the discriminator function $\Phi(R)$, this result can be restated as follows.

Theorem 16.6.1 (Rayleigh's Centrifugal Stability Theorem). In an inviscid fluid, the necessary and sufficient condition for a distribution of angular velocity $\Omega(R)$ to be stable is that $d(R^2\Omega)^2/dR > 0$ everywhere in the interval; and further, the distribution is unstable if $(R^2\Omega)^2$ should decrease anywhere inside the interval.

We have proved the above theorem only for axisymmetric perturbations. A discussion of stability of rotating flows to nonaxisymmetric perturbations is given by Chandrasekhar (1961, pp. 281–284).

Example 16.6.1 Rotating Couette Flow. We consider the flow within a cylindrical annulus generated by rotation of the inner cylinder relative to the outer cylinder, shown in Figure 16.8, where the inner cylinder at $R = R_1$ rotates at rate Ω_1 and the outer cylinder at $R = R_2$ rotates at rate Ω_2. A solution for this flow is obtained in Section 4.6, which when written in cylindrical polar coordinates is given by

$$V(R) = AR + \frac{B}{R}. \tag{16.6.9}$$

Because (16.6.9) is an exact solution of both the Euler and Navier-Stokes equations, the constants A and B can be specified to satisfy the no-slip conditions at the inner

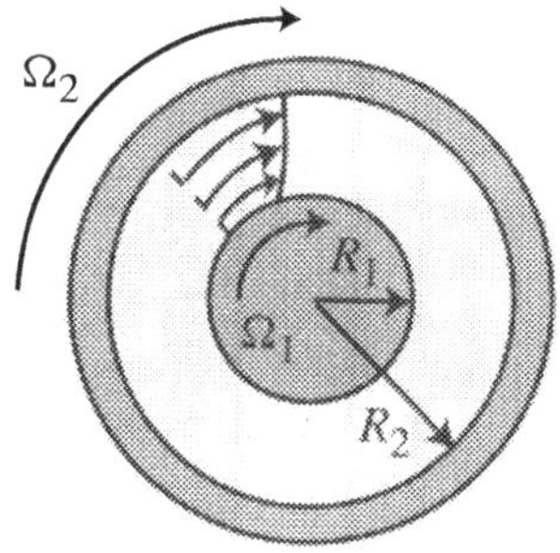

Figure 16.8 Rotating Couette flow in a cylindrical annulus.

and outer cylinders, giving

$$A = \frac{\Omega_2 R_2^2 - \Omega_1 R_1^2}{R_2^2 - R_1^2}, \qquad B = -R_1^2 R_2^2 \left(\frac{\Omega_2 - \Omega_1}{R_2^2 - R_1^2} \right). \tag{16.6.10}$$

The fluid angular rotation rate in the basic flow is $\Omega(R) = V/R$, so from (16.6.9) we can write

$$\frac{d}{dR}(R^2\Omega)^2 = 2(AR^2 + B)(2AR). \tag{16.6.11}$$

Substituting the expressions for the constants A and B into (16.6.11) yields

$$\frac{d}{dR}(R^2\Omega)^2$$
$$= -\frac{4\Omega_1^2 R_1^2 R_2^2 R}{(R_2^2 - R_1^2)^2} \left[\frac{\Omega_2}{\Omega_1}(R^2 - R_1^2) + \frac{R_1^2}{R_2^2}(R_2^2 - R^2) \right] \left(1 - \frac{\Omega_2 R_2^2}{\Omega_1 R_1^2} \right). \tag{16.6.12}$$

Applying the Rayleigh theorem yields the conditions for flow stability. When Ω_1 and Ω_2 are of the same sign, the first two groups of variables on the right-hand side of (16.6.12) are always positive, so the stability condition is determined from the sign of the last group, $1 - \Omega_2 R_2^2/\Omega_1 R_1^2$. The Rayleigh stability theorem thus indicates that the flow is stable whenever $\Omega_2 R_2^2 > \Omega_1 R_1^2$ and unstable when this condition is not satisfied. When Ω_1 and Ω_2 are of opposite sign, the last group of variables in (16.6.12) is always positive. The first term in the second variable group, $(\Omega_2/\Omega_1)(R^2 - R_1^2)$, is negative, but since this term vanishes as $R \to R_1$, the second group of variables cannot be negative for all R in the interval (R_1, R_2). Consequently, the flow is always unstable when Ω_1 and Ω_2 are not of the same sign.

Unstable rotating Couette flow is observed experimentally to develop a set of stationary vortices, called Taylor vortices, which wrap around the cylindrical annulus

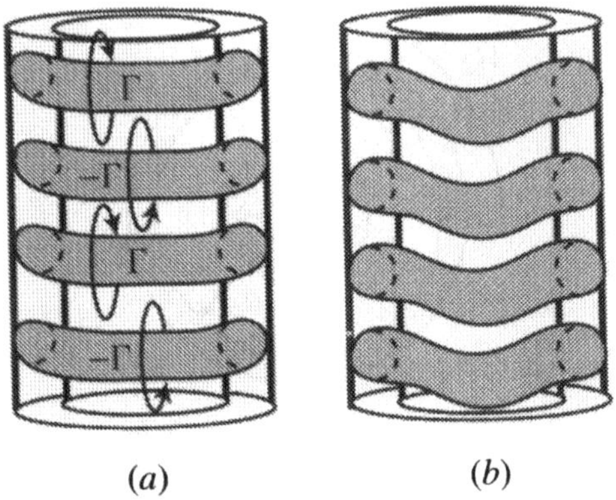

Figure 16.9 Schematic of the Taylor vortices that form due to centrifugal instability in a rotating cylindrical annulus: (*a*) axisymmetric and (*b*) nonaxisymmetric. Based on flow visualizations of Koschmieder (1979).

and are periodic in the axial direction (Figure 16.9*a*). As the difference between the cylinder rotation rates increases, the Taylor vortices themselves become unstable, which causes the vortices to develop nonaxisymmetric waves that propagate azimuthally around the cylinder (Figure 16.9*b*). Further increase in the difference between cylinder rotation rates introduces additional bifurcations, eventually leading a turbulent state.

We note that even though the basic rotating flow is a solution of the full Navier-Stokes equation, the above analysis requires the perturbations to satisfy only the Euler equation rather than the full viscous-flow equations. The viscous terms typically damp the centrifugal instability such that a flow that is stable according to the Rayleigh criterion would be expected to remain stable if viscous terms are included, but a flow that is unstable to the inviscid theory might be stabilized by inclusion of viscous terms. In other types of flows, however, inclusion of viscous terms can have the surprising effect of enhancing instability. For an introduction to viscous stability theory, the reader might refer to the books on hydrodynamic stability theory by Drazin and Reid (1981) or Chandrasekhar (1961).

16.7 STABILITY OF PARALLEL SHEAR FLOWS

Many fluid flows have velocity oriented primarily in one direction, with variation of velocity in an orthogonal direction. Such flows, generically referred to as *parallel shear flows*, are encountered in pipe and conduit flow problems and may sometimes be used to model stability of viscous boundary layers or shear layer flows. Several examples of parallel shear flow solutions are given in Section 4.6. The velocity and pressure fields for the basic inviscid flow solution under consideration are given by

$$\mathbf{U} = U(y)\mathbf{e}_x, \qquad P = \text{const.} \tag{16.7.1}$$

Since these flows are a solution of both the Euler and Navier-Stokes equations, the velocity field can be adjusted to satisfy no-slip boundary conditions of the form $U(x, y_1) = U_1$ and $U(x, y_2) = U_2$.

The perturbed velocity and pressure fields are given by

$$\mathbf{u} = U(y)\mathbf{e}_x + \mathbf{u}'(x, y, t), \qquad p = P + p'(x, y, t), \tag{16.7.2}$$

where since the perturbations are assumed to be periodic in the streamwise direction, we can write

$$\mathbf{u}' = [\hat{u}(y)\mathbf{e}_x + \hat{v}(y)\mathbf{e}_y]e^{ik(x-ct)}, \qquad p' = \rho\hat{p}(y)e^{ik(x-ct)}. \tag{16.7.3}$$

Only two-dimensional perturbations are considered since it is known (Squire, 1933) that parallel shear flows become unstable first to two-dimensional disturbances. The continuity and Euler equations for flow with no body force are

$$\frac{\partial u}{\partial x} + \frac{\partial v}{\partial y} = 0, \tag{16.7.4a}$$

$$\frac{\partial u}{\partial t} + u\frac{\partial u}{\partial x} + v\frac{\partial u}{\partial y} = -\frac{1}{\rho}\frac{\partial p}{\partial x}, \tag{16.7.4b}$$

$$\frac{\partial v}{\partial t} + u\frac{\partial v}{\partial x} + v\frac{\partial v}{\partial y} = -\frac{1}{\rho}\frac{\partial p}{\partial y}. \tag{16.7.4c}$$

Substituting (16.7.2) and (16.7.3) into (16.7.4) and omitting quadratic terms in the perturbation quantities give the perturbation evolution equations as

$$ik\hat{u} + \frac{d\hat{v}}{dy} = 0, \tag{16.7.5a}$$

$$ik\hat{u}(U - c) + \frac{dU}{dy}\hat{v} = -ik\hat{p}, \tag{16.7.5b}$$

$$ik\hat{v}(U - c) = -\frac{d\hat{p}}{dy}. \tag{16.7.5c}$$

Eliminating $\hat{u}$ and $\hat{p}$ from the above equations gives a second-order differential equation for $\hat{v}$ as

$$(U - c)\left(\frac{d^2\hat{v}}{dy^2} - k^2\hat{v}\right) - \frac{d^2U}{dy^2}\hat{v} = 0, \tag{16.7.6}$$

which is known as *Rayleigh's equation*.

In general, the constant c and eigenfunctions $\hat{v}$ of the Rayleigh equation are complex valued. The flow is stable when c is real valued and unstable when c is complex valued. Some information on parallel shear flows in general can be obtained by the following argument. Multiplying (16.7.6) by the complex conjugate $\bar{\hat{v}}$ of $\hat{v}$ and dividing by $U - c$ gives

$$\bar{\hat{v}}\frac{d^2\hat{v}}{dy^2} - k^2|\hat{v}|^2 - \frac{d^2U}{dy^2}\left[\frac{U-\bar{c}}{|U-c|^2}\right]|\hat{v}|^2 = 0, \tag{16.7.7}$$

where we note that $1/(U-c) = (U-\bar{c})/|U-c|^2$ and $|\hat{v}|^2 = \hat{v}\bar{\hat{v}}$. Taking the complex conjugate of (16.7.7) gives

$$\hat{v}\frac{d^2\bar{\hat{v}}}{dy^2} - k^2|\hat{v}|^2 - \frac{d^2U}{dy^2}\left[\frac{U-c}{|U-c|^2}\right]|\hat{v}|^2 = 0. \tag{16.7.8}$$

Subtracting (16.7.7) from (16.7.8) yields

$$\frac{d}{dy}\left(\bar{\hat{v}}\frac{d\hat{v}}{dy} - v\frac{d\bar{\hat{v}}}{dy}\right) = \frac{2ic_I}{|U-c|^2}\frac{d^2U}{dy^2}|\hat{v}|^2, \tag{16.7.9}$$

where c_I is the imaginary part of c. Integrating (16.7.9) over the interval (y_1, y_2), where $v(y_1) = v(y_2) = 0$, yields

$$c_I\int_{y_1}^{y_2}\frac{|\hat{v}|^2}{|U-c|^2}\frac{d^2U}{dy^2}\,dy = 0. \tag{16.7.10}$$

Either c_I must vanish, in which case the flow is stable, or the integral in (16.7.10) must vanish. In the latter case, it is necessary that the derivative d^2U/dy^2 must either vanish everywhere or change signs somewhere in the interval (y_1, y_2). This conclusion leads to the following theorem.

Theorem 16.7.1 (Rayleigh's Inflection Point Theorem). It is necessary for linear instability of a parallel shear flow that the velocity profile is either piecewise linear or have a point of inflection.

A stronger necessary condition can be developed by adding (16.7.7) and (16.7.8), integrating over (y_1, y_2) and using integration by parts to write

$$\int_{y_1}^{y_2}(U - c_R)\frac{|\hat{v}|^2}{|U-c|^2}\frac{d^2U}{dy^2}\,dy = -\int_{y_1}^{y_2}\left[\left|\frac{d\hat{v}}{dy}\right|^2 + k^2|\hat{v}|^2\right]dy < 0, \tag{16.7.11}$$

where c_R is the real part of c. If the flow is unstable, then the integral in (16.7.10) must vanish since $c_I \neq 0$ for unstable flow. Multiplying this integral by the constant factor $c_R - U_0$, where $U_0 \equiv U(y_0)$ and $d^2U/dy^2 = 0$ at $y = y_0$, we can add the term

$$(c_R - U_0)\int_{y_1}^{y_2}\frac{|\hat{v}|^2}{|U-c|^2}\frac{d^2U}{dy^2}\,dy$$

to the left-hand side of (16.7.11) to obtain

$$\int_{y_1}^{y_2} (U - U_0) \frac{|\hat{v}|^2}{|U - c|^2} \frac{d^2U}{dy^2}\, dy = -\int_{y_1}^{y_2} \left[\left| \frac{d\hat{v}}{dy} \right|^2 + k^2 |\hat{v}|^2 \right] dy < 0. \quad (16.7.12)$$

The restrictions imposed by this inequality lead to the following theorem.

Theorem 16.7.2 (Fjørtoft's Theorem). It is necessary for linear instability of a parallel shear flow that $(U - U_0)(d^2U/dy^2) < 0$ at some point in the flow field, where y_0 is a point at which $d^2U/dy^2 = 0$ and $U_0 \equiv U(y_0)$.

The two theorems stated above can be used to rule out the possibility of instability, but they do not definitively indicate whether or not a given flow is unstable. The stability of a particular parallel shear flow solution $U(y)$ can be determined by solving the Rayleigh equation (16.7.6) for the eigenfunctions and the value of the eigenvalue c for the specified boundary conditions. Several examples of parallel shear flows are illustrated in Figure 16.10, including the Couette flow generated by a plate sliding over another plate, the Pouisille flow driven by a pressure gradient in a fixed channel, and a shear layer. In Couette flow, the velocity profile is linear, so that d^2U/dy^2 everywhere vanishes. For this case, the inflection point theorem indicates that instability of the flow is possible. Detailed solution of Rayleigh's equation for Couette flow indicates that the flow is stable to linear disturbances (Drazin and Reid, 1981). In Pouisille flow, the velocity profile has the quadratic form $U(y) = C(y^2 - hy)$, where C and h are constants. The second derivative of U is constant, so the inflection point theorem requires that the flow must be stable. In a plane mixing layer, the velocity

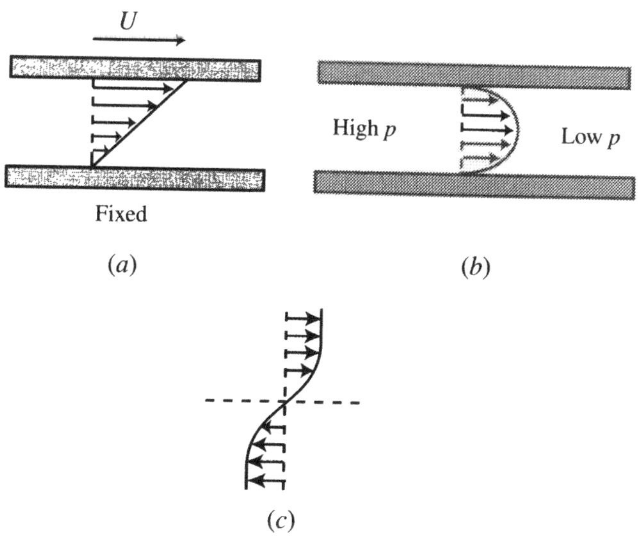

Figure 16.10 Examples of parallel shear flows: (*a*) Couette flow between a sliding plate and a fixed plate; (*b*) Pouisille flow in a channel; (*c*) shear layer flow.

profile has the form $U(y) = [(U_2 - U_1)/2]\tanh(y/h) + (U_2 + U_1)/2$. The second derivative of U vanishes at the midplane $y = 0$, so the inflection point theorem indicates that it is possible that the flow is unstable. Also, $(U - U_0)(d^2U/dy^2) < 0$ everywhere in the flow field, so the instability requirement of Fjørtoft's theorem is also satisfied. Detailed solution of the Rayleigh equation indicates that the shear layer flow is in fact unstable (Drazin and Reid, 1981). An example illustrating determination of flow stability by direct solution of the Rayleigh equation is given below.

Example 16.7.1 Stability of Piecewise Linear Shear Layer. In order to illustrate determination of flow stability by direct solution of the Rayleigh equation (16.7.6), we consider a shear layer with piecewise linear velocity profile, as shown in Figure 16.11. The velocity profile is given by

$$U(y) = \begin{cases} 1, & y > 1, \\ y, & -1 \leq y \leq 1, \\ -1, & y < -1. \end{cases} \tag{16.7.13}$$

The second derivative of U vanishes everywhere, so the inflection point theorem indicates that it is possible for the flow to be unstable. The Rayleigh equation for this problem reduces to

$$\frac{d^2\hat{v}}{dy^2} - k^2\hat{v} = 0. \tag{16.7.14}$$

This equation has solutions in terms of the exponential functions $\exp(ky)$ and $\exp(-ky)$. Upon imposing the boundary condition that the perturbation velocity vanishes as $y \to \pm\infty$, the general solution can be written as

$$\hat{v}(y) = \begin{cases} Ae^{-ky} & \text{for } y > 1, \\ Be^{-ky} + Ce^{ky} & \text{for } -1 \leq y \leq 1, \\ De^{ky} & \text{for } y < -1, \end{cases} \tag{16.7.15}$$

where A, B, C, and D are constants of integration. Continuity of vertical velocity v at $y = \pm 1$ gives

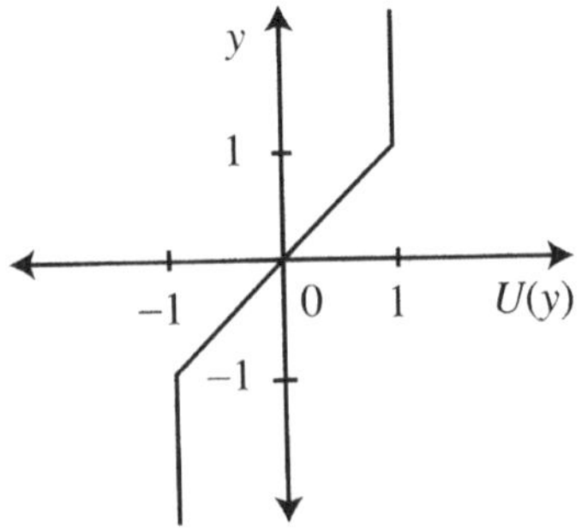

Figure 16.11 Shear layer with piecewise linear velocity profile.

$$A = B + Ce^{2k}, \qquad D = C + Be^{2k}. \tag{16.7.16}$$

The perturbation equations (16.7.5) can be used to obtain an equation for the pressure perturbation amplitude as

$$\hat{p}(y) = \frac{i}{k}\left[-(U - c)\frac{d\hat{v}}{dy} + \frac{dU}{dy}\hat{v}\right]. \tag{16.7.17}$$

Pressure continuity at $y = \pm 1$ gives

$$k(1 - c)A = k(1 - c)(B - Ce^{2k}) + B + Ce^{2k}, \tag{16.7.18a}$$

$$k(1 + c)D = k(1 + c)(-Be^{2k} + C) + Be^{2k} + C. \tag{16.7.18b}$$

Substituting the expression (16.7.16) for A and D into (16.7.18) gives a system of two equations for B and C as

$$\begin{pmatrix} 1 & [1 - 2k(1 - c)]e^{2k} \\ [1 - 2k(1 + c)]e^{2k} & 1 \end{pmatrix} \begin{pmatrix} B \\ C \end{pmatrix} = \begin{pmatrix} 0 \\ 0 \end{pmatrix}. \tag{16.7.19}$$

Setting the determinant of this matrix equation equal to zero gives an equation for the constant c as

$$c^2 = \frac{1}{4k^2}[(1 - 2k)^2 - e^{-4k}]. \tag{16.7.20}$$

A plot of c^2 versus k is given in Figure 16.12. This plot indicates that c is imaginary for wavenumbers in the interval $0 < k < 0.64$, and that the growth rate kc_I is greatest for $k \cong 0.40$. In comparison to the Kelvin-Helmholtz instability for a vortex sheet

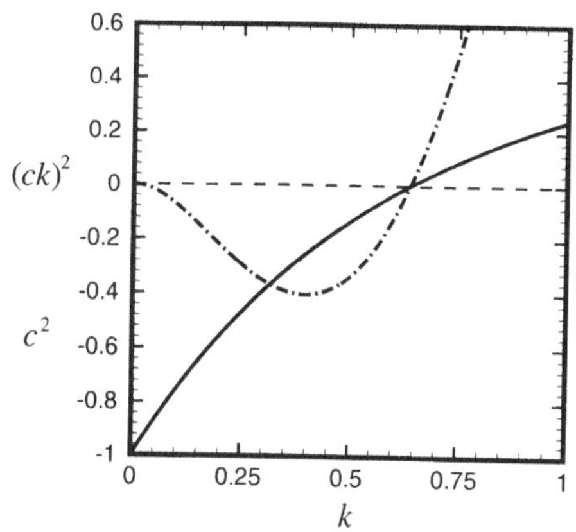

Figure 16.12 Plot of c^2 and $(ck)^2$ versus wavenumber k for the piecewise linear shear layer shown in Figure 16.11.

(Section 16.4), the presence of nonzero shear-layer thickness in the present example has the effect of stabilizing the short waves (for large k) and making the wavenumber of the fastest growing wave finite.

16.8 THREE-DIMENSIONAL INSTABILITY OF A VORTEX PAIR

Two antiparallel vortex tubes exhibit a long-wave instability due to the mutual and self interaction between the tubes when the distance separating the vortices is sufficiently small. The linear theory describing this instability for equal-strength vortices is usually credited to Crow (1970), although similar calculations are reported in the context of helium II by Raja Gopal (1963) using the local-induction approximation for the vortex self-induced velocity. This instability leads to the "cross-linking" of aircraft trailing vortices that can be observed on clear days with little ambient atmospheric turbulence.

The flow under consideration involves two vortices with strength Γ and $-\Gamma$, with equal core radius a and initial average separation distance b (Figure 16.13). In a Cartesian coordinate system (X, Y, Z), the two unperturbed vortices at the initial instant are located at $Y_1(0) = -b/2$ and $Y_2(0) = b/2$. The unperturbed vortices translate in the Z-direction with a constant speed $W = \Gamma/2\pi b$.

The position vector of the vortex filaments is denoted by $\mathbf{r}_N(\xi, t)$, where ξ specifies distance along the x coordinate axis and $N = \{1, 2\}$ identifies the vortex. The equation of motion for the perturbed vortices is

$$\frac{d\mathbf{r}_N}{dt} = \mathbf{u}_N, \tag{16.8.1}$$

where $\mathbf{u}_N$ is the sum of the self- and mutually induced velocity on the vortices. Perturbations $y_N(\xi, t)$ and $z_N(\xi, t)$ are introduced in the direction orthogonal to the unperturbed vortex axes, so that the vortex positions in the perturbed state are given by

$$\mathbf{r}_N(\xi, t) = \xi\mathbf{e}_x + \left[\left(\tfrac{1}{2}b\right)(-1)^N + y_N\right]\mathbf{e}_y + [Wt + z_N]\mathbf{e}_z. \tag{16.8.2}$$

The velocity field $\mathbf{u}_N$ can similarly be written as

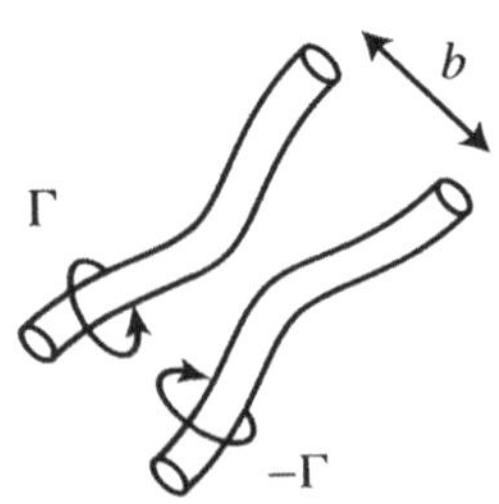

Figure 16.13 Antiparallel vortex pair with symmetric mode perturbations.

$$\mathbf{u}_N = u_{N2}\mathbf{e}_y + (u_{N3} + W)\mathbf{e}_z, \tag{16.8.3}$$

where $u_{N2}(\xi, t)$ and $u_{N3}(\xi, t)$ are velocity components arising from the vortex perturbations. Substituting (16.8.2) and (16.8.3) into (16.8.1) gives the vortex perturbation equations as

$$\frac{dy_N}{dt} = u_{N2}, \qquad \frac{dz_N}{dt} = u_{N3}. \tag{16.8.4}$$

The vortex perturbations are assumed to have the form

$$y_N(\xi, t) = \hat{y}_N e^{i(k\xi - \sigma t)}, \qquad z_N(\xi, t) = \hat{z}_N e^{i(k\xi - \sigma t)}, \tag{16.8.5}$$

where k is the wavenumber, σ is the frequency, and $(\hat{y}_N, \hat{z}_N)$ are constants. A dimensionless vortex separation distance β and a dimensionless vortex core radius δ are defined by $\beta \equiv bk$ and $\delta \equiv ka$, respectively. The vortex perturbation velocities are obtained by Crow (1970) in terms of the *mutual-induction functions* $\psi(\beta)$ and $\chi(\beta)$ and the *self-induction function* $\omega(\delta)$ as

$$u_{12} = \frac{\Gamma}{2\pi b^2}[-z_1 + \psi(\beta)z_2 - \beta^2\omega(\hat{\delta})z_1], \tag{16.8.6a}$$

$$u_{13} = \frac{\Gamma}{2\pi b^2}[-y_1 + \chi(\beta)y_2 + \beta^2\omega(\hat{\delta})y_1], \tag{16.8.6b}$$

$$u_{22} = \frac{\Gamma}{2\pi b^2}[-z_2 - \psi(\beta)z_1 + \beta^2\omega(\hat{\delta})z_2], \tag{16.8.6c}$$

$$u_{23} = \frac{\Gamma}{2\pi b^2}[-y_2 - \chi(\beta)y_1 - \beta^2\omega(\hat{\delta})y_2], \tag{16.8.6d}$$

where $\hat{\delta} \equiv \delta_C ka$ and $\delta_C = 0.5\exp(\frac{1}{4})$ is the Crow cut-off constant. This result requires that the amplitude A of the vortex displacement $[A \equiv \max_{-\pi/k<\xi<\pi/k}(y_N^2 + z_N^2)^{1/2}]$ is much smaller than the vortex separation distance ($A/b \ll 1$) and that the perturbed vortex has small slope ($Ak \ll 1$). The mutual-induction functions are given in terms of the modified Bessel function of the second kind by

$$\psi(\beta) = \beta^2 K_0(\beta) + \beta K_1(\beta), \qquad \chi(\beta) = \beta K_1(\beta). \tag{16.8.7}$$

The self-induction function is given for long-wave perturbations ($\delta \ll 1$) by the cut-off approximation as

$$\omega(\hat{\delta}) = \frac{1}{2}\left[\frac{\cos\hat{\delta} - 1}{\hat{\delta}^2} + \frac{\sin\hat{\delta}}{\hat{\delta}} - Ci(\hat{\delta})\right], \tag{16.8.8}$$

where $Ci(\cdot)$ is the cosine integral. In practice, (16.8.8) is a reasonable approximation for $\delta \leq 0.5$, and it deviates substantially from the exact solution for $\delta > 1$.

Symmetric perturbations (y_S, z_S) and antisymmetric perturbations (y_A, z_A) are defined by

$$y_S = \frac{y_2 - y_1}{b}, \qquad z_S = \frac{z_2 + z_1}{b}, \tag{16.8.9}$$
$$y_A = \frac{y_2 + y_1}{b}, \qquad z_A = \frac{z_2 - z_1}{b}.$$

Defining the dimensionless frequency by $\alpha \equiv 2\pi b^2\sigma/\Gamma$, the perturbation equations (16.8.4) can be written using (16.8.6) as

$$-i\alpha\hat{y}_S = F_2\hat{z}_S, \qquad -i\alpha\hat{z}_S = F_1\hat{y}_S, \tag{16.8.10}$$
$$-i\alpha\hat{y}_A = G_2\hat{z}_A, \qquad -i\alpha\hat{z}_A = G_1\hat{y}_A,$$

where

$$F_1 \equiv 1 + \chi(\beta) - \beta^2\omega(\hat{\delta}), \qquad F_2 \equiv 1 - \psi(\beta) + \beta^2\omega(\hat{\delta}), \tag{16.8.11}$$
$$G_1 \equiv 1 - \chi(\beta) - \beta^2\omega(\hat{\delta}), \qquad G_2 \equiv 1 + \psi(\beta) + \beta^2\omega(\hat{\delta}).$$

Solving the eigenvalue problem (16.8.10) yields two solutions for the growth rate α as

$$\alpha^2 = F_1F_2, \; G_1G_2. \tag{16.8.12}$$

A stability diagram is given in Figure 16.14, where the stable region corresponds to conditions where α^2 is positive and the unstable region to conditions where α^2 is negative. The fastest growing waves are indicated by a dashed line in Figure 16.14. The dimensionless perturbation growth rate is plotted as a function of β along the cut $\delta/\beta = 0.2$ in Figure 16.15.

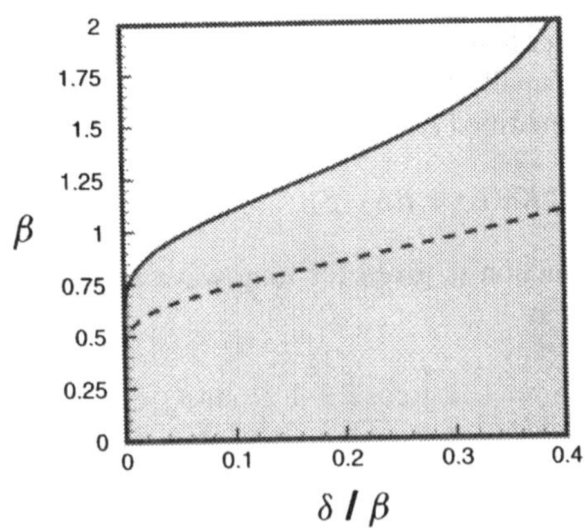

Figure 16.14 Stability diagram for an antiparallel vortex pair, showing dimensionless wavenumber β for stable (white) and unstable (shaded) perturbations. Conditions corresponding to the fastest growing wave are indicated by a dashed line.

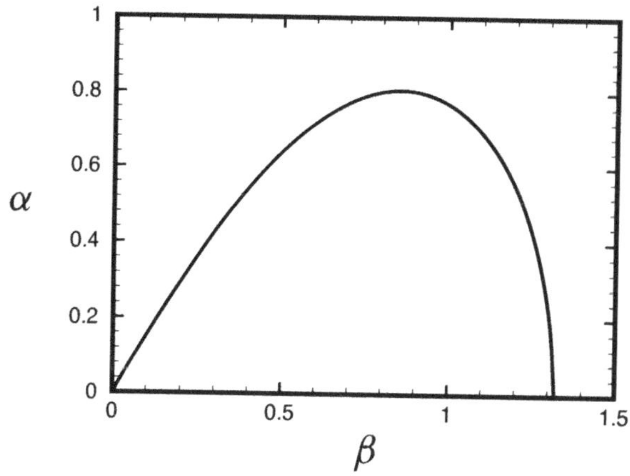

Figure 16.15 Dimensionless growth rate α of the vortex pair instability as a function of dimensionless wavenumber β.

The Crow stability analysis has been extended by a number of investigators. Moore (1972) used numerical computations (based on the filament model with cut-off) to show that the growth rate of unstable waves predicted by the linear stability theory remains approximately valid until nearly the point at which the vortices touch. Jimenez (1975) showed that a pair of parallel (corotating) vortices is unconditionally stable in the linear theory. Moore and Saffman (1972) and Widnall and Bliss (1971) examined the effect of uniform axial flow within the vortex core on the vortex stability. The effect of axial stretching and cross-flow shearing on the vortex stability is examined by Marshall (1992) and Marshall and Chen (1997), respectively. Stability of antiparallel vortices with unequal strength is examined by Klein et al. (1995). Stability of systems of parallel and antiparallel vortices is studied by Robinson and Saffman (1982).

BIBLIOGRAPHY

Chandrasekhar, S. (1961). *Hydrodynamic and Hydromagnetic Stability*, Clarendon Press, Oxford (reprinted Dover Publications, New York, 1981).

Crow, S.C. (1970). "Stability theory for a pair of trailing vortices," *AIAA Journal* **8**, 2172–2179.

Drazin, P.G. and W.H. Reid (1981). *Hydrodynamic Stability*, Cambridge University Press, Cambridge.

Fleischmann, S.T. and D.W. Sallet (1981). "Vortex shedding from cylinders and the resulting unsteady forces and flow phenomenon. Parts I and II," *Shock and Vibration Digest* **13**(10), 9–22, and **13**(11), 15–24.

Hooker, S.G. (1936). "On the action of viscosity in increasing the spacing ratio of a vortex street," *Proceedings of the Royal Society of London A* **154**, 67–89.

Jimenez, J. (1975). "Stability of a pair of co-rotating vortices," *Physics of Fluids* **18**(11), 1580–1581.

Kida, S. (1981). "Motion of an elliptical vortex in a uniform shear flow," *Journal of the Physical Society of Japan* **50**, 3517–3520.

Klein, R., A.J. Majda, and K. Damodaran (1995). "Simplified equations for the interaction of nearly parallel vortex filaments," *Journal of Fluid Mechanics* **288**, 201–248.

Koschmieder, E.L. (1979). "Turbulent Taylor vortex flow," *Journal of Fluid Mechanics* **93**, 515–527.

Kreyszig, E. (1999). *Advanced Engineering Mathematics*, 8th ed., John Wiley & Sons, New York.

Lamb, H. (1932). *Hydrodynamics*, Cambridge University Press, Cambridge (reprinted Dover Publications, New York, 1945).

Marshall, J.S. (1992). "The effect of axial stretching on the three-dimensional stability of a vortex pair," *Journal of Fluid Mechanics* **241**, 403–419.

Marshall, J.S., and H. Chen (1997). "Stability of a counter-rotating vortex pair immersed in cross-stream shear flow," *AIAA Journal* **35**(2), 295–305.

Moore, D.W. (1972). "Finite amplitude waves on aircraft trailing vortices," *Aeronautical Quarterly* **23**, 307–314.

Moore, D.W. (1979). "The spontaneous appearance of a singularity in the shape of an evolving vortex sheet," *Proceedings of the Royal Society of London A* **365**, 105–119.

Moore, D.W. and R. Griffith-Jones (1974). "The stability of an expanding circular vortex sheet," *Mathematika* **21**, 128–133.

Moore, D.W. and P.G. Saffman (1971). "Structure of a line vortex in an imposed strain," in *Aircraft Wake Turbulence and Its Detection*, edited by J.H. Olsen, A. Goldburg, and M. Rogers, Plenum Press, New York, pp. 339–354.

Moore, D.W. and P.G. Saffman (1972). "The motion of a vortex filament with axial flow," *Philosophical Transactions of the Royal Society of London A* **272**, 403–429.

Neu, J.C. (1984). "The dynamics of a columnar vortex in an imposed strain," *Physics of Fluids* **27**, 2397–2402.

Raja Gopal, E.S. (1963). "Motion and stability of vortices in a finite channel: application to liquid helium II," *Annals of Physics* **25**, 196–220.

Robinson, A.C. and P.G. Saffman (1982). "Three-dimensional stability of vortex arrays," *Journal of Fluid Mechanics* **125**, 411–427.

Rutland, D.F. and G.J. Jameson (1971). "A non-linear effect in the capillary instability of liquid jets," *Journal of Fluid Mechanics* **46**, 267–271.

Saffman, P.G. (1992). *Vortex Dynamics*, Cambridge University Press, Cambridge.

Squire, H.B. (1933). "On the stability of three-dimensional disturbances of viscous flow between parallel walls," *Proceedings of the Royal Society of London A* **142**, 621–628.

Stakgold, I. (1979). *Green's Functions and Boundary Value Problems*, John Wiley & Sons, New York.

Stuart, J.T. (1963). "Hydrodynamic stability," in *Laminar Boundary Layers*, ed. L. Rosenhead, Oxford University Press, Oxford, pp. 492–579 (reprinted Dover Publications, New York, 1988).

Van Dyke, M. (1982). *An Album of Fluid Motion*, Parabolic Press, Stanford, CA.

von Kármán, T. and H. Rubach (1912). "Über den mechanismus des Flüssigkeits und Luftwiderstandes," *Physikalische Zeitschrift* **13**(2), 49–59.

Widnall, S.E. and D.B. Bliss (1971). "Slender-body analysis of the motion and stability of a vortex filament containing an axial flow," *Journal of Fluid Mechanics* **50**, 335–353.
Wille, R. (1960). "Kármán vortex streets," *Advances in Applied Mechanics* **6**, 273–287.

PROBLEMS

1. Consider the following system of nonlinear ordinary differential equations for the unknowns $x(t)$ and $\theta(t)$:

$$\frac{dx}{dt} = A(x^2 - 3)\cos\theta, \qquad \frac{d\theta}{dt} = x^4 - 1,$$

where $0 < \theta < \pi$, $-\infty < x < \infty$, and A is a positive parameter. All quantities in this system are real valued.

   (a) Find two equilibrium solutions for this system.

   (b) Determine whether or not each of these equilibrium solutions are linearly stable.

2. Estimate the rate of strain induced on a two-dimensional vortex patch by another vortex patch, where both patches are initially circular with radius a and the vortex centers are separated by a distance b.

   (a) For the case where the vortices are of equal strength, use this estimate to (qualitatively) sketch the form of the deformed patches in a stable configura tion and determine whether this stable configuration can be maintained until the cores touch.

   (b) Repeat part (a) for the case where the other vortex has three times the strength of the first vortex. Estimate the value of the vortex separation distance for which the weaker patch will not be able to maintain an elliptical shape with constant aspect ratio.

3. Consider the problem of wave generation due to wind blowing over a nearly flat water surface. Assume the water to be infinitely deep and to have no ambient currents, and admit a nonzero surface tension at the interface.

   (a) Calculate the wind speed U_{crit} at which the interface first becomes unstable. [Note that the ratio $(\rho_2 - \rho_1)/(\rho_2 + \rho_1)$ is not small in this problem.]

   (b) At a wind speed $U_1 = 30$ m/s, what is the wavelength of the fastest growing wave? Based on your experience, does this seem to be a reasonable value for real water waves under this wind condition?

4. The oceans can be approximately modeled as consisting of an upper surface layer and a slightly denser and much thicker deep-water layer (Figure 16.16). A relatively sudden density transition between these two layers occurs in the pycnocline, which for purposes of long-wave internal wave motion can be idealized as a sharp material interface with no surface tension. A possible source of mixing

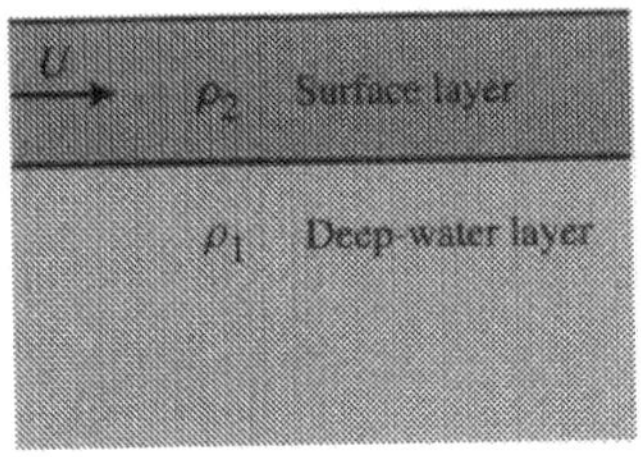

Figure 16.16 Approximate two-layer ocean model, consisting of a shallow surface layer covering a deep-water layer.

between these two layers occurs in response to the Kelvin-Helmholtz instability of the pycnocline due to wind-driven currents in the surface layer. We are given that $\rho_1 = 1010\ \text{kg/m}^2$, $\rho_2 = 1020\ \text{kg/m}^2$, $g = 9.8\ \text{m/s}^2$, and $h_1 = 150$ m. Solutions for the velocity potential ϕ and interface displacement η that satisfy the irrotational flow equations and the boundary conditions at $y = h_1$ and $y \to -\infty$ are

$$\phi_1 = B_1 \cosh[k(y - h_1)]e^{i(kx-\sigma t)}, \qquad \phi_2 = B_2 e^{ky} e^{i(kx-\sigma t)},$$
$$\eta = Ae^{i(kx-\sigma t)}.$$

(a) Write the linearized interfacial matching conditions, and state all assumptions that have been made. Using the above solutions, obtain the characteristic equation for the frequency $\sigma(k)$.

(b) The pycnocline thickness is about $h_p = 20$ m. Assume that unstable waves with wavelength λ less than h_p are damped out by the effects of finite interface thickness. Determine the minimum current speed U_1 for which there exist unstable waves with wavelength $\lambda \geq h_p$.

5. In Section 16.5, it is shown that inviscid theory predicts that all liquid jets immersed in a gas are unstable due to surface tension effects. Consider a water jet immersed in air at standard temperature and pressure. Write the ratio of the instability growth rate σ_I and the free-stream advection rate kW_0 as a function of the Weber number We $\equiv \rho a W_0^2/\gamma$. For a typical kitchen faucet, $a = 0.5$ cm and $W_0 = 0.1$ m/s. Estimate the Weber number and the distance that a perturbation would be advected downstream before its amplitude doubles due to the capillary instability.

6. Categorize each of the following fluid flows according to one of the following: (i) must be linearly stable, (ii) may be either linearly stable or unstable, or (iii) must be unstable. In each case, clearly explain the reason for your answer. Do not attempt to perform stability analyses from the basic Euler equations. In this problem, (r, θ) are polar coordinates, (x, y) are Cartesian coordinates, and a, Γ, and U are positive constants. All basic flows are two-dimensional, but the perturbations may be three-dimensional.

(a) $u_r = 0, \qquad u_\theta = \frac{\Gamma}{2\pi r}(1 - e^{-r^2/a^2})$

(b) $u_r = 0, \qquad u_\theta = \frac{\Gamma}{2\pi r}\left[1 - \left(1 - \frac{r^2}{a^2}\right) e^{-r^3/a^3}\right]$

(c) $u_x = U\left[1 - \left(1 - \frac{y}{a}\right) e^{-y/a}\right], \qquad v = 0 \text{ for } y > 0$

(d) $u_x = U(1 - e^{-y/a}), \qquad v = 0 \text{ for } y > 0$

7. The trailing vortices behind a jet aircraft have strength $\Gamma = 2000\ \text{m}^2/\text{s}$, core radius $a = 2$ m, and separation distance $b = 10$ m. Suppose that the initial perturbation amplitude of the trailing vortices at the rear of the aircraft is $A = 0.1$ m, due to turbulence in the aircraft wake.
   (a) What is the wavelength of the fastest growing perturbation?
   (b) Estimate the distance behind the aircraft at which the vortex pair perturbation amplitude grows so large that the trailing vortices touch.

APPENDIX A

COMMON EXPRESSIONS IN ORTHOGONAL CURVILINEAR COORDINATE SYSTEMS

We consider a system of orthogonal curvilinear coordinates with unit base vectors **a**, **b**, and **c** and related coordinates ξ_1, ξ_2, and ξ_3. A directed line segment $d\mathbf{x}$ of infinitesimal length can be expressed in terms of these base vectors as

$$d\mathbf{x} = \mathbf{a}h_1 d\xi_1 + \mathbf{b}h_2 d\xi_2 + \mathbf{c}h_3 d\xi_3, \tag{A.1}$$

where the *metrical coefficients* h_1, h_2, and h_3 are in general functions of ξ_1, ξ_2, and ξ_3. The product of a metrical coefficient and a coordinate increment is equal to an increment of arc length along the associated coordinate direction. The element of volume is then written in terms of the coordinate increments as

$$dv = h_1 h_2 h_3 d\xi_1 d\xi_2 d\xi_3. \tag{A.2}$$

The base vectors **a**, **b**, **c** are proportional to the derivative of $d\mathbf{x}$ in the direction of the respective coordinates. The above expression can be used together with the orthogonality of the base vectors to show that the derivatives of the base vectors along their respective coordinate directions are

$$\begin{aligned}
\frac{\partial \mathbf{a}}{\partial \xi_1} &= -\frac{1}{h_2}\frac{\partial h_1}{\partial \xi_2}\mathbf{b} - \frac{1}{h_3}\frac{\partial h_1}{\partial \xi_3}\mathbf{c}, \\
\frac{\partial \mathbf{b}}{\partial \xi_2} &= -\frac{1}{h_3}\frac{\partial h_2}{\partial \xi_3}\mathbf{c} - \frac{1}{h_1}\frac{\partial h_2}{\partial \xi_1}\mathbf{a}, \\
\frac{\partial \mathbf{c}}{\partial \xi_3} &= -\frac{1}{h_1}\frac{\partial h_3}{\partial \xi_1}\mathbf{a} - \frac{1}{h_2}\frac{\partial h_3}{\partial \xi_2}\mathbf{b}
\end{aligned} \tag{A.3}$$

TABLE A.1 Common Examples of Orthogonal Curvilinear Coordinate Systems and Associated Metrical Coefficients.

System	ξ_1	ξ_2	ξ_3	h_1	h_2	h_3
Cartesian	x	y	z	1	1	1
Cylindrical	R	α	z	1	R	1
Spherical	r	θ	α	1	r	$r \sin\theta$
Elliptical	ξ	η	z	$c(\sinh^2\xi + \sin^2\eta)^{1/2}$	$c(\sinh^2\xi + \sin^2\eta)^{1/2}$	1

and their derivatives in orthogonal directions are

$$\begin{aligned}
\frac{\partial \mathbf{a}}{\partial \xi_2} &= \frac{1}{h_1}\frac{\partial h_2}{\partial \xi_1}\mathbf{b}, & \frac{\partial \mathbf{a}}{\partial \xi_3} &= \frac{1}{h_1}\frac{\partial h_3}{\partial \xi_1}\mathbf{c},\\
\frac{\partial \mathbf{b}}{\partial \xi_1} &= \frac{1}{h_2}\frac{\partial h_1}{\partial \xi_2}\mathbf{a}, & \frac{\partial \mathbf{b}}{\partial \xi_3} &= \frac{1}{h_2}\frac{\partial h_3}{\partial \xi_2}\mathbf{c},\\
\frac{\partial \mathbf{c}}{\partial \xi_1} &= \frac{1}{h_3}\frac{\partial h_1}{\partial \xi_3}\mathbf{a}, & \frac{\partial \mathbf{c}}{\partial \xi_2} &= \frac{1}{h_3}\frac{\partial h_2}{\partial \xi_3}\mathbf{b}.
\end{aligned} \tag{A.4}$$

Common examples of orthogonal curvilinear coordinate systems are listed in Table A.1.

The gradient of a scalar function F is given by

$$\nabla F = \frac{\mathbf{a}}{h_1}\frac{\partial F}{\partial \xi_1} + \frac{\mathbf{b}}{h_2}\frac{\partial F}{\partial \xi_2} + \frac{\mathbf{c}}{h_3}\frac{\partial F}{\partial \xi_3}. \tag{A.5}$$

In taking derivatives of a vector quantity $\mathbf{f} = f_1\mathbf{a} + f_2\mathbf{b} + f_2\mathbf{c}$, it is necessary to account for the derivatives of both the scalar coefficients and the base vectors. The divergence and the curl operators acting on $\mathbf{f}$ can be written as

$$\nabla\cdot\mathbf{f} = \frac{1}{h_1h_2h_3}\left\{\frac{\partial(h_2h_3f_1)}{\partial\xi_1} + \frac{\partial(h_3h_1f_2)}{\partial\xi_2} + \frac{\partial(h_1h_2f_3)}{\partial\xi_3}\right\}, \tag{A.6}$$

$$\begin{aligned}
\nabla\times\mathbf{f} = {}& \frac{\mathbf{a}}{h_2h_3}\left\{\frac{\partial(h_3f_3)}{\partial\xi_2} - \frac{\partial(h_2f_2)}{\partial\xi_3}\right\} + \frac{\mathbf{b}}{h_3h_1}\left\{\frac{\partial(h_1f_1)}{\partial\xi_3} - \frac{\partial(h_3f_3)}{\partial\xi_1}\right\}\\
& + \frac{\mathbf{c}}{h_1h_2}\left\{\frac{\partial(h_2f_2)}{\partial\xi_1} - \frac{\partial(h_1f_1)}{\partial\xi_2}\right\}.
\end{aligned} \tag{A.7}$$

The Laplacian of a vector $\mathbf{f}$ can be obtained from the above expressions using either the definition $\nabla^2\mathbf{f} = \nabla\cdot\nabla\mathbf{f}$ or the vector identity $\nabla^2\mathbf{f} = \nabla(\nabla\cdot\mathbf{f}) - \nabla\times(\nabla\times\mathbf{f})$.

The directional derivative of a vector $\mathbf{f}$ along a direction $\mathbf{u}$ is given by

$$\begin{aligned}
&(\mathbf{u}\cdot\nabla)\mathbf{f}\\
&= \mathbf{a}\left\{\mathbf{u}\cdot\nabla f_1 + \frac{f_2}{h_1h_2}\left(u_1\frac{\partial h_1}{\partial\xi_2} - u_2\frac{\partial h_2}{\partial\xi_1}\right) + \frac{f_3}{h_1h_3}\left(u_1\frac{\partial h_1}{\partial\xi_3} - u_3\frac{\partial h_3}{\partial\xi_1}\right)\right\}
\end{aligned}$$

$$+\mathbf{b}\left\{\mathbf{u}\cdot\nabla f_2 + \frac{f_3}{h_2h_3}\left(u_2\frac{\partial h_2}{\partial \xi_3} - u_3\frac{\partial h_3}{\partial \xi_1}\right) + \frac{f_1}{h_2h_1}\left(u_2\frac{\partial h_2}{\partial \xi_1} - u_1\frac{\partial h_1}{\partial \xi_2}\right)\right\}$$

$$+\mathbf{c}\left\{\mathbf{u}\cdot\nabla f_3 + \frac{f_1}{h_3h_1}\left(u_3\frac{\partial h_3}{\partial \xi_1} - u_1\frac{\partial h_1}{\partial \xi_3}\right) + \frac{f_2}{h_3h_2}\left(u_3\frac{\partial h_3}{\partial \xi_2} - u_2\frac{\partial h_2}{\partial \xi_3}\right)\right\}. \quad \text{(A.8)}$$

The components D_{ij} of the rate of deformation tensor are given by the scalar product of $\mathbf{D}$ with the two base vectors indicated by the indices i and j. For instance, $D_{12} = \mathbf{a}\cdot\mathbf{D}\cdot\mathbf{b}$, and so forth. Using the definition (3.2.5) of $\mathbf{D}$, we can rewrite this expression in terms of the directional derivative of the velocity $\mathbf{u}$ as $D_{12} = \frac{1}{2}\{\mathbf{b}\cdot[(\mathbf{a}\cdot\nabla)\mathbf{u}] + \mathbf{a}\cdot[(\mathbf{b}\cdot\nabla)\mathbf{u}]\}$. Using (A.8), the components of the rate of deformation tensor are obtained as

$$\begin{aligned}
D_{11} &= \frac{1}{h_1}\frac{\partial u_1}{\partial \xi_1} + \frac{u_2}{h_1h_2}\frac{\partial h_1}{\partial \xi_2} + \frac{u_3}{h_1h_3}\frac{\partial h_1}{\partial \xi_3},\\
D_{22} &= \frac{1}{h_2}\frac{\partial u_2}{\partial \xi_2} + \frac{u_3}{h_2h_3}\frac{\partial h_2}{\partial \xi_3} + \frac{u_1}{h_2h_1}\frac{\partial h_2}{\partial \xi_1},\\
D_{33} &= \frac{1}{h_3}\frac{\partial u_3}{\partial \xi_3} + \frac{u_1}{h_3h_1}\frac{\partial h_3}{\partial \xi_1} + \frac{u_2}{h_3h_2}\frac{\partial h_3}{\partial \xi_2},\\
D_{12} &= \frac{h_2}{2h_1}\frac{\partial\,(u_2/h_2)}{\partial \xi_1} + \frac{h_1}{2h_2}\frac{\partial\,(u_1/h_1)}{\partial \xi_2}, \qquad \text{(A.9)}\\
D_{23} &= \frac{h_3}{2h_2}\frac{\partial\,(u_3/h_3)}{\partial \xi_2} + \frac{h_2}{2h_3}\frac{\partial\,(u_2/h_2)}{\partial \xi_3},\\
D_{13} &= \frac{h_3}{2h_1}\frac{\partial\,(u_3/h_3)}{\partial \xi_1} + \frac{h_1}{2h_3}\frac{\partial\,(u_1/h_1)}{\partial \xi_3}.
\end{aligned}$$

All of the terms in the equations of motion can be obtained in different coordinate systems by substitution of the appropriate values of the coordinates ξ_1, ξ_2, ξ_3 and the coefficients h_1, h_2, h_3 into the above expressions. Some examples in common orthogonal coordinate systems are given below.

A.1 Cylindrical Polar Coordinates

Common vector operations in cylindrical polar coordinates are

$$(\xi_1, \xi_2, \xi_3) = (R, \alpha, z), \qquad (\mathbf{a}, \mathbf{b}, \mathbf{c}) = (\mathbf{e}_r, \mathbf{e}_\alpha, \mathbf{e}_z), \qquad dv = R\,dR\,d\alpha\,dz,$$

$$h_1 = 1, \qquad h_2 = R, \qquad h_3 = 1, \qquad \mathbf{f} = f_R\mathbf{e}_R + f_\alpha\mathbf{e}_\alpha + f_z\mathbf{e}_z,$$

$$\frac{\partial \mathbf{e}_R}{\partial \alpha} = \mathbf{e}_\alpha, \qquad \frac{\partial \mathbf{e}_\alpha}{\partial \alpha} = -\mathbf{e}_R, \qquad \frac{\partial \mathbf{e}_z}{\partial \alpha} = 0,$$

$$\nabla F = \mathbf{e}_R\frac{\partial F}{\partial R} + \frac{\mathbf{e}_\alpha}{R}\frac{\partial F}{\partial \alpha} + \mathbf{e}_z\frac{\partial F}{\partial z},$$

$$(\mathbf{u}\cdot\nabla)\mathbf{f} = \mathbf{e}_R\left(\mathbf{u}\cdot\nabla f_R - \frac{u_\alpha f_\alpha}{R}\right) + \mathbf{e}_\alpha\left(\mathbf{u}\cdot\nabla f_\alpha + \frac{u_\alpha f_R}{R}\right) + \mathbf{e}_z(\mathbf{u}\cdot\nabla f_z),$$

$$\nabla\cdot\mathbf{f} = \frac{1}{R}\frac{\partial(Rf_R)}{\partial R} + \frac{1}{R}\frac{\partial f_\alpha}{\partial\alpha} + \frac{\partial f_z}{\partial z},$$

$$\nabla\times\mathbf{f} = \mathbf{e}_R\left\{\frac{1}{R}\frac{\partial f_z}{\partial\alpha} - \frac{\partial f_\alpha}{\partial z}\right\} + \mathbf{e}_\alpha\left\{\frac{\partial f_R}{\partial z} - \frac{\partial f_z}{\partial R}\right\} + \frac{\mathbf{e}_z}{R}\left\{\frac{\partial(Rf_\alpha)}{\partial R} - \frac{\partial f_R}{\partial\alpha}\right\},$$

$$\nabla^2 F = \frac{1}{R}\frac{\partial}{\partial R}\left(R\frac{\partial F}{\partial R}\right) + \frac{1}{R^2}\frac{\partial^2 F}{\partial\alpha^2} + \frac{\partial^2 F}{\partial z^2},$$

$$\nabla^2\mathbf{f} = \mathbf{e}_R\left\{\nabla^2 f_R - \frac{f_R}{R^2} - \frac{2}{R^2}\frac{\partial f_\alpha}{\partial\alpha}\right\} + \mathbf{e}_\alpha\left\{\nabla^2 f_\alpha - \frac{f_\alpha}{R^2} + \frac{2}{R^2}\frac{\partial f_R}{\partial\alpha}\right\}$$

$$+\mathbf{e}_z\left\{\nabla^2 f_z\right\}.$$

The components of the rate of deformation tensor can be written in terms of the components of the velocity vector $\mathbf{u}$ as

$$D_{RR} = \frac{\partial u_R}{\partial R}, \qquad D_{\alpha\alpha} = \frac{1}{R}\frac{\partial u_\alpha}{\partial\alpha} + \frac{u_R}{R}, \qquad D_{zz} = \frac{\partial u_z}{\partial z},$$

$$D_{R\alpha} = \frac{R}{2}\frac{\partial}{\partial R}\left(\frac{u_\alpha}{R}\right) + \frac{1}{2R}\frac{\partial u_R}{\partial\alpha}, \qquad D_{\alpha z} = \frac{1}{2}\frac{\partial u_\alpha}{\partial z} + \frac{1}{2R}\frac{\partial u_z}{\partial\alpha},$$

$$D_{zR} = \frac{1}{2}\frac{\partial u_R}{\partial z} + \frac{1}{2}\frac{\partial u_z}{\partial R}.$$

The Euler equations are

$$\frac{\partial u_R}{\partial t} + (\mathbf{u}\cdot\nabla)u_R - \frac{u_\alpha^2}{R} = -\frac{1}{\rho}\frac{\partial p}{\partial R},$$

$$\frac{\partial u_\alpha}{\partial t} + (\mathbf{u}\cdot\nabla)u_\alpha + \frac{u_R u_\alpha}{R} = -\frac{1}{\rho R}\frac{\partial p}{\partial\alpha},$$

$$\frac{\partial u_z}{\partial t} + (\mathbf{u}\cdot\nabla)u_z = -\frac{1}{\rho}\frac{\partial p}{\partial z}.$$

The components of the vorticity transport equation for an inviscid, uniform-density fluid are

$$\frac{\partial\omega_R}{\partial t} + (\mathbf{u}\cdot\nabla)\omega_R = \boldsymbol{\omega}\cdot\nabla u_R,$$

$$\frac{\partial\omega_\alpha}{\partial t} + (\mathbf{u}\cdot\nabla)\omega_\alpha + \frac{\omega_R u_\alpha}{R} = \boldsymbol{\omega}\cdot\nabla u_\alpha + \frac{\omega_\alpha u_R}{R},$$

$$\frac{\partial\omega_z}{\partial t} + (\mathbf{u}\cdot\nabla)\omega_z = \boldsymbol{\omega}\cdot\nabla u_z.$$

The equivalent expressions in two-dimensional polar coordinates can be obtained by substituting r, θ for R, α, respectively, and omitting derivatives and components in the z direction.

A.2 Spherical Coordinates

Common vector operations in spherical coordinates are

$$(\xi_1, \xi_2, \xi_3) = (r, \theta, \alpha), \qquad (\mathbf{a}, \mathbf{b}, \mathbf{c}) = (\mathbf{e}_r, \mathbf{e}_\theta, \mathbf{e}_\alpha), \qquad dv = r^2 \sin\theta dr d\theta d\alpha,$$

$$h_1 = 1, \qquad h_2 = r, \qquad h_3 = r\sin\theta, \qquad \mathbf{f} = f_r\mathbf{e}_r + f_\theta\mathbf{e}_\theta + f_\alpha\mathbf{e}_\alpha,$$

$$\frac{\partial \mathbf{e}_r}{\partial r} = 0, \qquad \frac{\partial \mathbf{e}_r}{\partial \theta} = \mathbf{e}_\theta, \qquad \frac{\partial \mathbf{e}_r}{\partial \alpha} = \mathbf{e}_\alpha \sin\theta,$$

$$\frac{\partial \mathbf{e}_\theta}{\partial r} = 0, \qquad \frac{\partial \mathbf{e}_\theta}{\partial \theta} = -\mathbf{e}_r, \qquad \frac{\partial \mathbf{e}_\theta}{\partial \alpha} = \mathbf{e}_\alpha \cos\theta,$$

$$\frac{\partial \mathbf{e}_\alpha}{\partial r} = 0, \qquad \frac{\partial \mathbf{e}_\alpha}{\partial \theta} = 0, \qquad \frac{\partial \mathbf{e}_\alpha}{\partial \alpha} = -\mathbf{e}_r \sin\theta - \mathbf{e}_\theta \cos\theta,$$

$$\nabla F = \mathbf{e}_r \frac{\partial F}{\partial r} + \frac{\mathbf{e}_\theta}{r}\frac{\partial F}{\partial \theta} + \frac{\mathbf{e}_\alpha}{r\sin\theta}\frac{\partial F}{\partial \alpha},$$

$$(\mathbf{u}\cdot\nabla)\mathbf{f} = \mathbf{e}_r\left(\mathbf{u}\cdot\nabla f_r - \frac{u_\theta f_\theta}{r} - \frac{u_\alpha f_\alpha}{r}\right) + \mathbf{e}_\theta\left(\mathbf{u}\cdot\nabla f_\theta - \frac{u_\alpha f_\alpha}{r}\cot\theta + \frac{u_\theta f_r}{r}\right)$$

$$+ \mathbf{e}_\alpha\left(\mathbf{u}\cdot\nabla f_\alpha + \frac{u_\alpha f_r}{r} + \frac{u_\alpha f_\theta}{r}\cot\theta\right),$$

$$\nabla\cdot\mathbf{f} = \frac{1}{r^2}\frac{\partial(r^2 f_r)}{\partial r} + \frac{1}{r\sin\theta}\frac{\partial(\sin\theta f_\theta)}{\partial \theta} + \frac{1}{r\sin\theta}\frac{\partial f_\alpha}{\partial \alpha},$$

$$\nabla\times\mathbf{f} = \frac{\mathbf{e}_r}{r\sin\theta}\left\{\frac{\partial(f_\alpha\sin\theta)}{\partial\theta} - \frac{\partial f_\theta}{\partial\alpha}\right\} + \frac{\mathbf{e}_\theta}{r}\left\{\frac{1}{\sin\theta}\frac{\partial f_r}{\partial\alpha} - \frac{\partial(r f_\alpha)}{\partial r}\right\}$$

$$+ \frac{\mathbf{e}_\alpha}{r}\left\{\frac{\partial(r f_\theta)}{\partial r} - \frac{\partial f_r}{\partial \theta}\right\},$$

$$\nabla^2 F = \frac{1}{r^2}\frac{\partial}{\partial r}\left(r^2\frac{\partial F}{\partial r}\right) + \frac{1}{r^2\sin\theta}\frac{\partial}{\partial\theta}\left(\sin\theta\frac{\partial F}{\partial\theta}\right) + \frac{1}{r^2\sin^2\theta}\frac{\partial^2 F}{\partial\alpha^2},$$

$$\nabla^2\mathbf{f} = \mathbf{e}_r\left\{\nabla^2 f_r - \frac{2f_r}{r^2} - \frac{2}{r^2\sin\theta}\frac{\partial(f_\theta\sin\theta)}{\partial\theta} - \frac{2}{r^2\sin\theta}\frac{\partial f_\alpha}{\partial\alpha}\right\}$$

$$+ \mathbf{e}_\theta\left\{\nabla^2 f_\theta + \frac{2}{r^2}\frac{\partial f_r}{\partial\theta} - \frac{f_\theta}{r^2\sin^2\theta} - \frac{2\cos\theta}{r^2\sin^2\theta}\frac{\partial f_\alpha}{\partial\alpha}\right\}$$

$$+ \mathbf{e}_\alpha\left\{\nabla^2 f_\alpha + \frac{2}{r^2\sin\theta}\frac{\partial f_r}{\partial\alpha} + \frac{2\cos\theta}{r^2\sin^2\theta}\frac{\partial f_\theta}{\partial\alpha} - \frac{f_\alpha}{r^2\sin^2\theta}\right\}.$$

The components of the rate of deformation tensor can be written in terms of the components of the velocity vector *u* as

$$D_{rr} = \frac{\partial u_r}{\partial r}, \qquad D_{\theta\theta} = \frac{1}{r}\frac{\partial u_\theta}{\partial \theta} + \frac{u_r}{r}, \qquad D_{\alpha\alpha} = \frac{1}{r\sin\theta}\frac{\partial u_\alpha}{\partial \alpha} + \frac{u_r}{r} + \frac{u_\theta \cot\theta}{r},$$

$$D_{r\theta} = \frac{r}{2}\frac{\partial}{\partial r}\left(\frac{u_\theta}{r}\right) + \frac{1}{2r}\frac{\partial u_r}{\partial \theta},$$

$$D_{\theta\alpha} = \frac{\sin\theta}{2r}\frac{\partial}{\partial \theta}\left(\frac{u_\alpha}{\sin\theta}\right) + \frac{1}{2r\sin\theta}\frac{\partial u_\theta}{\partial \alpha}, \qquad D_{\alpha r} = \frac{1}{2r\sin\theta}\frac{\partial u_r}{\partial \alpha} + \frac{r}{2}\frac{\partial}{\partial r}\left(\frac{u_\alpha}{r}\right).$$

The Euler equations are

$$\frac{\partial u_r}{\partial t} + (\mathbf{u}\cdot\nabla)u_r - \frac{u_\theta^2}{r} - \frac{u_\alpha^2}{r} = -\frac{1}{\rho}\frac{\partial p}{\partial r},$$

$$\frac{\partial u_\theta}{\partial t} + (\mathbf{u}\cdot\nabla)u_\theta + \frac{u_r u_\theta}{r} - \frac{u_\alpha^2 \cot\theta}{r} = -\frac{1}{\rho r}\frac{\partial p}{\partial \theta},$$

$$\frac{\partial u_\alpha}{\partial t} + (\mathbf{u}\cdot\nabla)u_\alpha + \frac{u_r u_\alpha}{r} + \frac{u_\theta u_\alpha \cot\theta}{r} = -\frac{1}{\rho r\sin\theta}\frac{\partial p}{\partial \alpha}.$$

The components of the vorticity transport equation for an inviscid, uniform-density fluid are

$$\frac{\partial \omega_r}{\partial t} + (\mathbf{u}\cdot\nabla)\omega_r = \boldsymbol{\omega}\cdot\nabla u_r,$$

$$\frac{\partial \omega_\theta}{\partial t} + (\mathbf{u}\cdot\nabla)\omega_\theta + \frac{\omega_r u_\theta}{r} = \boldsymbol{\omega}\cdot\nabla u_\theta + \frac{\omega_\theta u_r}{r},$$

$$\frac{\partial \omega_\alpha}{\partial t} + (\mathbf{u}\cdot\nabla)\omega_\alpha + \frac{\omega_r u_\alpha}{r} + \frac{\omega_\theta u_\alpha \cot\theta}{r} = \boldsymbol{\omega}\cdot\nabla u_\alpha + \frac{\omega_\alpha u_r}{r} + \frac{\omega_\alpha u_\theta \cot\theta}{r}.$$

INDEX

Added mass, 251–252
Airfoil
  cambered, 179–180
  flat–plate, 175–177
  Joukowski, 177–178
  vortex sheet representation, 209–212
Analytic function, 123
Atwood ratio, 327
Axisymmetric flow, 220–221, 260–265

Baroclinic, *see* Vorticity, baroclinic generation
Barotropic fluid, 75, 93
Bernoulli coefficient, 115, 118–120
Bernoulli theorem, 115, 213
Bifurcation, 332
  Hopf, 332–333
  pitchfork, 332–333
  turning point, 332–333, 336
Biot-Savart equation, 86–89
Birkhoff-Rott equation, 207
Blasius
  force law, 168–169
  moment law, 170
Boundary-integral method
  doublet sheet, 242–243
  for interfacial waves, 325–327
  for pressure, 116–119
  source sheet, 238–242
  vortex sheet, 243–245
Boundary layer, 4
Bragg-Hawthorne equation, 279
Bubble
  oscillations of, 229–230
  pressure difference over surface of, 72

Cauchy-Goursat theorem, 116
Cauchy-Riemann equations, 123
Cauchy stress formula, 47–48
Channel flow, through a narrow slit, 153–155
Centrifugal instability, 348–352
Circle theorem, 133–135
Circulation, 39, 90, 105, 108–109
Complex lamellar, 78–79
Configuration
  current, 24
  reference, 24
Conformal transformation, 124, 139–162
  of complex potential, 142
  of complex velocity, 143
  critical point, 139–140
  of vortex and source strength, 143–144
  of a vortex patch, 186–187

Constitutive equation, 52
Constraint, 53–54
  force due to, 54
Contact line, 74
Continuity equation, *see* Incompressibility condition
Continuum, 24
Contour dynamics method, 191–194
  axisymmetric, 273–277
Contour surgery, 192
Contraction, 11
Control volume, 41–42, 45, 49
Couette flow
  plane, 56–57, 355
  rotating, 57–58, 350–352
Creeping flow, 3
Crow cut-off, 300
Crow instability, 358–361
Curl, 14
Cylindrical polar coordinates, 220, 368–370

D'Alembert theorem, 173–174
Deformation
  gradient, 28–29
  Jacobian of, 28
  rate of, 29
Del operator, 14
Derivative
  directional, 14
  material, 26–27
  partial, 26
Determinant, 13
Dilatation, 34
  rate of, 34, 86–87
Dirac delta, 82
Direction cosine tensor, 15
Discontinuity jump condition
  energy, 75
  mass, 66–67
  momentum, 67–68, 70–72
Discontinuity surface, 63–72
Discrete vortex method, 193–203
  axisymmetric, 288–291
Dissipation rate, 5, 53
Divergence
  of a vector, 14, 79
  theorem, 20–21
Doublet
  axisymmetric, 223–224
  two-dimensional, 128–131
Dyadic product, 11

Elliptic coordinates, 140–142, 368
Elliptic integrals, 263, 276–277
Elliptical vortex, 188–191, 193
  stability in straining flow, 332–336
Enstrophy, 103–105
Equipotential
  line, 78
  surface, 78
Euler-Cauchy equation, 57
Euler equation, 55
Eulerian viewpoint, 26

Fjørtoft theorem, 355
Fredholm equation, second kind, 118, 242
Free streamline theory, 157–162
Föppl vortices, 181
Force
  added mass, 252–253
  body, 47
  buoyancy, 253–255
  contact, 46, 69–70
  due to a doublet near a body, 248–249
  due to a source near a body, 246–248
  on a circular cylinder with nonzero circulation, 170–171
  on a small particle near a line vortex, 254–255

Gradient, 14
Green function, 82, 87
  axisymmetric flows, 262–263
Green second identity, 21
Green theorem, generalized, 19

Hamilton equations, 186
Harmonic function, 81, 115, 125, 142
Hasimoto soliton, 306–307
Helicity, 102–103, 109
  axisymmetric, 264–265
Helmholtz representation, 79–80, 83–86
Helmholtz vortex laws, 40, 97
Hill spherical vortex, 271–273, 279
Hodograph plane, 158

Image set, 136–139
Impulse
  angular, 99–101, 106, 109, 185
  axisymmetric, 263–264
  linear, 99–101, 106, 109, 184
Incompressibility condition, 34–35, 53–54, 79–83, 115
Incompressible fluid, 34
Index
  dummy, 10
  free, 10
  notation, 10
Integral
  identities, 18–22
  length scale, 5
Interaction energy, 107, 185–186
Interface matching conditions
  kinematic, 73, 315
  dynamic, 73, 316
Invariants, *see* Vorticity, invariants
Irrotational flow, 7, 76–79, 81–83, 95

Joukowski transformation, 148–151

Kaden spiral, 207–208
Kelvin-Helmholtz instability, 344–346
Kelvin circulation theorem, 97–98, 262
Kelvin waves, *see* Vortex tube, bending waves *and* Vortex tube, area–varying waves
Kinematics, 24
Kinetic energy, 51–53, 83, 124
  axisymmetric, 264
  invariance of, 101–102, 107, 109
  transport equation, 51
Kirchhoff elliptic vortex, 188–191
Kolmogorov scale, 5
Kronecker delta, 11
Kutta condition, 176
Kutta-Joukowski theorem, 171, 173–175

Lagally theorem, 172–173, 248
Lagrangian viewpoint, 26
Lamb transformation, 268
Lamellar flow, *see* Irrotational flow
Laplace
  equation, 81–82
  formula, 72
Laplacian operator, 14
Laurent series, 167–168
Limit cycle, 332
Line doublet, 237–238
Line source, 224–225, 235–237
Liquid jet, capillary instability, 346–348
Local corrections, method of, 199
Local-induction approximation, 303–306

Martensen theorem, 244–245
Mass conservation, 44–46
Material
  derivative, *see* Derivative, material
  point, 28
  region, 28
  with respect to an arbitrary vector field, 64–65
Mechanical power, 51–53
Metrical coefficients, 367
Moment of momentum conservation, 49–51
Momentum conservation, 46–49
Momentum jump condition, *see* Discontinuity jump condition, momentum
Monge potentials, 86
Multipole expansion method, 199–203

Navier-Stokes equation, 55
No-penetration condition, 55, 73
Normal modes, 332
No-slip condition, 55–56, 74

Orifice, liquid flow through, 159–162
Orthogonal tensor, 14–15

Pathline, 36–37
Permutation symbol, 11
Poiseuille flow
  axisymmetric, 59
  plane, 58–59, 355
Polygon, simple closed, 151
Potential
  complex, 123
  flow, *see* Irrotational flow
  vector, 79–80, 86–87
  velocity, 77–79
Pressure
  boundary-integral equation, 116–119
  Poisson equation, 114–115

Pressure (*cont.*)
variation in steady two-dimensional flows, 119–120

Radius of gyration, 106, 109
Rankine
circular vortex, 187–188
half-body, 226–227
solid, 227–228
Rayleigh centrifugal stability theorem, 350
Rayleigh equation, for oscillations of a bubble, 230
Rayleigh equation, for stability of a parallel shear flow, 353
Rayleigh inflection point theorm, 354
Rayleigh-Taylor instability, 343–344
Residue theorem, 167
Reynolds number, 3
Reynolds transport theorem, 40–42
alternative form, 45–46
extended for discontinuity surface, 63–66
Rodrigues formula, 71–72
Rosenhead cut-off, 301

Scalar,
product of tensors, 12
product of vectors, 12
triple product, 14
Schrödinger equation, 305–306
Schwarz-Christoffel equation, 151–157, 160–161
Schwarz functions, 186–191
Serret-Frenet equations, 298
Sharp edge, flow around, 147–148
Shear flow
parallel, 56–59, 352–358
past a circular cylinder, 213–215
piecewise linear, 356–358
Slender body theory, 235–238
Solenoidal flow, *see* Incompressible flow
Source
axisymmetric, 222–223
in corner, 145–146
near circular cylinder, 137–138
near flat plate, 136–137
near sphere, 232–233, 248
two-dimensional, 126–127

Space curve
curvature, 298
torsion, 298
unit vectors, 297
Sphere theorem
Butler, 231–233
Weiss, 233–235
Spherical coordinates, 220, 371–372
Stability, definition, 331
asymptotic, 331
Stagnation-point flow, *see* Straining flow, planar
Steady flow, 26, 37, 115, 119–120, 277–279
Stokes stream function, 220–221, 261
Stokes theorem, 21–22
Strain, rate of, *see* Deformation, rate of
Straining flow
axisymmetric, 94, 222
past a circle, 134–135
planar, 125–126, 334
Streakline, 36–37
Stream function, 80–81, 88
Streamline, 37, 78, 80–81
Stream tube, 37–39
strength, 38, 80–81
Stress
tensor, 48
vector, 47–48
Stretch, 30, 96
rate of, 30
Sturm-Liouville equation, 331–332, 349–350
Summation convention, 10
Superposition principle, 128
Surface
closed, 18
normal, 18
orientable, 19
piecewise smooth, 18
simply connected, 18
smooth, 18
Surface tension, 68–72, 319–320, 346–348
Swirling sink flow, 132–133

Tangent projection, 18
Tapered body, *see* Uniform flow, past a tapered body
Taylor series, 88, 168, 317

Tensor
product, 10–11
skew-symmetric, 15–16
symmetric, 15–16
Trace, 13
Transport theorem, *see* Reynolds transport theorem
Transpose, 13
Turbulent flow, 5

Uniform flow, 125, 211–222
past a circle, 131–132, 134, 170–171
past a compound cylinder, 135
past an ellipse, 150
past a flat plate, 150–151
past a semi-infinite rectangular body, 155–157
past a sphere, 225–226
past a tapered body, 228–229

Vector
axial, 16
identities, 16–18
product, 12
Velocity
asymptotic limit at infinity, 88–90
complex, 123–124
gradient, 29
Velocity maximum theorem, 82–83
Viscosity
bulk, 52
dynamic, 52
implications of the second law of thermodynamics, 52
Viscous flow
exact solutions, 55–59
length scales, 4
vorticity invariants in, 107–109
Volume
composite, 19–20
elementary, 19–20
Vortex
bathtub, 132–133
breakdown, 286–288
critical condition, 286–287
in corner, 146–147
cutting by a blade, 96–97
filament, 296–298
induced force on small particle, 254–255
merger, 196–197
near circular cylinder, 138–139
near flat plate, 137
pair, 129–133, 146–147, 358–361
patch, 186–193
reconnection, 5, 96
shock, 287–288
street, 337, 342
stretching, 93–95
stripping, 291–292
swirl ratio, 287
two-dimensional, 127–128
Vortex-in-cell method, 197–199
Vortex lines, 39–40
knottedness of, 103
motion of, 95–97
Vortex ring, 265–271
elliptical, 302
impulse, 266
kinetic energy, 271
leapfrogging motion, 295
self-induced velocity, 271
Vortex sheet
equation of motion, 206–207
equivalence to doublet sheet, 204–206
roll-up, 207–208
singularity, 208
stability of, 344–346
strength, 203
velocity jump, 204
Vortex systems, 183–186
stability of, 337–342
Vortex tube, 39–40, 296–311
area-varying waves, 279–286
bending waves, 306–311
cut-off model, 298–303
plug-flow model, 283–303
solitary waves, 279, 306–307
Vorticity
baroclinic generation, 93
center of, 106, 109, 265
flux, 108
invariants, 98–109, 263–265
tensor, 29
transport equation, 92–95
vector, 32

Wave, interfacial, 314–327
capillary, 319–320

Wave, interfacial (*cont.*)
deep-water approximation, 318
dispersion equation, 318
energy, 322–323
energy transport rate, 323–324
free-surface, 314
group velocity, 324–325
internal, 314–316
linear theory, 316–319
particle displacement by, 320–321
phase velocity, 318
shallow-water approximation, 318
stability, 343–346
Weber number, 364
Wedge, flow past, 144–145

Printed and bound by CPI Group (UK) Ltd, Croydon, CR0 4YY
16/09/2026
14963457-0003